T5-CSA-008

Current Topics in Bioenergetics

Volume 9

Contributors to This Volume

Richard S. Criddle

D. W. Deamer

Daniel V. DerVartanian

R. J. Gillies

Richard John Guillory

Richard F. Johnston

Jean LeGall

Thomas J. Murphy

Harry D. Peck, Jr.

Adil E. Shamoo

Robert J. Stack

Haruhiko Takisawa

Yuji Tonomura

K. Van Dam

H. V. Westerhoff

Taibo Yamamoto

Current Topics in Bioenergetics

Edited by
D. RAO SANADI

Boston Biomedical Research Institute
Boston, Massachusetts

VOLUME 9

1979

ACADEMIC PRESS
NEW YORK SAN FRANCISCO LONDON

A Subsidiary of Harcourt Brace Jovanovich, Publishers

ACADEMIC PRESS, INC.
111 Fifth Avenue, New York, New York 10003

United Kingdom Edition published by
ACADEMIC PRESS, INC. (LONDON) LTD.
24/28 Oval Road, London NW1 7DX

Library of Congress Catalog Card Number: 66-28678

ISBN 0-12-152509-0

PRINTED IN THE UNITED STATES OF AMERICA

79 80 81 82 9 8 7 6 5 4 3 2 1

Contents

Irreversible Thermodynamic Description of Energy Transduction in Biomembranes

H. V. WESTERHOFF AND K. VAN DAM

Intracellular pH: Methods and Applications

R. J. GILLIES AND D. W. DEAMER

Mitochondrial ATPases

RICHARD S. CRIDDLE, RICHARD F. JOHNSTON, AND ROBERT J. STACK

Ionophores and Ion Transport Across Natural Membranes

ADIL E. SHAMOO AND THOMAS J. MURPHY

Reaction Mechanisms for ATP Hydrolysis and Synthesis in the Sarcoplasmic Reticulum

TAIBO YAMAMOTO, HARUHIKO TAKISAWA, AND YUJI TONOMURA

Flavoproteins, Iron Proteins, and Hemoproteins as Electron-Transfer Components of the Sulfate-Reducing Bacteria

JEAN LEGALL, DANIEL V. DERVARTANIAN, AND HARRY D. PECK, JR.

Applications of the Photoaffinity Technique to the Study of Active Sites for Energy Transduction

RICHARD JOHN GUILLORY

List of Contributors

Numbers in parentheses indicate the pages on which the authors' contributions begin.

RICHARD S. CRIDDLE (89), *Department of Biochemistry and Biophysics, University of California, Davis, California 95616*

D. W. DEAMER (63), *Department of Zoology, University of California, Davis, California 95616*

DANIEL V. DERVARTANIAN (237), *Department of Biochemistry, University of Georgia, Athens, Georgia 30602*

R. J. GILLIES (63), *Department of Zoology, University of California, Davis, California 95616*

RICHARD JOHN GUILLORY (267), *John A. Burns School of Medicine, University of Hawaii, Honolulu, Hawaii 96822*

RICHARD F. JOHNSTON (89), *Department of Biochemistry and Biophysics, University of California, Davis, California 95616*

JEAN LEGALL (237), *Laboratoire de Chimie Bacterienne, CNRS, 13274 Marseille Cedex 2, France*

THOMAS J. MURPHY (147), *Department of Radiation Biology and Biophysics, University of Rochester School of Medicine and Dentistry, Rochester, New York 14642*

HARRY D. PECK, JR. (237), *Department of Biochemistry, University of Georgia, Athens, Georgia 30602*

ADIL E. SHAMOO (147), *Department of Radiation Biology and Biophysics, University of Rochester School of Medicine and Dentistry, Rochester, New York 14642*

ROBERT J. STACK (89), *Department of Biochemistry and Biophysics, University of California, Davis, California 95616*

HARUHIKO TAKISAWA (179), *Department of Biology, Faculty of Science, Osaka University, Toyonaka, Osaka, Japan*

YUJI TONOMURA (179), *Department of Biology, Faculty of Science, Osaka University, Toyonaka, Osaka, Japan*

K. VAN DAM (1), *Laboratory of Biochemistry, B. C. P. Jansen Institute, University of Amsterdam, Plantage Muidergracht 12, 1018 TV Amsterdam, The Netherlands*

H. V. WESTERHOFF (1), *Laboratory of Biochemistry, B. C. P. Jansen Institute, University of Amsterdam, Plantage Muidergracht 12, 1018 TV Amsterdam, The Netherlands*

TAIBO YAMAMOTO (179), *Department of Biology, Faculty of Science, Osaka University, Toyonaka, Osaka, Japan*

Preface

Consistent with our goals of fostering cross-fertilization between the different branches of bioenergetics, this volume has articles ranging from a theoretical, thermodynamic perspective of energy transducing reactions to the detailed kinetic analysis of a specific aspect of an ion pump. The underlying theme of membrane-dependent bound reactions runs through all of the articles, including that on sulfate reduction which requires ATP generated from respiratory reactions. The elaborate treatment of photoaffinity labeling as a tool for the study of bioenergetics may prove valuable for a considerable length of time. The discussion of probes for intracellular pH determination should stimulate development of additional techniques and their application in pharmacology and pathology.

Studies with isolated proteins or protein complexes derived from the membranes have advanced knowledge about individual reactions in the overall processes of energy transduction. However, their operation within the membrane phase is still obscure and is limited by insufficient development of technique. It would seem that each advance in membrane technology produces a burst in our knowledge about energy transduction, which is followed by a hiatus. We now appear to be poised for a final attack, particularly on the chemistry of photosynthesis and oxidative phosphorylation.

D. RAO SANADI

Contents of Previous Volumes

Volume 1

Volume 2

Volume 3

Volume 4

Volume 5

Volume 6

Volume 7
Photosynthesis: Part A

Volume 8
Photosynthesis: Part B

Irreversible Thermodynamic Description of Energy Transduction in Biomembranes

H. V. WESTERHOFF and K. VAN DAM
Laboratory of Biochemistry
B.C.P. Jansen Institute
University of Amsterdam
Amsterdam, The Netherlands

ISBN 0-12-152509-0

I. Introduction

The biologically important processes of respiratory-chain phosphorylation and of photophosphorylation involve the transduction of redox or light energy, respectively, for the synthesis of ATP from ADP and P_i against an energy gradient. The mechanism of this energy transduction has been hotly debated, but it seems that on many points a consensus among the workers in the field is approaching (Boyer *et al.*, 1977). A central role in reaching this consensus has been played by the chemiosmotic concept, formulated by Mitchell (1961, 1968). It rationalizes the fact that these energy transductions occur at membranes and, at the same time, establishes a relationship between them and so-called active transport across biomembranes.

In the past few years, isolation of membrane proteins and their reconstitution into realistic model membranes has made great progress. Many of the components involved in respiratory-chain phosphorylation and photophosphorylation as well as components of other membrane-linked bioenergetic processes have been purified and recombined into functioning parts of the overall process (Kagawa, 1972; Racker and Stoeckenius, 1974; Crane *et al.*, 1976; Goldin, 1977).

It seems an appropriate moment, therefore, to attempt to devise a quantitative description of biological energy transduction. The main purpose of such a description should be to allow us to formulate testable predictions, just as in the case of enzyme kinetics. Furthermore, it should be possible to refine the description as our knowledge of the system increases.

We chose the basis for our description in physics by applying fundamental equations of linear irreversible thermodynamics (de Groot, 1952; de Groot and Mazur, 1962). It may be hoped that building of such bridges between physics, chemistry, and biology is to the benefit of workers in these disciplines. Applications of nonequilibrium thermodynamics to biological systems have been reported by other authors (see, e.g., Rottenberg *et al.*, 1970; Oster *et al.*, 1973; Katchalsky and Curran, 1974; Paterson, 1970). The equations derived were phenomenological in nature.

A central feature of our description is that we take certain molecular mechanisms as our starting point (see also Lagarde, 1976). By doing this, we arrive at equations in which the constants have a mechanistically defined meaning. In this respect, a comparison with enzyme kinetics is again appropriate: other mechanisms would lead to other equations. It is the testing of the experimental soundness of such equations that determines the strategy for further experimentation.

II. From Theoretical Physics to Membrane Biochemistry

A. The Dissipation Function

In chemical or physical processes, a continuous entropy production occurs. It is this entropy change that may be considered as the "irreversible component" of such processes. A commonly used equation to describe the entropy production is the following (Katchalsky and Curran, 1974):

$$\sigma = j_u \nabla(1/T) + \sum_j j_j \nabla(-\tilde{\mu}_j/T) + \sum_r j_{\text{chem. r}}(A_r/T) \tag{1}$$

This equation states that the rate of entropy production of a system is the sum of a number of forces multiplied by their associated fluxes. The energy flux is associated with a temperature gradient, diffusion is associated with an (electro) chemical gradient of a compound, and the chemical reaction rate is associated with the affinity of that reaction.

σ always has a positive value (second law of thermodynamics), but the sign of each of the flux-force couples is not *a priori* defined. Thus, the negative contribution of a diffusion flux against a concentration gradient may be compensated by the positive contribution of a chemical reaction proceeding at a high affinity. For this the two fluxes must be coupled. Any independent (not coupled) set of fluxes and forces must conform to the criterion of positive entropy production (Section III).

In the usual isothermal systems there is no significant temperature gradient. Under such circumstances, the dissipation function $\Phi(=T\sigma)$ can be integrated, and it is found that Φ equals the specific loss of Gibbs free energy, due to irreversibility.

A flux j_i in a system can be considered to be a function of all the forces within that system. Close to equilibrium, it can be derived (see, e.g., Sauer, 1977) that in general the relation between fluxes and forces is

$$j_i = L_{ij} X_j \tag{2}$$

Thus

$$L_{ij} = (\partial j_i / \partial X_j)_{X_{k \neq j}} \tag{3}$$

The full set of equations relating j and X can be written in a matrix form and called the phenomenological equations (Onsager, 1931). The matrix of constants L_{ij} will be independent of fluxes and forces, but can certainly depend on the values of local parameters, since these can change with equilibrium conditions (see Section II,D).

The validity of the linear (even proportional) phenomenological equations has been proved only for conditions close to equilibrium. However, experiments indicate that the equations remain valid for large deviations from equilibrium in the case of diffusion processes, whereas for chemical reactions their validity is confined to conditions in which the affinity is not much larger than $RT\ln 2$. Under certain conditions enzyme-catalyzed reactions have larger regions of linearity (Rottenberg, 1973).

It has been shown by Onsager (1931) that the matrix of phenomenological coefficients is symmetrical:

$$L_{ij} = L_{ji} \tag{4}$$

This fortunately reduces the number of independent variables that one has to measure to establish the values of all the constants in the phenomenological equations. The correctness of the Onsager reciprocal relation has been verified in many different experimental systems (see, e.g., Dunlop and Gosting, 1959; Miller, 1960; Blumenthal *et al.*, 1967).

B. Steady-State Criteria

Under steady-state conditions the entropy production is minimal (Prigogine, 1955). Since, with linear phenomenological equations

$$\sigma = \sum_{i,j} X_i L_{ij} X_j \tag{5}$$

σ is a quadratic function of the forces. Also, σ is always positive, so that the absolute minimum is reached if $\sigma = 0$, which occurs if all forces are zero.

If all forces are variable, there is only one possible steady state: equilibrium. Since in the steady state σ must be minimal, the derivatives with respect to all variable forces must be equal to zero. In the case that one of the forces (X_k) is kept constant, use of the Onsager reciprocal relation allows us to state:

$$0 = (\partial\sigma/\partial X_j)_{X_i, X_k} = 2\sum_i X_i L_{ij} = 2j_j \qquad \text{for} \qquad j \neq k \tag{6}$$

The entropy production then reduces to

$$\sigma = j_k X_k = \sum_i L_{ik} X_i X_k \tag{7}$$

Steady states reached by systems if one of the forces (the input force) is kept constant are static-head conditions (Rottenberg *et al.*, 1970), as the output fluxes are zero. An example of a static-head condition is the open-circuit situation of an electrical cell with finite capacity in the wiring. Sometimes one of the forces in a system will change faster than the others. Generally

$$dX_i/dt = j_i/C_i = -[(L_i X_i)/C_i] \quad (8)$$

in which C_i is the capacity of the particular process. For the purpose of simplicity, i is assumed not to depend on any other process either through its capacity or through cross coefficients. In systems to which this assumption does not apply, one should try to rearrange flows and forces (by linear combinations) so that independence is achieved. If this is not possible, then comparison of capacities will be very complicated, and, in practice, only qualitative reasoning should lead to assumptions that are checked experimentally. For an electrical current, the usual meaning can be attached to the term capacity; in other cases, it may be the buffer capacity of the system, etc.

Assuming that L_i and C_i are independent of time, it is easily seen that

$$X_i(t) = X_i(0)\exp[-(L_i/C_i)t] \quad (9)$$

For such a two-fluxes/two-forces system we can write

$$\sigma = \sigma_1 + \sigma_2 = L_1 X_1^2 + L_2 X_2^2 = L_1 X_1(0)^2 \cdot \exp[-2t(L_1/C_1)] + L_2 X_2(0)^2 \cdot \exp[-2t(L_2/C_2)] \quad (10)$$

If now the initial forces and L's are equal, but $C_2 \gg C_1$, then σ_1 will change much faster than σ_2 and at $t \gg L_1/C_1$, j_1 will be negligible compared with j_2. This near-steady state condition is comparable with the case where X_2 is kept constant, X_1 is variable, and $j_1 = 0$ in the steady state (static head). The difference between a near-steady state and a real steady state lies in the degree of constancy of the parameters. In practical systems where all fluxes and forces are variable, there will occur steady-state-like situations if L_i/C_i terms differ markedly.

C. Adaptation of the Dissipation Function to a Vesicular System

1. *Integration*

The processes of energy transduction, considered in this paper, are occurring in vesicular structures surrounded by a membrane. Since the normal dissipation function is written for local values of parameters, we have to adapt the equation to such vesicular systems.

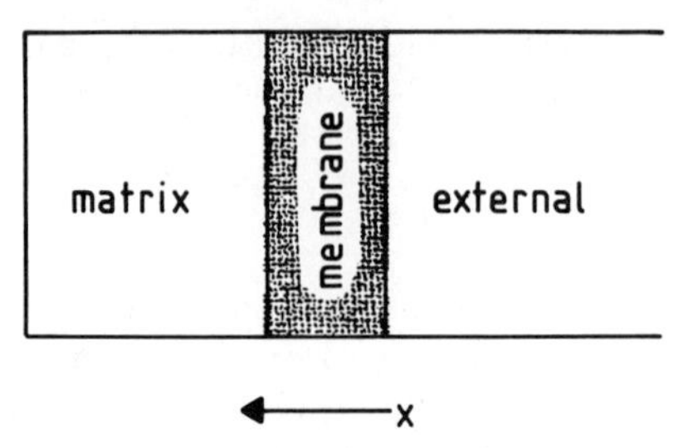

FIG. 1. The three-compartment idealization of energy-transducing vesicles.

The following assumptions will be made. Each vesicle, or representative part thereof, can be considered as consisting of a juxtaposition of three compartments (Fig. 1):

1. The matrix space, in which $\nabla(-T) = \nabla(-\tilde{\mu}_i) = 0$
2. The membrane space, in which $\partial j_u/\partial x = \partial j_i/\partial x = j_{\text{chem. r}}$
3. The external space, in which $\nabla(-T) = \nabla(-\tilde{\mu}_i) = 0$

If the effective surface area of the vesicle membrane and the effective volume of the matrix space are called O and V, respectively, then integration of the dissipation function yields:

$$\Sigma = \iiint_{\text{vesicle}} \sigma dV = -\left(\iint_O j_u dO\right)\Delta\left(\frac{-1}{T}\right) - \sum_i \left(\iint_O j_i dO\right)\Delta\left(\frac{\tilde{\mu}_i}{T}\right) + \sum_{\text{r}} \frac{A_{\text{r}}}{T} \iiint_V J_{\text{chem. r}} dV \quad (11)$$

The integrals of the flows in this expression are the net flows usually considered and will be denoted by capital J's:

$$\Sigma = J_u \Delta\left(\frac{-1}{T}\right) + \sum_i J_i \Delta\left(\frac{\tilde{\mu}_i}{T}\right) + \sum_{\text{r}} J_{\text{chem. r}}^{\text{in}} \frac{A_{\text{r}}^{\text{in}}}{T^{\text{in}}} + \sum_{\text{r}} J_{\text{chem. r}}^{\text{ex}} \frac{A^{\text{ex}}}{T^{\text{ex}}} \quad (12)$$

The units in which the flows have to be expressed then are joule per second per unit vesicle, and three times moles per second per unit vesicle. Unit vesicle can be replaced by milligrams of protein or milligrams dry weight.

2. *Discarding the Entropy Flow*

The entropy production can be considered to be a sum of separate other functions:

$$\Sigma = \Sigma_1 + 2\Sigma_{12} + \Sigma_{12} \quad (13)$$

in which

$$\Sigma_1 = L_{11}X_1^2, \Sigma_2 = L_{22}X_2^2, \quad \text{and} \quad \Sigma_{12} = X_1X_2(L_{12} + L_{21})/2$$

If we now further introduce the parameter

$$\sigma_{12} = L_{12}/L_{11} \cdot X_2/X_1 \tag{14}$$

and use Onsager's reciprocal relation, we can rewrite the entropy production

$$\Sigma = \Sigma_1(1 + 2\sigma_{12}) + \Sigma_2 \tag{15}$$

Another way of looking at σ_{12} follows from inspection of the equation:

$$J_1 = L_{11}X_1 + L_{12}X_2 = L_{11}X_1(1 + \sigma_{12}) \tag{16}$$

From both equations it can be concluded that if one studies the process (J_1, X_1) and $|\sigma_{12}| \ll 1$, the process (J_2, X_2) can be left out of consideration, as it influences neither the entropy production nor the flow of the first process.

This approach proves useful in discarding the temperature gradient-linked processes in most biological systems (see also Spanner, 1954). Since $\Delta(\tilde{\mu}_i)$ also depends on $\Delta(T)$, we rewrite the expression (12) for the entropy production, using

$$\Delta_T(\tilde{\mu}_i) = \Delta\tilde{\mu}_i + \bar{S}_i\Delta T \tag{17}$$

$$J'_{\mathrm{q}} = J_{\mathrm{u}} - \sum_i J_i\bar{H}_i \tag{18}$$

$$\Delta\left(\frac{1}{T}\right) \approx \frac{-1}{(\bar{T})^2}\Delta T \tag{19}$$

$$\Delta\left(\frac{\tilde{\mu}_{\mathrm{K}}}{\bar{T}}\right) \approx -\frac{\tilde{\mu}_{\mathrm{K}}}{(\bar{T})^2}\Delta T + \frac{\Delta\tilde{\mu}_{\mathrm{K}}}{(\bar{T})} \tag{20}$$

$$\Sigma = \frac{J'_{\mathrm{q}}}{(\bar{T})^2}\Delta(T) + \sum_i \frac{J_i}{\bar{T}}\Delta_T(\tilde{\mu}_i) + \sum_{\mathrm{r}} J_{\mathrm{chem.\,r}}\frac{A_{\mathrm{r}}}{T} \tag{21}$$

The relevance of considering the first term in this equation in a description of the processes centered in the other terms will now be discussed more quantitatively for the example of valinomycin-catalyzed (Ovchinnikov *et al.*, 1974) potassium flow across biological membranes:

$$\Sigma = \frac{J'_{\mathrm{q}}}{\bar{T}}\frac{\Delta(T)}{\bar{T}} + \frac{J_{\mathrm{K}}}{\bar{T}}\Delta_T(\tilde{\mu}_{\mathrm{K}}) \tag{22}$$

In the way described above we define σ_{KT}

$$\sigma_{KT} = \frac{L_{TK}}{L_{KK}} \frac{\Delta(T)}{\overline{T} \cdot \Delta_T(\tilde{\mu}_K)} \tag{23}$$

Using Onsager symmetry of the phenomenological equations and the definition

$$S_K^* = \left(\frac{J_s}{J_K}\right)_{\Delta T = 0} \qquad Q_K^* = \left(\frac{J'_q}{J_K}\right)_{\Delta T = 0} \tag{24}$$

one can derive

$$\sigma_{KT} = \frac{(S_K^* - \bar{S}_K)\Delta T}{\Delta_T(\tilde{\mu}_K)} = \frac{Q_K^*}{\overline{T}} \frac{\Delta T}{\Delta_T(\tilde{\mu}_K)} \tag{25}$$

Now $(S^* - \bar{S})$ is equal to the amount of entropy carried by the potassium ions that move across the membrane less the partial entropy of the average potassium ion in solution. The difference exceeds zero only because the potassium ions that are able to take the barrier of the membrane have a higher than average (e.g., kinetic) energy.

An estimation of $(S^* - \bar{S})$ is obtained by considering that the main entropy change upon moving K^+ ions from their hydration shell into the cage of valinomycin is the release of the hydrating water molecules. Assuming a shell of 6 water molecules around a K^+ ion in solution (Lakshminarayanaiah, 1969), we find

$$S^* - \bar{S} \approx 6\bar{S}_{H_2O} - 6\bar{S}_{H_2O}\,\text{hydrating} < 7\bar{S}_{H_2O} \approx 500\ \text{J} \cdot \text{mol}^{-1}\ K^{-1}$$

If $\Delta_T(\tilde{\mu}_K) = 500\ \text{J} \cdot \text{mol}^{-1}$ (equivalent to 5 mV only) and $\sigma_{KT} = 10^{-2}$, $\Delta(T)$ would have to be $\geq 0.01\ K$. Such a temperature drop across a biomembrane of the usual thickness of less than 100 Å is equivalent to a gradient of more than $10^6\ K \cdot m^{-1}$, which is much too large to be compatible with the heat conductance of such membranes (Davies, 1965).

As to the correctness of the estimation, we may note that from the temperature dependence of valinomycin-catalyzed potassium movement across lipid bilayers (Krasne *et al.*, 1971) we can approximate that Q_K^*/T slightly exceeds $60\ \text{J} \cdot \text{mol}^{-1}\ K^{-1}$.

The conclusion is that σ_{TK} must be (much) smaller than 10^{-2}, and this in turn means that coupling between heat flow and flow of matter is negligible.

The important corollary of this conclusion is that the description of biological processes can treat the part of the dissipation function not containing the heat processes separately as:

$$\Sigma = \sum_i \frac{J_i}{\overline{T}} \Delta_{P,T}(\tilde{\mu}_i) + \sum_r J_{\text{chem. r}} \frac{A_r}{\overline{T}} \tag{26}$$

where the condition $\Delta P \approx 0$ was used in $\Delta_{P,T}(\tilde{\mu}_i) = \Delta_T(\tilde{\mu}_i)$.

Neglecting ΔP is equivalent to assuming that the membrane is not stressed: it is comparable to a paper bag that is not fully inflated. The highly folded inner membrane of the mitochondion will fit this description. Also other vesicles in isosmotic media will approximate this situation, if the internal osmolarity is not increased by chemical reactions or the action of pumps. In cases where swelling does occur, one should be careful in using the equation derived below. As soon as elastic forces come into play, $\Delta\tilde{\mu}_i$ will include a term $\overline{V}_i \Delta P$, and even if L_w is high, $\Delta\Pi$ will not necessarily be equal to zero (see below). However, since $\overline{V}_i$ is usually in the order of $0.2 \cdot 10^{-4}$ $m^3 \cdot \text{mol}^{-1}$, the elastic forces must reach $2.5 \cdot 10^7\ N \cdot m^{-2}$ (250 atm) to introduce a $\overline{V}_i \Delta P$ term of 0.5 $\text{kJ} \cdot \text{mol}^{-1}$. Since a transmembrane concentration difference of an impermeant substance of one molar induces a pressure difference of 22 atm only, the $\overline{V}_i \Delta P$ will seldom be relevant (see also Mitchell, 1968).

In the following we will use Δ for $\Delta_{P,T}$ and d for $d_{P,T}$. In this paper we will concentrate on the dissipation function $\mathbf{\Phi} = \overline{T}\Sigma$. Here $\overline{T}$ is the average temperature in the system, but, since in the type of biological systems we will be concerned with temperature gradients will be small:

$$T \approx \overline{T}$$

so that we can write

$$\mathbf{\Phi} = \sum_i J_i \Delta_{P,T}(\tilde{\mu}_i) + \sum_{\mathrm{r}} J_{\mathrm{chem.\,r}} A_{\mathrm{r}} \tag{27}$$

3. *Discarding the Water Flow*

Since water is one of the substances for which a gradient might exist across the biomembrane, we should write the dissipation function as follows:

$$\mathbf{\Phi} = J_{\mathrm{w}}\, \Delta(\tilde{\mu}_{\mathrm{w}}) + \sum_{i=1}^{n-1} J_i \Delta(\tilde{\mu}_i) + \sum_{\mathrm{r}} J_{\mathrm{chem.\,r}} A_{\mathrm{r}} \tag{28}$$

The permeability of biomembranes to water is very high (cf. Dainty and Ginzburg, 1964; Massari *et al.*, 1972) and comparable to the ability of water to go into the vapor phase. Ions do not have this ability. Therefore, it will be assumed that $J_{\mathrm{w}}\, \Delta(\tilde{\mu}_{\mathrm{w}})$ does not couple to any other process, and we can consider these other processes separately:

$$\mathbf{\Phi} = \sum_{i=1}^{n-1} J_i \Delta(\tilde{\mu}_i) + \sum_{\mathrm{r}} J_{\mathrm{chem.\,r}} A_{\mathrm{r}} \tag{29}$$

It remains interesting to examine the water flow due to the existence of the Gibbs–Duhem relation:

$$\sum_{i=1}^{n} c_i d\tilde{\mu}_i = 0 \tag{30}$$

The condition of electroneutrality ($\sum_{i=1}^{n} c_i z_i = 0$) reduces this equation to

$$\sum_{i=1}^{n} c_i d\mu_i = 0 \tag{31}$$

and integration of this equation across the membrane yields

$$\sum_{i=1}^{n} \bar{c}_i^m \Delta\mu_i \approx 0 \tag{32}$$

Here $\bar{c}_i^m$ represents the average concentration of substance i in the membrane. $\Delta\mu_i$ now contains only the concentration-dependent part of the molar free-energy difference (see also above).

As we mentioned above, the permeation of water through the membrane is very high, so that quickly

$$\Delta\mu_{\mathrm{w}} = 0 \tag{33}$$

This combines to the equation

$$\sum_{i=1}^{n-1} \bar{c}_i^m \Delta\mu_i \approx 0 \tag{34}$$

This equation shows that the remaining forces in the dissipation function are not independent. It can be calculated that, if changes in an external substance, small in comparison to the sum of concentrations inside the vesicle, occur, their influence on other forces will be small. In any case, possible mistakes caused by this effect can always be eliminated by considering all forces, never supposing that other flows that have not been considered remain constant.

D. Significance of the Proportionality Constants

In principle, one can now write phenomenological equations for any system, making use of the Onsager reciprocal relation to reduce the number of proportionality constants. This useful approach was taken by Rottenberg *et al.* (1970) and seems simpler than the one we use. However, the real advantage of our method will appear to be that the proportionality constants will receive a meaning that can be easily translated into mechanistic terms.

To illustrate the importance of this point, we will derive the equation relating the flux and affinity of a very simple chemical reaction

$$A \underset{k_{-1}}{\overset{k_1}{\rightleftharpoons}} B$$

From chemical kinetics, the net rate of this reaction is given by

$$v = k_1[A] - k_{-1}[B] \tag{35}$$

At equilibrium $v = 0$, which can be used to derive the relationship

$$k_{-1} = k_1 \frac{[A]_{eq}}{[B]_{eq}} \tag{36}$$

Inserting this into the kinetic equation leads to

$$v = k_1[A]_{eq}\left(\frac{[A]}{[A]_{eq}} - \frac{[B]}{[B]_{eq}}\right) \tag{37}$$

From thermodynamics we take the expression for the chemical potential

$$\mu_i = \mu_i^o + RT \ln[i] \tag{38}$$

which leads upon rearrangement to

$$[i] = \exp\left(\frac{\mu_i - \mu_i^o}{RT}\right) \tag{39}$$

Applying this to the reactants, we obtain the kinetic equation

$$V = k_1[A]_{eq}\left\{\exp\left(\frac{\mu_A - \mu_A^{eq}}{RT}\right) - \exp\left(\frac{\mu_B - \mu_B^{eq}}{RT}\right)\right\} \tag{40}$$

The normal approximation in linear irreversible thermodynamics is that the conditions are close to equilibrium. In the present case, this leads to the approximation (using also $\mu_A^{eq} - \mu_B^{eq} = 0$):

$$V = k_1 \frac{[A]_{eq}}{RT} (\mu_A - \mu_B) \tag{41}$$

This last equation has the expected form of a flux equation. The flux through the reaction is proportional to the affinity of that reaction ($\mu_A - \mu_B$).

$$J_{chem.\,1} = L_{chem.\,1}(\mu_A - \mu_B)$$

Thus the mechanistic meaning of $L_{chem.\,1}$ is given by

$$L_{chem.\,1} = k_1[A]_{eq}/(RT)$$

L is proportional to the amount of catalyst present.

The important fact to note, however, is that the proportionality constant still contains the equilibrium concentration of one of the reactants. An experimental consequence of this is that one must be careful in calculating the value of the proportionality constant from experiments in which variations in the affinity are effected by changing the concentrations of reactants.

A second aspect of the above derivation is that one can now easily understand the reliability of the approximation taken in linear irreversible thermodynamics: for the exponential term to be linear, it must be $\ll RT$. One can derive that simple enzyme catalyzed reactions have linear relations between flux and force for more than 50% of the velocity range (Van der Meer *et al.*, in preparation).

III. Development of the Description of a Biological System by Building from Thermodynamically Independent Units

A. Fully Coupled Processes

As stressed above, for a description of any system we may consider the dissipation function as a sum of independent dissipation functions, each relating to such a part of the system that the fluxes and forces are coupled.

First we will consider a fully coupled process (see Fig. 2), such as the reaction

$$\text{Glucose} + \text{ATP} \rightleftharpoons \text{glucose 6-phosphate} + \text{ADP (affinity } A_{hk})$$

which is catalyzed by an enzyme. (In the following we neglect the complication that enzymes show saturation.) We could imagine that this reaction was built up of two partial reactions

$$\text{ATP} \rightleftharpoons \text{ADP} + \text{phosphate (affinity } A_P)$$

and

$$\text{Glucose} + \text{phosphate} \rightleftharpoons \text{glucose 6-phosphate (affinity } A_{gp})$$

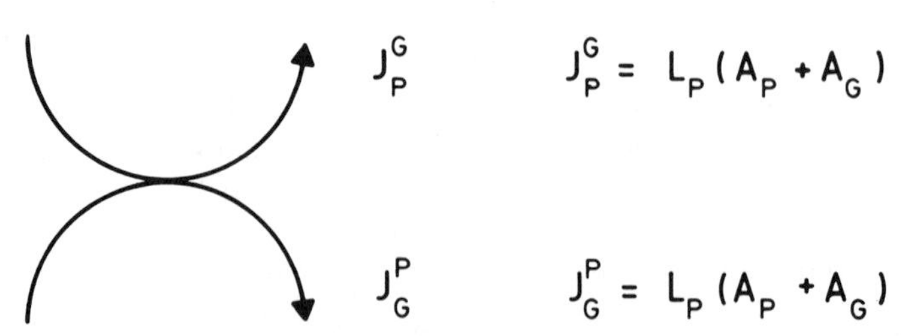

FIG. 2. A fully coupled system: Two chemical reactions with affinities A_P and A_G, rates J_P^G and J_G^P, and proportionality constant L_P.

If, however, no catalysts for these reactions are present, they proceed at a very slow rate compared to the first reaction.

We now have the possibility to describe the first reaction in two ways: as it is written or as the sum of two fully coupled partial reactions. It is obvious that the latter description might be advantageous if, for instance, the system contains other reactions in which phosphate is involved. It turns out that the two descriptions are equivalent.

Thus, independent of the affinities of the partial reactions, the flux through them should be equal. The dissipation function will be

$$\Phi = J_P A_P + J_{gp} A_{gp} = J_P(A_P + A_{gp}) = J_P A_{hk} \tag{42}$$

and the phenomenological equations can be written either with a 1×1 or a 2×2 matrix.

$$\begin{pmatrix} J_P \\ J_G \end{pmatrix} = \begin{pmatrix} L_{hk} & L_{hk} \\ L_{hk} & L_{hk} \end{pmatrix} \begin{pmatrix} A_P \\ A_{gp} \end{pmatrix} \tag{43}$$

Linear algebra would tell us here, from the fact that the determinant

$$\begin{pmatrix} L_{hk} & L_{hk} \\ L_{hk} & L_{hk} \end{pmatrix} = 0$$

that the two flows are fully dependent and that the system could be described as

$$J_P = J_{gp} = L_P(A_P + A_{gp}) \tag{44}$$

B. Passive Transport

Passive transport is the flow of a solute down its electrochemical gradient across a membrane. In principle, if this transport is catalyzed (facilitated diffusion) we still consider it passive. In any case, the description of such a process is straightforward (see Fig. 3) and related to the diffusion down an electrochemical gradient in free solution

$$J_S = L_S \Delta\tilde{\mu}_S \tag{45}$$

It should be remembered that, as in the case of diffusion (Katchalsky and Curran, 1974), the proportionality constant will depend on the absolute concentration of the solute.

Certain catalysts are able to facilitate the diffusion of more than one solute across a membrane. For instance, the protein involved in transport of lactose across bacterial membranes can supposedly move across the membrane only if loaded with both lactose and a proton, or if empty (West and Mitchell,

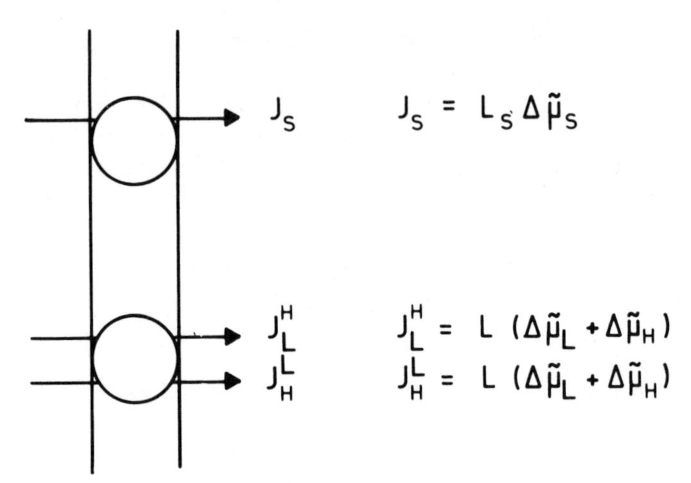

FIG. 3. Two examples of passive transport. The first, facilitated diffusion, involves only one substance. In the second example, two substances move in a strictly coupled manner. J_L^H the flux of substance L in as much as it is coupled to the flux of protons J_H^L.

1973; for review, see Simoni and Postma, 1975). Thus, we have another example of fully coupled fluxes and the considerations of Section III,A hold. We can describe the system as two equal fluxes, each being not linearly related to its own force only, or as a flux of two solutes, linearly related to the force of the overall process (see also Fig. 3):

$$\Phi = J_L^H \Delta\tilde{\mu}_L + J_H^L \Delta\tilde{\mu}_H = J_L^H(\Delta\tilde{\mu}_L + \Delta\tilde{\mu}_H) \tag{46}$$

The phenomenological equation is

$$J_L^H = L(\Delta\tilde{\mu}_L + \Delta\tilde{\mu}_H) \tag{47}$$

It may be noted that a flux of lactose opposite to its (electro) chemical gradient could occur if $\Delta\tilde{\mu}_H$ is sufficiently negative.

A more detailed theoretical discussion about the validity of the postulated linearity of certain transport processes will not be given here (but see, e.g., Geck and Heinz, 1976).

C. Active Transport

One definition of active transport is: the flux of solutes against their electrochemical gradient coupled to a chemical reaction that runs down its affinity.

In terms of the description given up to now, we may consider the mito-

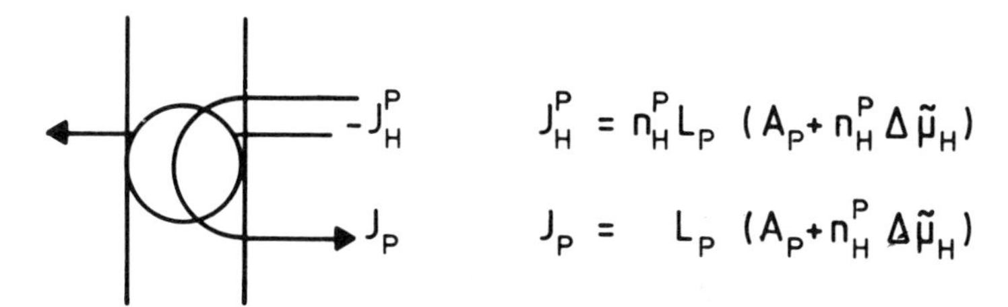

FIG. 4. Active transport: A transmembrane movement of a substance (in this example H^+) is strictly coupled to the flow through a chemical reaction (in this example the ATPase reaction).

chondrial ATPase complex as an ideal example of active transport. It couples the hydrolysis of ATP to transmembrane movement of H^+ ions (Mitchell, 1968; Kagawa, 1972). At present, the stoichiometry between the two processes is being disputed, but we will assume full coupling.

If the hydrolysis of one molecule of ATP leads to transport of n_H^P protons (see Fig. 4), we can immediately write (since $J_H = n_H^P J_P$):

$$\Phi = J_P A_P + J_H \Delta\tilde{\mu}_H = J_P(A_P + n_H^P \Delta\tilde{\mu}_H) \tag{48}$$

$$J_P = L_P(A_P + n_H^P \Delta\tilde{\mu}_H) \quad \text{and} \quad J_H = n_H^P L_P(A_P + n_H^P \Delta\tilde{\mu}_H) \tag{49}$$

Again, it is useful to point out that ATP hydrolysis could be reversed if $\Delta\tilde{\mu}_H$ is sufficiently negative.

We can also write the phenomenological equations in a matrix form

$$\begin{pmatrix} J_P \\ J_H \end{pmatrix} = \begin{pmatrix} L_P & n_H^P L_P \\ n_H^P L_P & (n_H^P)^2 L_P \end{pmatrix} \begin{pmatrix} A_P \\ \Delta\tilde{\mu}_H \end{pmatrix} \tag{50}$$

The validity of the Onsager reciprocal relation follows quite naturally from this treatment.

D. GROUP TRANSLOCATION

In some transport systems the movement of a solute across the membrane is obligatorily coupled to its conversion. Such systems are called group translocation systems, and one of the best characterized is the phosphoenolpyruvate phosphotransferase system. It catalyzes the movement of sugars across a bacterial membrane with simultaneous conversion of the sugar to sugar phosphate. The phosphate group is derived from phosphoenolpyruvate (PEP).

The unusual feature of this type of system is that (in a first approximation) no free sugar is found within the cell and no sugar phosphate outside the

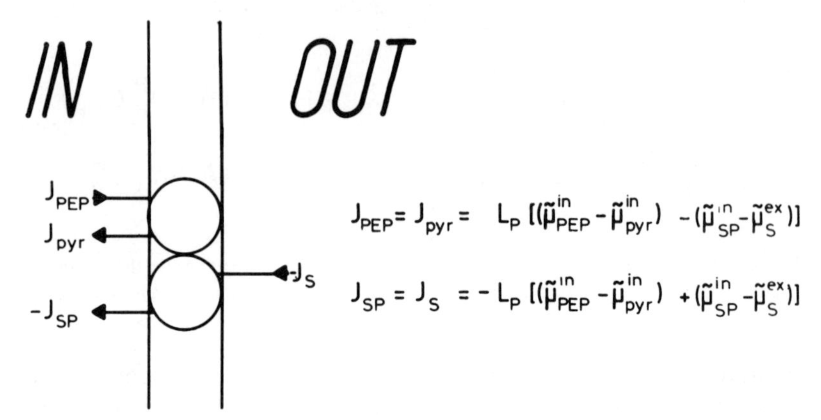

FIG. 5. Group translocation: Transmembrane movement of a substance is strictly coupled to its own chemical modification in a reaction with other substances.

cell. Therefore, the gradient of the sugar per se cannot be considered in the equations.

The driving force of the reaction must be the ΔG of the process that actually occurs:

$$\text{sugar}_{\text{ex}} + \text{PEP}_{\text{in}} \rightleftharpoons \text{sugar-P}_{\text{in}} + \text{pyruvate}_{\text{in}}$$

Applying the linear approach to this process, we obtain the following description of the system (see also Fig. 5):

$$\begin{pmatrix} J_{\text{PEP}} \\ J_{\text{S}} \end{pmatrix} = \begin{pmatrix} L_{\text{PEP}} & L_{\text{PEP}} \\ L_{\text{PEP}} & L_{\text{PEP}} \end{pmatrix} \begin{pmatrix} \mu_{\text{PEP}}^{\text{in}} - \mu_{\text{pyr}}^{\text{in}} \\ \mu_{\text{SP}}^{\text{in}} - \mu_{\text{S}}^{\text{ex}} \end{pmatrix} \tag{51}$$

This description considers only the group translocation process itself, not other reactions that may lead to conversion of the reactants in this process. In a more complete description, such other reactions are added in a linear combination of processes.

E. LINEAR COMBINATION OF PROCESSES

In practice, one often finds situations where the solutes under consideration participate in more than one process. As an example, we will consider the simultaneous operation of the proton-translocating ATPase and a passive permeation of the membrane by protons (see Fig. 6, upper part).

The dissipation function will be a linear combination of these processes

$$\Phi = J_{\text{P}} A_{\text{P}} + J_{\text{H}}^{\text{P}} \Delta\tilde{\mu}_{\text{H}} + J_{\text{H}}^{1} \Delta\tilde{\mu}_{\text{H}} \tag{52}$$

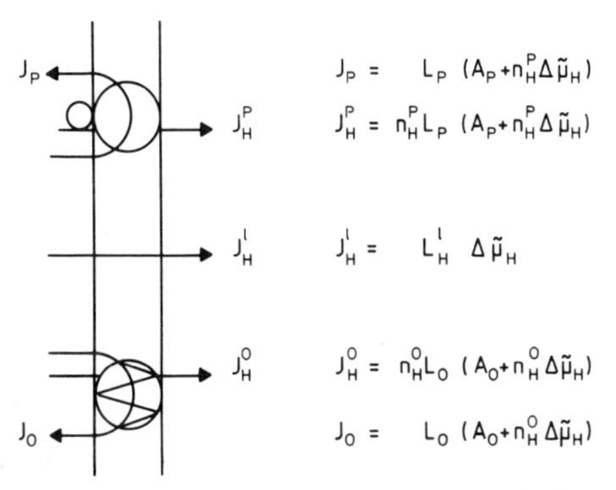

FIG. 6. Linear combination of processes as applied to oxidative phosphorylation. The upper system represents the ATPase–proton pump (stoichiometry n_H^P); the lower system represents the oxidase–proton pump. The arrow in the middle indicates the proton leak.

We have defined now two fluxes of H^+ ions: J_H^P through the ATPase and J_H^1 through the leak. The proportionality constant L^1 is a measure of the conductivity of the membrane for protons. Also, we assume that $\Delta\tilde{\mu}_H$ is the same at the site of the leak and at the site of the ATPase.

Of course, macroscopically only the total proton flux (J_H) is measurable. It is practical to consider the form

$$\Phi = J_P A_P + J_H \Delta\tilde{\mu}_H \tag{53}$$

Using the relations of Section III,C, we can write

$$\begin{pmatrix} J_P \\ J_H \end{pmatrix} = \begin{pmatrix} L_P & n_H^P L_P \\ n_H^P L_P & (n_H^P)^2 L_P + L_H^1 \end{pmatrix} \begin{pmatrix} A_P \\ \Delta\tilde{\mu}_H \end{pmatrix} \tag{54}$$

Experimentally, we cannot distinguish by which of the two pathways the protons are moving. A measure of the relative importance of the two pathways is the coupling coefficient as defined by Rottenberg *et al.* (1970):

$$q = \left(\frac{L_{PH}^2}{L_{PP} L_{HH}}\right)^{1/2} = \left(\frac{(n_H^P)^2 L_P}{(n_H^P)^2 L_P + L_H^1}\right)^{1/2} \tag{55}$$

It is evident that at $L_H^1 = 0$ the coupling is complete ($q = 1$). With increasing leakiness of the membrane toward protons, q decreases. In the limit $L_H^1 \to \infty$, $q = 0$, i.e., the ATP hydrolysis is completely uncoupled from proton transport.

Deviations of q from 1 or 0 give us the phenomenological information that the coupling is not complete. However, under such circumstances, no further information concerning mechanistic details can be derived from the system without a description that further increases understanding of the relation between thermodynamic parameters and molecular events.

IV. Oxidative Phosphorylation in Mitochondria

A. The Idealized Mitochondrion

Mitochondria are the subcellular organelles that catalyze the synthesis of ATP from ADP and phosphate, coupled to the oxidation of substrates by molecular oxygen. They consist of an outer membrane, which allows free permeation of all low molecular weight solutes, an extensively folded inner membrane, which functions as the energy transducer, and an inner (matrix) space. For the purpose of our paper the outer membrane can be neglected.

The inner-membrane surface-to-matrix volume ratio is quite high: per gram protein approximately 1 ml of matrix and 40 m^2 surface are present (Mitchell, 1968). For this reason, relatively small fluxes per unit surface cause relatively large concentration changes in the matrix space. Also, the electrical capacity of the membrane relative to the buffer capacity of the matrix space is not insignificant (Mitchell, 1968).

The mitochondria themselves are small vesicles, with maximal diameter of the order of 1 μm. Thus, the volume enclosed in each vesicle is very small, of the order or 10^{-15} l or less. For certain components, such as H^+ ions, this poses statistical problems. At a matrix pH of 9 (which occurs under certain circumstances), there must be only a few free protons per mitochondrion. But even at pH 7.5, the number of H^+ and OH^- ions is too small to justify the usual statistical treatment as a homogeneous volume element. This problem has been noted and has been circumvented by considering the time average on top of the local composition average (Mitchell, 1966).

These statistical problems are aggravated by the fact that mitochondria are not uniform in size and that—conceivably—their membrane may also vary in composition. In view of the large surface-to-volume ratio, the latter problem is probably not serious; for instance, one heart mitochondrion contains more than 50,000 respiratory-chain complexes (Lehninger, 1975).

To be able to derive any equations we have to idealize the mitochondrion to a certain extent. We will consider the mitochondria as a homogeneous population as far as surface-to-volume ratio is concerned (see also Section II,C), and the inner membrane is postulated to have a two-dimensionally homogeneous structure. Formally speaking, the sum of all the mitochondria in a certain experiment will be taken equivalent to one large mitochondrion having the total surface and internal volume.

B. The Chemiosmotic Hypothesis

Although other models can also be described, the chemiosmotic model of oxidative phosphorylation is ideally suited for the type of thermodynamic treatment we have adopted in this paper (Van Dam and Westerhoff, 1977). According to this hypothesis, the coupling between redox reactions and ATP synthesis occurs via the intermediate of an electrochemical gradient of protons across the mitochondrial inner membrane (Mitchell, 1961).

The H^+-translocating ATPase has already been considered in Section III,C. We can now add a similar H^+-translocating respiratory chain. Analogous to the ATPase, we can derive for the coupled fluxes of protons and redox equivalents.

$$J_O = L_O A_O + n_H^O L_O \Delta\tilde{\mu}_H \tag{56}$$

$$J_H = n_H^O L_O A_O + (n_H^O)^2 L_O \Delta\tilde{\mu}_H = n_H^O J_O \tag{57}$$

Taking the ATPase, redox reactions, and the possibility of an H^+ leak (see Section III,D and Fig. 6) together, the total set of equations describing oxidative phosphorylation according to the chemiosmotic model is:

$$\begin{pmatrix} J_P \\ J_H \\ J_O \end{pmatrix} = \begin{pmatrix} L_P & n_H^P L_P & 0 \\ n_H^P L_P & (n_H^P)^2 L_P + (n_H^O)^2 L_O + L_H^1 & n_H^O L_O \\ 0 & n_H^O L_O & L_O \end{pmatrix} \begin{pmatrix} A_P \\ \Delta\tilde{\mu}_H \\ A_O \end{pmatrix} \tag{58}$$

The lack of direct coupling between oxidation and phosphorylation, as indicated by the zeros for the respective coupling constants, is a direct result of the model we chose to describe. In fact, the disappearance of these cross-coefficients is the most notable difference between these equations and the ones that can be derived by the same procedures, starting from a model of energy transduction, involving chemical high-energy intermediates.

It can be quickly verified that if $L_H^1 = 0$, i.e., there is no leak of protons through the membrane, then in the steady state where there is no net proton flux across the membrane ($J_H = 0$), the following equation holds:

$$n_H^P J_P + n_H^O J_O = 0 \tag{59}$$

It precisely describes the relation between oxidation and phosphorylation under conditions of complete coupling. Only in this case the empirical P/O ratio ($-J_P/J_O$) gives us information about the mechanistic coupling ratios n_H^P and n_H^O. In the normal case, where the mitochondria are slightly uncoupled (and *a fortiori* under conditions where protonophorous uncoupler has been added), no such simple relation between J_P and J_O exists.

C. The Problem of Substrate Transport

1. *The Redox Substrates*

The reactive sites for most mitochondrial redox substrates lie at the matrix side of the inner membrane, so that reduced and oxidized substances have to move across this membrane and even other reactions have to occur (see Fig. 7, lower part). Furthermore, the A_O is defined at the reactive site, whereas we can more accurately measure substances outside the mitochondria. Both these factors can be introduced into our equations.

The often used substrate succinate moves across the membrane in a 1 : 1 exchange for malate and is oxidized primarily to fumarate (Chappell and Haarhoff, 1967). We can add the transport reaction and the (de)hydration reactions to the dissipation function:

$$\Phi = J_O A_O^{in} + J_H^O \Delta\tilde{\mu}_H + J_1 A_1^{in} + J_m \Delta\tilde{\mu}_{mal} + J_s \Delta\tilde{\mu}_{suc} + J_2 A_2^{ex} \quad (60)$$

with

$$A_1^{in} = (\tilde{\mu}_{fum}^{in} - \tilde{\mu}_{mal}^{in}) \quad \text{and} \quad A_2^{ex} = (\tilde{\mu}_{fum}^{ex} - \tilde{\mu}_{mal}^{ex})$$

The 1 : 1 exchange of substrates across the membrane can be accounted for by taking

$$J_m = -J_S \quad (61)$$

Furthermore, $J_H^O = n_H^O J_O$, so that we obtain

$$\Phi = J_O(A_O^{in} + n_H^O \Delta\tilde{\mu}_H) + J_1 A_1^{in} + J_m(\Delta\tilde{\mu}_{mal} - \Delta\tilde{\mu}_{suc}) + J_2 A_2^{ex} \quad (62)$$

Each of the four terms in this dissipation function is independent of any of the others, so that in the flow-force equations only the straight coefficients differ from zero.

We can define a steady-state situation in which the intramitochondrial substrates no longer change in concentration. This is reached when

$$J_O = J_1 = J_m = -J_2 \quad (63)$$

This further simplifies the dissipation function to

$$\begin{aligned}\Phi &= J_O\{(A_O^{in} + n_H^O \Delta\tilde{\mu}_H) + A_1^{in} + (\Delta\tilde{\mu}_{mal} - \Delta\tilde{\mu}_{suc}) - A_2^{ex}\} \\ &= J_O\{A_O^{ex} + n_H^O \Delta\tilde{\mu}_H\}\end{aligned} \quad (64)$$

Since each of the fluxes depends only on its own force, we can also write

$$\Phi = J_O[J_O/L_O + J_O/L_1 + J_O/L_m + J_O/L_2] = J_O(J_O/L_O^*) \quad (65)$$

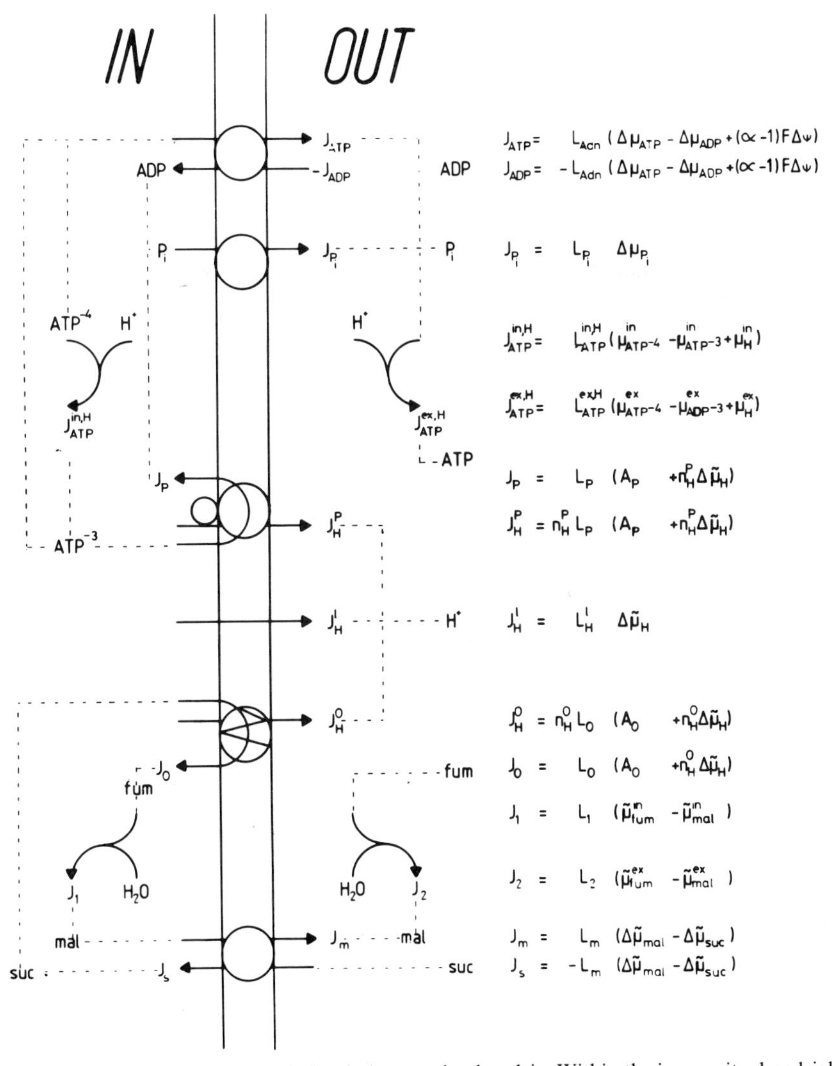

FIG. 7. Oxidative phosphorylation in intact mitochondria. Within the inner mitochondrial membrane the adenine nucleotide translocator, the phosphate translocator, the ATPase–proton pump, the proton leak, the oxidase–proton pump, and the dicarboxylic acid translocator are depicted. Furthermore, acid–base reactions of ATP and hydration of fumarate may occur.

in which

$$1/L_{\mathrm{O}}^{*} \doteq 1/L_{\mathrm{O}} + 1/L_{1} + 1/L_{\mathrm{m}} + 1/L_{2} \tag{66}$$

so that

$$J_{\mathrm{O}} = L_{\mathrm{O}}^{*}[A_{\mathrm{O}}^{\mathrm{ex}} + n_{\mathrm{H}}^{\mathrm{O}}\Delta\tilde{\mu}_{\mathrm{H}}] \tag{67}$$

The only difference between this equation and the one with the internal affinity lies in the modified proportionality constant. This proportionality constant is obtained as though the resistance of the additional reactions were added in series with that of the oxidation reaction itself.

2. *The Substrates for the ATPase*

In principle, the substrate transport problems for the ATPase, which also has its active site at the matrix side of the membrane, can be treated in an analogous way as described above for the redox reaction (see Fig. 7, upper part).

One extra complication that arises in this case is that the 1 : 1 exchange of ATP for ADP is electrogenic to an unknown extent (Klingenberg and Rottenberg, 1977). It can be easily verified that if $(1 - \alpha)$ excess negative charges were moving together with the ATP and P_i were moving electroneutrally, this would formally be equivalent to the movement of $(1 - \alpha)$ H^+ ions in the opposite direction in the steady state. These H^+ ions have to be added to the stoichiometric number of H^+ ions transported during hydrolysis of one molecule of ATP$(n_{\mathrm{H}}^{\mathrm{P}})$.

The movement of phosphate is electroneutral and can be described as a movement of phosphoric acid (Chappell and Crofts, 1966).

Looking again at the steady state in which the intramitochondrial substrate concentrations are constant, we derive (Van Dam and Westerhoff, 1977)

$$\Phi = J_{\mathrm{P}}\{A_{\mathrm{P}}^{\mathrm{ex}} + (n_{\mathrm{H}}^{\mathrm{P}} + 1 - \alpha)\Delta\tilde{\mu}_{\mathrm{H}}\} \tag{68}$$

and

$$J_{\mathrm{P}} = L_{\mathrm{P}}^{*}\{A_{\mathrm{P}}^{\mathrm{ex}} + (n_{\mathrm{H}}^{\mathrm{P}} + 1 - \alpha)\Delta\tilde{\mu}_{\mathrm{H}}\} \tag{69}$$

in which

$$1/L_{\mathrm{P}}^{*} = (1/L_{\mathrm{P}} + 1/L_{\mathrm{Pi}} + 1/L_{\mathrm{AdN}}) \tag{70}$$

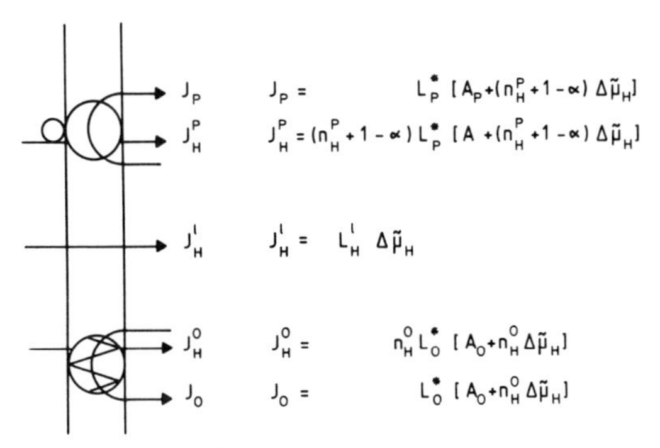

FIG. 8. Black box description of mitochondrial oxidative phosphorylation. The equations in terms of extramitochondrial affinities and flows describe the steady state of the processes depicted in Fig. 7. $(1 - \alpha)$ represents the degree of electrogenicity of the adenine nucleotide translocator.

The total set of equations describing oxidative phosphorylation in intact mitochondria can now be given

$$\begin{pmatrix} J_O \\ J_H \\ J_P \end{pmatrix} = \begin{pmatrix} L_O^* & n_H^O L_O^* & 0 \\ n_H^O L_O^* & (n_H^O)^2 L_O^* + (n_H^P + 1 - \alpha)^2 L_P^* + L_H^1 & (n_H^P + 1 - \alpha) L_P^* \\ 0 & (n_H^P + 1 - \alpha) L_P^* & L_P^* \end{pmatrix} \times \begin{pmatrix} A_O^{ex} \\ \Delta\tilde{\mu}_H \\ A_P^{ex} \end{pmatrix} \tag{71}$$

These equations should hold under conditions of constant intramitochondrial composition. We conclude that in the appropriate steady states the mitochondrial system with its substrate concentrations can be summarized as the black box shown in Fig. 8 (compare to Fig. 7).

D. EXPERIMENTS WITH SUBMITOCHONDRIAL PARTICLES

Submitochondrial particles are derived from mitochondria, usually through breakage by ultrasonic irradiation. An important characteristic of these mitochondrial fragments is that they form closed vesicular structures, but with the orientation of the membrane inverted (Lee and Ernster, 1966): the active sites of ATPase and dehydrogenases are now directly accessible from the external medium. This circumvents the problems, connected with substrate transport, discussed in Section IV,C.

Although for this reason submitochondrial particles would be the material of choice for testing the derived equations, they have certain disadvantages that have limited their use in this respect. In the first place, submitochondrial particles are relatively leaky toward protons, which results in poor coupling between phosphorylation and oxidation and the absence of strong respiratory control. Second, the usual procedures for tracing intravesicular probes for pH or ψ, involve centrifugation or filtration; both techniques are more difficult to apply to submitochondrial particles than to intact mitochondria because of the relatively small size of the former. The quantitative correctness of the use of fluorescent probes for pH measurement (Rottenberg and Lee, 1975) can be questioned.

One type of experiment concerns the static-head situation in which oxidation has led to buildup of a maximal affinity of the phosphorylation reaction. In this situation (see Section III,C)

$$J_P = n_H^P L_P \Delta\tilde{\mu}_H + L_P A_P = 0 \tag{72}$$

from which follows

$$(-A_P/\Delta\tilde{\mu}_H)^{*p} = n_H^P \tag{73}$$

(in submitochondrial particles we can use the proportionality constants without asterisk and the factor $(1 - \alpha)$ because adenine nucleotide transport is absent).

Most measurements of A_P and $\Delta\tilde{\mu}_H$ available in the literature, are not usable here, as they were not obtained under identical conditions. In those cases where they can be used, the ratio $-(-A_P/\Delta\tilde{\mu}_H)$ varies between 2 and 5 (Rottenberg and Lee, 1975; Van Dam *et al.*, 1978b; Azzone *et al.*, 1978c; Sorgato *et al.*, 1978).

It should be stressed that under conditions where there truly is no net flux through the ATPase, $A_P/\Delta\tilde{\mu}_H$ must be constant independent of the leakiness of the rest of the membrane. The fact that in practice this ratio is not constant and differs from 2 as predicted by the current model of oxidative phosphorylation will be discussed below.

Another interesting example of experiments that can be done to test the equations for oxidative phosphorylation can be found in the work of Rottenberg and Gutman (1977). These authors measured ATP-driven reversed electron transport in submitochondrial particles. The coupled transfer of electrons from succinate to NAD^+ and ATP hydrolysis appears to be completely reversible: reduction of fumarate by NADH leads to ATP synthesis.

The rate of the reaction turned out to be linear with both the free energy of the redox reaction (A_O) and the free energy of the ATPase reaction (A_P).

This can be explained only within the framework of the chemiosmotic hypothesis, if the $J_H = 0$ steady state had been reached in the 3–5 seconds after addition of ATP, in which the authors measured the rate of conversion of NAD^+ to NADH. If we accept this steady-state condition, then from Eq. (58)

$$J_O^{*h} = \frac{[(n_H^P)^2 L_P + L_H^1] L_O A_O - n_H^O L_O n_H^P L_P A_P}{(n_H^P)^2 L_P + L_H^1 + (n_H^O)^2 L_O} \tag{74}$$

Apart from the linearity, this equation also predicts that in the plot of J_O versus A_P (if A_O is taken relatively constant) the intersection of the line with the A_P axis occurs at higher A_P, if a protonophore (which increases L_H^1) is added. In the same plot the slope of the line should decrease with the addition of protonophore. Both these predictions are confirmed by experiments described in the article of Rottenberg and Gutman (1977). These authors also plotted A_O versus A_P for the cases where $J_O = 0$. The above equation predicts that

$$\left(\frac{A_O}{A_P}\right)^{*h*o} = \frac{n_H^O}{n_H^P} \cdot \frac{1}{1 + [L_H^1/(n_H^P)^2 L_P]} \tag{75}$$

This prediction is a bit more precise than the prediction emerging straightaway from the second law of thermodynamics

$$\left(\frac{A_O}{A_P}\right) \leqslant \left(\frac{n_H^O}{n_H^P}\right) \tag{76}$$

Although the slope of A_O versus A_P approximately equals 0.8, the ratio of these two forces varies between 1.25 and 1.5. Thus, as the authors conclude, the experiments seem to falsify both chemical and chemiosmotic models in which the "P/O per site" is postulated to equal 1. However, the authors did not prove that appearance of NADH and fumarate were stoichiometric; a small leak across the cyanide block could account for an increase in A_O/A_P:

$$\left(\frac{A_O}{A_P}\right)^{*o*h} = \frac{n_H^O}{n_H^P} \cdot \frac{1}{1 + [L_H^1/(n_H^P)^2 L_P] + \{[(n_H^O)']^2 L_O'/(n_H^P)^2 L_P\}} + \frac{n_H^O (n_H^O)' L_O'}{(n_H^P)^2 L_P + L_H^1 + [(n_H^O)']^2 L_O'} \cdot \frac{A_O'}{A_P} \tag{77}$$

Here the primes indicate the leakage reaction. Additional experiments checking this point and the point of attainment of the $J_H = 0$ steady state should be carried out.

E. Experiments with Intact Mitochondria

In intact mitochondria much more experimental material is available to test the equations (Rottenberg, 1970; Padan and Rottenberg, 1973; Nicholls, 1974, 1977; Küster *et al.*, 1976; Van Dam *et al.*, 1978; Azzone *et al.*, 1978a,b).

We first look at the condition of maximal affinity of the phosphorylation reaction (state IV; Chance and Williams, 1955):

$$J_P = (n_H^P + 1 - \alpha)L_P^* \Delta\tilde{\mu}_H + L_P^* A_P^{ex} = 0 \tag{78}$$

or

$$\left(\frac{-A_P^{ex}}{\Delta\tilde{\mu}_H}\right)^{*p} = (n_H^P + 1 - \alpha) \tag{79}$$

We must be careful to relate only parameters that have been obtained under identical conditions. Yet several examples of such data can be found in the literature. The value of $n_H^P + (1 - \alpha)$ deduced from these comparisons is always greater than 2 and usually smaller than 3. A notable point of dispute is that some authors claim that this value increases with the degree of uncoupling of the membrane (Van Dam *et al.*, 1978b; Azzone *et al.*, 1978b), whereas others find no such variation (Nicholls, 1977).

Another steady-state situation that can be approached relatively rapidly, especially in the absence of ionophores, is the one in which there is no net flux of protons across the membrane ($J_H = 0$). In this situation, we can eliminate $\Delta\tilde{\mu}_H$ from the equations and derive the following relation between rates of oxidation and phosphorylation:

$$(J_P)^{*h} = -\left(\frac{n_H^O}{n_H^P} + \frac{L_H^1}{n_H^O \cdot n_H^P \cdot L_O^*}\right) \cdot J_O + \frac{L_H^1}{n_H^O \cdot n_H^P} A_O \tag{80}$$

The slope of the line relating J_P and J_O at constant A_O should depend (among others) on the value of L_O^*. If $L_O^* \gg L_H^1$, the slope will approach the value n_H^O/n_H^P, which may be related to the "theoretical P/O ratio." In experiments with rat liver mitochondria where L_O^* was varied by adding inhibitors of the oxidation, it was found that at the highest rates of oxidation the slope approached 2 for succinate and 3 for β-hydroxybutyrate (Tsou and Van Dam, 1969). Later experiments (Van Dam *et al.*, 1978a) confirmed the linearity of the relation between J_P and J_O with a slope that increased with uncoupler. Furthermore, if a protonophorous uncoupler was added to increase L_H^1, as predicted a higher rate of oxidation (larger L_O^*) was required before these values of the slope were reached.

In intact mitochondria also the more complicated situation, where finite fluxes of both oxidation and phosphorylation occur, has been tested.

It was first found by Rottenberg (1973) that the Onsager reciprocal relation was valid, i.e., the dependence of oxidation rate on A_P was equal to the dependence of phosphorylation rate on A_O. In these experiments, however, it also appeared that the fluxes were linear with the forces but not proportional to them.

In a later series of experiments, Van Dam *et al.*, (1978b) tested the Eqs. (71) while varying the flux through the phosphorylation reaction by addition of glucose and hexokinase. They concluded that a reasonable fit between experimental results and equations could be obtained with certain proportionality constants. The stoichiometric number of protons, moving across the membranes per oxidation of one molecule of succinate or the synthesis of one molecule of ATP were slightly above 4 and 2, respectively. Recalculation of the data shows that n_H^O was just above eight.

F. Localized Chemiosmosis?

Overall, the predictions of the equations describing oxidative phosphorylation on the basis of a chemiosmotic model, are borne out by the experiments designed to test them. However, some disturbing discrepancies remain, such as the variability of $A_P/\Delta\tilde{\mu}_H$ at $J_P = 0$.

One possible explanation of this discrepancy is that the measurements of $\Delta\tilde{\mu}_H$ are wrong. Indeed, each of the experimental methods used (centrifugation, filtration, flow dialysis, ion-selective electrodes) has its possible artifacts. However, in those instances where these methods have been directly compared, they led to not significantly different results.

We are faced with the choice to reject the chemiosmotic model or to adopt it so as to accommodate these anomalies. We chose the latter and suggested the introduction into the model of a region of limited accessibility for protons around the respiratory-chain and ATPase supercomplexes (Van Dam *et al.*, 1978b; cf. Azzone *et al.*, 1978a,b; and, for chloroplasts, Ort *et al.*, 1976). Phenomenologically this results in an apparent partial direct coupling between oxidation and phosphorylation (see the parallel coupling model of Padan and Rottenberg, 1973). The $\Delta\tilde{\mu}_H$ measured between the bulk aqueous phases inside and outside the mitochondria is then an underestimate as compared to the local $\Delta\tilde{\mu}_H$ experienced by the ATPase and respiratory chain. For this reason the adopted model could be called localized chemiosmosis. It differs fundamentally from the localized proton hypothesis of Williams (see review, Williams, 1978) in which 2 phases instead of 3 (4) are postulated to play a role in the mechanism of energy transduction.

V. Light-Energized Systems

A. Bacteriorhodopsin Liposomes

1. Properties of the System

Bacteriorhodopsin is a protein occurring in the cytoplasmic membrane of the extremely halophilic bacterium *Halobacterium halobium* (Oesterhelt and Stoeckenius, 1971). It can be easily extracted from this membrane and recombined with added lipid to form small unilamellar vesicles (Oesterhelt and Stoeckenius, 1974).

In these vesicles, bacteriorhodopsin is able to transport protons across the membrane at the expense of absorbed light quanta (Kayushin and Skulachev, 1974). This is the same process it catalyzes in the intact cell, where the gradient of protons, formed in this way, can be utilized to drive endergonic processes, such as ATP synthesis (Hartmann *et al.*, 1977).

The reconstituted bacteriorhodopsin vesicles are an almost ideal system for the study of energy transduction from light to $\Delta\tilde{\mu}_H$. They are composed of precisely defined building blocks, are stable, can be obtained in large quantities, and are easy to manipulate, since energization is effected by illumination and does not involve disturbance of the composition of the suspension. For these reasons, the amount of work done on this system in the past few years has been substantial.

As will be shown below, the bacteriorhodopsin liposomes also lend themselves to a description within the framework of linear irreversible thermodynamics. This description and the outcome of experiments to test it may lead to new insights into the functioning of this light-dependent H^+ pump.

2. Development of a Model from Building Blocks

As discussed in Section II, we need consider only the fluxes of solutes across the membrane and (in this case) the light-dependent proton flux, to describe the system of energy-transducing vesicles (Westerhoff *et al.*, 1979). As an extra complication in the case of reconstituted bacteriorhodopsin vesicles, we have to account for the possibility that within one vesicle the pumps are oriented in opposite directions with respect to the membrane in which they are embedded (Gerber *et al.*, 1977). We will neglect the possibility of the presence of two populations of vesicles, each with all the pumps orientated in one direction or the other (Hellingwerf *et al.*, 1978b).

The building blocks required for our model have been depicted in Fig. 9. Each of the units functions independent of the others.

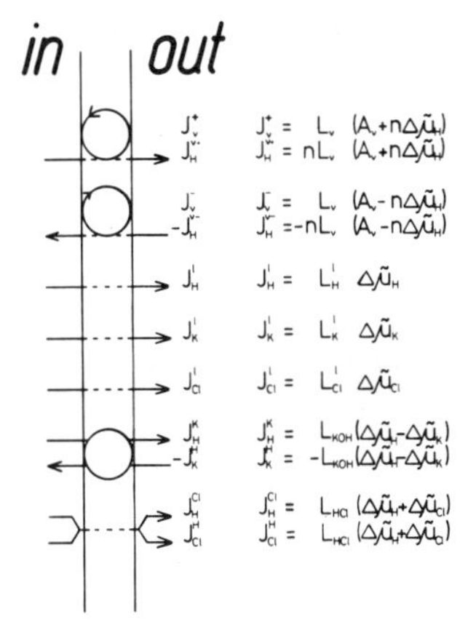

FIG. 9. Light-driven ion movements in a vesicular system. Within the membrane, an outward proton pump, an inward proton pump, a proton, potassium and chloride leak, a K^+/H^+ exchange, and a Cl^-/OH^- exchange are depicted.

Several important elements are introduced in this figure. Since light absorption causes proton movement, the force across the pump should at least contain a component exerted by this photon absorption. We give this part of the force the name A_v. Furthermore, at least near equilibrium, J_H^v should be sensitive to $\Delta\tilde{\mu}_H$ back pressure. The total force across the pump is described by $(A_v + n\Delta\tilde{\mu}_H)$ (cf. Section III,C).

We will confine our treatment to those reactions of bacteriorhodopsin that result in proton pumping ($nJ_v = J_H^v$). The dissipation of part of the photon force not coupled to H^+ movement is not taken into consideration (cf. Section II,C,2).

The dissipation function describing bacteriorhodopsin vesicles can then be written as

$$\Phi = J_v A_v + J_H \Delta\tilde{\mu}_H + J_K \Delta\tilde{\mu}_K + J_{Cl} \Delta\tilde{\mu}_{Cl} \tag{81}$$

in which

$$J_v = (1 - \alpha)J_v^+ + \alpha J_v^- \tag{82}$$

$$J_H = (1 - \alpha)J_H^{v+} + \alpha J_H^{v-} + J_H^1 + J_H^K + J_H^{Cl} \tag{83}$$

$$J_K = J_K^1 + J_K^H \tag{84}$$

$$J_{Cl} = J_{Cl}^1 + J_{Cl}^H \tag{85}$$

We can combine the relations of Fig. 9 with these flux equations and obtain the following set of equations:

$$\begin{pmatrix} J_v \\ J_H \\ J_K \\ J_{Cl} \end{pmatrix} = \begin{pmatrix} L_v & n(1-2\alpha)L_v & 0 & 0 \\ n(1-2\alpha)L_v & n^2L_v + L_H^1 + L_{KOH} + L_{HCl} & -L_{KOH} & -L_{HCl} \\ 0 & -L_{KOH} & L_K^1 + L_{KOH} & 0 \\ 0 & L_{HCl} & 0 & L_{Cl}^1 + L_{HCl} \end{pmatrix} \times \begin{pmatrix} A_v \\ \Delta\tilde{\mu}_H \\ \Delta\tilde{\mu}_K \\ \Delta\tilde{\mu}_{Cl} \end{pmatrix} \quad (86)$$

The great advantage of this set of equations over the generalized set of phenomenological equations is that it visualizes how each of the proportionality constants reflects the underlying mechanisms. Thus, for example, the effect of degree of orientation of the bacteriorhodopsin in the membrane can be quantitatively evaluated by the parameter α, which can change between 0 and 1 (if $\alpha = 0.75$, 75% of the bacteriorhodopsin molecules pump inward and 25% pump outward; here n is a positive integer, probably 1). Variations in α change the proportionality constant relating J_v and $\Delta\tilde{\mu}_H$ only between $n \cdot L_v$ and $-n \cdot L_v$.

3. *Light as a Thermodynamic Force*

It is not *a priori* clear how the absorption of photons by bacteriorhodopsin must be translated into thermodynamic variables. The number of photons per unit time met by each bacteriorhodopsin molecule could be accommodated either in A_v or in L_v. In our model (Westerhoff *et al.*, 1979) we have adopted the following approach. In each photochemical cycle a bacteriorhodopsin molecule absorbs only one photon, and therefore A_v is expected to be independent of the light intensity (of course A_v may depend on the energy of the photon, i.e., the wavelength of the incident light, though for a given chromophore such a dependence would be limited owing to the discrete nature of light absorption). If each bacteriorhodopsin molecule undergoes a photochemical cycle driven by a certain thermodynamic force only after absorption of a photon, then in the absence of saturation effects the total flux through all photochemical cycle is proportional to the number of bacteriorhodopsin molecules that absorb a photon: L_v will be proportional to the bacteriorhodopsin concentration as well as to the intensity of illumination. Every single molecule will still be subject to the balance between A_v and $\Delta\tilde{\mu}_H$.

In mechanistic terms we can consider bacteriorhodopsin in a membrane equivalent to an ideal voltage source, in which absorption of a photon at the same time opens the pathway for protons across the membrane and exposes the proton to the thermodynamic force $(A_v + n\Delta\tilde{\mu}_H)$.

4. *Experiments with Bacteriorhodopsin Liposomes*

Because of the complexity of the equations describing ion fluxes and coupled light absorption in bacteriorhodopsin vesicles, we will have to content ourselves in the beginning with testing predictions under well defined steady-state conditions. One such steady state is the one in which net fluxes of ions have ceased.

Specifically, if electrical and proton flux are zero, we can derive

$$(\Delta\mu_H + \Delta\mu_{Cl})^{*e*h} = \frac{-n(1-2\alpha)L_e L_v A_v + \{L_{eg}L_{KOH} + L_K^1(n^2L_v + L_H^1)\}(\Delta\mu_K + \Delta\mu_{Cl})^{*e*h}}{L_n L_{eg} + L_e(n^2L_v + L_H^1)} \tag{87}$$

If the medium contains 150 mM KCl, it can be calculated that the value of $(\Delta\mu_K + \Delta\mu_{Cl})^{*e*h}$ must remain very small. Also $(\Delta\mu_{Cl})^{*e*h}$ itself will be negligible as compared to $(\Delta\mu_H)^{*e*h}$ so that we can approximate the steady-state pH gradient by

$$(\Delta\mu_H)^{*e*h} = \frac{-n(1-2\alpha)L_v A_v}{(n^2L_v + L_H^1)[1 + (L_n/L_e)] + L_n} \tag{88}$$

Similarly

$$F(\Delta\psi)^{*e*h} = \frac{-n(1-2\alpha)L_v A_v}{(n^2L_v + L_H^1)[1 + (L_e/L_n)] + L_e} \tag{89}$$

and thus

$$(\Delta\tilde{\mu}_H)^{*e*h} = \frac{-n(1-2\alpha)L_v A_v}{(n^2L_v + L_H^1) + (L_n L_e)/(L_n + L_e)} \tag{90}$$

We will now look at the effects of the K^+ ionophore valinomycin and the K^+/H^+ exchanger nigericin (Ovchinnikov *et al.*, 1974) on the steady-state proton uptake. For simplicity, we assume

$$L_{HCl} \ll L_{KOH} \quad \text{and} \quad L^1_{Cl} \ll L^1_K$$

which reduces the above equation to

$$(\Delta\mu_H)^{*e*h} = \frac{-n(1-2\alpha)L_v A_v}{(n^2 L_v + L^1_H)(1 + L_{KOH}/L^1_K) + L_{KOH}} \tag{91}$$

$$F(\Delta\psi)^{*e*h} = \frac{-n(1-2\alpha)L_v A_v}{(n^2 L_v + L^1_H)(1 + L^1_K/L_{KOH}) + L^1_K} \tag{92}$$

and

$$(\Delta\tilde{\mu}_H)^{*e*h} = \frac{-n(1-2\alpha)L_v A_v}{(n^2 L_v + L^1_H) + L_{KOH}L^1_K/(L_{KOH} + L^1_K)} \tag{93}$$

If now the proton conductivity of bacteriorhodopsin and the membrane in parallel is larger than that of K^+/H^+ exchange and K^+ leak in series, i.e., if

$$L_{KOH}L^1_K/(L_{KOH} + L^1_K) \ll n^2 L_v + L^1_H$$

then we can further simplify the equations to

$$(\Delta\mu_H)^{*e*h} = \frac{(\Delta\tilde{\mu}_H)^{*e*h}_{max}}{1 + L_{KOH}/L^1_K} \tag{94}$$

$$F(\Delta\psi)^{*e*h} = \frac{(\Delta\tilde{\mu}_H)^{*e*h}_{max}}{1 + L^1_K/L_{KOH}} \tag{95}$$

and

$$(\Delta\tilde{\mu}_H)^{*e*h} = (\Delta\tilde{\mu}_H)^{*e*h}_{max} = \frac{-n(1-2\alpha)L_v A_v}{n^2 L_v + L^1_H} \tag{96}$$

These equations state that, at moderately low valinomycin and nigericin concentrations, their effect on $(\Delta\tilde{\mu}_H)^{*e*h}$ will be negligible, since their effect on $(\Delta\mu_H)^{*e*h}$ will be balanced by their effect on $F(\Delta\psi)^{*e*h}$. At higher concentrations of the ionophores, the approximation no longer holds and their addition will result in a decrease in $(\Delta\tilde{\mu}_H)^{*e*h}$.

These predictions were verified with bacteriorhodopsin liposomes, using the flow dialysis technique to determine the distributions across the vesicle membrane of CNS^- and methylamine, indicators of $\Delta\psi$ and ΔpH, respectively (Hellingwerf *et al.*, 1979).

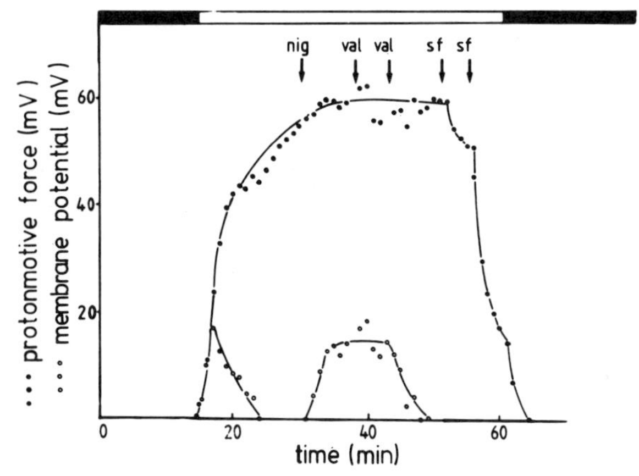

FIG. 10. Proton-motive force (●——●) and membrane potential (○——○) in illuminated vesicles reconstituted from bacteriorhodopsin and purified lipids. (For details, see Hellingwerf *et al.*, 1979.) Illumination period indicated by white area in the bar above the figure. Nig (nigericin), val (valinomycin), and sf (sf 6847, a protonophore) were added at the times indicated.

As can be seen in Fig. 10, the predictions are borne out by the experimental results. Addition of valinomycin after nigericin results in an increase in ΔpH at the expense of $\Delta\psi$, so that $\Delta\tilde{\mu}_H$ remains constant.

Another important question that must be raised in connection with our description of bacteriorhodopsin action is whether there is indeed a back-pressure of $\Delta\tilde{\mu}_H$ on the light-dependent pump. Two different lines of evidence suggest that this is the case.

In the first place, it was shown by Hellingwerf *et al.*, (1978a) that the kinetics of the change in concentration of one of the intermediates of the photocycle induced by flash excitation are influenced by $\Delta\tilde{\mu}_H$ (manipulated by background illumination and addition of ionophores) in a manner compatible with a back pressure control of $\Delta\tilde{\mu}_H$ on the pump.

A second piece of evidence is derived from a model system in which bacteriorhodopsin vesicles are attached to lipid-impregnated Millipore filters (Hellingwerf *et al.*, 1979; see also Drachev *et al.*, 1974). In this system a light-induced electrical potential can be measured directly. If an external voltage is applied to the filter in the dark, the magnitude of the subsequent light-induced change in potential varies linearly with the applied voltage. This proves that the bacteriorhodopsin pump is at least regulated by the $\Delta\psi$ component of $\Delta\tilde{\mu}_H$ [see also Herrmann and Rayfield (1978)]. Furthermore, since the applied potential at which the light-induced change in potential is reduced to zero turns out to be independent of light intensity (Hellingwerf *et al.*, 1979), we obtain evidence from this system that A_v does not depend on light intensity

$$A_v = -n(\Delta\tilde{\mu}_H)_{J^v_H=0}$$

It has been suggested that the stoichiometry of the bacteriorhodopsin proton pump is not a constant and approaches zero at high $\Delta\tilde{\mu}_H$. At present we cannot fully exclude a variable stoichiometry. However, if the variable stoichiometry (m) is related to $\Delta\tilde{\mu}_H$ via the equation

$$m = n[1-(n\Delta\tilde{\mu}_H/A_v)] \qquad (\alpha=1;\text{ all pumps assumed to pump inward}) \qquad (97)$$

then the flux through the pump

$$J^v_H = -mL_vA_v \qquad (98)$$

will be identical to the one given by our description with constant stoichiometry (n).

B. Chloroplasts

1. *Thylakoid Membranes*

Energy transduction from light to ATP in chloroplasts is catalyzed by the membranes of the grana stacks or thylakoids (for review, see Avron, 1977). These thylakoid membranes contain light-harvesting pigments, reaction centers (photosystems I and II), carotenoids, redox components, and an ATPase complex comparable to that of mitochondria.

In a first approximation, the photosynthetic electron-transport chain from water to $NADP^+$, including the two photosystems, might be considered analogous to the mitochondrial respiratory chain. In both, the movement of reducing equivalents along the chain is coupled to generation of a transmembrane $\Delta\tilde{\mu}_H$. This gradient of protons can be used to drive the synthesis of ATP (Jagendorf and Uribe, 1966). Here we will not work out a set of equations for photophosphorylation. Rather, we take advantage of the property of chloroplasts that they are able to generate very quickly a measurable membrane potential and, thus, a $\Delta\tilde{\mu}_H$ upon illumination. We will examine an experiment in which the effect of flash-induced $\Delta\tilde{\mu}_H$ on the activity of the chloroplast ATPase complex was investigated.

2. *Experiments with Chloroplasts*

The equations derived for the bacteriorhodopsin liposomes can be used to a large extent for describing the uptake and release of protons by chloroplasts in the absence of ADP and phosphate. As an example, we would predict similar antagonistic effects of valinomycin and nigericin on light-induced H^+ uptake. Such effects have indeed been described by Fiolet (1975), who demonstrated that the decrease in proton uptake, caused by nigericin, could be almost completely overcome by subsequent addition of valinomycin. Fully parallel changes were observed in the quenching of the fluorescence of acridines, which are supposedly recording changes in ΔpH.

A much more complicated situation arises if we also allow for activity of the phosphorylating enzymes, especially because the chloroplast ATPase seems to be strongly regulated by dissociation or alterations in binding of its naturally occurring inhibitor protein (Bakker–Grunwald and Van Dam, 1974; see also van de Stadt *et al.*, 1973.) Under energized conditions the ATPase appears to be active, whereas deenergized conditions lead to inactivation of the ATPase. An interesting question is whether the inhibitor dissociation is promoted by the pH gradient, by the membrane potential, or by $\Delta\tilde{\mu}_H$. Some experiments reported by Harris and Crofts (1978) might answer this question.

In these experiments, illumination was not continuous but consisted of repetitive saturating flashes. The authors found that ATP synthesis in chloroplasts did not always start immediately upon illumination: the duration of the lag decreased with the flash frequency. The relevant equations describing the necessary parameters are analogous to those used for the bacteriorhodopsin liposomes (under conditions of electroneutral flow, which will be reached after a few flashes).

$$\Delta\tilde{\mu}_H^{*e} = \Delta\mu_H^{*e} + F\Delta\psi^{*e} \tag{99}$$

$$F\Delta\psi^{*e} = \frac{-nL_v A_v - (n^2 L_v + L_H^1)\Delta\mu_H^{*e} - L_K^1 \Delta\mu_K^{*e}}{n^2 L_v + L_H^1 + L_K^1 + L_{Cl}^1} \tag{100}$$

$$\Delta\tilde{\mu}_H^{*e} = \frac{-nL_v A_v + (L_K^1 + L_{Cl}^1)\Delta\mu_H^{*e} - L_K^1 \Delta\mu_K^{*e}}{n^2 L_v + L_H^1 + L_K^1 + L_{Cl}^1} \tag{101}$$

Here $\Delta\mu_{Cl}$ has been neglected in view of the relatively large chloride concentrations.

The development of $\Delta\mu_H^{*e}$ as a function of the number of flashes did not depend on the flash frequency, i.e., the back leakage of protons during the dark period was not significant ($L_H^1 \ll L_v$). The magnitude of the "membrane potential," deduced from the shift in absorbance of the endogenous carotenoids, did depend on the flash frequency. Since the lag in onset of phosphorylation also increases with increasing interval between flashes, the authors conclude that $\Delta\psi$ effects the inhibitor dissociation. However, at this point the total proton-motive force ($\Delta\tilde{\mu}_H$) is an equally suited candidate for the activation signal. A second experiment makes this latter possibility the more likely one.

In the presence of 20 m*M* KCl and valinomycin, the lag phase in phosphorylation is prolonged (see also Ort *et al.*, 1976). During the preequilibration period the intravesicular concentration of K^+ must also have reached 20 m*M*, and it can be calculated that at these concentrations $\Delta\mu_K$ cannot change significantly during the first 125 flashes and, therefore [see Eq. (100)], $\Delta\psi$ cannot have attained a high value (L_K^1 is large). Yet, after this number of flashes, phosphorylation is found. This may be interpreted as the consequence of an increase in $\Delta\tilde{\mu}_H^{*e}$, since Harris and Crofts showed that even in the absence of valinomycin a significant $\Delta\mu_H^{*e}$ developed within 125 flashes. In the formula for $\Delta\tilde{\mu}_H^{*e}$, the importance of $\Delta\mu_H^{*e}$ is stressed since it is multiplied by the (now large) L_K^1.

A definitive experiment to test our interpretation would be to determine the effect of nigericin on the lag phase in photophosphorylation. In this case the equations tell us that, because $\Delta\mu_H^{*e}$ will increase more slowly, the development of $\Delta\tilde{\mu}_H^{*e}$ will be retarded whereas $\Delta\psi^{*e}$ will remain at a higher value (see also above). According to Harris and Crofts' interpretation this should result in a short lag phase, whereas we would predict a prolonged lag.

The relevance of the distinction between $\Delta\tilde{\mu}_H$ and $\Delta\psi$ as signals for the ATPase activation is clear: during steady-state phosphorylation $\Delta\tilde{\mu}_H$ is high but $\Delta\psi$ in chloroplasts is low. Mechanistically, implying $\Delta\tilde{\mu}_H$ as a regulating factor probably means that the first protons that move across the membrane have to do an extra amount of work, resulting in displacement of the inhibitor.

This section merely serves as an example of how our description can be used to discuss and outline experiments. Thus, some remaining possibilities to explain the activation of the chloroplast ATPase by the membrane potential will not be discussed. Also, the implications of localized chemiosmosis (Ort *et al.*, 1976) for this system will not be examined here.

C. Chromatophores

From the plasma membrane of certain photosynthetic bacteria small vesicles can be derived in which most of the enzymes involved in photophosphorylation are present. These vesicles, called chromatophores, are able to synthesize ATP at the expense of absorbed photons by a mechanism that presumably involves a transmembrane electrochemical gradient of protons (see, e.g., Crofts, 1974).

The proton gradient is formed by a cyclic electron transfer coupled to a vectorial proton transfer. The mechanism of this process has not yet been fully elucidated, but a recently proposed scheme (Petty *et al.*, 1977) is as follows. If the reaction center pigment is excited, it can pick up an electron from a donor at one side of the membrane (a cytochrome *c*) and transfer it to an acceptor at the other side of the membrane (a quinone). The reduced quinone picks up a proton and in this form is able to reduce the cytochrome *c* again with intermediate participation of cytochrome *b* and another redox component. Since cytochrome *c* releases the proton it obtained from the reduced quinone, and because the involvement of the other two redox compounds effectuates the transport of a second proton, the net effect of the cycle is a light-induced transfer of two protons across the membrane. (A similar cyclic electron transport occurs in chloroplasts.)

It is conceivable that some of the partial reactions of this process are short-circuited; for instance, the quinone may reduce cytochrome *c* without prior protonation. Such side reactions should make chromatophores very interesting objects of study, particularly to compare with the bacteriorhodopsin proton pump in which no such redox intermediates occur. Equations to describe the consequences of such side reactions can be derived by procedures outlined above.

The chromatophores are also useful to study the relation between $\Delta\tilde{\mu}_H$ and ATP synthesis, and such studies have been reported (Baccarini–Melandri *et al.*, 1977; Schuldiner *et al.*, 1974; Kell *et al.*, 1978). Just as in the case of mitochondria, it can be shown that in chromatophores (where the active site for ATP synthesis is located on the external surface of the vesicles) the relation

$$(-A_P/\Delta\tilde{\mu}_H)^{*P} = n_H^P \tag{102}$$

should hold. Initial reports by Casadio *et al.* (1974) indicated a very high value for $\Delta\tilde{\mu}_H$, so that the calculated stoichiometric number $-n_H^P$ was smaller than 2. However, later experiments have shed some doubt on the validity of the use of the carotenoid shift as an indicator for bulk $\Delta\psi$ across the membrane (although it may register a local gradient close to the electron-transfer chain and ATPase). The magnitude of $\Delta\tilde{\mu}_H$ may have to be corrected downward, and, consequently, the calculated $-n_H^P$ will be larger than 2. Interestingly, also in experiments with chromatophores it appears that n_H^P is variable (Baccarini–Melandri *et al.*, 1977).

VI. Transport Across Plasma Membranes

A. Bacterial Transport

Bacterial membranes contain many transport systems for the uptake of nutrients from the surrounding medium. In this section emphasis will be on those systems that catalyze solute $-H^+$ symport. One such a symport is the lac permease of *Escherichia coli*, which catalyzes the coupled movement of a proton with either a lactose or a thiomethylgalactoside molecule (West and Mitchell, 1973).

In analogy with the descriptions of other systems, we will consider (see Fig. 11) the presence in the bacterial membrane of a redox proton pump, the translocator for the sugar (S) and several permeation processes of ions or weak acids (HA). The bulk anion and cation are represented by Cl^- and K^+, respectively. The important assumptions are again that all defined processes are mutually independent and that flow-force relations have a proportional form.

The following set of equations results (see Fig. 11; $n_O^H = n_H^O$)

$$\begin{pmatrix} J_O \\ J_H \\ J_S \\ J_A \\ J_K \\ J_{Cl} \end{pmatrix} = \begin{pmatrix} L_O & n_H^O L_O & 0 & 0 & 0 & 0 \\ n_H^O L_O & L_{HH} & n_H^S L_S & n_H^A L_A & -L_{KOH} & L_{HCl} \\ 0 & n_H^S L_S & L_S & 0 & 0 & 0 \\ 0 & n_H^A L_A & 0 & L_A & 0 & 0 \\ 0 & -L_{KOH} & 0 & 0 & L_K^1 + L_{KOH} & 0 \\ 0 & L_{HCl} & 0 & 0 & 0 & L_{Cl}^1 + L_{HCl} \end{pmatrix} \times \begin{pmatrix} A_O \\ \Delta\tilde{\mu}_H \\ \Delta\mu_S \\ \Delta\tilde{\mu}_A \\ \Delta\tilde{\mu}_K \\ \Delta\tilde{\mu}_{Cl} \end{pmatrix} \quad (103)$$

$$L_{HH} = (n_H^O)^2 L_O + (n_H^S)^2 L_S + (n_H^A)^2 L_A + L_H^1 + L_{KOH} + L_{HCl}$$

A series of increasing restrictions arise in the system, so that it passes through a series of near steady states (see Section II,B). Such restrictions are needed to reduce the number of variables to an experimentally manageable amount; if their validity is provided for an experimental system, the predictive value of the description is sharpened considerably.

The first restriction is that of electroneutral flow, a condition that will

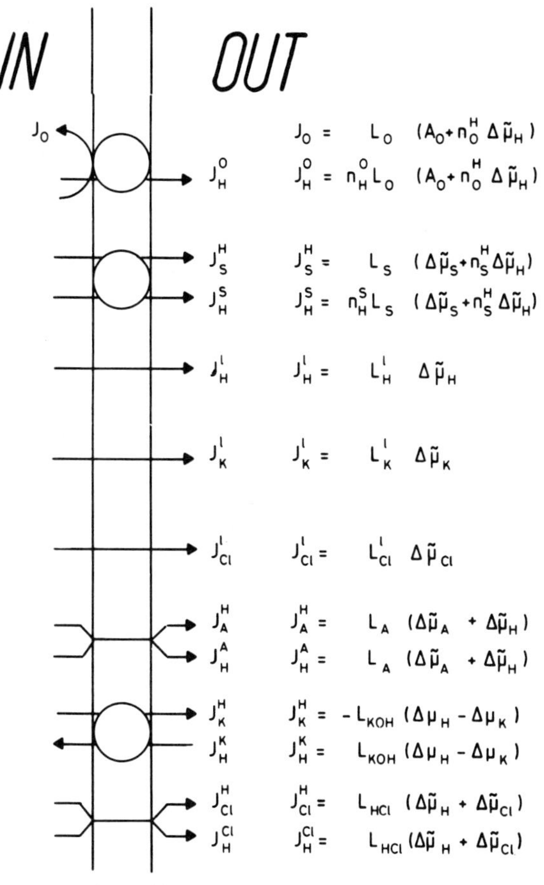

FIG. 11. Substrate uptake and ion movements in bacterial systems. Within the membrane an oxidase–proton pump, a sugar–proton symport, proton, potassium, and chloride leaks, permeation of an undissociated acid, K^+/H^+, and Cl^-/OH^- exchange are depicted.

arise in this system within seconds (see Section VI,B). By elimination we can derive the following equation:

$$\begin{pmatrix} J_O \\ J_H \\ J_S \\ J_A \\ J_K \end{pmatrix}^{*e} = \lambda_e \begin{pmatrix} A_O \\ \Delta\mu_H + \Delta\mu_{Cl} \\ \Delta\mu_S \\ \Delta\mu_A - n_H^A \Delta\mu_{Cl} \\ \Delta\mu_K + \Delta\mu_{Cl} \end{pmatrix} \tag{104}$$

where λ_e is a matrix with elements, some of which are complicated functions of the parameters defined in the first equation (103).

A further reduction in variables can be reached by looking at the steady state, where no net proton movement occurs ($J_H = 0$, measurable with a pH electrode):

$$\begin{pmatrix} J_O \\ J_A \\ J_S \\ J_K \end{pmatrix}^{*e*h*} = \lambda_{eh} \begin{pmatrix} A_O \\ \Delta\mu_A - n_H^A \Delta\mu_{Cl} \\ \Delta\mu_S \\ \Delta\mu_K + \Delta\mu_{Cl} \end{pmatrix} \quad (105)$$

Finally, a steady state is reached in which the sugar uptake has become maximal ($J_S = 0$), and the equation becomes

$$\begin{pmatrix} J_O \\ J_A \\ J_K \end{pmatrix}^{*e*h*s*} = \lambda_{ehs} \begin{pmatrix} A_O \\ \Delta\mu_A - n_H^A \Delta\mu_{Cl} \\ \Delta\mu_K + \Delta\mu_{Cl} \end{pmatrix} \quad (106)$$

Of course, also other steady-state conditions could have been chosen. If, for instance, the concentration of the acid (A) is low, the $J_A = 0$ steady state will be reached very quickly. If oxygen concentration is low, anaerobiosis ($J_O = 0$) will set in. In our description K^+ represents the bulk cation and, therefore, the $J_K = 0$ steady state may be further away. Since in most cases $\Delta\mu_K + \Delta\mu_{Cl} = C\Delta\pi \approx 0$, we can leave out the right-hand column of the matrices and the lowest line of the equation (see Section II,C,2), unless we are interested in the flux of the bulk cation across the membrane.

If one measures the initial rate of lactose uptake upon energization of *E. coli* in the absence of an acid HA, the approximation of electroneutral flow can be used. Initially $\Delta\mu_S$ and $\Delta\mu_H$ will be negligible and the approximation

$$-(J_S)^{*e}_{\text{initial}} = \frac{n_H^O L_O n_H^S L_S}{(n_H^O)^2 L_O + (n_H^S)^2 L_S + L_H^1 + L_K^1 + L_{Cl}^1} A_O \quad (107)$$

will hold. This equation predicts that the initial rate of lactose uptake will decrease with increasing L_H^1 (protonophore) or L_K^1 (valinomycin). Furthermore, a hyperbolic dependence on number of carriers (L_S) or number of respiratory chains (L_O) in the membrane is expected.

For the steady-state accumulation of lactose we can use the simultaneous conditions $J_H = 0$ and $J_S = 0$ on top of the electroneutral flux. We arrive at (L_A is taken equal to zero)

$$(\Delta\mu_S)^{*e*h*s} = \frac{n_H^S \cdot n_H^O L_O}{(n_H^O)^2 L_O + L_H^1 + L_n L_e/(L_n + L_e)} A_O \quad (108)$$

Thus, the number of sugar carriers should have no effect on the steady-state sugar gradient. Protonophores or inhibitors of respiration will inhibit the accumulation. Effects of valinomycin and nigericin are only of second order, analogous to the situation in bacteriorhodopsin liposomes.

Under the same assumptions we can also derive

$$(\Delta\mu_S)^{*e*h*s} = -n_H^S(\Delta\tilde{\mu}_H)^{*e*h*s} \tag{109}$$

Experiments relevant to the above equations have been reported by the group of Kaback, using coli membrane vesicles depleted of the endogenous soluble proteins (Ramos *et al.*, 1976; Ramos and Kaback, 1977). The results are shown in Fig. 12.

In accordance with Eq. (108), nigericin has hardly any effect on lactose accumulation. The effect of valinomycin is small, but significant. Since under the experimental conditions used, the intravesicular lactose concentration was 100 mM, which will cause significant osmotic effects (possibly $L_n \neq 0$; 100 mM potassium phosphate was present), similar data concerning proline

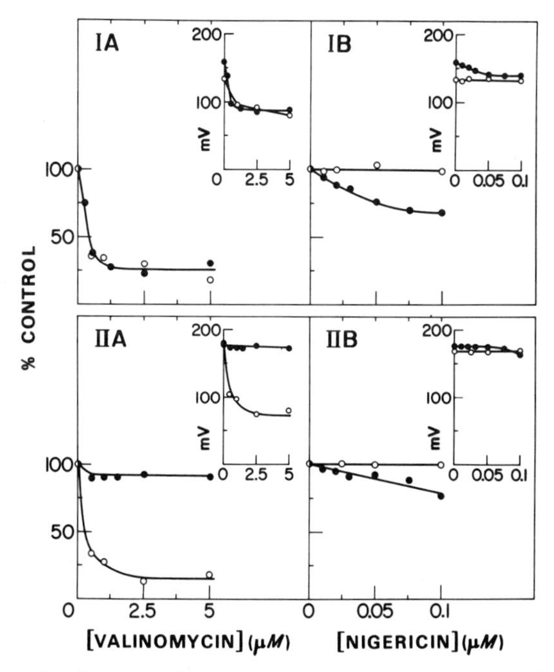

FIG. 12. Effect of valinomycin (A) and nigericin (B) on steady-state levels of accumulation of lactose (top) and proline (bottom) in *Escherichia coli* ML 308–225 membrane vesicles at pH 5.5 (●) and pH 7.5 (○). Insets show $58.8 \cdot {}^{10}\log$ (concentration gradient) = $\Delta\mu_S$. From Ramos and Kaback (1977); reproduced by permission.

uptake (see Fig. 12, which resulted in an intravesicular proline concentration of only 8 mM) are even more convincing (cf. Section II,C,2).

The prediction of Eq. (109) that the ratio $(-\Delta\mu_S/\Delta\tilde{\mu}_H)^{*e*h*s}$ will be constant (n_H^S) seems to be falsified by experiments of Ramos and Kaback (1977) in which they varied the extravesicular pH. From pH 5.5 to 7.0, the ratio changes from 1.0 to 1.8 in the case of proline. Closer inspection of the data shows that $\Delta\psi$ is rather constant over this range, but that ΔpH decreases by about 1 unit per pH unit. The measurement of ΔpH is based on distribution of acetate (HA), which even at pH 5.5 is relatively slow in entering and leaving the vesicles, as shown by the authors themselves. We may suspect that with increasing pH, the penetration of acetic acid becomes increasingly more difficult and that at higher external pH the $J_A = 0$ steady state was not reached. Experiments to verify this are clearly needed.

Some interesting older observations by the same group (Kaback and Barnes, 1971) on lactose movement in *E. coli* and *Staphylococcus aureus* vesicles might find an explanation through our description. It was found that oxamate, a competitive inhibitor of D-lactate dehydrogenase in *E. coli*, inhibits the initial uptake of lactose but does not induce efflux if added to vesicles maximally loaded with the sugar (see Fig. 13). Cyanide, on the other hand, both inhibits initial uptake and induces efflux. In the steady state of maximal lactose uptake, the rate of sugar efflux caused by respiratory inhibition should be, from Eq. (105):

$$J_S^{\text{efflux}} = \frac{-n_H^S L_S\{(n_H^A)^2 L_A + L_n + L_e\}}{L'_{Hn}} n_H^O L'_O A'_O + \left\{L'_{eg} - (n_H^S)^2 L_S - \frac{(n_H^S)^2 L_S L_e^2}{L'_{Hn}}\right\} \frac{L_S}{L'_{eg}} \Delta\mu_S - \frac{n_H^S L_S L_e n_H^A L_A}{L'_{Hn}} (\Delta\mu_A - n_H^A \Delta\mu_{Cl}) \tag{110}$$

with

$$L'_{Hn} = \{(n_H^A)^2 L_A + L_n + L_e\} L'_{eg} - L_e^2$$
$$L'_{eg} = (n_H^O)^2 L'_O + (n_H^S)^2 L_S + L_H^1 + L_e$$

Primes indicate values possibly altered owing to inhibition. Possibly oxamate enters the vesicles in its neutral form; in the case of oxamate there is a significant L_A. The effect of L_A on the rate of efflux does not become clear straightaway from the above equation: Through the terms containing A'_O and $\Delta\mu_S$, oxamate is predicted to increase efflux, but through the third term it is expected to inhibit efflux. In the experiments discussed here the inhibitors were quite effective in abolishing respiration: $L'_O \approx 0$. Therefore,

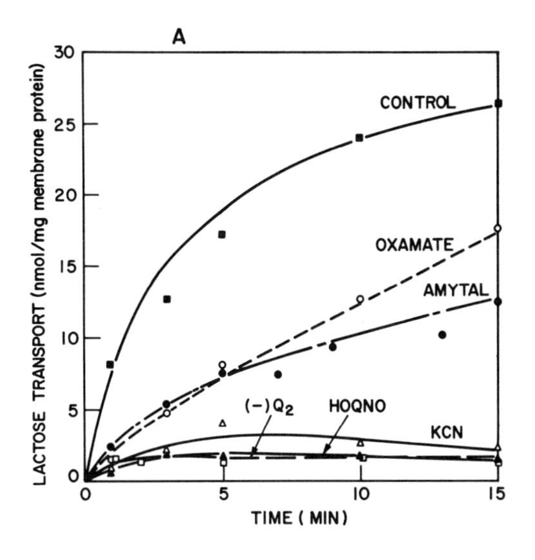

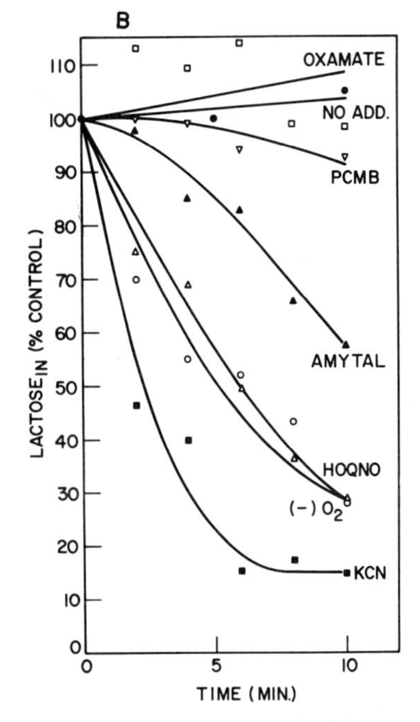

FIG. 13. A. Effect of oxamate, cyanide, and other inhibitors on initial uptake of lactose into *Escherichia coli* ML 308–225 membrane vesicles. (B) Inhibitor-induced lactose efflux from *E. coli* ML 308–225 membrane vesicles, preincubated for 15 minutes in the presence of D-lactate (20 mM) and labeled lactose. A: ○, oxamate, △, cyanide. B: □, oxamate; ■, cyanide. (A and B from Kaback and Barnes (1971); reproduced by permission.)

at relatively high values of L_A, efflux is expected to be retarded. The effect will be stronger at higher values of $\Delta\mu_A$. Particularly if oxamate reaches its steady state distribution before it effectively inhibits, one can derive that a near steady state may be attained in which (assuming $L'_O = 0$)

$$\frac{\Delta\mu_S^{\text{after efflux}}}{A_O^{\text{before efflux}}} = \frac{C}{1 + \left\{\dfrac{L_n + [L_e L_H^1/(L_e + L_H^1)]}{(n_H^A)^2 L_A}\right\}} \tag{111}$$

Here C is a complicated expression not containing L_A. Clearly such a near steady state can be reached only if $L_A \neq 0$. This explanation can be put to an experimental test: Under conditions of large L_n (nigericin + K^+) or large $L_e L_H^1/(L_e + L_H^1)$ (valinomycin + protonophore acting in series) the difference in effect on efflux between oxamate and cyanide should be less pronounced.

Apart from coupled solute–cation translocators, bacterial membranes contain several other types of transport systems. A very complicated but intriguing one is the so-called phosphoenolpyruvate phosphotransferase system (for review, see Postma and Roseman, 1976). It catalyzes the inward translocation of sugars coupled to their phosphorylation by phosphoenolpyruvate. In this case we cannot define a gradient of the free sugar across the membrane since intravesicular free sugar may not be present. Nevertheless, using the procedures developed above, we can describe such a system by a set of equations. Such a description, choosing appropriate values for the constants, can successfully explain phenomena such as competition for uptake between different phosphotransferase-system sugars and "inducer exclusion" (Postma and Roseman, 1976).

B. Eukaryotic Plasma Membranes

1. *Characteristics of the Eukaryotic Plasma Membrane*

The eukaryotic cell differs from the structures considered so far in that it is divided in subcompartments. For studies of transport across the plasma membrane this only brings the complication that we have to define precisely the forces at the cytoplasmic side. The affinity of the ATPase reaction, experienced by the plasma membrane, will, for instance, not be equal to that calculated from total cellular concentrations of ATP, ADP, and phosphate.

Methods have been developed to measure concentrations in subcellular compartments, but another approach is to allow the intracellular processes to come to a steady state. This is possible, because intracellular capacities are usually much smaller than the extracellular ones.

An example of a cellular steady state is that of electroneutrality of the transmembrane flux. The passive electrical resistance of an eukaryotic plasma membrane has been estimated at $0.3\ \Omega \cdot m^2$ (Reuss and Finn, 1975) and the electrical capacity at $100\ mF \cdot m^{-2}$ (Montal, 1976), so that the electrical relaxation time will be of the order of 0.03 second. Of course, the internal resistance of the enzymes functioning as the molecular voltage source will be connected in parallel to this, so that the relaxation time of the electrical potential rise is expected to be shorter than 0.03 second.

In organized multicellular systems, transcellular potential differences may exist in addition to transmembrane potentials. In epithelial tissue, for instance, the transcellular resistance is approximately $10^{-2}\ \Omega \cdot m^2$ (as a result of shunt pathways between cells, see Schultz, 1977). The transepithelial capacity may be estimated at $20\ \mu F \cdot m^{-2}$. Thus, transepithelial relaxation times are expected and found (Frömter, 1977) to be shorter than 1 second.

The time required to reach steady-state conditions of flux of solutes is more difficult to evaluate. In pure phospholipid vesicles an RC-time of 1.2 days can be calculated from the potassium permeability ($10^{-4}\ m \cdot sec^{-1}$; Lakshminarayanaiah, 1969). Since eukaryotic cells have a much smaller surface-to-volume ratio, their relaxation on the basis of a simple calculation would last longer. However, the presence of proteins, some of which have a specific function in transmembrane ion movements, will strongly reduce this relaxation time. It is clear that under most conditions it will be necessary to verify that steady-state conditions have been reached, before the corresponding simplified description is allowed.

2. *An Example of Indirect Active Transport*

As an example of the functioning of eukaryotic plasma membranes, we will discuss the transepithelial transport of glucose. Transport across epithelial tissue has been described using linear irreversible thermodynamics via phenomenological equations (Essig and Caplan, 1968). In certain ranges of the forces, the equations were shown to apply and coupling coefficients could be calculated (Lang *et al.*, 1977a). However, the authors themselves (Lang *et al.*, 1977b) emphasize the lack of understanding about the relation between phenomenological coefficients and mechanistic parameters. Our method elaborates such relations. The model depicted in Fig. 14 seems to be in accordance with most experimental observations (see, e.g., Kimmich and Randles, 1977: Frömter, 1977). It includes an ATP-driven Na^+/K^+

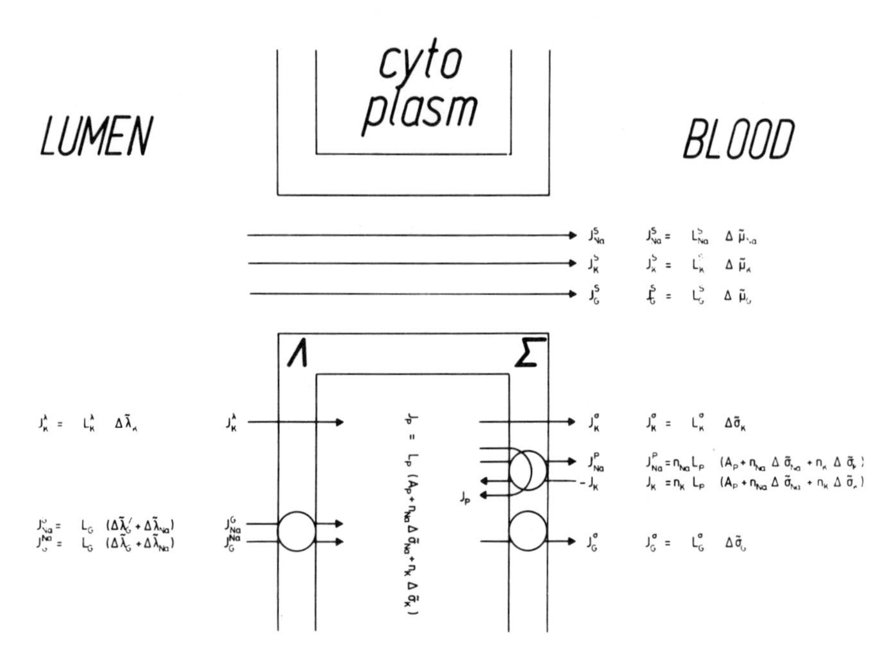

FIG. 14. Transport across epithelial cells. In the luminal membrane of an epithelial cell, a potassium leak and a glucose–sodium symport are depicted. In the serosal membrane, another potassium leak, the (Na + K) ATPase, and facilitated glucose diffusion are depicted. Through the junction also glucose, potassium, and sodium may flow.

pump, a glucose carrier in the membrane adjacent to the blood stream, and a coupled sodium–glucose translocator in the luminal part of the membrane.

The total set of equations is a 10 × 10 system. It can already be reduced to a 7 × 7 system by use of the boundary condition:

$$\Delta\tilde{\mu}_i = \Delta\tilde{\lambda}_i + \Delta\tilde{\sigma}_i \tag{112}$$

where $\Delta\tilde{\mu}$, $\Delta\tilde{\lambda}$, and $\Delta\sigma$ represent the transepithelial, the transluminal membrane, and the transserosal membrane electrochemical potential differences, respectively. The reduction is carried out by rewriting the dissipation function for each solute as in the following example:

$$\mathbf{\Phi}_{\mathrm{g}} = J_{\mathrm{g}}^{\lambda}\Delta\tilde{\lambda}_{\mathrm{g}} + J_{\mathrm{g}}^{\sigma}\Delta\tilde{\sigma}_{\mathrm{g}} + J_{\mathrm{g}}^{\mathrm{s}}\Delta\tilde{\mu}_{\mathrm{g}} = J_{\mathrm{g}}^{\mathrm{tot}}\Delta\tilde{\mu}_{\mathrm{g}} + J_{\mathrm{g}}^{\delta}\Delta\tilde{\sigma}_{\mathrm{g}} \tag{113}$$

with

$$J_{\mathrm{g}}^{\mathrm{tot}} = J_{\mathrm{g}}^{\mathrm{s}} + J_{\mathrm{g}}^{\lambda} \quad \text{and} \quad J_{\mathrm{g}}^{\delta} = J_{\mathrm{g}}^{\sigma} - J_{\mathrm{g}}^{\lambda} \tag{114}$$

The set of equations, fully describing the system, becomes

$$\begin{pmatrix} J_P \\ J_g^\delta \\ J_{Na}^\delta \\ J_K^\delta \\ J_g^{tot} \\ J_{Na}^{tot} \\ J_K^{tot} \end{pmatrix} =$$

$$\begin{pmatrix} L_P & 0 & n_{Na}L_P & n_K L_P & 0 & 0 & 0 \\ 0 & L_g^\sigma + L_g^\lambda & L_g^\lambda & 0 & -L_g^\lambda & -L_g^\lambda & 0 \\ n_{Na}L_P & L_g^\lambda & n_{Na}^2 L_P + L_g^\lambda & n_{Na}n_K L_P & -L_g^\lambda & -L_g^\lambda & 0 \\ n_K L_P & 0 & n_{Na}n_K L_P & (n_K^2 L_P + L_K^\sigma + L_K^\lambda) & 0 & 0 & -L_K^\lambda \\ 0 & -L_g^\lambda & -L_g^\lambda & 0 & L_g^s + L_g^\lambda & L_g^\lambda & 0 \\ 0 & -L_g^\lambda & -L_g^\lambda & 0 & L_g^\lambda & L_{Na}^s + L_g^\lambda & 0 \\ 0 & 0 & 0 & -L_K^\lambda & 0 & 0 & L_K^s + L_K^\lambda \end{pmatrix}$$

$$\times \begin{pmatrix} A_P \\ \Delta\sigma_g \\ \Delta\tilde{\sigma}_{Na} \\ \Delta\tilde{\sigma}_K \\ \Delta\mu_g \\ \Delta\tilde{\mu}_{Na} \\ \Delta\tilde{\mu}_K \end{pmatrix} \quad (115)$$

These equations are too complicated to tackle experimentally. Further boundary conditions to reduce the number of equations are welcome. Interestingly, the condition of electroneutral flux across each of the membranes separately does not apply: net charge flux through the circuit luminal membrane–serosal membrane shunt may occur. One, therefore, must look for other forces or flows that are likely to come to a steady state rapidly. By a suitable linear combination of forces we can then eliminate such highly variable forces.

A useful set of such forces is

$$\frac{\Delta\tilde{\mu}_{Na} - \Delta\tilde{\mu}_K}{2} = \frac{\Delta\mu_{Na} - \Delta\mu_K}{2} \quad (116)$$

$$\frac{\Delta\tilde{\sigma}_{Na} - \Delta\tilde{\sigma}_K}{2} = \frac{\Delta\sigma_{Na} - \Delta\sigma_K}{2} \quad (117)$$

$$\frac{\Delta\tilde{\mu}_{\mathrm{Na}} + \Delta\tilde{\mu}_{\mathrm{K}}}{2} = \frac{\Delta\mu_{\mathrm{Na}} + \Delta\mu_{\mathrm{K}}}{2} + F\Delta\psi^{s} \tag{118}$$

$$\frac{\Delta\tilde{\sigma}_{\mathrm{Na}} + \Delta\tilde{\sigma}_{\mathrm{K}}}{2} = \frac{\Delta\sigma_{\mathrm{Na}} + \Delta\sigma_{\mathrm{K}}}{2} + F\Delta\psi^{\sigma} \tag{119}$$

of which the first two are relatively constant and the last two are highly variable.

It is relatively simple to find the flows conjugate to these forces, as illustrated for a part of the system:

$$\begin{aligned}
\mathbf{\Phi}_{\mathrm{NaK}} &= J^{\delta}_{\mathrm{Na}}\Delta\tilde{\sigma}_{\mathrm{Na}} + J^{\delta}_{\mathrm{K}}\Delta\tilde{\sigma}_{\mathrm{K}} \\
&= J^{\delta}_{\mathrm{Na}}\left[\left(\frac{\Delta\tilde{\sigma}_{\mathrm{Na}} - \Delta\tilde{\sigma}_{\mathrm{K}}}{2}\right) + \left(\frac{\Delta\tilde{\sigma}_{\mathrm{Na}} + \Delta\tilde{\sigma}_{\mathrm{K}}}{2}\right)\right] \\
&\quad + J^{\delta}_{\mathrm{K}}\left[\left(\frac{\Delta\tilde{\sigma}_{\mathrm{Na}} + \Delta\tilde{\sigma}_{\mathrm{K}}}{2}\right) - \left(\frac{\Delta\tilde{\sigma}_{\mathrm{Na}} - \Delta\tilde{\sigma}_{\mathrm{K}}}{2}\right)\right] \\
&= (J^{\delta}_{\mathrm{Na}} - J^{\delta}_{\mathrm{K}})\frac{(\Delta\sigma_{\mathrm{Na}} - \Delta\sigma_{\mathrm{K}})}{2} + (J^{\delta}_{\mathrm{Na}} + J^{\delta}_{\mathrm{K}})\left(F\Delta\psi^{\sigma} + \frac{\Delta\sigma_{\mathrm{Na}} + \Delta\sigma_{\mathrm{K}}}{2}\right)
\end{aligned} \tag{120}$$

It follows that the first steady states to be reached are at

$$J^{\delta}_{\mathrm{Na}} + J^{\delta}_{\mathrm{K}} = 0 \tag{121}$$

and

$$J^{\mathrm{tot}}_{\mathrm{Na}} + J^{\mathrm{tot}}_{\mathrm{K}} = 0 \tag{122}$$

By use of these conditions the $\Delta\psi$ terms can be eliminated from Eq. (115), which then can be reduced to a 5×5 system. Further introduction of, for instance, the $J^{\delta}_{\mathrm{g}} = 0$ and $J^{\delta}_{\mathrm{Na}} = 0$ steady-state conditions will result in a final 3×3 equation of the form

$$\begin{pmatrix} J_{\mathrm{P}} \\ J^{\mathrm{tot}}_{\mathrm{g}} \\ J^{\mathrm{tot}}_{\mathrm{Na}} \end{pmatrix}^{*\mathrm{e}*\mathrm{g}^{\delta}*\mathrm{Na}^{\delta}} = L^{*\mathrm{e}*\mathrm{g}^{\delta}*\mathrm{Na}^{\delta}} \begin{pmatrix} A_{\mathrm{P}} \\ \Delta\mu_{\mathrm{g}} \\ \Delta\mu_{\mathrm{Na}} - \Delta\mu_{\mathrm{K}} \end{pmatrix} \tag{123}$$

The constants in these equations are very complicated combinations of the mechanistic constants. Nevertheless, it is possible to calculate these constants, and we can quantitatively predict the variation of each of them upon changing the partial processes involved. Let us now consider some experiments underlining this claim.

One of these concerns the uptake of 3-*O*-methyl-D-glucose in isolated epithelial cells from the intestine (Kimmich and Randles, 1977). This allows for quite a few simplifications, since the two parts of the membrane are so to speak short-circuited. The steady-state accumulation of glucose in these cells will be described by

$$(\Delta\sigma_g)^{*e*g} = \frac{L_g^\lambda L_P A_P + L_g^\lambda(2L_P - L_K^\sigma - L_K^\lambda)(\Delta\sigma_{Na} - \Delta\sigma_K)}{L_g^\sigma(L_P + L_g^\lambda + L_K^\lambda + L_K^\sigma) + L_g^\lambda(L_P + L_K^\sigma + L_K^\lambda)} \tag{124}$$

where $\Delta\sigma_g$, $\Delta\sigma_{Na}$, and $\Delta\sigma_K$ are the chemical potential differences across the membrane of the sugar, Na^+ and K^+, respectively. L_K^λ and L_K^σ represent the K^+ permeability of the luminal and the serosal part of the cell membrane. L_g^λ and L_g^σ determine the activity of the glucose transport systems. L_P and A_P are the proportionality constant and the affinity of the (Na/K)-ATPase. n_{Na} and n_K have been taken $+3$ and -2, respectively (Goldin, 1977).

A net accumulation of the sugar will result if the activity of the ATPase outweighs the probably outwardly directed $(\Delta\sigma_{Na} - \Delta\sigma_K)$. An inhibitor of the mucosal glucose transport (phlorizin) will lower L_g^λ and should lower the accumulation of sugar. Conversely, the inhibitor of serosal glucose transport phloretin lowers L_g^σ and, therefore, increases the accumulation. These predictions are borne out by the experiments (Kimmich and Randles, 1977) (see also Fig. 15).

In different experiments the electrical potential across the epithelial cell membrane and across the whole epithelium was measured (Frömter, 1977). The expression for the potential across the serosal membrane is

$$\begin{aligned} -L_\square(\Delta\psi^\sigma)^{*e} = {} & (L_{Na}^s + L_K^s + L_g^\lambda + L_K^\lambda)L_P A_P - L_g^\lambda(L_{Na}^s + L_K^s)\Delta\lambda_g \\ & + [3L_P(L_{Na}^s + L_K^s + L_g^\lambda + L_K^\lambda) + L_g^\lambda(L_{Na}^s + L_K^s)]\Delta\sigma_{Na} \\ & + [(-2L_P + L_K^\sigma)(L_{Na}^s + L_K^s + L_g^\lambda + L_K^\lambda) \\ & + L_K^\lambda(L_{Na}^s + L_K^s)]\Delta\sigma_K + (L_{Na}^s L_K^\lambda - L_K^s L_g^\lambda)(\Delta\mu_{Na} - \Delta\mu_K) \end{aligned} \tag{125}$$

with

$$L_\square = (L_g^\lambda + L_{Na}^s + L_K^\lambda + L_K^s)(L_P + L_K^\sigma) + (L_K^\lambda + L_g^\lambda)(L_{Na}^s + L_K^s) \tag{126}$$

In experiments of Frömter (1977) (see Fig. 16), a rapid depolarization (0.3 second) of the serosal membrane upon addition of glucose on the luminal side of the intestinal epihelium was measured under conditions of $\Delta\mu_{Na} = \Delta\mu_K = 0$. The initial situation is such that A_P outweighs $\Delta\sigma_{Na}$ $(-\Delta\psi^\sigma = 81$ mV). Before addition of the sugar, L_g^λ can effectively be considered zero (see Section II,D). After its addition, there will be a sharp increase in $\Delta\lambda_g$ and L_g^λ will become nonzero. If the shunt pathway is operative $(L_{Na}^s + L_K^s > 0)$, a decrease in $-\Delta\psi$ will result, as found experimentally by Frömter (1977) (see Fig. 16).

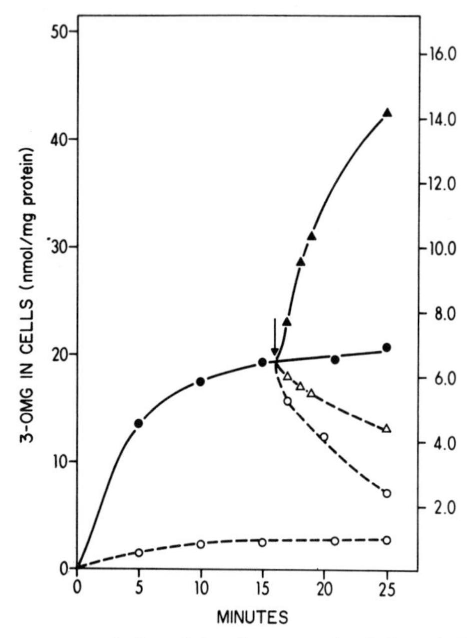

FIG. 15. Steady state accumulation of the glucose analog 3-*O*-methyl-D-glucose (3-OMG) by isolated intestinal epithelial cells. Righthand ordinate indicates the 3-OMG distribution ratio. ●, Control; ▲, phloretin; △, phlorizin and phloretin; ○, phlorizin. From Kimmich and Randles (1977); reproduced by permission.

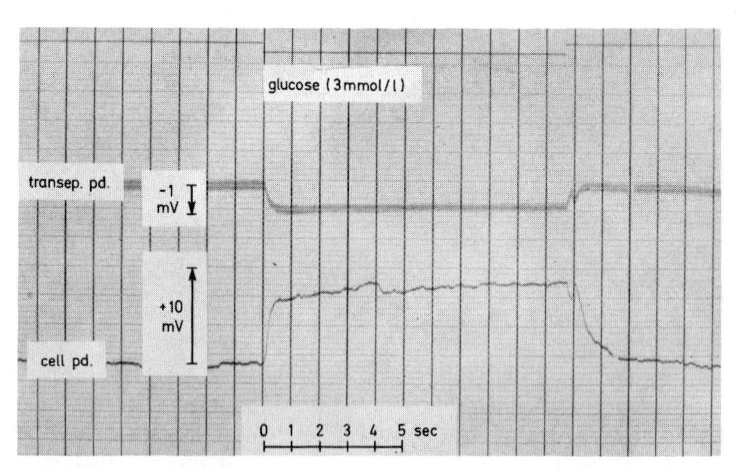

FIG. 16. Electric potential changes resulting from addition of glucose to the luminal side of an epithelial cell layer. Upper trace: glucose concentration; middle trace: transepithelial potential (blood–lumen); lower trace: potential across the serosal membrane (plasma–blood). Before glucose application, its value was −81 mV. From Frömter (1977); reproduced by permission.

The transepithelial glucose-evoked potential can also be estimated from the equations. For simplicity, we calculate

$$\frac{\partial(+\Delta\psi^{s})^{*e}}{\partial(\Delta\lambda_{g})_{\Delta_{\mu Na}, \Delta_{\mu K}, \Delta_{\sigma Na}, \Delta_{\sigma K}, \Delta_{\sigma g}, \Delta_{P}}} = \frac{-L_{g}^{\lambda}(L_{P} + L_{K}^{\sigma})}{L_{\square}} \tag{127}$$

and, since this term is negative under all conditions, the addition of glucose to the luminal side of the epithelium should decrease the transepithelial potential (luminal-serosal). This is also found experimentally (Frömter) (see Fig. 16).

The effect of the shunt pathway on the transport process can be evaluated in different ways. One way is to look at the impact of the (Na/K)-ATPase on transepithelial glucose transport. If we assume that $\Delta\sigma_{g}$ is not influenced by the shunt, we calculate

$$\left(\frac{\partial J_{g}^{tot}}{\partial A_{P}}\right)_{\substack{(\Delta\mu_{Na} - \Delta\mu_{K}), \Delta\mu_{g}, \\ \Delta\sigma_{g}, \Delta\sigma_{Na}, \Delta\sigma_{K}}}$$

$$= \frac{L_{g}^{\lambda} L_{P}(L_{Na}^{s} + L_{K}^{s})}{(L_{Na}^{s} + L_{K}^{s})(L_{P} + L_{g}^{\lambda} + L_{K}^{\lambda} + L_{K}^{\sigma}) + (L_{g}^{\lambda} + L_{K}^{\lambda})(L_{P} + L_{K}^{\sigma})} \tag{128}$$

Clearly, opening of the shunt (i.e., increasing $L_{Na}^{s} + L_{K}^{s}$) will increase the dependence of the glucose flux on the affinity of the ATPase. Under most conditions increased permeability of the shunt pathway will lead to increased transepithelial glucose transport.

VII. Intermediary Metabolism and Biomembranes

A. The Liver Cell

The isolated liver cell is a useful experimental system to study intermediary metabolism at the level of an eukaryotic cell (Zuurendonk *et al.*, 1976). A special extra complexity of this cell, as compared to a prokaryotic cell, is its subcompartmentation into cytosol, mitochondria, nucleus etc. It is the cooperation between these subcompartments in metabolism that introduces an extra dimension in the description.

For the purpose of our description, we consider the contents of each of the subcompartments as a homogeneous phase. Thus, for instance, there will be no significant concentration gradients of solutes except across membranes.

The problem of the statistical significance of solute concentrations within a small organelle, such as a mitochondrion, was touched upon in Section

IV,A. This problem is accentuated in the case of a whole cell, since each of these cells will not contain a very large number of mitochondria. For the time being we will assume that it is allowed to transfer the reasoning of Section IV,A to the large population of whole cells, so that a suspension of liver cells can be treated as the sum of a large number of average and equal cells.

The metabolism within each of the cellular compartments can of course be treated as an independent unit. In the metabolic pathways, such as gluconeogenesis or ureogenesis, parts of these series of reactions occur in one compartment and parts in another. Thus, interaction between pathways and also regulation of their rate at the level of movement of the substrates across the compartment membranes may be expected.

B. A Complex of Reactions

A complicated series of chemical reactions may be difficult to describe if no limiting conditions are set. Fortunately, however, the usual condition in cellular metabolism is one of a steady state in which the concentration of most of the intermediates changes only very slowly.

In such a situation, which we already met in connection with the transport of mitochondrial metabolites (see Section IV,C), the rate of flux through the series of reactions will be proportional to the affinity of the overall reaction

$$J_{\text{overall}} = L_{\text{overall}} \times A_{\text{overall}} \tag{129}$$

The proportionality constant will be composed of all the proportionality constants of the partial reactions, as though all partial reactions acted as resistances in series

$$\frac{1}{L_{\text{overall}}} = \sum \frac{1}{L_{\text{partial}}} \tag{130}$$

It is immediately evident that all partial reactions having a large L will contribute little to the overall L. Changing one of the L's from zero to infinity will result in an asymptotic increase of L_{overall} from zero to a certain maximal value.

If parallel reactions occur, and the force across each of these is equal, then

$$L_{\text{overall}} = \sum L_{\text{partial}} \tag{131}$$

In practice, many parallel and serial reactions occur in cellular metabolism and the complete description will remain complicated even with the two general rules given here.

C. Gluconeogenesis in Isolated Liver Cells

Gluconeogenesis is an example of intricate coupling between two metabolic sequences of reactions. On the one hand, the mitochondria catalyze the process of oxidation of substrates by oxygen coupled to the synthesis of ATP. On the other hand conversion of, for instance, lactate to glucose is coupled to the hydrolysis of ATP. Since the cell contains a limited amount of adenine nucleotides, a steady state must be reached in which hardly any change with time in the concentration of ATP occurs, but where continuous oxidation of substrates leads to continuous conversion of lactate to glucose.

We are now in a position to describe such a system to various degrees of sophistication. The simplest is to relate the rate of flux through the overall reaction to the overall affinity. Although in this way we may loose sight of the mechanistic contribution of each of the partial reactions to the overall proportionality constant, it will nevertheless turn out to give useful information. First, however, we will consider each of the component reaction sequences separately.

The mitochondrial oxidative phosphorylation was already treated extensively in Section IV. Here, we use only the condition in which no net flux of protons across the mitochondrial membrane occurs ($J_H = 0$). In this steady state the following relation holds [derived from Eq. (71)]:

$$J_O = \frac{L_O^* L_P^* (n_H^P)^2}{(n_H^P)^2 L_P^* + (n_H^O)^2 L_O^* + L_H^1} \left[A_O - \frac{n_H^O}{n_H^P} A_P + \frac{L_H^1}{(n_H^P)^2 L_P^*} A_O \right] \qquad (132)$$

(For simplicity, we have neglected the possibly electrogenic nature of adenine nucleotide transport.) In this steady state the oxygen uptake must be a linear function of A_P, if A_O is kept constant. For isolated mitochondria this was shown to be the case experimentally by Rottenberg (1973).

In intact liver, A_P will change, if the demand for ATP varies, for instance with changing rates of gluconeogenesis. Experiments under such conditions were reported by Van der Meer *et al.* (1978). They measured A_P and A_O in intact rat-liver cells and plotted J_O against $(A_O - 3A_P)$. This treatment of the data makes implicit assumptions about the value of the stoichiometric constants n_H^P and n_H^O and about the size of L_H^1. The fact that a good fit of the data with a straight line was obtained, lends some support to the validity of these assumptions. In these experiments some variation in A_O occurred, which made it necessary to make the plot as described, not as a plot of J_O against A_P.

From the form of the equation it is clear that at constant A_O increasing leaks of the mitochondrial membrane for protons (L_H^1) will cause an increase in oxygen uptake. A linear relation should always exist between J_O and

$$A_O - \frac{n_H^O}{n_H^P} \cdot \left(1 + \frac{L_H^1}{(n_H^P)^2 \cdot L_P^*}\right)^{-1} A_P$$

The deviation of this from the form used ($A_O - 3A_P$) will be large if L_H^1 is large; the data given by Van der Meer *et al.* (1978) are not accurate enough for detection of the absence of such a deviation with certainty (see also Van Dam *et al.*, 1978a).

The reactions of gluconeogenesis can also be considered separately. A special complication that arises here is that, generally speaking, the forward and backward reactions of gluconeogenesis occur via different pathways. Conversion of 1 mol of glucose to 2 mol of lactate is coupled to synthesis of 2 mol of ATP, but the reverse reaction requires hydrolysis of 6 mol of ATP. (In liver, conversion of pyruvate to phosphoenolpyruvate goes via intermediate formation of oxaloacetate and costs two ATP.) Thus, we have two parallel sets of reactions, involving the same end products. Their coupling constant differs. As we have indicated in Section VI,C, there are ways to reduce the description of a network of metabolic interconversions to a relatively simple set of equations. Dissecting the pathways from glucose to lactate into three common and three parallel metabolic pathways, one can write formulas for each of these nine parts by summing $1/L^i$ and the stoicheiometric numbers n_P^{Gi} of the metabolic conversions occurring in that part. The reduction of the three sets of equations occurs as is shown below for one of them (see also Fig. 17):

$$\begin{pmatrix} J_G \\ J_P \end{pmatrix} = \begin{pmatrix} L_G & n_P^G L_G \\ n_P^G L_G & (n_P^G)^2 L_G + L_P^{12} \end{pmatrix} \begin{pmatrix} A_G \\ A_P \end{pmatrix} \tag{133}$$

in which

$$L_G = L_G^1 + L_G^2 \tag{134}$$

$$n_P^G = \frac{n_P^{G1} L_G^1 + n_P^{G2} L_G^2}{L_G^1 + L_G^2} \tag{135}$$

and

$$L_P^{12} = \frac{L_G^1 \cdot L_G^2 (n_P^{G1} - n_P^{G2})^2}{L_G^1 + L_G^2} \tag{136}$$

No data are available to check these equations, but they are certainly experimentally testable. Of course, the effect of the presence of two pathways will be noticeable only if $n_P^{G1} L_G^1 \approx n_P^{G2} L_G^2$. Furthermore, if $n_P^{G1} = n_P^{G2}$, L_P^{12} will be zero.

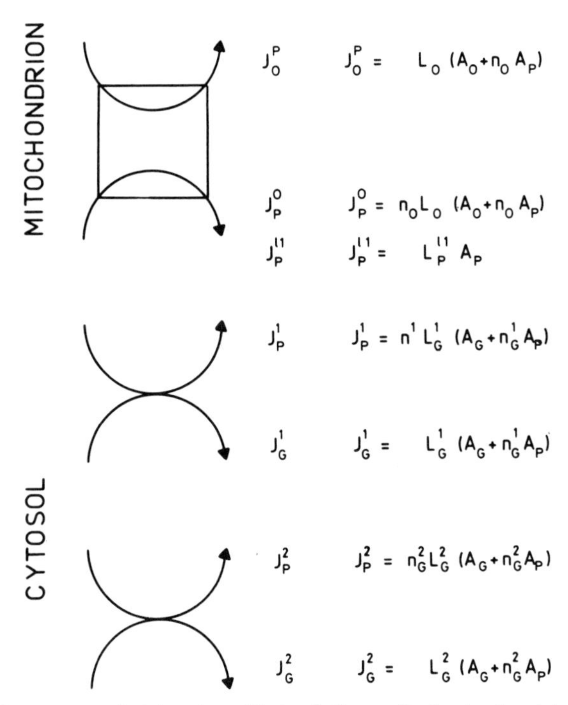

FIG. 17. Gluconeogenesis driven by oxidation in liver cells. A mitochondrion and two pathways from 2 lactate to glucose and vice versa are depicted (Note: in the text n_G^1 and n_G^2 are represented by n_P^{G1} and n_P^{G2}.)

The next step is that of a reduction of six sets of equations for serial reactions to one set of equations. Again the resulting proportionality coefficient L is given by Eq. (130), whereas the overall stoichiometric number is obtained by summing the stoichiometric numbers of the partial reactions. The L_P^1 coefficient is equal to the sum of the partial L_P^1's. In these derivations it has been assumed that there is only one value of the phosphorylation potential, that all intermediate metabolites are present in amounts small enough to reach their steady-state distribution, and that there are no important branch pathways from the glycolytin/gluconeogenic considered. The resulting overall equation will then be fully analogous to Eq. (133).

We can combine oxidative phosphorylation and gluconeogenesis into an overall set of equations describing conversion of oxidative energy into glucose synthesis from smaller molecules. Such a set is as follows:

$$\begin{pmatrix} J_O \\ J_P \\ J_G \end{pmatrix} = \begin{pmatrix} L_O^\dagger & n_P^O L_O^\dagger & 0 \\ n_P^O L_O^\dagger & (n_P^O)^2 L_O^\dagger + L_P^1 + (n_P^G)^2 L_G & n_P^G L_G \\ 0 & n_P^G L_G & L_G \end{pmatrix} \begin{pmatrix} A_O \\ A_P \\ A_G \end{pmatrix} \tag{137}$$

in which

$$L_O^{\dagger} = L_O^* \frac{1}{1 + (n_H^O)^2 L_O^*/[(n_H^P)^2 L_P^* + L_H^1]} \tag{138}$$

$$n_P^O = -\frac{n_H^O}{n_H^P} \frac{1}{1 + L_H^1/[(n_H^P)^2 L_P^*]} \tag{139}$$

$$L_P^{11} = \frac{1}{1/L_P^* + (n_H^P)^2/L_H^1} \tag{140}$$

$$L_P^1 = L_P^{11} + L_P^{12} \tag{141}$$

In practice both n_P^G and n_P^O are negative.

To this set of equations we can set a constraint using a steady-state condition in which there is no net change in the adenine nucleotides in the cell ($J_P = 0$). This gives

$$(J_G)^{*P} = -\left(\frac{n_P^O}{n_P^G} + \frac{L_P^1}{n_P^O n_P^G L_O^{\dagger}}\right) J_O + \frac{L_P^1}{n_P^O n_P^G} A_O \tag{142}$$

This equation predicts a linear relationship between rate of gluconeogenesis and rate of oxygen uptake (at constant A_O). Furthermore, the slope of the line relating those two parameters will vary, depending on the relative values of the constants. Notably, if $-n_P^G$ increases, the slope will decrease. Lines for different n_P^G, but the same L_P^1 intersect the J_O axis at the same point. Such an increase in $-n_P^G$ could be the result of an increase in the value of the L_G^i connected with the largest $-n_P^{Gi}$. In metabolic terms this could occur by regulatory influences on the enzymes involved in gluconeogenesis and the effect is referred to as futile cycling or substrate cycling. Our equations give a quantitative description of this effect.

Linear relationships between J_O and J_G have indeed been found experimentally (Van Dam *et al.* 1978a). The difference in slope between different incubation conditions can, according to the above reasoning, be ascribed to different rates of substrate cycling under those conditions. However, the slope also depends on various mitochondrial parameters, such as the proton permeability of the inner mitochondrial membrane and the activity of the adenine nucleotide translocator.

VIII. Conclusion

In this article, we have tried to show that it is possible to derive quantitative relations between measurable parameters of energy-transducing systems, starting from mechanistically independent units and using linear irreversible thermodynamics. The main advantage of this approach over the "black box" approach is that the constants in our equations have a mechanistically defined meaning.

This advantage is not only theoretical, as the increased insight in sharper formulations of hypotheses leads directly to a better design of experiments to test the hypotheses. In several instances, such tests have already led to either rejection or reformulation of the original description of the system under study. Although it has been shown for certain experimental systems that the presented approach is successful in predicting the outcome of some experiments, space was too limited to present a critical review of the applicability of the approach to all available test systems. Many relevant reports in the literature have not been discussed.

According to its theoretical basis, the description is still limited, since it describes only conditions so close to equilibrium that the use of proportionality can be founded theoretically. Also, it neglects the complications of enzyme kinetics. Nevertheless, we could apply it to a number of examples, chosen arbitrarily from the field of membrane bioenergetics.

The fact that our treatment can cover a wide range of experiments gives us confidence that it is worthwhile to pursue and that it will serve both as a catalyst and as a regulator of further investigations.

ACKNOWLEDGMENTS

We thank Henk Albers, Klaas Hellingwerf, Wim Hermens, Bob Scholte, Jan Siegenbeek van Heukelom, Roelof van der Meer, and Ton Wiechmann for clarifying discussions and critical reading of the manuscript. We are grateful to Cor Brugman for discussing with us aspects of transmembrane heat flow. Furthermore, we thank Drs. Barnes, Frömter, Kaback, Kimmich, Ramos, and Randles, as well as Excerpta Medica Foundation, the publishers of the *Journal of Biological Chemistry*, and the American Chemical Society for permission to reproduce Figs 12, 13, 15, and 16 from their publications. The help of the authors providing the figures in manuscript form is greatly appreciated. Last but not least, we acknowledge the technical assistance of Ans Brouwer, Hans van der Meijden, and Dirk-Jan Reijngoud.

List of Symbols

A_i	Affinity negative free energy change $(-\Delta G)$ of chemical reaction i
C_i	Capacity for i
c_i	Concentration of i
F	Faraday constant $= 9.65 \times 10^4$ C mol^{-1}
H_i	Partial molar enthalpy of component i
$J_{\text{chem.r}}$	Flux through chemical reaction r per unit vesicle
J_i^{a}	Flux of component i per unit vesicle, coupled to process a
$J_i^{\text{a*b}}$	Flux of component i per unit vesicle coupled to process a, at steady state of process b
J'_{q}	Outward "reduced" heat flow $= J_{\text{u}} - \sum_i H_i J_i$
J_{s}	Outward flow of entropy per unit vesicle
J_{u}	Outward flow of energy per unit vesicle

$j_{\text{chem. r}}$	Flux through chemical reaction r per unit volume
j_i	Flow of i per unit surface area
j_u	Flow of energy per unit surface area
L_i	Proportionality constant relating the flow through an autonomous process i to the force across the process
$L_O^\dagger$	Phenomenological constant relating mitochondrial oxidation to the extramitochondrial A_O, under conditions of zero transmembrane proton flow
L_O^*	Proportionality constant for mitochondrial oxidation relating externally measured oxidation to the external A_O and the proton-motive force
L_P^*	Proportionality constant relating ATP synthesis by mitochondria to the extramitochondrial A_P and the proton-motive force
L_{ij}	Phenomenological constant relating flux of i to force of j
$\ln x$	$e_{\log x}$
n_i^j	Mechanistic stoichiometric coupling constant relating flow i to flow j. $= J_i^j/J_j^i$
P	Pressure
P_i	Inorganic phosphate
Q	Heat
Q_i^*	Heat of transfer of substrate $i = (J_q'/J_i)_{\Delta T=0}$
q	Coupling coefficient
R	Gas constant $= 8.31$ J mol^{-1} K^{-1}
S_i	Partial molar entropy of i
S_i^*	Entropy of transport of substrate $i = (Js/J_i)_{\Delta T=0}$
T	Absolute temperature
t	Time
X_i	Thermodynamic force across process i
Z_i	Electrical charge of i of Faradays per mole
α	Fraction of a process
ΔX	$X_{\text{in}} - X_{\text{out}}$
$\Delta_T X$	ΔX minus its temperature-dependent part
$\Delta_{P,T} X$	$\Delta_T X$ minus its pressure-dependent part
$\Delta\tilde{\lambda}_i$	Electrochemical potential difference of i across luminal membrane (lumen-plasma)
Π	Osmotic pressure
$\Delta\tilde{\sigma}_i$	Electrochemical potential difference of i across serosal membrane (plasma-blood)
μ_i	Chemical potential of i
$\tilde{\mu}_i$	Electrochemical potential of i
$\sum X_i$	Summation of all X_i's
σ	Rate of entropy production per unit volume
Φ	Dissipation function; rate of Gibbs free energy dissipation per unit volume
ψ	Electrical potential
Σ	Rate of entropy production per unit vesicle
$\mathbf{\Phi}$	Dissipation function; rate of Gibbs free energy dissipation per unit vesicle
$\bar{X}$	Average value of X
∇	$=$ grad $= (\partial/\partial x, \partial/\partial y, \partial/\partial z) =$ vectorial space derivative
σ	Serosal membrane
∂	Partial differential

Subscripts and Superscripts

A	Acid
Cl	Chloride or bulk anion
e	Electrical; $L_e = L_K^1 + L_{Cl}^1$

eg	Electrogenic, $L_{eg} = n^2 L_v + (n_H^O)^2 L_O + (n_H^S)^2 L_S + L_H^1 + L_K^1 + L_{Cl}^1$
ex	Extravesicular
fum	Fumarate
G	Gluconeogenic and glycolytic; 2 lactate $\rightleftharpoons$ glucose
g	Glucose
gp	Glucose-6-phosphatase; Glucose + $P_i \rightleftharpoons$ glucose 6-phosphate
H	Proton
hk	Hexokinase; glucose + ATP $\rightleftharpoons$ glucose 6-phosphate + ADP
in	Intravesicular
K	Potassium, or bulk cation
L	Lactose
l	Passive leak
m	Malate
mal	Malate
Na	Sodium
n	Electroneutral proton flow; $L_n = L_{KOH} + L_{HCl}$
O	Redox
P	ATPase; ATP + $H_2O \rightleftharpoons$ ADP + P_i
P_i	Inorganic phosphate
q	Heat
S	Sugar
s	Referring to shunt pathways between epithelial cells
s	Entropy
suc	Succinate
w	Water
δ	Difference of flows through luminal–serosal membrane
λ	Luminal membrane
ν	Photon-, light absorption

Matrix Notation

$$\begin{pmatrix} J_1 \\ J_2 \end{pmatrix} = \begin{pmatrix} L_{11} & L_{12} \\ L_{21} & L_{22} \end{pmatrix} \begin{pmatrix} X_1 \\ X_2 \end{pmatrix}$$

is equivalent to

$$\begin{cases} J_1 = L_{11} X_1 + L_{12} X_2 \\ J_2 = L_{21} X_1 + L_{22} X_2 \end{cases}$$

REFERENCES

Avron, M. (1977). *Annu. Rev. Biochem.* **46**, 143–155.

Azzone, G. F., Pozzan, T., Massari, S., and Bragadin, M. (1978a). *Biochim. Biophys. Acta* **501** 296–306.

Azzone, G. F., Pozzan, T., and Massari, S. (1978b). *Biochim. Biophys. Acta* **501**, 307–316.

Azzone, G. F., Pozzan, T., Viola, E., and Arslan, P. (1978c). *Biochim. Biophys. Acta* **501**, 317–329.

Baccarini-Melandri, A., Casadio, R., and Melandri, B. A. (1977). *Eur. J. Biochem.* **78**, 389–402.

Bakker-Grunwald, T., and Van Dam, K. (1974). *Biochim. Biophys. Acta* **347**, 290–298.

Blumenthal, R., Caplan, S. R., and Kedem, O. (1967). *Biophys. J.* **7**, 735–757.

Boyer, P. D., Chance, B., Ernster, L., Mitchell, P., Racker, E., and Slater, E. C. (1977). *Annu. Rev. Biochem.* **46**, 955–1026.

Casadio, R., Baccarini-Melandri, A., Zannoni, D., and Melandri, B. A. (1974). *FEBS Lett.* **49**, 203–207.

Chance, B., and Williams, G. R. (1955). *J. Biol. Chem.* **217**, 409–427.

Chappell, J. B., and Crofts, A. R. (1966). *In* "Regulation of Metabolic Processes in Mitochondria" (J. M. Tager, S. Papa, E. Quagliariello and E. C. Slater, eds.), Biochim. Biophys. Acta Library, Vol. 7, pp. 293–314. Elsevier, Amsterdam.

Chappell, J. B., and Haarhoff, K. N. (1967). *In* "Biochemistry of Mitochondria," (E. C. Slater, Z. Kaniuga, and L. Wojtczak, eds.), pp. 75–91. Academic Press, New York.

Crane, R. K., Malathi, P., and Preiser, H. (1976). *Biochem. Biophys. Res. Commun.* **71**, 1010–1016.

Crofts, A. R. (1974). *In* "Perspectives in Membrane Biology" (G. Estrada-O and C. Gilter, eds.), pp. 373–412. Academic Press, New York.

Dainty, J., and Ginzburg, B. Z. (1964). *Biochim. Biophys. Acta.* **79**, 102–111.

Davies, M. (1965). *Biophys. J.* **5**, 651–654.

De Groot, S. R. (1952). "Thermodynamics of Irreversible Processes." North-Holland Publ., Amsterdam.

De Groot, S. R., and Mazur, P. (1962). "Non-Equilibrium Thermodynamics." North-Holland Publ., Amsterdam.

Drachev, L. A., Jasaitis, A. A., Kaulen, A. D., Kondrashin, A. A., Liberman, E. A., Nemecek, I. B., Ostroumov, S. A., Semenov, A. Y., and Skulachev, V. P. (1974). *Nature* (*London*) **249**, 321–324.

Dunlop, P. J., and Gosting, L. J. (1959). *J. Phys. Chem.* **63**, 86–93.

Essig, A., and Caplan, S. R. (1968). *Biophys. J.* **8**, 1434–1457.

Fiolet, J. W. T. (1975). Ph.D. Thesis, Univ. of Amsterdam, Amsterdam.

Frömter, E. (1977). *In* "Intestinal Permeation" (M. Kramer and F. Lauterbach, eds.), pp. 393–405. Excerpta Med. Found., Amsterdam.

Geck, P., and Heinz, E. (1976). *Biochim. Biophys. Acta* **443**, 49–53.

Gerber, G. E., Gray, C. P., Wildenauer, D., and Khorana, H. G. (1977). *Proc. Natl. Acad. Sci. U.S.A.* **74**, 5426–5430.

Goldin, S. M. (1977). *J. Biol. Chem.* **252**, 5630–5642.

Harris, D. A., and Crofts, A. R. (1978). *Biochim. Biophys. Acta* **502**, 87–102.

Hartmann, R., Sickinger, H. D., and Oesterhelt, D. (1977). *FEBS Lett.* **82**, 1–6.

Hellingwerf, K. J., Schuurmans, J. J., and Westerhoff, H. V. (1978a). *FEBS Lett.* **92**, 181–186.

Hellingwerf, K. J., Arents, J. C., Scholte, B. J., and Westerhoff, H. V. (1979). *Biochim. Biophys. Acta*, in press.

Hellingwerf, K. J., Tegelaers, F. P. W., Westerhoff, H. V., Arents, J. C., and Van Dam, K. (1978b). *In* "Halophilism" (S. R. Caplan and M. Ginsburg, eds.), pp. 283–290. Elsevier/North-Holland, Amsterdam.

Herrman, T. R., and Rayfield, G. W. (1978). *Biophys. J.* **21**, 111–125.

Jagendorf, A. T., and Uribe, E. (1966). *Brookhaven. Symp. Biol.* **19**, 215–241.

Kaback, H. R., and Barnes, E. M., Jr. (1971). *J. Biol. Chem.* **246**, 5523–5531.

Kagawa, Y. (1972). *Biochim. Biophys. Acta* **265**, 297–338.

Katchalsky, A., and Curran, P. F. (1974). "Non-Equilibrium Thermodynamics in Biophysics." Harvard Univ. Press, Cambridge, Massachusetts.

Kayushin, L. P., and Skulachev, V. P. (1974). *FEBS Lett.* **39**, 39–42.

Kell, D. B., Ferguson, S. J., and John, P. (1978). *Biochim. Biophys. Acta* **502**, 111–126.

Kimmich, G. A., and Randles, J. (1977). *In* "Intestinal Permeation" (M. Kramer and F. Lauterbach, eds.), pp. 94–106. Excerpta Med. Found., Amsterdam.

Klingenberg, M., and Rottenberg, H. (1977). *Eur. J. Biochem.* **73**, 125–130.
Krasne, S., Eisenman, G., and Szabo, G. (1971). *Science* **174**, 412–415.
Küster, U., Bohnensack, R., and Kunz, W. (1976). *Biochim. Biophys. Acta* **440**, 391–402.
Lagarde, A. E. (1976). *Biochim. Biophys. Acta* **426**, 198–217.
Lakshminarayanaiah, N. (1969). "Transport Phenomena in Membranes." Academic Press, New York.
Lang, M. A., Caplan, S. R., and Essig, A. (1977a). *Biochim. Biophys. Acta* **464**, 571–582.
Lang, M. A., Caplan, S. R., and Essig, A. (1977b). *J. Membr. Biol.* **31**, 19–29.
Lee, C. P., and Ernster, L. (1966). *In* "Regulation of Metabolic Processes in Mitochondria" (J. M. Tager, S. Papa, E. Quagliariello, and E. C. Slater, eds.), Biochim. Biophys. Acta Library, Vol. 7, pp. 218–234. Elsevier, Amsterdam.
Lehninger, A. L. (1975). "Biochemistry." Worth, New York.
Massari, S., Frigeri, L., and Azzone, G. F. (1972). *J. Membr. Biol.* **9**, 57–70.
Miller, D. G. (1960). *Chem. Rev.* **60**, 15–37.
Mitchell, P. (1961). *Nature (London)* **191**, 144–148.
Mitchell, P. (1966). *Biol. Rev.* **41**, 445–502.
Mitchell, P. (1968). "Chemiosmotic Coupling and Energy Transduction." Glynn Res. Ltd., Bodmin, Cornwall, England.
Montal, M. (1976). *Annu. Rev. Biophys. Bioeng.* **5**, 119–175.
Nicholls, D. G. (1974). *Eur. J. Biochem.* **50**, 305–315.
Nicholls, D. G. (1977). *Biochem. Soc. Trans.* **5**, 200–203.
Oesterhelt, D., and Stoeckenius, W. (1971). *Nature (London) New Biol.* **233**, 149–152.
Oesterhelt, D., and Stoeckenius, W. (1974). *In* "Biomembranes," Part A (S. Fleischer and L. Packer, eds.), Methods in Enzymology, Vol. 31, pp. 667–678. Academic Press, New York.
Onsager, L. (1931). *Phys. Rev.* **37**, 405–426.
Ort, D. R., Dilley, R. A., and Good, N. E. (1976). *Biochim. Biophys. Acta* **449**, 108–124.
Oster, G. F., Perelson, A. S., and Katchalsky, A. (1973). *Q. Rev. Biophys.* **6**, 1–134.
Ovchinnikov, Y. A., Ivanov, V. T., and Shkrob, A. M. (1974). "Membrane-Active Complexones," Biochim. Biophys. Acta Library, Vol. 12. Elsevier, Amsterdam.
Padan, E., and Rottenberg, H. (1973). *Eur. J. Biochem.* **40**, 431–437.
Paterson, R. (1970). *In* "Membranes and Ion Transport" (E. E. Bittar, ed.), Vol. 1. pp. 123–164. Wiley (Interscience), New York.
Petty, K. M., Jackson, J. B., and Dutton, P. L. (1977). *FEBS Lett.* **84**, 299–303.
Postma, P. W., and Roseman, S. (1976). *Biochim. Biophys. Acta* **457**, 213–257.
Prigogine, I. (1955). "Introduction to Thermodynamics of Irreversible Processes." Thomas, Springfield, Illinois.
Racker, E., and Stoeckenius, W. (1974). *J. Biol. Chem.* **249**, 662–663.
Ramos, S., and Kaback, H. R. (1977). *Biochemistry* **16**, 848–859.
Ramos, S., Schuldiner, S., and Kaback, H. R. (1976). *Proc. Natl. Acad. Sci. U.S.A.* **73**, 1892–1896.
Reuss, L., and Finn, A. L. (1975). *J. Membr. Biol.* **25**, 115–139.
Rottenberg, H. (1970). *Eur. J. Biochem.* **15**, 22–28.
Rottenberg, H. (1973). *Biophys. J.* **13**, 503–511.
Rottenberg, H., and Gutman, M. (1977). *Biochemistry* **16**, 3220–3227.
Rottenberg, H., and Lee, C.-P. (1975). *Biochemistry* **14**, 2675–2680.
Rottenberg, H., Caplan, S. R., and Essig, A. (1970). *In* "Membranes and Ion Transport" (E. E. Bittar, ed.), Vol. 1, pp. 165–191. Wiley (Interscience), New York.
Sauer, F. A. (1977). *In* "Intestinal Permeation" (M. Kramer and F. Lauterbach, eds.), pp. 320–321. Excerpta Med. Found., Amsterdam.

Schuldiner, S., Padan, E., Rottenberg, H., Gromet-Elhanan, Z., and Avron, M. (1974). *FEBS Lett.* **49**, 174–177.

Schultz, S. G. (1977). *In* "Intestinal Permeation" (M. Kramer and F. Lauterbach, eds.), pp. 382–392, Excerpta Med. Found., Amsterdam.

Simoni, R. D., and Postma, P. W. (1975). *Annu. Rev. Biochem.* **44**, 523–554.

Sorgato, M. C., Ferguson, S. J., Kell, D. B., and John, P. (1978). *Biochem. J.* **174**, 237–256.

Spanner, D. C. (1954). *Symp. Soc. Exp. Biol.* **8**, 76–93.

Tsou, C. S., and Van Dam, K. (1969). *Biochim. Biophys. Acta* **172**, 174–176.

Van Dam, K., and Westerhoff, H. V. (1977). *In* "Structure and Function of Energy-Transducing Membranes" (K. Van Dam and B. F. Van Gelder, eds.), pp. 157–167. Elsevier, Amsterdam.

Van Dam, K., Casey, R. A., Van der Meer, R., Groen, A. K., and Westerhoff, H. V. (1978a). *In* "Frontiers of Biological Energetics," Vol. 1, pp. 430–438.

Van Dam, K., Wiechmann, A. H. C. A., Hellingwerf, K. J., Arents, J. C., and Westerhoff, H. V. (1978b). *Proc. FEBS Meet., 11th, Copenhagen, 1977* **45**, 121–132.

Van der Meer, R., Akerboom, T. P. M., Groen, A. K., and Tager, J. M. (1978). *Eur. J. Biochem.* **84**, 421–428.

Van de Stadt, R. J., De Boer, B. L., and Van Dam, K. (1973). *Biochim. Biophys. Acta* **292**, 338–349.

West, I. C., and Mitchell, P. (1973). *Biochem. J.* **132**, 587–592.

Westerhoff, H. V., Scholte, B. J., and Hellingwerf, K. J. (1979). *Biochim. Biophys. Acta*, in press.

Williams, R. J. P. (1978). *Biochim. Biophys. Acta* **505**, 1–44.

Zuurendonk, P. F., Akerboom, T. P. M., and Tager, J. M. (1976). *In* "Use of Isolated Liver Cells and Kidney Tubules in Metabolic Studies" (J. M. Tager, H. D. Söling, and J. R. Williamson, eds.), pp. 17–27. North-Holland Publ., Amsterdam.

Intracellular pH: Methods and Applications

R. J. GILLIES and D. W. DEAMER
Department of Zoology
University of California
Davis, California

I. Introduction

The role of hydrogen ions in cytoplasmic function is attracting increasing interest among biochemists and cell biologists. Mitchell (1961, 1966) was among the first to propose that proton activity gradients are involved in bioenergetic processes at the subcellular level. It is now generally accepted that protons are actively transported by phosphorylating membranes such as mitochondria and chloroplasts, and that the energy of the resulting electrochemical gradient can be coupled to the synthesis of ATP. An important result of this research was that new techniques were developed to measure transmembrane proton gradients, and these are now being applied to whole tissues and cells.

ISBN 0-12-152509-0

Although the existence of electrochemical proton gradients across subcellular organelle membranes is established, there is considerably less confidence that proton activity (intracellular pH) plays an active role at the cellular level. As pointed out by Waddell and Bates (1969) in their comprehensive review of this area, it is difficult to define intracellular pH. If there are indeed membranous compartments within the cell that can produce and maintain proton gradients, it is not clear that intracellular pH corresponds to the bulk cytoplasmic pH. Furthermore, even in simple systems, pH varies near charged surfaces. An example is the dramatic alteration in the measured p*K* of carboxyl groups in fatty acid monolayers, which can increase as much as 4 pH units over the expected value owing to strong field effects (Patil *et al.*, 1975). The cell has many such charged surfaces, which presumably induce pH gradients in their neighborhood.

Despite these reservations, it should still be possible to learn something about the approximate range of cytoplasmic pH within a given level of accuracy. If several different methods indicate that the intracellular pH of a given cell type is in the range of 7.0 under defined conditions, it is reasonable to accept this as an operational value. Waddell and Bates (1969) summarized values for a number of different animal cells, and pH_i values ranged around 7.0 ± 0.5, depending on cell type and conditions. Although it is useful to have estimates of absolute pH_i values, even more significant is the relationship between intracellular pH variation and various cell functions.

The purpose of this review is to discuss aspects of intracellular pH that have attracted current interest. We will use a number of terms that are conveniently applied to specific parameters of pH in relation to cell morphology and function. The terms pH_i and pH_e will refer to the measured intracellular and extracellular bulk-phase pH values. We will also use the term ΔpH to refer to the measured pH gradient across a cell membrane, and the term δpH will designate a change in the magnitude of ΔpH.

There are several basic questions that can be approached with present techniques for monitoring pH_i, and these form the basis of this review:

1. What methods are appropriate for studying pH_i in different systems, and what are their relative strengths and weaknesses?
2. Is the cytoplasmic pH regulated by the cell?
3. Does pH_i vary during specific cell functions, such as the cell cycle, fertilization, and protein synthesis?
4. If pH_i is found to vary with a given cell function, is the variation simply correlated with that function, or can it be shown to be causal? For instance, it is possible that cells use pH_i to regulate metabolism or as a signal to initiate specific events.

5. In damage processes, such as those initiated by ischemia in heart disease, do changes in pH_i contribute to cell dysfunction?

Recent work permits at least partial answers to these questions. We will discuss three methods for measuring pH_i in tissues and cellular systems. These include weak acid and weak base distribution across membranes, refinements of pH-sensitive glass microelectrodes, and ^{31}P nuclear magnetic resonance (NMR) spectroscopy. Each will be discussed in terms of theory, application, and relative strengths and weaknesses, and an example of current research involving each method will be described in detail.

II. Weak Acids

A. Introduction

Intracellular pH can be estimated by the equilibrium distribution of weak acids or bases across the cell membrane. The most commonly used weak acid indicator is 5,5-dimethyloxazolidine-2,4-dione (DMO). DMO was introduced by Waddell and Butler (1959) who showed that it distributes across muscle cell membranes in response to pH gradients. DMO has been used to estimate the internal pH of whole tissues (Waddell and Butler, 1959), mitochondria (Addanki *et al.*, 1968), bacteria (Harold *et al.*, 1970), fungi (Leguay, 1977), protozoa (Gillies and Deamer, 1977, 1978, 1979), human leukocytes (Levin *et al.*, 1976; Baron and Ahmed, 1969; Ahmed and Baron, 1971; Zieve *et al.*, 1967), and erythrocytes (Tomodal *et al.*, 1977).

Ideally, only the uncharged form of a weak acid or weak base probe is freely permeable to the membrane and is present in equal concentrations on either side. The ratio of charged to uncharged form is given by a Henderson–Hasselbach relationship in either compartment, and the equilibrium distribution is therefore pH dependent (Fig. 1).

pH_i is estimated by the relationship:

$$pH_i = pK + \log[R(10^{(pH_e - pK)} - 1) - 1] \tag{1a}$$

for acids, and

$$pH_i = pK + \log[R(10^{(pK - pH_e)} + 1) - 1] \tag{1b}$$

for bases, where

$$R = (TA_i/TA_e)(V_e/V_i) \tag{2}$$

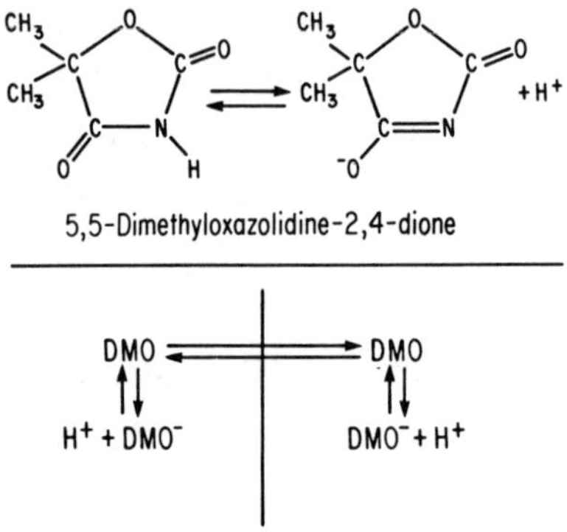

5,5-Dimethyloxazolidine-2,4-dione

FIG. 1. Diagrammatic equilibration of 5,5-dimethyloxazolidine-2,4-dione (DMO) across a biological membrane. DMO is a weak acid, with a pK of approximately 6.2. As shown, the protonated form freely crosses the membrane and establishes an equilibrium with its charged form in either compartment, which is determined by a Henderson–Hasselbach relationship.

The subscripts i and e refer to internal and external compartments; TA represents total amount of indicator in either compartment, and V is the volume of that compartment. Estimates of these volumes are necessary, and the external pH is measured during an experiment. The pK_a of DMO has been tabulated as a function of temperature and ionic strength by Boron and Roos (1976).

Equations (1) and (2) can be simplified for acids to

$$\mathrm{pH_i} = \mathrm{pH_e} + \log[\mathrm{R}] \tag{3}$$

and for bases to

$$\mathrm{pH_i} = \mathrm{pH_e} - \log[\mathrm{R}] \tag{4}$$

if the pK of the indicator is more than 2 pH units away from the $\mathrm{pH_e}$. The value ($\log[R]$) represents the ΔpH.

Another relationship that follows from these arguments is the difference between the $\mathrm{pH_i}$ of two similar samples; it is defined by

$$\delta\mathrm{pH} = \mathrm{pH_i}(1) - \mathrm{pH_i}(2) \tag{5}$$

If the volumes, pK and $\mathrm{pH_e}$ are identical between the two samples (1) and (2), substituting in Eq. (4) gives:

$$\delta\mathrm{pH} = [\mathrm{TA_i}(1) \cdot \mathrm{TA_e}(2)]/[\mathrm{TA_i}(2) \cdot \mathrm{TA_e}(1)] \tag{6}$$

This equation is quite useful, since it allows calculation of δpH without knowing V_i and V_e. It is generally applicable only when using weak bases.

B. Theoretical Considerations

For a compound to be sufficiently sensitive to transmembrane pH gradients, its pK must be in the correct range. In general, the farther the pK is from the measured pH, the more sensitive it will be. If the pH_i is 7.0, DMO (pK_a = ca. 6.2) is approximately 85% dissociated and therefore relatively insensitive. Weak-base indicators are more sensitive than DMO because of this phenomenon. On the other hand, DMO is more generally applicable and can be used in systems with either acidic or alkaline interiors because of its moderate pK (Fig. 2). DMO therefore sacrifices sensitivity in favor of applicability.

A slight permeability to the charged species can affect the estimated pH_i since the charged form will respond to membrane potentials, if present. Boron and Roos (1976) have calculated that if the ratio of charged to uncharged DMO permeability was 1 : 200, the estimated pH_i would be 0.03 pH unit too low, given a ΔpH of 0.46 and membrane potential of −50 mV.

C. Strengths and Weaknesses of DMO Technique

Although there is considerable evidence that DMO behaves ideally in estimating pH_i, certain errors can occur as a consequence of the techniques used. A major source of error arises during separation of internal (V_i) from external (V_e) volume. Separation is necessary in order to measure the distribution of DMO between V_i and V_e. Usually, filtration or centrifugation are used for separating V_i and V_e. Filtration of cells and organelles can introduce substantial background radioactivity on the filter, which must be controlled (Nichols, 1974). Separation by centrifugation is slower and adds the risk of oxygen deprivation. These possibilities are discussed in a recent review by Rottenburg (1979).

Determination of the internal volume is a prerequisite to estimating the pH_i. Measurement of this volume is usually accomplished by incubating the system in $[^3H]H_2O$ to mark total volume and a ^{14}C-labeled external volume marker such as sorbitol, manitol, or sucrose. Since the exchangeable water volume can change with the metabolic state of a cell (Beall *et al.*, 1976), internal volume should be determined under each condition to be tested. External volume markers usually need a substantial time to uniformly diffuse (>1 hour), which makes the volume determination difficult and at times impractical (Boron and Roos, 1976). If the volume is determined in this way however, pH_i can be expressed in terms of exchangeable V_i, which can be significant if the system has a variable percentage of its water in a bound state. Volume can be related linearly to another parameter, such a protein,

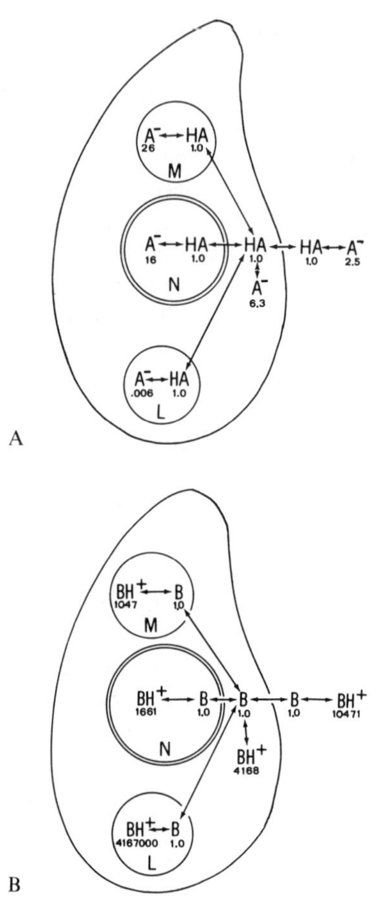

FIG. 2. Expected equilibrations of 5,5-dimethyloxazolidine-2,4-dione (DMO) (Fig. 2A) and methylamine (Fig. 2B) into compartments of eukaryotic cell. Size and pH of compartments are presented in Table I. Compartments are cytosol (C), nucleus (N), mitochondria (M), and lysosomes (L). Note the large amount of amine accumulation (Fig. 2B) in lysosomes due to their very low pH.

which allows an indirect estimation of cell volume at each experimental condition.

In controlled studies DMO has been shown to behave almost ideally (Waddell and Butler, 1959). It does not bind to fragments of muscle (Boron and Roos, 1976), bacteria (Kashket and Wong, 1969), and protozoa (Gillies and Deamer, 1978, 1979). DMO equilibrates rapidly in many systems (Addanki *et al.*, 1968; Ahmed and Baron, 1971; Tomodal *et al.*, 1977), suggesting that it is neither metabolized nor transported. The nontoxicity of DMO has also been established in both prokaryotic (Leguay, 1977; Kashket and Wong, 1969) and eukaryotic (Gillies and Deamer, 1978, 1979) systems. These reports indicate that DMO equilibrates with cells and tissues in response to a pH gradient and is not artifactual. Further support of this comes from studies in which results from the DMO technique are directly compared to those from either microelectrodes (Boron and Roos, 1976) or nuclear magnetic resonance (NMR) (Paddan *et al.*, 1976; Navon *et al.*, 1977a,b). In both cases, DMO correlated very highly with the results from the other two techniques.

A major strength of the DMO technique is its general applicability. It can be used to determine the pH_i of systems as small as mitochondria (Addanki *et al.*, 1968) or as large as whole organisms (Waddell and Butler, 1959). This range of applicability is a distinct advantage over the microelectrode, which necessitates using large ($>30\ \mu m$) cells. The systems to be studied with DMO can also be maintained in conditions optimal for growth or function. For instance, with cell suspensions this is an advantage over the NMR technique, in which cells must be maintained at a very high density (Navon *et al.*, 1977a). Finally, intracellular pH determinations using DMO are inexpensive and require only basic laboratory equipment, such as a centrifuge and scintillation counting equipment.

D. Application of DMO to Eukaryotic Cells

A recent application of the DMO technique has been to estimate the intracellular pH of eukaryotic cells. The first workers to do so were Zieve *et al.* (1967), who studied transmembrane acid–base regulation in human leukocytes. More recently, a pH dependency of glycolysis *in situ* was determined by Tomodal *et al.* (1977) using DMO with human erythrocytes and was corroborated in *Escherichia coli* independently by Shulman and co-workers using ^{31}P NMR (Navon *et al.*, 1977b). A major interest in our laboratory is the relationship between intracellular pH and cellular growth control. We have studied these parameters in *Tetrahymena*, a ciliated protozoan whose cell cycle is readily synchronized by a variety of methods (Zeuthen, 1971; Cameron and Jeter, 1970). Application of the DMO technique to eukaryotic cells poses some new practical and theoretical considerations. These cells contain several membrane-bound compartments, which can maintain a different internal pH.

We have approached the problem of compartmentation in *Tetrahymena* according to the following rationale. The major cytoplasmic compartments of *Tetrahymena* include the nucleus, mitochondria, cytosol, and food vacuoles. Food vacuoles are very acidic and would exclude DMO. They

TABLE I

CALCULATION OF pH FROM EQUILIBRIUM DISTRIBUTIONS OF BOTH WEAK ACID, DMO (pK = 6.20), AND WEAK BASE, METHYLAMINE (pK = 10.62)[a,b]

Compartment	pH	Volume	Concentration	Total	Concentration	Total
External	6.6	100	3.50	3.50	10,472	10,472
Cytosol	7.0	60	7.30	4.38	4,169	2,501
Nucleus	7.4	15	16.80	2.52	1,669	249
Lysosome	4.0	15	1.01	0.15	4,167,000	625,050
Mitochondria	7.6	10	27.00	2.70	1,048	105

Indicator	External conc. (Ex.)	Internal conc. (In.)	In. : Ex. ratio	Apparent pH_i
DMO	3.50	9.75	2.786	7.14
MA	10,472.00	627,905.00	59.960	4.82

[a] DMO = 5,5-dimethyloxazolidine-2,4-dione; MA = methylamine.
[b] Data are from case presented in Figs. 2A and 2B.

comprise 15% of the total cell volume in a log-phase culture. We have therefore run our experiments on log-phase cells and have corrected the volume estimation by 15%. Our reported pH_i for these cells presumably represents contributions from cytosol, nucleus, and mitochondria (Table I). More detailed resolution of compartmentalized pH_i cannot be expected using the DMO technique.

The protocol for pH_i determinations using DMO in *Tetrahymena* is identical in theory to that used in other systems. The labeled tracer is incubated with the cells until equilibrium is reached. The cells are then separated from the medium, and aliquots are taken from each and assayed for radioactivity. Finally, the intracellular volume is estimated either directly or indirectly.

In a typical experiment, *Tetrahymena* are incubated for 5 minutes in the presence of [^{14}C]DMO and [^{3}H]H_2O, a volume marker. This system comes to equilibrium within 1 minute and remains stable for at least 90 minutes (Gillies and Deamer, 1978, 1979). The suspensions are then centrifuged at 800 g for 2–5 minutes. An aliquot of the supernatant is assayed for radioactivity, and the rest is removed by aspiration. The sides of the tubes are wiped dry, and the pellet is solubilized in 0.5 ml of 0.1 N NaOH and assayed for radioactivity and protein. Supernatant and pellet radioactive samples are equally quenched. The protein in the pellet sample is related to cell volume by a factor of 5.47×10^{-3} $\mu l/\mu g$ for exponential, asynchronous cells. An advantage of this method is that it allows a volume determination for each sample.

Resolution of the raw data of supernatant counts, pellet counts, and pellet protein into a pH value is done by a Fortran program written and executed on a Burroughs 6700. The calculations involve:

1. Resolving the counts per minute (cpm) to disintegrations per minute (dpm) by correcting for background, spillovers, and efficiencies.
2. Calculating internal volume (V_i) from pellet protein.
3. Calculating total volume (V_t) from pellet tritium.
4. Calculating external volume (V_e) by subtracting V_i from V_t.
5. Calculating external [^{14}C](TA$_e$) by multiplying V_e by the ratio of ^{14}C to volume in the supernatant.
6. Calculating internal [^{14}C](TA$_i$) by subtracting DMO$_e$ from total ^{14}C in pellet.
7. Calculating pH$_i$ using Eq. (1a).

We are currently using this technique to monitor the pH$_i$ of *Tetrahymena* under different growth conditions and during the cell cycle. The first series of experiments to be described determine pH$_i$ and growth rates at different external pH. As shown in Fig. 3, *Tetrahymena* maintain their internal pH near 7.2 in the external pH range 5.0 to 7.3. Determinations were performed on two strains (GL and DN$_3$), which were either grown at or titrated to the appropriate pH, with similar results. Maintenance of the internal pH against a gradient was also found to require a convenient energy source, such as

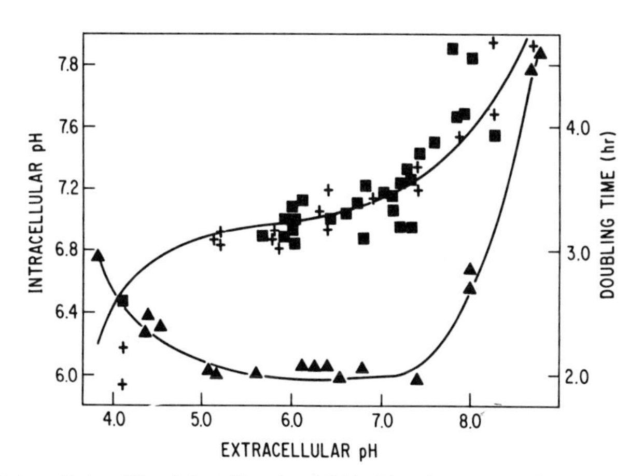

FIG. 3. Intracellular pH and doubling time (▲) in *Tetrahymena* as a function of external pH. pH$_i$ was determined by the DMO method on two strains of *T. pyriformis*; GL (■) and DN$_3$ (+), which were either grown at or titrated to appropriate pH$_e$. The results from both types of experiments are pooled. Doubling time was determined as cell number at times after initial inoculum of 2000 cells/ml. Cell number was determined using an electronic particle counter.

glucose (data not shown). The ability of the cells to maintain a constant internal pH is correlated with the external pH range of maximum growth.

These results suggest that (1) *Tetrahymena* do not require a proton gradient across the plasma membrane for growth and (2) regulation of pH_i is probably not by simple buffering. The first point follows from the observation that the cell growth rate does not correlate with the magnitude of the pH gradient. The second follows from the fact that these cells maintain an interior that is alkaline with respect to the environment. If the intracellular buffer was only chemical, one would not expect a gradient to be maintained at all. The direction of the gradient suggests the presence of a vectorial "proton pump" at the cell surface. Such structures have been described previously for muscle (Aickin and Thomas, 1977), and prokaryotes (Walker and Smith, 1975; Harold *et al.*, 1970; Oesterhelt and Stoeckenius, 1973).

To examine possible changes in pH_i during the cell cycle in *Tetrahymena*, we synchronized the cells either by starvation and refeeding (Cameron and Jeter, 1970) or by heat shock (Zeuthen, 1971). In the first case, cells were either allowed to grow to stationary phase or were placed in starvation buffer overnight (Wolfe, 1975). Upon refeeding, the cells exhibit good (70%)

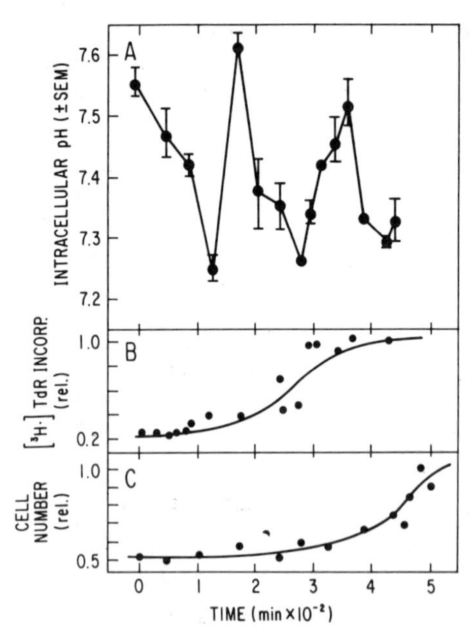

FIG. 4. Intracellular pH, thymidine (TdR) incorporation, and cell number as a function of time after refeeding in *Tetrahymena pyriformis* GL. pH_i is expressed as mean ±SEM of results from 3 different experiments. TdR incorporation and cell number are expressed as percentage relative to values at 500 minutes.

DNA synthetic synchrony (S phase) at 170 minutes and relatively poor (12 %) division synchrony (D phase) at 300 minutes. During this generation, there are two distinct alkaline shifts of ca. 0.35 pH unit. These shifts are temporally related to within 30 minutes of the onset and end of S phase (Fig. 4). The second peak also occurs just prior to the beginning of cell division.

Heat shock synchrony yields a culture that synchronously divides (division index = 90 %) 85 minutes after the sixth and subsequent heat shocks of 33.8°C. DNA synthesis occurs immediately after cytokinesis and involves up to 75 % of the cells at any one time. In this system also there are two alkaline shifts of 0.4 pH unit per cell cycle. The shifts are temporally related to within 15 minutes of the beginning and end of either S or D phase (Fig. 5).

Comparison of these data from both synchronization procedures yields some interesting correlations. The greatest difference in the cell cycle kinetics of the two populations is the timing of the cytokinesis. DNA synthesis in both cultures proceeds at about the same (relative) time and to the same degree. The shifts in intracellular pH seen in the two are surprisingly similar. This has led us to conclude that the shifts in pH are associated with DNA synthetic events and are not affected by the time, duration, or magnitude of

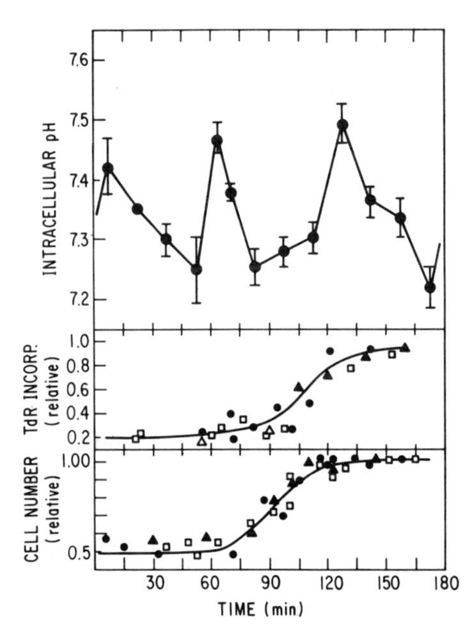

FIG. 5. pH_i, thymidine (TdR) incorporation, and cell number as a function of time after heat shock, as described in text. All results are pooled from 3 experiments and expressed as described for Fig. 4.

cytokinesis. Whether these shifts are common to other cell types and their involvement in regulation remains to be established.

In conclusion, DMO has been shown by many workers to be an effective indicator of pH_i. It is suitable for a wide range of applications from tissues to subcellular organelles. Under controlled conditions, it has been shown to behave almost ideally, although some resolution may be lost when it is applied to multicompartmented systems. Application of this technique to eukaryotes has shown pH_i to be affected by agents such as CO_2 (Levin *et al.*, 1976), and trypsin (Ahmed and Baron, 1971). Changes in pH_i correlate highly with some parameters of cell growth, such as development (Carter and Nijhout, 1977), glycolysis (Tomodal *et al.*, 1977), and the cell cycle (Gillies and Deamer, 1978, 1979).

III. Weak Bases

A. Introduction

The distribution of weak bases between two membrane-bounded compartments is governed by the same considerations described earlier for weak acids. However, the distribution will be in the opposite direction, so that weak bases are concentrated in the relatively acidic volume.

The most common choices for weak base pH probes have been the monoamines methylamine (MA) and 9-aminoacridine (9-AA). MA is measured as the ^{14}C-labeled compound, and the experimental protocol used to measure DMO distribution is directly applicable to it, including the requirement for separation of V_0 from V_i by centrifugation or filtration. 9-AA has the advantage of simplicity of quantification, in that it is a fluorescent compound that can be measured by fluorometry. Most investigators have assumed that 9-AA fluorescence is quenched when it becomes concentrated in a membrane-enclosed space. This offers the advantage that V_e and V_i need not be separated, since the amount of amine inside can be related to the amount of quenching. If half the fluorescence disappears, it is assumed that half the amine is in the internal volume. An important advantage of 9-AA is that its distribution within a cell can be visualized by fluorescence microscopy. Despite the fact that much of the fluorescence is quenched, enough remains to provide contrast in a fluorescence microscope. This effect has been used by Meizel and Deamer (1978) to visualize the 9-AA distribution in hamster sperm, as will be discussed later.

A disadvantage of 9-AA is that it can bind to cell structures by mechanisms other than pH gradients. The amount of 9-AA that appears to be accumulated by a membrane-bound space can therefore be a mixture of nonspecific binding and accumulation related to a pH gradient. It is possible to control for nonspecific binding by adding compounds, such as nigericin or ammonium, which permit pH gradients to decay. Any remaining binding of 9-AA is assumed to be independent of pH gradients and can be subtracted from the total amount accumulated to provide an estimate of the pH-dependent binding.

Other possible artifacts include incomplete quenching of fluorescence and leakage of the charged form of the amine through cation channels in the membrane. The most practical controls for such effects are to compare results from separate methods for monitoring pH_i, to calibrate the system with pH_e shifts of known magnitude, and to test for ideal behavior of the amine.

In any new system to which 9-AA is to be applied, there are a number of simple tests that will determine whether it is behaving as an ideal weak base. For instance, one may alter the volume ratio to see whether the distribution of the amine follows the relationship:

$$\log(A_i/A_e) = \log(V_e/V_i) - K \quad (7)$$

where K is the constant ΔpH. This relationship and others have been tested in liposome systems in which pH gradients of known magnitude were established (Deamer *et al.*, 1972; Casadio and Melandri, 1977), and it was found that within limits 9-AA behaved as an ideal weak base and could be used to estimate ΔpH and pH_i with reasonable accuracy.

B. Application of 9-Aminoacridine to Hamster Sperm

In order to show how 9-AA may be used in eukaryotic cells, we will describe a current application in which the amine was monitored by fluorescence microscopy to localize the site of accumulation within a cell and by fluorometry to derive information about the pH_i. The cell type used was hamster sperm, and the structure of interest was the acrosome, a cap-like organelle located in the anterior region of the sperm head. The acrosome contains hydrolytic enzymes involved in fertilization, and it has been suggested that acrosome pH may control activation of the enzymes (Meizel and Mukerji, 1975). Therefore it was desired to estimate the pH_i of the acrosome, and the most suitable method was weak-base distribution.

The rationale for this work was to use fluorescence microscopy to visualize the sperm compartment in which the major fraction of the dye was accumulating. The pH dependence of the accumulation could be tested by conditions known to discharge pH gradients, as described earlier. For instance, ammonium could be added as NH_4Cl in relatively high concentrations. As NH_3 diffuses into an acidic volume, protons are used to form NH_4^+, thereby

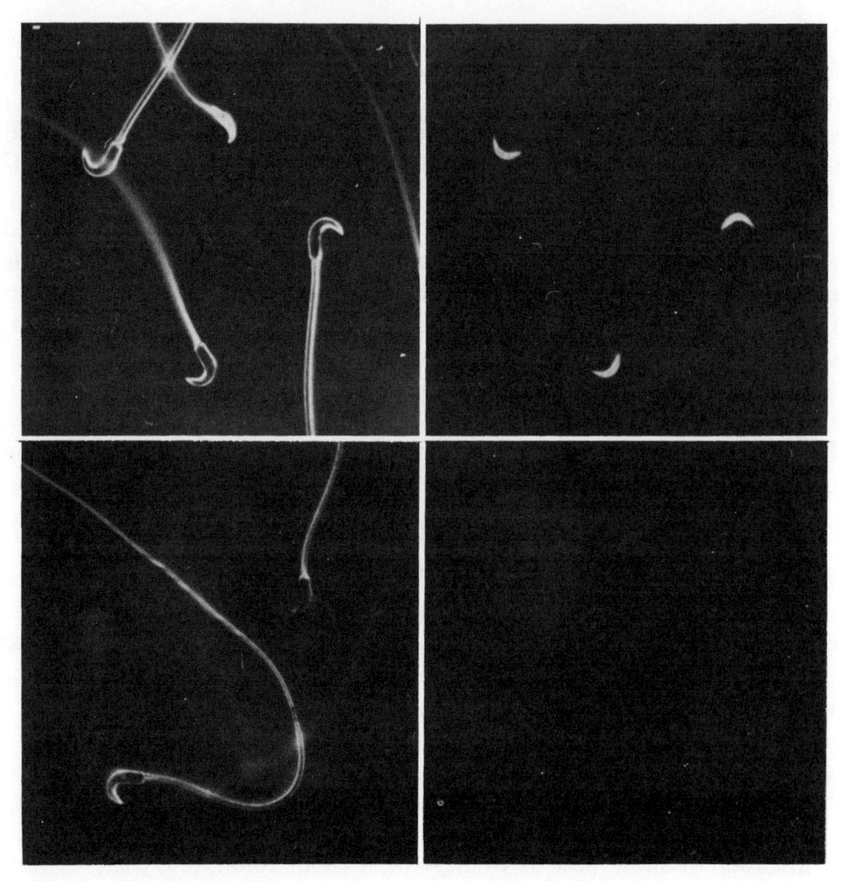

FIG. 6. Fluorescence of hamster sperm in the presence of 9-aminoacridine. The figures at left show hamster sperm under dark-field microscopy; and figures at right, with fluorescence microscopy. The figures at top are untreated controls, and the lower ones are in the presence of 5 m*M* NH_4Cl. Note the concentration of fluorescence in the acrosomal region of sperm at upper right. In the presence of NH_4Cl, fluorescence is completely eliminated. (See text for details.)

discharging the pH gradient. A second check was addition of the antibiotic nigericin, which permits exchange of potassium or sodium for H^+ and again causes pH gradients to decay rapidly.

Figure 6 shows results from such an experiment. The 9-AA concentrated in the acrosomes of motile sperm so that individual acrosomes were clearly visible in fluorescent contrast against the dark background. Addition of either ammonium or nigericin caused 9-AA to disappear from the acrosome, and it was concluded that the acrosome was the major intracellular compartment involved in the 9-AA response.

The ΔpH can be estimated if V_i is known. In these studies, acrosomal volume was estimated by a simple geometric analysis of electron micrographs of sperm and was found to be 0.4 μm^3. From knowledge of sperm number in the sample, and from the amount of pH-dependent binding of 9-AA (determined by the change in fluorescence before and after ammonium or nigericin addition), the pH_i of the acrosome was estimated to be 3.2–3.4.

It is clear that combined microscopic and fluorometric analysis with 9-AA can be a useful tool in working with intracellular pH gradients of relatively large magnitude. However, the initial applications to whole cells suggest that nonspecific binding makes amines less useful than DMO in measurements of overall cytoplasmic pH. Also, amines may accumulate in a small, but highly acidic, intracellular compartment, thereby biasing any estimate of average cytoplasmic pH. This is clearly shown in Table I and Fig. 2B, in which this effect causes a 2 pH unit discrepancy between results from a weak-acid and a weak-base probe.

IV. Microelectrodes

A. Introduction

Glass microelectrodes offer the most direct way of estimating pH_i. Over the last decade, a number of laboratories have applied this technique to a variety of cells. The apparent simplicity is deceptive, however. Considerable care must be taken in the interpretation of results, as well as in experimental procedures. This method for monitoring pH_i has recently been used to study cytoplasmic pH changes in sea urchin eggs following fertilization. The results offer an interesting comparison with equivalent data from other methods, as will be discussed later.

There are two basic types of microelectrodes that have been used to monitor intracellular pH, the basic difference being an exposed or recessed pH-sensitive glass tip. The former type was used by Carter *et al.* (1967) to study muscle cells. The observed pH_i was surprisingly low, near 6.0, with a direct correlation between membrane potential and the apparent pH_i. Since the exposed tip could introduce artifacts related to conductance around the tip through the puncture in the cell, Aicken and Thomas (1977) repeated this work using electrodes with recessed tips. The measured pH_i was markedly higher, near 7.0, and the correlation between pH and membrane potential was not observed. This suggested that recessed-tip electrodes directly sense the pH_i of the cell and are the method of choice.

B. Limitations of the Microelectrode Method

There are several limitations of the microelectrode technique that are related to instrumentation. In an ion-sensitive glass electrode, the output of the electrode is a function of the ion selectivity, the potential developed by the ion-selective glass, and the membrane potential of the cell. Therefore, a membrane-potential measurement is required together with the glass-electrode measurement so that the membrane potential can be subtracted from the glass-electrode potential to give the potential due to pH. In order to assure that both electrodes are in the same cell, a polarizing current is passed between electrodes while observing the change in membrane potential.

A second limitation concerns the time required for the electrode to reach a stable readout. The response time is usually on the order of 15–30 seconds, so that any changes in pH_i with shorter time constants cannot be monitored. A related problem is that the cytoplasm can change in the vicinity of the electrode over time. For instance, the cell after a few minutes may form a membranous vesicle around the electrode tip which masks the true pH.

C. Intracellular pH and Fertilization

One of the most exciting possible functions of pH_i is that the cell uses intracellular pH as a signal or regulating factor. This was first suggested for the activation of sea urchin eggs by Mazia and co-workers (Steinhardt and Mazia, 1973; Epel *et al.*, 1974). This was supported by further investigation in which eggs were homogenized at various time intervals following fertilization, and the pH of the homogenate was measured (Johnson *et al.*, 1976). It was assumed that changes in the pH of the homogenate would reflect changes in the pH_i. If this assumption is correct, the results suggested that pH_i increases markedly between 1 and 4 minutes after fertilization. This work was confirmed by Lopo and Vacquier (1977), who also observed an apparent decrease in pH_i over the next 120 minutes to original levels. These workers used the detergent Triton X-100 to release the pH gradients as well as homogenization, with similar results.

Although the implications of this work are exciting, homogenization is a crude method and this important result should be confirmed by other techniques if possible. Shen and Steinhardt (1978) have used microelectrodes to study pH_1 after fertilization and initiation of protein synthesis in sea urchin eggs. Gillies and Deamer (unpublished) have used methylamine and DMO to monitor pH_i in the same system after fertilization. These results and those of the earlier homogenization method are shown in Fig. 7. It is clear that there is qualitative agreement in the three methods, but that DMO and microelectrodes give somewhat higher initial values for pH_i. It is reasonable to conclude that pH_i increases in sea urchin eggs after fertilization.

What is the significance of this result? Is there simply a correlation, or could the pH change represent a signal? If the pH change does regulate some cell function after fertilization, then experimental manipulation of the pH may also be expected to cause a similar effect. Ammonium chloride has

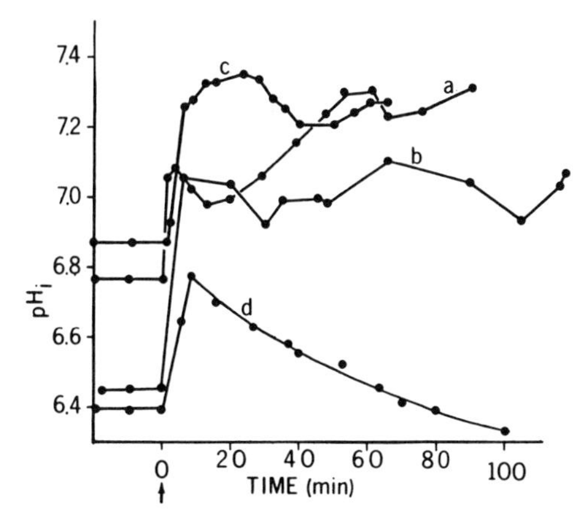

FIG. 7. Intracellular pH in sea urchins as determined by (curve a) the DMO method, (curve b) methylamine equilibration, (curve c) microelectrodes, and (curve d), homogenization. Eggs of *Strongylocentrotus purpuratus* (a, b, and d) and *Lytechinus pictus* (c) were fertilized by the addition of sperm at 0 minute. a, b: unpublished observations; c: Shen and Steinhardt (1978); d: Johnson *et al.* (1977); Lopo and Vacquier (1977).

long been known to parthenogenetically activate sea urchin eggs, and Shen and Steinhardt (1978) have demonstrated that this treatment can produce intracellular pH shifts. As the ammonia molecule diffuses into the cell, it utilizes internal protons and increases pH_i by a mechanism similar to that in Fig. 1 (Winkler and Grainger, 1978). In later experiments, pH_i was experimentally varied by ammonium addition while monitoring both the pH_i by microelectrode and amino acid incorporation into protein, a major event in the latter phases of sea urchin egg development. It was found that an increase of 0.01 pH unit produced an immediate 3- to 4-fold increment in the rate of protein synthesis. The increase was shown to be specific to the cytoplasmic phase, rather than reflecting an increased rate of amino acid transport (Grainger *et al.*, 1979). This remarkable result clearly shows that at least one cell function, protein synthesis, can be controlled by small changes in the pH of the cytoplasm.

V. Nuclear Magnetic Resonance Method

A. INTRODUCTION

The most recent technique developed for estimating pH_i involves ^{31}P nuclear magnetic resonance (NMR) spectra. This method utilizes the resonance frequency of intracellular phosphate, which is affected by the pH of its environment. Thus, as pH shifts from the mid-7 range to the mid-6 range, NMR spectra show a shift in the phosphorus peak as phosphate is titrated. The shift can be readily calibrated and thereby provides information about the pH of the environment around the phosphate.

The major advantage of the NMR method is that it is noninvasive and can be run on cells and tissue samples as large as perfused rabbit hearts. The disadvantage is that one must know where phosphate is located within the cell and keep track of it during any metabolic changes that are imposed on the cell or tissue.

The NMR method has been successfully applied to cell organelles (Casey *et al.*, 1977; Ogawa *et al.*, 1978a, b), bacteria (Ugurbil *et al.*, 1978), rat liver cells (Cohen *et al.*, 1978), and heart tissue (Jacobus *et al.*, 1977). (For review, see Burt *et al.*, 1977.) The application to heart tissue offers a clear example of the power of the method, and an application will be described here in which NMR was used to relate myocardial performance to changes in intracellular pH.

B. Intracellular pH and Ischemia

The correlation between pH_i and myocardial performance is a major research area in molecular cardiology. This stems from the fact that during the ischemic transition in the heart, lactic acid is produced by anaerobic metabolism and protons are also generated from the hydrolysis of ATP. As a result, pH_i decreases. Since the acidic pH_i associated with ischemia may inhibit glycolysis (Revetto *et al.*, 1975), ischemia not only limits ATP availability from oxidative phosphorylation, but the resulting acidosis also reduces energy production from glycolysis.

The net result is decreased ATP availability and turnover, followed by a very rapid and significant fall in ventricular function during the first minute of ischemia. Although the decline in ventricular performance has been under intense study for many years, the relationship between availability of high-energy phosphate, pH_i, and contractile performance is not understood. In order to define the role of intracellular pH during normal as well as ischemic states, it is necessary to have reliable techniques for monitoring intracellular pH.

Neely *et al.* (1975) first investigated the relationship between myocardial metabolism and pH_i. These investigators characterized the changes in lactic acid produced during ischemia and hypoxia and proposed that intracellular acidosis may inhibit glycolysis at glyceraldehyde-3-phosphate dehydrogenase. Steenbergen *et al.* (1977), using the DMO method, have correlated respiratory acidosis with the fall in ventricular function. Despite the wealth of correlative evidence, it is still a difficult problem to provide direct or even indirect measurements of intracellular pH in functioning myocardium.

C. Application of NMR to Perfused Heart

^{31}P NMR is clearly a powerful tool to apply to this problem. The specific advantages of the NMR method are its sensitivity and applicability to the perfused, working heart. However, there are certain complications with the NMR technique in relation to myocardial tissue. Among these is the unknown distribution of phosphate in the cells. For instance, it is possible that most of the phosphate in the heart is sequestered in the mitochondrial matrix so that the NMR signal provides information primarily about the mitochondrial volume. A second limitation of the technique as applied to myocardium is that during global ischemia there may be membrane permeability changes. If phosphate leaks into the interstitial space, the NMR spectra may reflect a composite of mitochondrial, cytoplasmic, and interstitial pH.

The first NMR estimates of the pH_i of the perfused myocardium were conducted with phosphate-containing buffer that was aerated either with pure oxygen and buffered by Tris or with 95% oxygen, 5% carbon dioxide and buffered with sodium bicarbonate (Jacobus *et al.*, 1977). Under either of these conditions, the estimated pH_i of the myocardium was approximately 7.4.

This work was done with a 23-kilogauss electromagnet, and under these conditions phosphate resonates at 42.5 megahertz. With this instrument only small rat hearts could be perfused, limiting both mass of tissue as well as NMR resolution. More recently a superconducting Bruckert WX 180 instrument was used, which had a field of 40.5 kilogauss with phosphate resonating at 72.89 megahertz. With this instrumentation rabbit hearts could be perfused with phosphate-free buffer. Under these conditions the pH_i estimated from the phosphate peak was 7.18 under several perfusion conditions (Jacobus *et al.*, 1978).

An important application of the NMR method has been to assay the relationship between ventricular performance and intracellular pH during steady state, transient respiratory acidosis and total global ischemia in the perfused rabbit heart. The rabbit heart has a significant advantage over the rat heart in that it does not vasoconstrict during respiratory acidosis. Therefore the possible contribution of localized ischemia during respiratory acidosis may be minimized in the rabbit heart model. Respiratory acidosis was induced in two ways. In the first, the perfusate was aerated with 95% O_2, 5% CO_2, together with variable amounts of 65% O_2, 35% CO_2.

With the change in CO_2 at constant bicarbonate, the pH of the buffer could be shifted from a normal pH of 7.4 down to pH 6.6. Under these conditions, the heart was stable and changes in pH_i could be related to myocardial function. The second technique used was to induce partial hypoxia with a gas mixture of 65% O_2, 30% N_2, and 5% CO_2. This gas was then mixed with 65% O_2, 35% CO_2, so that the pO_2 was constant while N_2 and CO_2 content were changing. Under both conditions, the phosphate signal indicated a decrease in pH_i of 0.25 unit when the extracellular pH dropped from 7.5 to 6.7.

A significant finding in these studies was the correlation of pH_i with ventricular performance, which was monitored by inserting a fluid-filled latex balloon into the left ventricle. The fluid in the balloon was conducted to a pressure transducer so that ventricular performance could be monitored in the isovolumic rabbit heart. During the condition of respiratory acidosis, either normal or hypoxic, there was a fall in left ventricular developed pressure from 100% down to approximately 34% at pH 6.7. When the depression of ventricular performance was plotted versus pH_i, it was possible to calculate the change in intracellular pH required to depress ventricular performance in half. For either the condition of respiratory acidosis or hypoxic acidosis, a pH change of 0.3 unit was sufficient. This remarkable result suggests that pH_i and ventricular performance are more tightly coupled than anticipated from previous intracellular pH studies.

NMR studies of the heart have clearly opened questions about the role of intracellular pH and tissue function. The experiments relating function, metabolism and pH_i suggest that there is a tight coupling between these parameters in the heart.

VI. Conclusions

In this review, we have described several techniques to monitor pH_i and discussed a current application of each. We can now attempt to answer the questions originally posed in the introduction.

1. What methods are appropriate for studying pH_i in different cell systems? It should now be clear that each technique has strengths and limitations that require an informed choice when they are to be used in a given system. The DMO method is most generally applicable and relatively inexpensive, but does not provide the resolution of the other methods. Methylamine distribution is well suited for determining the pH_i of single-compartment systems with acidic interiors, such as chloroplasts and isolated lysosomes; and 9-aminoacridine is useful for visualizing acidic compartments

within multicompartmented systems. Microelectrodes provide the highest resolution of pH_i (0.01 pH unit), but can be applied only to single cells with diameters greater than 20–30 μm. NMR is the most expensive but offers the best possibility for resolving compartmentalized pH_i under noninvasive conditions.

It is interesting to ask whether these diverse techniques provide similar values for pH_i. Table II summarizes measured pH_i values for several kinds of cells and organelles in which two or more of the techniques have been applied. In general there is surprisingly good agreement. When there are

TABLE II

REPRESENTATIVE pH_i VALUES FROM DIFFERENT TECHNIQUES

System	External pH	DMO[r]	MA[r]	Electrode	NMR[r]
Barnacle muscle[a]	7.5	7.39	7.28	7.35	—
Mammalian heart					
Aerobic	7.29	7.03[b,c]	7.02[n]	7.02[n]	7.15[d]
Ischemic[o]	6.8	6.8[b], 6.7[c]			5.7[e,p]
Erlich ascites cells	6.57	6.75[f]	—	—	6.71[g]
Sea urchin eggs					
Unfertilized	7.8	6.76[h]	6.45[h]	6.78[i]	—
Fertilized (10 min)	7.4	7.08[h]	7.04[h]	7.22[i,q]	—
Mitochondria state 4	6.95	7.4[j]	—	—	7.4[k]
Escherichia coli	7.9	7.75[l]	7.73[l]	—	—
(energized)	6.6	—	—	—	7.50[m]

[a] Boron and Roos (1976).
[b] Neely *et al.* (1975).
[c] Steenbergen *et al.* (1977).
[d] Jacobus *et al.* (1978).
[e] Jacobus *et al.* (1977).
[f] Poole *et al.* (1972).
[g] Navon *et al.* (1977a).
[h] Gillies (unpublished).
[i] Shen and Steinhardt (1978).
[j] Addanki *et al.* (1968).
[k] Ogawa *et al* (1978a, b).
[l] Paddan *et al.* (1976).
[m] Navon *et al.* (1977b).
[n] Ellis and Thomas (1976).
[o] Caused by low perfusate flow.
[p] Caused by no perfusate flow.
[q] Extracellular pH not reported.
[r] DMO = 5,5-dimethyloxazolidine-2,4-dione; MA = methylamine; NMR nuclear magnetic resonance.

differences, they can be qualitatively understood from the expected equilibrations presented in Table I and Figs. 2A and B. For instance, DMO and methylamine indicate pH_i values slightly higher (DMO) and lower (MA) than those from microelectrodes and NMR when applied to multicompartmented systems.

We conclude that different methods give consistent results when applied to a given system, and that measured pH_i values are meaningful estimates of the actual pH_i.

2. Is the cytoplasmic pH regulated by the cell? Results with *Tetrahymena* and many other cell types show that pH_i is maintained at a fairly constant level when pH_e is varied within a physiological range. Furthermore, it has been shown in both *Tetrahymena* and sea urchin eggs that pH_i undergoes reproducible changes at specific stages of cellular growth.

We conclude from these results that pH_i can be regulated by cells within limits.

3. Does pH_i vary during specific cell functions, and perhaps contribute to the regulation of that function? We have presented evidence that intracellular pH does change during specific functions of at least three types of eukaryotic cells. These changes are highly correlated with processes such as cell cycle in *Tetrahymena*, derepression in sea urchin eggs, and ischemia in rabbit heart. Furthermore, in sea urchin eggs external manipulation of pH_i can markedly affect the rate of protein synthesis *in situ*.

These findings are consistent with the possibility that cells have mechanisms in which pH_i plays a regulatory role or acts as a signal.

4. Does pH_i contribute to cell dysfunction during damage processes? The work on ischemic myocardium which we have reviewed here strongly suggests that alterations in pH_i are among the primary events that contribute to loss of performance in heart tissue.

Finally, we can briefly address two further questions: How might a cell manipulate pH_i, and what are some possible mechanisms by which pH_i could act to regulate a metabolic pathway? There are only a few mechanisms that might contribute to changes in pH_i, and they are presented in Fig. 8. Note that some of the processes are cyclic (i.e., the NTP cycle) during steady-state metabolism, so that no net pH change results. The remaining pathways include cytoplasmic CO_2 accumulation, anaerobic lactate production, and active or passive transport of protons across the plasma membrane. Cells can apparently use all these mechanisms to produce pH_i changes. The best illustration of this is in myocardium, where pH_i variation under different conditions appears to be produced by different mechanisms. Under conditions of respiratory acidosis, small pH decreases occur mainly as a consequence of CO_2 accumulation. During global ischemia, the pH_i

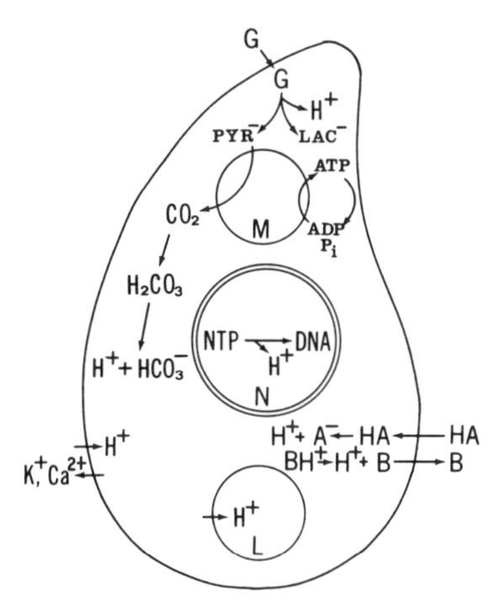

FIG. 8. Reactions involving protons in eukaryotic cells. Compartments are identical to those in Figs. 2A and 2B. Abbreviations: G, glucose; NTP, nucleoside triphosphate; HA, weak acid; B, weak base. Note that pH changes could be produced by anaerobic glycolysis, respiration, nucleic acid synthesis, transmembrane antiport or symport, and weak acid–weak base equilibria. Lysosomes also maintain acidic interiors, and under certain conditions hydrolysis of high-energy phosphate compounds could decrease cytoplasmic pH.

falls more than 1.5 pH units. This dramatic alteration in pH_i presumably reflects a combination of both lactate formation and net proton production by complete hydrolysis of cellular high-energy phosphate stores.

We conclude that cells change pH_i by diverse mechanisms, and that the "choice" of mechanisms depends on the metabolic state of the cell.

How might a change in intracellular pH produce a regulatory effect? The most likely possibility is that protons associate with buffering sites on enzymes and substrates and produce direct or indirect effects on enzyme activity. This can serve to regulate the overall activity of an enzymic sequence such as glycolysis. A typical enzyme in a sequence may be relatively insensitive to pH effects such that small changes in pH produce little alteration in its activity. However, if one enzyme in a sequence is either markedly activated or inhibited by small pH changes, we would clearly have a mechanism for governing the overall rate of that sequence. It is interesting to note that phosphofructokinase has an unusually high slope of activity against pH, and that this enzyme can govern glycolytic rates *in situ*.

ACKNOWLEDGMENTS

The authors would like to thank Drs. W. J. Jacobus, R. A. Steinhardt, and S. Shen for their helpful discussions during preparation of this manuscript.

REFERENCES

Addanki, S., Cahill, F., and Sotos, J. (1968). *J. Biol. Chem.* **243**, 2337–2348.

Ahmed, S., and Baron, D. (1971). *Clin. Sci.* **40**, 487–495.

Aickin, C., and Thomas, R. C. (1977). *J. Physiol.* (*London*) **267**, 791–810.

Baron, D., and Ahmed, S. (1969). *Clin. Sci.* **37**, 205–219.

Beall, P. T., Hazelwood, C., and Rao, P. N. (1976). *Science* **192**, 904–907.

Boron, W. F., and Roos, A. (1976). *Am. J. Physiol.* **231**, 799–809.

Burt, C. T., Glonek, T., and Barany, M. (1977). *Science* **195**, 145–149.

Cameron, I. L., and Jeter, E. J. (1970). *J. Protozool.* **17**, 429–431.

Carter, N. W., Rector, F. C., Campion, D. S., Seldin, D. W., Nunn, A. C., and Howard, W. (1967). *J. Clin. Invest.* **46**, 920–933.

Carter, R., and Nijhout, M. (1977). *Science* **195**, 407–409.

Casadio, R., and Melandri, B. A. (1977). *J. Bioenerg. Biomembr.* **9**, 17–29.

Casey, R. P., Njus, D., Radda, G. K., and Sehr, P. A. (1977). *Biochemistry* **16**, 972–975.

Cohen, S. M., Ogawa, S., Rottenburg, H., Glynn, P., Yamane, T., Brown, T., Shulman, R. G., and Williamson, J. (1978). *Nature* (*London*) **273**, 554–555.

Deamer, D. W., Prince, R., and Crofts, A. R. (1972). *Biochim. Biophys. Acta* **274**, 323.

Ellis, D., and Thomas, R. L. (1976). *Nature* (*London*), **262**, 224–225.

Epel, D., Steinhardt, R., Humphreys, T., and Mazia, D. (1974). *Dev. Biol.* **40**, 245–255.

Gillies, R. J., and Deamer, D. W. (1977). *J. Cell Biol.* **75**, 82a.

Gillies, R. J., and Deamer, D. W. (1978). *J. Supramol. Struct.* **52**, 320.

Gillies, R. J., and Deamer, D. W. (1979). *J. Cell. Physiol.* (in press).

Grainger, J., Winkler, M., Shen, S., and Steinhardt, R. (1979). *Develop. Biol.* **68**, 396–406.

Harold, F. M., Paulasova, E., and Baarda, J. (1970). *Biochim. Biophys. Acta* **196**, 235–244.

Hollis, D. P., Nunnally, R., Jacobus, W., and Taylor, G. (1977). *Biochem. Biophys. Res. Commun.* **75**, 1086–1090.

Jacobus, W. E., Pores, I. H., Taylor, G. J., Nunnaly, R. L., Hollis, D. P., and Weisfeldt, M. L. (1978). *J. Mol. Cell. Card.* **10** (Suppl. 1), 39.

Jacobus, W. E., Taylor, G., Hollis, D., and Nunnally, R. (1977). *Nature* (*London*) **265**, 756–758.

Johnson, J., Epel, D., and Paul, M. (1976). *Nature* (*London*) **262**, 661–664.

Kashket, E. R., and Wong, P. (1969). *Biochim. Biophys. Acta* **193**, 212–214.

Leguay, J. (1977). *Biochim. Biophys. Acta* **497**(1), 329–333.

Levin, G., Collinson, P., and Baron, D. (1976). *Clin. Sci. Mol. Med.* **50**, 293–299.

Lopo, A., and Vacquier, V. (1977). *Nature* (*London*) **269**, 590–592.

Meizel, S., and Deamer, D. W. (1978). *J. Histochem. Cytochem.* **26**, 98–105.

Meizel, S., and Mukerji, S. K. (1975). *Biol. Reprod.* **13**, 83.

Mitchell, P. (1961). *Nature* (*London*) **191**, 144.

Mitchell, P. (1966). *Biol. Rev. Cambridge Philos. Soc.* **41**, 445.

Navon, G., Ogawa, S., Shulman, R. G., and Yamane, T. (1977a). *Proc. Natl. Acad. Sci. U.S.A.* **74**, 87–91.

Navon, G., Ogawa, S., Shulman, R. G., and Yamane, T. (1977b). *Proc. Natl. Acad. Sci. U.S.A.* **74**, 888–891.

Neely, J. R., Whitmer, J., and Rovetto, M. (1975). *Circ. Res.* **37**, 733–741.

Nichols, O. G. (1974). *Eur. J. Biochem.* **50**, 305.

Oesterhelt, D., and Stoeckenius, W. (1973). *Proc. Natl. Acad. Sci. U.S.A.* **70**, 2853–2857.

Ogawa, S., Rottenberg, H., Brown, T., Shulman, R. G., Castillo, C., and Glynn, P. (1978a). *Proc. Natl. Acad. Sci. U.S.A.* **75**, 1796–1808.

Ogawa, S., Shulman, R. G., Glynn, P., Yamane, T., and Navon, G. (1978b). *Biochim. Biophys. Acta* **502**, 45–50.

Paddan, E., Zilberstein, D., and Rottenberg, H. (1976). *Eur. J. Biochem.* **63**, 533–541.
Patil, G. S., Matthews, R. H., and Cornwell, D. G. (1975). *Adv. Chem. Ser.* No. 144, 44–66.
Poole, D., Butler, T., and Williams, M. (1972). *Biochim. Biophys. Acta* **266**, 463–470.
Revetto, M. J., Lamberton, W. F., and Neely, J. R. (1975). *Circ. Res.* **37**, 742–751.
Rottenberg, H. (1979). *Methods in Enzymology* **55**, 547–569.
Shen, S., and Steinhardt, R. A. (1978). *Nature* (*London*) **272**, 253–254.
Steenbergen, C., Deleeuw, G., Rich, T., and Williamson, J. (1977). *Circ. Res.* **41**, 849–858.
Steinhardt, R., and Mazia, D. (1973). *Nature* (*London*) **241**, 400–401.
Tomodal, A. S., Tsuda-Hirota, S., and Minakami, S. (1977). *J. Biochem.* (*Tokyo*) **81**, 697–701.
Ugurbil, K., Rottenberg, H., Glynn, P., and Shulman, R. G. (1978). *Proc. Natl. Acad. Sci. U.S.A.* **75**, 2244–2248.
Waddell, W. J., and Bates, R. G. (1969). *Physiol. Rev.* **49**, 285–329.
Waddell, W. J., and Butler, T. (1959). *J. Clin. Invest.* **38**, 720–729.
Walker, N. A., and Smith, F. (1975). *Plant Sci. Lett.* **4**, 125–132.
Winkler, M., and Grainger, J. (1978). *Nature* (*London*) **273**, 536–538.
Wolfe, J. (1975). *J. Cell. Physiol.* **85**, 73–86.
Zeuthen, E. (1971). *Exp. Cell Res.* **68**, 49–60.
Zieve, P., Haghshenass, M., and Krevens, J. (1967). *Am. J. Physiol.* **212**, 1099–1102.

Mitochondrial ATPases

RICHARD S. CRIDDLE, RICHARD F. JOHNSTON, and ROBERT J. STACK
Department of Biochemistry and Biophysics
University of California
Davis, California

ISBN 0-12-152509-0

I. Introduction

Since the last general reviews on mitochondrial ATPases were published (Senior, 1973; Pedersen, 1975; Tzagoloff, 1976; Kozlov and Skulachev, 1977; Penefsky, 1974; Boyer *et al.*, 1977; Sebald, 1977), virtually every aspect of structure and function of this enzyme has become a focus for intensive study by workers in various laboratories. The first portions of this review will attempt a general presentation of available data obtained largely in the past two to three years, with little interpretative comment. Later we will deal with questions of where these results have led in seeking an interpretation of the involvement of the ATPase enzyme in coupled ATP synthesis.

The terminology used to describe the ATPase enzyme complex and its various subfractions varies from laboratory to laboratory. No logically based system of terminology that relates enzyme activities to subfractions has been agreed upon. Even the title of this review, referring to mitochondrial ATPases, fails to recognize that the major goal of most workers in the field is to relate this enzyme complex to its ATP synthase activity rather than to the hydrolysis reaction.

All mitochondrial ATPases (EC 3.6.1.3) (as well as chloroplast and representative bacterial ATPases involved in oxidative and photophosphorylation) can be separated to yield a soluble unit with ATP hydrolytic activity and a "membrane" sector lacking this activity but capable of modifying the catalytic properties of the soluble ATPase. These fractions were first recognized for their roles as factors required for the coupling of oxidation to phosphorylation and were called factors F_1 and F_0, respectively (Racker, 1970). Subsequent partial clarification of their respective roles in ATPase activity has not yet prompted a change in this terminology, and probably should not until more exact roles can be described. Therefore, for the purpose of this review and for uniformity in presentation, we will refer to the catalytically active soluble ATPase unit as F_1-ATPase or simply as F_1. The membrane sector of the complex will still be referred to as F_0, and the intact complex (or the membrane-bound form of the enzyme) will be called ATPase.

The individual polypeptide subunits of ATPase have generally been identified on the basis of their electrophoretic mobilities on SDS[1] gels. Arabic numbers, Roman numerals, or Greek letters have been used to indicate subunits with differing mobilities; assigning for example, 1 or α to the largest (slowest migrating component) and 2 or β to the second largest, etc. This system was the most logical available when few enzymes had been studied and little was known about the function of individual protein subunits. However, this makes discussion of comparative properties of subunits from different enzyme sources very difficult. The major emphasis of this review will be on summarizing the many new findings that relate ATPase enzyme structure to function. This information comes from studies of many different enzyme preparations and has required us to impose a standard subunit nomenclature for our discussion. As indicated in Table I, we have maintained the standard F_1 and F_0 designations and further defined the peptides numerically into subgroups which we feel have common functional properties as well as reasonably constant size. Justification of these groupings will, it is hoped, become apparent in the discussion that follows.

II. Subunit Composition and Structure of F_1

A. Composition

Soluble F_1-ATPases have been isolated from several different mitochondrial sources (Pullman *et al.*, 1960; Lambeth and Lardy, 1971; Schatz *et al.*, 1967), from bacterial membranes (Abrams, 1965; Davies and Bragg, 1972; Hachimori *et al.*, 1970), and also from the membranes of chloroplasts (Vambutas and Racker, 1965). Remarkable similarities in molecular weights of the intact F_1 and of the individual subunits are noted in comparisons among enzyme preparations from these widely different sources (for reviews, see Senior, 1973; Pedersen, 1975; Kozlov and Skulachev, 1977; Tzagaloff, 1976; Penefsky, 1974). All are complexes of high molecular weight, 3.2 to 3.8×10^5, and have subunits of five different sizes. Table I indicates F_1 subunit size for representative studies of F_1 from several enzyme preparations. Muller *et al.* (1977) have estimated that purified beef heart F_1 may also contain as much as 8% by weight nonprotein material. Recent reports that

[1] Abbreviations: ANP, azidonitrophenol; CCP, *m*-chlorocarbonylcyanide-*m*-chlorophenylhydrazone; CMB, *p*-chloromercuribenzoate; DCCD, dicylohexylcarbodiimide; DNP, dinitrophenol; NBD, 4-chloro-7-nitrobenzo-2-oxo-1,3-diazole; PCP, pentachlorophenol; S-13, *S*-chloro-3-*t*-butyl-2′-chloro-4′-nitro salicylamide; SDS, sodium dodecyl sulfate; 1799, bis-hexafluoroacetonyl acetone.

TABLE I

SUBUNIT COMPOSITIONS AND MOLECULAR WEIGHTS OF ATPase PREPARATIONS AND SUBUNIT DESIGNATIONS TO BE USED FOR UNIFORMITY IN THIS REVIEW

Subunit designations	Yeast, *Saccharomyces cerevisiae* Tzagoloff and Meagher (1971)	Yeast, *S. cerevisiae* Ryrie (1977)	Yeast, *S. pombe* Goffeau *et al.* (1973)	Beef heart Stiggall *et al.* (1978)	*Escherichia coli* Foster and Fillingame (1978)	Thermophilic bacterium Sone *et al.* (1976)	Liver Catterall and Pedersen (1971)	Chloroplast Nelson *et al.* (1973)
F_1-1	58[a]	60	60	53	58	56	62	59
F_2-2	54	54	58	50	52	53	57	56
	38.5	—	—	—	—	—	—	—
F_1-3	31	32	32	33	28	32	36	37
F_1-4	12	11.5	13.5	15	20	15	12	17
F_1-5	—	—	8	8.9	12	11	7	13
	29	—	—	31	—	—	—	—
F_0-1	21	23	20	24	26	19	—	—
OSCP[b]	19	21.5	17	22.5	—	—	—	—
	—	17	—	—	—	—	—	—
F_0-2	12	13	13.5	13	15	13.5	—	—
	11	11.5	9	—	—	—	—	—
F_0-3	7.5	?	8	8–9	8	5.4	—	—

[a] Numbers refer to molecular weights $\times 10^{-3}$.
[b] OSCP, oligomycin sensitivity-conferring protein.

ATPase is a glycoprotein may account for some of this nonprotein material (Andreu *et al.*, 1973, 1978).

The stoichiometry of the five subunits in F_1 has been discussed in earlier reviews (Senior, 1973; Pedersen, 1975). Most estimates of stoichiometry have been based on the somewhat questionable method of analyzing relative staining intensities of protein bands following SDS disc gel electrophoresis, combined with molecular weight analysis of the subunits and of intact F_1. Differences in color yields for different proteins may cause inaccuracies. Also the nature of the reported 8% contaminant and its effect on total protein molecular weight make calculations quite uncertain. The most common stoichiometry arrived at by this method is that the five subunits exist in the ratio (from largest to smallest) of 3 : 3 : 1 : 1 : 1 for liver, beef heart, and yeast mitochondria (Catterall *et al.*, 1973; Senior, 1973) although the ratio of 2 : 2 : 2 : 1 : 1 has also been proposed for beef heart based on dye staining. Aurovertin binding studies were used in proposing a stoichiometry of 2 : 2 : 2 : 2 : 2 for beef heart enzyme (Verschour *et al.*, 1977). Analysis of thiol groups reactive with *N*-ethylmaleimide also led to a similar stoichiometry (2 : – : 2 : – : 2) (Senior, 1975). This latter stoichiometry is in agreement with that reported for *Escherichia coli* F_1 (Bragg and Hou, 1972, 1975; Vogel and Steinhart, 1976). Chloroplast F_1 was proposed to have the composition 2 : 2 : 1 : 1 : 1 (Nelson, 1976) and a molecular weight about the same as the mitochondrial enzyme (Farron, 1970). Enns and Criddle (1977b) showed that dimers of subunits F_1-1, F_1-2 and F_1-3 could be produced by reaction of yeast F_1 with a bifunctional protein cross-linking reagent. This observation appears to require that at least two of each of these subunits be present in each F_1 complex.

The best measurement to date of F_1 subunit stoichiometry from any source of enzyme comes from studies of a thermophilic bacterial F_1. Computer filtering of electron micrographs of F_1 crystals suggests that these contain six large subunits (F_1-1 and F_1-2) per molecule rather than four and suggest stoichiometry ratios of 3 : 3 : 1 : 1 : 1 (Wakabayashi *et al.*, 1977). Amzel and Pedersen (1978) have recently reported crystallization of liver mitochondrial F_1 containing the standard 5 subunit types. X-Ray analysis of these crystals should clearly define subunit stoichiometry.

The existence of five types of subunits in F_1 from so many sources, including chloroplasts and bacteria, has been used to argue that all must serve essential roles in the integrated enzymic formation of ATP during oxidative phosphorylation (Senior, 1973). It is becoming increasingly clear, however, that stoichiometric amounts of the F_1-associated subunits are not required for ATP hydrolytic activity. Studies defining which subunits are required for activity are most advanced with bacterial ATPases. The *Micrococcus lysodeikticus* enzyme has been isolated in active form lacking the smallest subunit (Andreu *et al.*, 1973). Based on subunit reconstitution studies, Yoshida *et al.* (1975) reported that assembly of only the F_1-1, F_1-2, and F_1-3 subunits of thermophilic bacterium PS3 is required for reconstitution of active F_1-ATPase. Futai (1977) obtained similar results in reconstitution of active *E. coli* ATPase during only the three largest subunits. Some limited ATPase activity was also obtained by these reconstitution studies with only subunits F_1-1 and F_1-2. Nelson *et al.* (1974) isolated active *E. coli* F_1-ATPase with only four subunits (lacking F_1-4). Trypsin treatment of this preparation yielded active enzyme with only the two largest subunits remaining. Sternweiss (1977, 1978) isolated active F_1-ATPase with only subunits F_1-1, F_1-2, and F_1-3. He further indicated that both F_1-4 and F_1-5 were required to bind this ATPase to F_1-depleted membranes for reconstitution of energy functions.

The latent ATPase activity of chloroplasts that is expressed following heat treatment (Farron and Racker, 1970) or limited tryptic hydrolysis of chloroplast membrane preparations (Vambutas and Racker, 1965) has the customary five F_1 peptides; however, prolonged tryptic digestion yields active enzyme containing only subunits F_1-1 and F_1-2 (Nelson, 1976). The effects of trypsin are difficult to evaluate, and some important peptide fragments from other subunits with an effect on activity could still be present in this preparation.

In contrast to these studies with the chloroplast enzyme, Ryrie (1977) has recently demonstrated a requirement for the continued presence of protease inhibitors (particularly *p*-aminobenzamidine) during purification of yeast mitochondrial F_1 to yield a highly active preparation containing only subunits, F_1-1, F_1-2, and F_1-3. An additional 42,000 molecular weight subunit was found in F_1 only when protease inhibitor concentrations were decreased. While this may be an indirect effect on extraction rather than the implied direct effect of proteolysis, it is clear that a three-subunit enzyme preparation from yeast is also an active ATPase.

The sizes of F_1 subunits appear to be amazingly well conserved over a wide range of organisms when lined up as in Table I. The counterpart of the 38,500 molecular weight component commonly isolated with yeast (*Saccharomyces cerevisiae*)-F_1 is lacking in the other F_1 preparations but, based on Ryrie's (1977) studies, it is clearly not needed for activity. Comparison with the various active preparations of the other F_1 enzymes makes it probable that the fourth largest subunit of yeast F_1 and the third largest from the other enzymes, designated F_1-3 in the table, are functionally identical.

Reconstitution of a fully active ATP hydrolase in all stidies to date requires at least F_1-1, F_1-2, and F_1-3. The minimum subunit requirement for activity of an already assembled enzyme may be only the two largest subunits as noted in the two cases of protease-digested chloroplast (Nelson *et al.*, 1973) and *E. coli* (Nelson, 1976) ATPases, which could, however, still have some small fragments of other subunits bound. The F_1-3 peptide may be required only in a structural role during assembly of F_1-1 and F_1-2 into an active form, since marginal ATPase activity was reconstituted from two subunits of *E. coli* enzyme indicating that these units do contain the hydrolytic site. The remaining subunits of F_1 may function in binding F_1 to F_0 (at least in the bacterial system) as suggested by Sternweiss (1977).

As functional roles for individual subunits of the ATPase complex become further defined, it will be necessary to recognize that size and functional roles may differ when comparing enzymes from various sources, and adjustments in Table I may be required. A recent example of such problems has been described in studies of the RNA polymerase complex. Size analysis of *E. coli* enzyme led to designation of a large β' and a smaller β subunit. β was found to bind rifampicin. In *Bacillus subtilis*, however, the larger subunit binds rifampicin. Functionally, both rifampicin binding units should be termed β, abandoning the size designation (Doi, 1978).

B. Structure

Electron microscopic detection of knoblike structures on negatively stained membranes (Fernandez-Moran, 1962) and their subsequent identification as F_1 (Racker *et al.*, 1965) yielded the first indications that there is a tightly packed structural arrangement of subunits in the F_1 complex to form generally spherical 80–100 Å particles. More careful electron microscopic studies have subsequently indicated a hexagonal arrangement of large subunits within the complex, promoting the suggestion that this corresponds to three alternating subunits 1 and 2 in the beef heart (Kozlov and Mikelsaar, 1974) and the rat liver (Catterall and Pedersen, 1974) F_1 ATPases. Some of the early models proposed for ATPase have been reviewed by Pedersen (1975). Image filtering of electron micrographs of crystalline F_1 from thermophilic bacteria PS3, provides a much clearer picture indicating that these particles have a considerable 6-fold and 3-fold symmetry with the length of each side of a hexagon about 38 Å (Wakabayashi *et al.*, 1977).

The nearest-neighbor associations of individual subunits within the ATPase complex have been investigated in several laboratories using chemical cross-linking agents. F_1 preparations from beef heart (Satre *et al.*, 1976; Wingfield and Boxer, 1975) and yeast (Enns and Criddle, 1977b) mitochondria and from chloroplasts (Baird and Hammes, 1976) and *E. coli* (Bragg and Hou, 1975) have been cross-linked, and the subunit patterns have been analyzed by gel electrophoresis. Satre *et al.* (1976) indicate formation of cross-linked pairs of the types F_1-1 : F_1-1 and F_1-1 : F_1-2 but not F_1-2 : F_1-2. In contrast, Enns and Criddle (1977b) indicated that yeast F_1 is able to form F_1-1 : F_1-1, F_1-2 : F_1-2, and F_1-3 : F_1-3 cross-links, but no F_1-1 : F_1-2, F_1-1 : F_1-3 or F_1-2 : F_1-3 subunit pairs. In neither study were cross-linked products containing F_1-4 or F_1-5 noted. Studies with *E. coli* F_1 (Bragg and Hou, 1975) indicated formation of F_1-1 : F_1-2 subunits but no F_1-1 : F_1-1 or F_1-2 : F_1-2, whereas chloroplast F_1 yielded predominantly F_1-1 : F_1-2 cross-linking (Baird and Hammes, 1976). Several reasons might be postulated to account for the differences noted here, in addition to the obvious possibility that all these enzymes have differing structures. The structural arrangement of F_1 proteins may depend upon isolation conditions. As discussed above, subunit composition of F_1, and probably also the structure, varies with the isolation procedures employed. Also studies of nucleotide, activator, and inhibitor binding are consistent with structural differences induced by the presence or the absence of the effector molecules (see Section VII, on ATPase kinetics). The levels of these effector molecules bound to isolated F_1 and, therefore, their effects on structure probably vary with the isolation procedures. The cross-linking reagents used have differing lengths separating the functional groups and can also give different results. Further structural analyses by cross-linking studies will have to include a consideration of these parameters.

Again, X-ray crystallographic analysis promises to yield the most definitive information on arrangement of subunits in F_1 and may eventually lead to identification of the structural differences suggested by kinetic analysis.

III. Composition and Structure of the Intact ATPase Complex

A. Composition

The intact ATPase has not been well characterized from as many sources as has F_1 owing to difficulties in solubilization of the F_1 with a bound-membrane sector. At the time of the last general review (Pedersen, 1975) only the yeast ATPase had been extensively purified (Tzagoloff and Meagher, 1971). Recently, however, purified ATPases have been prepared from beef heat mitochondria (Stiggall *et al.*, 1978; Serrano *et al.*, 1976; Sadler *et al.*, 1974), *E. coli* (Foster and Fillingame, 1978), and thermophilic bacteria (Sone *et al.*, 1975). In general, a mild detergent treatment (deoxycholate or Triton X-100) has been used to solubilize the enzyme from heart (Stekhoven *et al.*, 1972; Tzagoloff *et al.*, 1968; Capaldi, 1973) and from yeast (Tzagoloff and Meagher, 1971; Ryrie, 1977) mitochondria and bacterial membranes (Bragg and Hou, 1975). F_0 is best prepared in purified form by first isolating ATPase, then fractionating by treatment with salts (Tzagoloff *et al.*, 1968) or chloroform (Ryrie, 1977). Direct isolation of F_0 from F_1-depleted membranes has not yielded highly purified preparations (Ragan and Racker, 1973; Ryrie, 1977). Subunit compositions of F_0 are, therefore, generally studied as differences between F_1 and ATPase. A postulated subunit stoichiometry for the yeast enzyme put into the nomenclature of this review is

$$[(F_1\text{-}1)_3(F_1\text{-}2)_3(F_1\text{-}3)_2(F_1\text{-}4)_1(F_0\text{-}1)_2(F_0\text{-}2)_1(F_0\text{-}3)_6(\text{OSCP})_1 + 38{,}500 \text{ peptide} + 11{,}000 \text{ peptide}]$$

(Enns and Criddle, 1977b). This estimate is similar to that arrived at by Pedersen (1975). It is based on the same considerations as noted for F_1 (except for subunit F_0-3) and therefore suffers from the same limitations. Subunit F_0-3 stoichiometry is estimated from apparent size of a polymer form as discussed later. X-Ray crystallographic analysis to aid in resolution of the stoichiometry questions does not seem so near in this case.

Some differences are noted in the numbers and sizes of the F_0 subunits associated with ATPase from various sources (Table I). However, all mitochondrial ATPases studied have both soluble and membrane-sector complexes, and most of the major subunits of F_0 appear to have recognizable counterparts in each system. Currently, the predominant assay for a functional F_0 is the ability to confer cold stability and oligomycin (or DCCD) sensitivity to F_1-ATPase (Kagawa and Racker, 1966). Alternatively, F_0 has been characterized by its ability to increase proton permeability in a lipid membrane, a property that is moderated by F_1 (Kagawa *et al.*, 1973).

The active unit of F_0 as isolated contains at least four different polypeptides. Again it is difficult to decide at present which of these are essential to integrated ATP synthesis. It appears that at least the polypeptides F_0-1, F_0-2, F_0-3, and OSCP (in mitochondria) plus some 3 to 30% by weight phospholipids are important for functions studied to date. It does not appear possible at this time to relate the careful reconstitution studies of thermophilic bacterial ATPase (Sone *et al.*, 1976) directly to the mitochondrial F_0 structure. However, the three F_0 peptides from thermophilic bacteria and *E. coli* do have size counterparts in mitochondrial F_0. The peptide with molecular weights near 30,000 in yeast and beef heart ATPase is not designated as an essential F_0 peptide in Table I because Ryrie's (1977) preparation of ATPase from *S. cerevisiae* and the Goffeau *et al.* (1973) preparation from *S. pombe* lack this subunit. The approximately 11,000 molecular weight (MW) component is not present in bacterial enzyme and may be absent in the beef heart preparation of Stiggall *et al.* (1978). Analysis of these small peptides is difficult, and the corresponding protein from heart mitochondria could be migrating with another band on SDS gels.

B. STRUCTURE

The general arrangement of subunits in the intact ATPase was first studied by electron microscopy. The enzyme contained knoblike F_1 structures attached to the inner mitochondrial membrane, presumably by F_0. Studies of isolated ATPase structure are complicated by the detergent extraction procedures utilized in solubilization as well as the possible structural alterations caused by removal of the F_0 sector from its membrane environment. Still, Tzagoloff (1976) and also Racker (1970) have been able to obtain electron micrographs of detergent-extracted yeast and beef ATPases that show F_1 connected to a regularly shaped F_0 unit. Aggregation of these particles occurs in the absence of detergent, and reconstitution of vesicles with stalks and knobs is noted. Reconstitution of ATPase with lipid or lipid vesicles also yields the now standard knob and stalk arrangement of ATPase observable by negative-staining techniques of electron microscopy.

The debate continues as to whether F_1 actually protrudes from the membrane to the extent noted by negative staining under any other conditions (Fernandez-Moran, 1962). Other methods of sample preparation or extensive protein cross-linking before negative staining indicate that it probably does not (Packer, 1978). Nevertheless, these studies have shown clear evidence of an asymmetrical organization of enzyme subunits and confirmed that the biochemist's separation of F_1 from F_0 also separates units with structural and functional counterparts in the mitochondria. This is extremely important in consideration of the link between the membrane-bound enzyme structure and vectorial processes that occur during ATP synthesis.

More-detailed analysis of the protein subunit arrangement in ATPase has been initiated by chemical cross-linking studies (Enns and Criddle, 1977b). These are subject to the same questions regarding the effects of detergents, isolation, etc., but have led to some useful conclusions. In general, proteins of F_1 become cross-linked predominantly to F_1 proteins. However, a link to F_0 appears to be through subunit F_0-1. This subunit becomes linked to subunits F_1-1, F_1-2, and F_1-3 as well as to subunit F_0-3. This is consistent with a stuctural model in which F_1 as a unit is associated with the F_0 sector proteins via subunit F_0-1, which must be located at the junction between the two.

IV. Chemical and Physical Properties of ATPase Subunits

The isolated individual subunits of ATPase with the exception of F_0-3 have not been studied extensively. Subunits F_1-1, F_1-2, and F_0-3 are the only ones readily isolated in large yields; subunits F_1-1 and F_1-2 because of their size and relative abundance, and subunit F_0-3 because of its peculiar solubility properties. Amino acid compositions of all the ATPase peptides have been analyzed and show that F_1-1 and F_1-2 are similar to each other but quite distinct from the complex as a whole or from the other peptides. Structural and chemical information known about the individual peptides is presented below.

A. Subunits F_1-1 and F_1-2

Together subunits F_1-1 and F_1-2 comprise 65–80% of the mass of F_1. They probably appear in equal molar ratios. The similarities in amino acid compositions of these subunits (Tzagoloff, 1976) have prompted an analysis of peptide homology between these two, but it has been reported that extensive homology was not observed (Senior, 1973). This question has now been reexamined by separation of yeast enzyme subunits F_1-1 and F_1-2 by SDS gel electrophoresis, iodination with ^{125}I, and tryptic hydrolysis. ^{125}I-labeled peptides were then analyzed by autoradiography. Subunits F_1-1 and F_1-2 each yield about 25 peptides labeled by this method, as many as 15 being common to both peptides (Satav and Criddle, 1979). This similarity is confirmed with chymotryptic peptides. It appears, therefore, that regions with common primary amino acid sequence do exist in these two subunits.

Subunit F_1-2 has been identified as the binding site for the ATPase inhibitor aurovertin. Verschour *et al.* (1977) dissociated beef heart F_1 with 0.85*M* LiCl and separated the subunits by DEAE chromatography. Only the isolated subunit F_1-2 was able to bind aurovertin. Their measurements indicated two noninteracting aurovertin binding sites (i.e., 2 subunit F_1-2) per ATPase when measured with or without ATP present. This agrees with data of Chang and Penefsky (1973), who reported 2 sites per F_1 in the presence of ATP but only one in its absence, and the data of Lardy *et al.* (1975), who showed 2 sites per F_1 on enzyme from rat liver and about 1.25 per F_1 in their beef heart preparations.

Subunit F_1-2 was confirmed to be the aurovertin-binding site in studies of yeast ATPase by Douglas *et al.* (1977). In addition these workers demonstrated that in an aurovertin-resistant mutant (nuclear) yeast, subunit 2 was unable to bind aurovertin.

Chemical modification of F_1-ATPase using ATP analogs has also generally resulted in reaction with subunit F_1-2. For example, 4-chloro-7-nitrobenzo-2-oxo-1,3 diazole (NBD) binds specifically to F_1-2 (Ferguson *et al.*, 1975; Deters *et al.*, 1975; Nelson *et al.*, 1974). A photoactive derivative of ATP, 8 azido-ATP, was shown by Wagenvoord *et al.* (1977) to also bind specifically to the subunit F_1-2, suggesting an ATP binding site on this subunit. Azidonitrophenolaminopropionyl-ATP was bound to ATPase upon photoactivation, but the site of binding of this analog was not determined (J. Russell *et al.*, 1976). Subsequently, Lunardi *et al.* (1977) used the butyryl derivative of this compound to show binding to both subunits F_1-1 and F_1-2 of beef heart F_1 with a corresponding loss in enzyme activity. Stoichiometry of binding was not clear, but about 0.95 mol of inhibitor per mole of F_1 caused about 80% inactivation. Quercetin, which mimics effects of the natural ATPase inhibitor (Lang and Racker, 1974), blocks binding to both subunits. Lunardi *et al.* (1977) suggested that the ATP analog may bind initially to subunit F_1-2 followed by reaction of the functional group with protein sites. The length of the carbon bridge in this case is sufficiently large that reaction may link the nucleotide to any adjacent subunits. They further suggest that varying this carbon bridge length may be used as a probe for structure analysis of the distance to neighboring peptides.

Additional chemical reagents used to modify ATPase have indicated the role of an essential tyrosine. By studying the reaction with 4-chloro-7-nitrobenzofurazan, Ferguson *et al.* (1976) suggested that modification of just one of the F_1-2 peptides produces an inactive enzyme. Esch and Allison (1978) labeled the nucleotide binding site of nucleotide-free beef heart F_1 with *p*-fluorosulfonyl-[^{14}C]benzoyl-5′-adenosine. Both subunits F_1-1 and F_1-2 were labeled. Inactivation of enzyme was proportional to the labeling on F_1-2. An amino acid sequence of the labeled peptide from subunit 2 containing substituted tyrosine residues was presented.

Additional information on structure of ATPase and binding sites on subunit F_1-2 was obtained by Hanstein and Hatefi (1974a) by affinity labeling with a dinitrophenol derivative (see section on uncouplers). The primary binding site for this molecule was a 29,000 MW protein. However, at higher pH considerable binding to F_1-2 was also noted. This suggests that the uncoupler binding site may be located between the 29,000 MW uncoupler protein and F_1-2. Thus a substantial body of information is available indicating that subunit F_1-2 must play a key catalytic or regulatory role in the hydrolysis of the γ-phosphate bond of ATP.

Specific labeling of F_1-1 has not been so readily obtained. Senior (1975) showed, however, that reaction with low levels of the thiol reagent *N*-ethylmaleimide with F_1 was quite specific for the largest subunit.

Other reagents used to modify and inactivate F_1-ATPase indicate a reactive arginine residue important for activity, but the peptide containing this reactive site has not been determined. Using butanedione and phenylglyoxal as arginine-specific reagents, Marcus *et al.* (1976) reported one arginine binding site that is protected by ATP. Frigeri *et al.* (1977) reported two sites: modification of one blocks ATPase activity; the other, more susceptible, site appears to be near the uncoupler binding site and is blocked by dinitrophenol.

It seems somewhat unexpected that all chemical modification studies to date would focus attention on subunit F_1-2 as the hydrolytic site on ATPase, while providing virtually no information on a role for subunit F_1-1 even though subunit F_1-1 appears to be essential for activity. Most of the chemical modification reactions have been run using isolated F_1. It may be important to look more closely at such reactions with intact or membrane-bound ATPases, particularly those in the process of ATP synthesis, to see whether alterations in the patterns of modification are noted.

B. Subunit F_1-3

Chemical modifications, analog binding, and inhibitor binding studies have not implicated subunit F_1-3 in the catalytic site for ATP hydrolysis and offer no indication of its role. Structural analysis of active enzyme preparations and reconstitution studies do, however, suggest an important role related to hydrolytic activity. As discussed above, all active F_1-ATPases purified to date have subunits F_1-1, F_1-2, and F_1-3 in stoichiometric amounts. Subunit F_1-3 is lost during protease digestion in two cases, leaving an active enzyme (Nelson *et al.*, 1974; Nelson, 1976), but reconstitution studies with thermophilic bacterial F_1 and *E. coli* F_1 both show a requirement for F_1-1, F_1-2, and F_1-3 (Nelson, 1976; Nelson *et al.*, 1973; Yoshida *et al.*, 1975; Futai, 1977). Chemical cross-linking studies of isolated ATPase complex suggest that subunit F_1-3 is centrally located adjacent to F_1-1 and F_1-2 in the complex and at the interface between F_1 and F_0 (Enns and Criddle, 1977b).

C. Subunits F_1-4 and F_1-5

Little is known regarding the function or structural location of F_1-4. It has been suggested that a subunit that may correspond to F_1-4 or F_1-5 in bacteria aids in binding F_1 to F_0 in bacteria (Sternweiss, 1977), but may not play this role in the mitochondrial enzyme. It has not been shown to be coupled to other F_1 peptides in protein cross-linking studies and is not modified by any of the substrate analog studies. Subunit F_1-5 has at times been considered to correspond to the ATPase inhibitor protein, but other workers consider the inhibitor protein to be yet a sixth F_1 peptide (for reviews, see Senior, 1973; Pedersen, 1975). As subunits F_1-4 and F_1-5 can be removed from the bacterial enzyme with little change in activity (Bragg *et al.*, 1973), it seems that F_1-5 is not a protein inhibitor in this case and a role for this peptide in ATP synthesis is yet to be determined.

D. SUBUNIT F_0-1

Every preparation of ATPase has a component with molecular weight near 20,000 that is a major component of F_0. Attention was focused on this subunit initially as one of the protein components synthesized on motochondrial ribosomes (Tzagoloff and Meagher, 1971; Enns and Criddle, 1977b; Groot Obbink *et al.*, 1976). It has been somewhat more difficult, however, to actually demonstrate whether this protein is a mitochondrial gene product. A major step toward the conclusion that it is indeed mitochondrial comes from the studies of Trembath *et al.*, (1975a, b; Monk *et al.*, 1979), using a class A oligomycin-resistant yeast strain that was in addition cold sensitive. This mitochondrial gene mutation caused a decreased level of subunit F_0-1 synthesis and also affects assembly of F_1 with F_0, particularly below the critical temperature. A tentative link was established between class A oligomycin resistance (which also offers coresistance to venturicidin) and F_0-1.

In recent studies, ^{3}H-labeled pantothenic acid has been added to pantothenate-starved yeast cells and incorporation of ^{3}H into ATPase has been noted (Criddle *et al.*, 1977b). SDS-gel electrophoretic analysis showed that the label was associated with a 20,000 MW polypeptide that comigrated with subunit F_0-1. It was concluded that pantothenic acid is covalently attached to this subunit of the ATPase complex. The role of this F_0-1-bound pantothenic acid moiety in ATPase function is not yet clear, but a venturicidin-sensitive thioesterase activity that may be dependent on this unit has been observed and will be discussed later (Scharf *et al.*, 1979).

E. SUBUNIT F_0-2

SDS gel electrophoresis of most ATPase preparations indicates one or sometimes several bands with molecular weights in the range of 12,000 to 15,000. Resolution of these components is often poor and separation of F_0

peptide from F_1-5 is frequently not clear. As a result, characterization of the F_0-2 peptide and even estimation of its stoichiometry in the complex has not progressed far in most cases. No known function can be assigned specifically to this peptide. It has been reported that an ATPase subunit this size may be synthesized in the mitochondria (Tzagoloff and Meagher, 1971) but other studies have created some doubt about this conclusion (Enns and Criddle, 1977b; Monk, 1977).

F. Subunit F_0-3

The best studied of the mitochondrial ATPase polypeptide subunits is the proteolipid component F_0-3. This is due in large part to its relative ease of isolation by extraction into neutral chloroform : methanol (Tzagoloff *et al.*, 1973). Neutral chloroform : methanol extraction and ether precipitation yields the yeast and *Neurospora* proteins in pure form. Proteolipids from *E. coli*, beef heart mitochondria, and chloroplasts are still contaminated at this step and require further purification by chromatography (Sebald and Wachter, 1978).

F_0-3 from five sources has been analyzed. All contain between 76 and 81 amino acids. All exhibit low polarity, as more than 75% of the total amino acid side chains are nonpolar. Amino acid sequence studies for the yeast, *Neurospora*, and *E. coli* proteins exhibit considerable sequence homology, characterized by small clusters of polar amino acids separated by very nonpolar segments (Machleidt *et al.*, 1976; Sebald *et al.*, 1977).

F_0-3 is the site of DCCD inhibition of oxidative phoshorylation (Catall *et al.*, 1971; Stekhoven *et al.*, 1972; Sebald and Wachter, 1978). DCCD becomes covalently linked to a glutamic acid side chain near the C-terminal end of the yeast and *Neurospora* proteins (Machleidt *et al.*, 1976; Wachter and Sebald, 1977; Sebald *et al.*, 1977). This subunit also contains the site for binding oligomycin and can be specifically labeled by reductive alkylation of the subunit F_0-3-oligomycin complex with $NaBT_4$ (Enns and Criddle, 1977a). Moreover, ethidium bromide appears to block ATPase activity by binding at this subunit (Johnston *et al.*, 1979b). A photoaffinity derivative of ethidium bromide also appears to bind specifically to subunit F_0-3 (Mahler and Bastos, 1974).

F_0-3 from yeast containing the class B oligomycin resistance marker is found to be altered (Tzagoloff *et al.*, 1976; Groot Obbink *et al.*, 1976). The protein is isolated in monomeric form from these strains under conditions that yield protein oligomers from wild-type strains (Tzagoloff *et al.*, 1976; Groot Obbink *et al.*, 1976). A change in primary amino acid sequence has been noted with at least two of these mutant strains (Sebald *et al.*, 1976; Wachter and Sebald, 1977). Sensitivity to oligomycin both of *in vivo* oxidative phosphorylation and of isolated ATPase are reduced in the mutant strains (Shannon *et al.*, 1973; Wakabayashi, 1972). The O^R class B mutants also commonly have ATPase activities with reduced sensitivities to ethidium bromide and DCCD (Johnston *et al.*, 1979b). Oligomycin-resistant mutants of *Neurospora*, which also appear to affect this subunit, are of nuclear origin (Sebald *et al.*, 1976, 1977; Edwards and Unger, 1978). The properties of subunit F_0-3 will be discussed in greater detail in Section X,A on proton translocation by ATPase.

G. Oligomycin Sensitivity-Conferring Protein (OSCP)

OSCP was first described as a component of the beef heart mitochondrial ATPase by MacLennan and Tzagoloff (1968). It was described as a soluble coupling factor (MacLennan and Tzagoloff, 1968; Knowles *et al.*, 1971) that can confer oligomycin sensitivity (Tzagoloff, 1970) and DCCD sensitivity (Knowles *et al.*, 1971) to ATPase activity of F_1 in the presence of urea-extracted or trypsin-treated submitochondrial particles (Datta and Penefsky, 1970; Knowles *et al.*, 1971). OSCP has been prepared from beef heart (Datta and Penefsky, 1970), rat liver (L. Russell *et al.*, 1976), and yeast (Tzagoloff, 1970). The yeast protein was shown to be a nuclear gene product (Tzagoloff, 1970).

The subunit designations of Table I do not assign OSCP as either an F_1 or F_0 peptide, and its functional role is not clear. No analogous peptide is found with bacterial ATPases. The OSCP protein was classically thought to be the stalk portion of the ATPase linking F_1 to F_0 in the negative-staining electron micrographs (Senior, 1973; Pedersen, 1975). In addition, as the name implies, it was thought by many to be not only a factor conferring inhibitor sensitivity (Kagawa and Racker, 1966; MacLennan and Tzagoloff, 1968), but the site of action of oligomycin as well.

More recent findings raise doubts concerning both of these conclusions. L. Russell *et al.* (1976) show that OSCP preparations contain more than one protein. One of these was shown to effect binding of F_1 to trypsin-urea treated mitochondrial particles in the absence of OSCP, indicating that OSCP may not be the "stalk" of earlier studies. Oligomycin sensitivity was not regained by this association until OSCP was added. However, OSCP is certainly not the site of action of oligomycin inhibition of ATPase activity as established by affinity-binding studies (Enns and Criddle, 1977a) and proton-permeability studies (Criddle *et al.*, 1977b). Hinkle and Horstman (1971) and Ernster (1975) indicated that OSCP was not required for oligomycin to inhibit proton permeability of urea-extracted submitochondrial particles. In fact, OSCP alone could partially inhibit that function. Thus it appears that OSCP may act somewhere near the F_1–F_0 interface in fully coupled mitochondrial ATP synthesis but is not the "stalk" or the oligomycin-binding protein.

H. Other Proteins Associated with ATPase

SDS-gel electrophoresis of purified mitochondrial ATPase preparations routinely yields protein bands in addition to those of F_1, and those noted here as F_0-1, F_0-2, F_0-3, and OSCP. Of these, the proteins with molecular weights in the range of 10,000–11,000 and 29,000–31,000 shown in Table I are the ones most commonly considered to be part of the F_0 membrane sector. No information is available at this time regarding function of proteins of lower molecular weight, and, as discussed earlier, they may not be an integral part of the ATPase complex. Tzagoloff and Meagher (1971) did, however, report that a protein with molecular weight in this range isolated with yeast mitochondrial ATPase preparations was synthesized in the mitochondria. Resolution of peptides of low molecular weight on the SDS gels commonly used for ATPase analysis has not been very good, making it difficult to define with certainty the size or even the number of polypeptides unique to the ATPase in this molecular weight range.

SDS gels of most ATPase preparations also contain two or more bands with molecular weights in the range of 70,000–90,000. These high molecular weight bands have generally been viewed as contaminants or incompletely dissociated oligomers of ATPase subunits, since they occur in variable, nonstoichiometric amounts. This has yet to be demonstrated clearly. If partial dissociation is the answer, a study of these bands could lead to information on pairing of subunits into the complex.

It is tempting to speculate that the 28,000–30,000 MW peptides shown in Table I for yeast (*S. cerevisiae*) and beef heart ATPase may correspond to the 29,000 MW uncoupler binding protein reported by Hanstein (1976), but as yet no demonstration of this has been reported.

ATPases of mitochondria, chloroplast, and *E. coli* membranes (Pullman and Monroy, 1963; Nelson *et al.*, 1972; Nieuwenhuis *et al.*, 1974; Klein *et al.*, 1977) are known to be associated with an endogenous protein inhibitor that, as suggested above, has often been considered to be part of F_1. The beef heart inhibitor was the first to be isolated (Pullman and Monroy, 1963). It, like other inhibitors subsequently isolated, is a small ($\sim$10,000 MW), heat-stable protein. It strongly suppresses ATPase activity of isolated F_1 and in submitochondrial particles effectively blocks ATP-driven energy transfer reactions, but possibly not ATP synthesis (Asami *et al.*, 1970). The physiological role played by inhibitor protein is still not clear, but it is postulated to control the back flow of energy from ATP to the mitochondrial ion transport and electron transport systems (Van de Stadt *et al.*, 1973; Asami *et al.*, 1970). Inhibitor protein has recently been purified from yeast, and the amount of inhibitor protein in two ATPase mutant strains has been directly correlated with the different levels of the mitochondrial ATPase complex in these strains to further indicate some physiological role for the inhibitor (Ebner and Maier, 1977). The relation between inhibitor protein and individual bands observed on SDS gels for various ATPase preparations has not been clarified appreciably since the reviews of Senior (1973) and Pedersen (1975). A portion of the protein staining in the low molecular weight region on SDS-gel electrophoretic examination of most mitochondrial ATPase preparations is probably inhibitor protein.

A close association between the adenine nucleotide translocase of the inner membrane and mitochondrial ATPase has been suggested (for review of the adenine translocase, see Vignais, 1976). It now seems clear that ATPase can be isolated free from the translocase, ruling out any direct involvement of translocase in the steps of ATP hydrolysis (Serrano *et al.*, 1976). However, this does not rule out a close *in vivo* association of the two. Kinetic studies of ^{14}C-labeled ATP efflux suggest that equilibrium is not obtained between newly synthesized ATP and the adenine nucleotide pool within the mitochondria, suggesting a direct transfer of ATP from the synthetase to translocase (Out and Kemp, 1974; Kemp and Out, 1975). Studies of the type reported by Bertagnolli and Hanson (1973) also may be interpreted to support this close association. Arsenate-loaded mitochondria were shown to leak that molecule when provided with reduced substrate. This has been postulated to occur via formation of an adenine diphosphoarsenate complex that is immediately transferred across the membrane and hydrolyzed in the external aqueous medium.

The arguments for an integral involvement of these two enzymes are not conclusive at this time, and further work must be done to resolve this important question. Chemical cross-linking studies of the inner membrane coupled with radioactive inhibitor labeling may lead to further elucidation of the ATPase-translocase interaction.

V. Biosynthesis of ATPase

Subunits of the mitochondrial ATPase can be divided into two major categories based on the genetic source of the information coding for these proteins (for reviews, see Tzagoloff *et al.*, 1973; Sebald, 1977). The peptides of one set are nuclear gene products and are synthesized in the extramitochondrial cytoplasm; the others are mitochondrial gene products synthesized in the mitochondria. A number of difficulties encountered in careful definition of the genetic source of mitochondrial components have been reviewed earlier (Schatz and Mason, 1974). However, for some of the ATPase proteins the results are clear. It appears that in every system studied to date, all the components of F_1 are of nuclear origin. Probably the clearest indication of this comes from demonstration of F_1 synthesis in petite strains of yeast which were altered in or lacking mitochondrial DNA, and were therefore incapable of mitochondrial protein synthesis (Perlman and Mahler, 1970; Schatz, 1968; Kovác and Weissová, 1968). Integration of F_1 depends upon mitochondrial protein synthesis (Schatz, 1968). More recent studies have now clearly shown that a nuclear aurovertin-resistant mutation can be related directly to an alteration in subunit F_1-2 of the yeast ATPase establishing clearly the genetic origin of this peptide (Agsteribbe *et al.*, 1976). This type of combined genetic and biochemical approach will ultimately be needed to clarify the genetic source of the rest of the ATPase peptides.

Analysis of the gene source coding for peptides of the F_0 membrane sector of the ATPase is complicated by their low aqueous solubility and accompanying problems of purification. It is now quite generally accepted, however, that at least two mitochondrial gene products are a part of the yeast ATPase. Tzagoloff and Meagher (1971) reported that, in yeast, F_1-1, F_1-2, F_1-3, and the 29,000 MW subunit are all of mitochondrial origin. This conclusion was based on results from differential inhibition of cytoplasmic and mitochondrial protein synthesis by cycloheximide and chloramphenicol. Later, Enns and Criddle (1977b) extended this type of analysis and found that higher levels of cycloheximide prevented labeling in two of the subunits but had no effect on the labeling of subunits F_0-1 and F_0-3. In addition, pulse-chase experiments were employed to demonstrate that cycloheximide was probably not causing either assembly or processing blocks to the labeling of other peptides. Finally, Trembath *et al.* (1975a) studied incorporation of ^{14}C-labeled amino acid into antibody precipitable yeast ATPase with added cycloheximide and observed label incorporation only into 29,000 MW and F_0-1 and F_0-3 subunits. When the antibody precipitate was washed with 0.9% saline prior to examination on SDS gels, the band corresponding in size to the 29,000 MW subunit was lost. This leads again to the conclusion that subunits F_0-1 and F_0-3 may be the only tightly integrated proteins of yeast ATPase that are mitochondrial gene products.

Using chloroamphenicol–cycloheximide differential inhibition studies to define mitochondrial gene products has major experimental limitations (Mahler, 1973). However, genetic and biochemical analyses of yeast mitochondrial gene mutants affecting ATPase generally support the conclusions from inhibitor studies. In particular, two classes of oligomycin-resistant mutants of yeast mitochondria have been mapped on the mitochondrial genome and are related to changes in subunit proteins of the ATPase. One of these (class B oligomycin resistant) has been shown to alter the aggregation state of F_0-3 in the isolated ATPase (Tzagoloff *et al.*, 1976; Groot Obbink *et al.*, 1976) and to result in a change in the amino acid composition of this subunit (Wachter and Sebald, 1977). One class A oligomycin-resistant mutant of yeast, which also has the property of being cold sensitive, was shown by workers in Linnane's laboratory to be deficient in F_0-1 when grown at restrictive temperatures (Trembath *et al.*, 1975b). This appears to to be a deficiency in synthesis rather than in assembly, providing further evidence that F_0-1 is a mitochondrial gene product (Monk, 1977). Genetic analysis also indicates a third mitochondrial gene locus conferring oligomycin resistance in yeast, but the relation of this locus to specific mitochondrial proteins is not yet clear (Trembath *et al.*, 1976).

The cycloheximide-resistant, chloramphenicol-sensitive criterion for defining the site of synthesis of mitochondrial proteins has been applied to several cell sources (for review, see Schatz and Mason, 1974). In addition, other studies have been made of protein synthesis in isolated mitochondria. In general, the numbers and sizes of mitochondrially synthesized proteins is similar for each of these systems. However, differences are noted, and extrapolation of the results from studies with yeast mitochondrial ATPase to other less defined sources should be made with caution. Recent analysis of the site of synthesis of the small proteolipid from *Neurospora*, analogous to F_0-3 of the yeast ATPase, shows that this peptide in *Neurospora* is a nuclear gene product (Sebald and Wachter, 1978; Edwards and Unger, 1978).

VI. Reactions Catalyzed by Isolated ATPase and Its Component Subunits

Pedersen (1975) has summarized details of the catalytic properties of mitochondrial ATPases including substrate specificities for the reactions and the effects of activators and inhibitors on the kinetics of the reactions catalyzed. It is now possible to detail the catalytic properties further with a list of individual reactions that have recently been reported to be catalyzed by preparations of ATPase and subfractions of the enzyme. This list has become quite extensive with recent studies and leads to some interesting considerations about possible integrated modes of action of enzyme subunits in ATP synthesis. Several of these reactions have been reviewed extensively elsewhere and will not be discussed in detail here except to remind the reader of some key points important for further consideration of enzyme action.

Most, however, represent new developments that need to be evaluated for their *in vivo* importance.

A. ATP Hydrolysis

The ATP hydrolysis activity (reaction 1; see Table II) has been used as the standard reaction for following enzyme purification and physical and kinetic properties. Hydrolysis at physiological pH ranges is accompanied by the liberation of a proton, an important consideration in the overall function of oxidative phosphorylation. This reaction may be catalyzed either by isolated F_1 or by the intact ATPase complex. The reaction with F_1 is cold sensitive (unless stabilized with methanol, Ryrie, 1977) and is specifically inhibited by "F_1 inhibitors," such as aurovertin, 1799, and the protein inhibitor (Racker, 1970). More general inhibitors, such as orthophosphate and arsenate, and stimulation by bicarbonate and bisulfite have been discussed elsewhere (Pedersen, 1975) or will be described in more detail with reaction kinetics. Hydrolysis catalyzed by the intact ATPase maintains sensitivity to F_1-inhibitors and gains additional sensitivity to inhibitors that bind to the F_0 position of the complex. Thus, ATP hydrolysis by intact ATPase is sensitive to inhibitors such as oligomycin (rutamycin), venturicidin, trialkyltins, and ethidium bromide (see reviews in Pedersen, 1975; Senior, 1973).

B. Proton Translocation Linked to ATP Hydrolysis

Mitochondria, chloroplasts, and bacteria are known to contain systems capable of acting as proton pumps as indicated in reaction 2. The role of ATPase in mitochondrial H^+ transport across membranes was first proposed by Mitchell (Mitchell, 1961; Mitchell and Moyle, 1965, 1967) and demonstrated by studies showing H^+ movement resulting from ATP hydrolysis in mitochondria with blocked respiratory components and in reconstituted lipoprotein vesicles. This work has been thoroughly reviewed (Kozlov and Skulachev, 1977), and we will consider only specific involvement of ATPase components here.

To date, it appears that the subunit requirements for enzyme preparations catalyzing hydrolysis-driven proton translocation are essentially identical to those studied more carefully for ATP-P_i exchange (see below). An additional technical requirement for measurement of activity is that enzyme must be embedded into a lipid membrane with low H^+ permeability so that the vectorial movement of protons may be determined. Both the F_1 and F_0

TABLE II

REACTIONS THAT CAN BE CATALYZED BY ATPASE AND ITS COMPONENTS

Reaction	Catalytic agent	Inhibitor
1. $ATP + H_2O \rightarrow ADP + P_i + H^+$	F_1	Aurovertin, 1799, Inhibitor protein, Dio-9, Efrapeptin } F_1 inhibitors
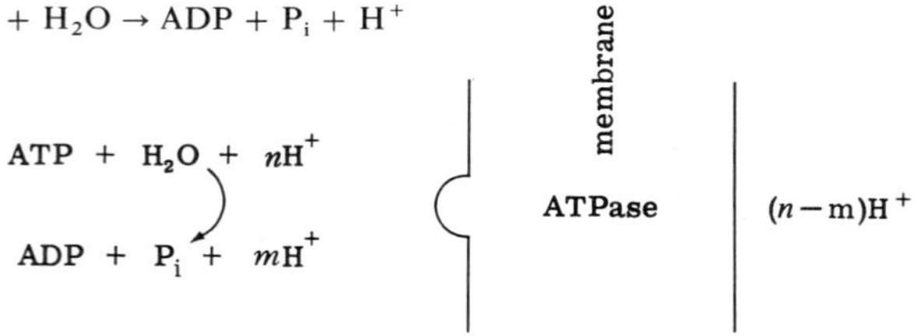 $ATP + H_2O + nH^+ \rightarrow ADP + P_i + mH^+$ (membrane, ATPase, $(n-m)H^+$)	$(F_1 + F_0)$-ATPase	F_1 inhibitors Oligomycin Venturicidin Trialkyltin derivatives Ethidium bromide DCCD
2. A. 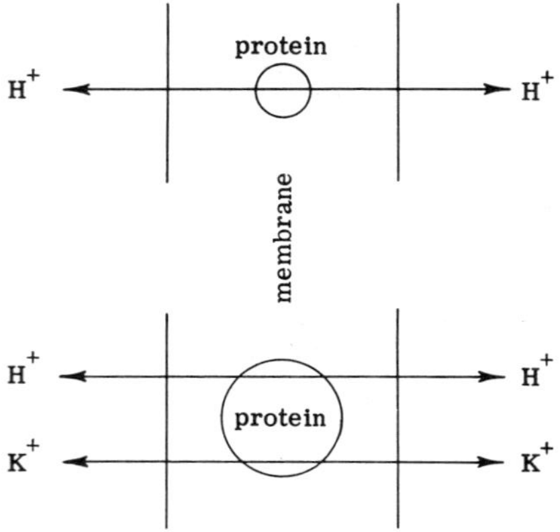 $H^+ \leftarrow$ protein (membrane) $\rightarrow H^+$	ATPase + lipid membrane	F_1 inhibitors Oligomycin Venturicidin DCCD Trialkyltin
B. $H^+ \leftarrow$ protein (membrane) $\rightarrow H^+$	F_0-3	Oligomycin DCCD Ethidium bromide
$K^+ \leftarrow$ protein (membrane) $\rightarrow K^+$	F_0-3	As in A
3. $2ADP \rightleftarrows ATP + AMP$	F_1 (mitochondria and chloroplast)	No Mg^{2+} requirement

4. $R{-}C(=O){-}O{-}P(=O){-}O^- + ADP \rightarrow ATP + RCO_2^-$	F_1 or ATPase	F_1 inhibitors
($R{-}C(=O){-}OH$ = oleic acid)		
5. $R{-}C(=O){-}O{-}P(=O)(O){-}O^- \rightarrow R{-}C(=O){-}O^- + HO{-}P(=O)(O^-){-}O^- + H^+$	F_1 or ATPase	Efrapeptin Aurovertin Stimulated by: DNP 1799
6. $CoA{-}S{-}C(=O){-}R + H_2O \rightarrow CoA{-}SH + R{-}C(=O){-}O^- + H$	F_0 (subunit F_0-1?)	Venturicidin Oligomycin Hg^{2+}
7. $R{-}C(=O){-}O{-}P(=O)(O){-}O^- + CoA{-}SH \rightarrow R{-}C(=O){-}O^-{-}S{-}CoA + P_i$	ATPase	—
8. Lipoic acid-linked reactions		
A. Dihydrolipoate + ADP + $P_i \rightarrow$ ATP + lipoate	Submitochondrial particles	F_1 inhibitors Uncouplers DCCD Venturicidin Oligomycin
B. Dihydrolipoate + ADP + $P_i \xrightarrow[\text{oleate}]{\text{oleoyl-CoA}}$ ATP + lipoate	ATPase	As in A
C. Oleoyl-S-lipoate + ADP + $P_i \rightarrow$ ATP + oleate + lipoate	ATPase	As in A

Continued

TABLE II—*Continued*

Reaction	Catalytic agent	Inhibitor
9. Exchange reactions		
A. $ATP + {}^{32}P_i \rightleftarrows ATP(\gamma\text{-}{}^{32}P) + P_i$	ATPase + phospholipid	F_1 inhibitors Uncouplers Oligomycin Venturicidin Trialkyltin DCCD
(a) ATP-P_i exchange stimulated by oleoyl phosphate	ATPase	F_1 inhibitors Other acyl phosphates
(b) ATP-P_i exchange stimulated by oleoyl-CoA	ATPase	As in (a) plus Oligomycin Venturicidin DCCD Other acyl CoAs
(c) ATP-P_i exchange stimulated by oleoyllipoate	ATPase	As in (b) plus Uncouplers
(d) ATP-P_i exchange stimulated by dihydrolipoate	ATPase	As in (c)
B. ${}^{14}C{-}ADP + ATP \rightleftarrows {}^{14}C{-}ATP + ADP$	ATPase plus phospholipid	As in A
C. $ATP + H_2{}^{18}O \rightleftarrows ATP({}^{18}O) + H_2O$	ATPase (lipid membranes?)	F_1 inhibitors Oligomycin DCCD Uncouplers
D. $P_i + H_2{}^{18}O \rightleftarrows P_i({}^{18}O) + H_2O$	ATPase (lipid membranes?)	F_1 inhibitors Oligomycin DCCD Uncouplers inhibit $\frac{1}{2}$ reaction

portions of the complex are required. The evidence suggests that the membrane sector of ATPase plays a major role in H^+ movement (Mitchell and Moyle, 1967). Inhibition of ATP hydrolysis by membrane-sector specific inhibitors, such as oligomycin, venturicidin, trialkyltins, and DCCD probably results from preventing the coupled movement of H^+ either to an acceptor in F_0 and/or through the membrane sector (and through the lipid membrane when ATPase is bound to membranes).

Evidence is accumulating to indicate that reaction 2 is reversible, i.e., imposition of a H^+ gradient can be used to drive ATP synthesis (Racker and Stoeckenius, 1974; Sone *et al.*, 1975). The number of H^+ moved per ATP hydrolyzed therefore becomes a critical question in linking the H^+ translocation reaction of this enzyme to the proposed chemiosmotic mechanisms for ATP synthesis.

A means of investigating this question has been to study the amount of H^+ transported across membranes during hydrolysis of ATP by mitochondrial particles with electron transport blocked (Mitchell and Moyle, 1965; Moyle and Mitchell, 1973; Brand *et al.*, 1976; Brand and Lehninger, 1977). This H^+ : ATP ratio is now considered to be 2–3. Accurate determination of this ratio, without questions concerning other proton-consuming reactions of mitochondrial particles, awaits careful measurement in a reconstituted ATPase–liposome system. It is clear from reaction 1 that simple hydrolysis of ATP in aqueous buffer at a physiological pH results in liberation of only about 1 H^+ per ATP hydrolyzed. Therefore, the observed larger stoichiometry of H^+ transported per ATP requires that protons from the aqueous solution on the F_1 side of the mitochondrial vesicles must somehow be associated with the ATPase and pumped through the membrane, coincident with ATP hydrolysis. This reaction is reminiscent of Na^+-K^+ ATPases of erythrocyte membranes, which in analogous fashion must bind Na^+ on the ATPase side of the membrane and exchange $3\,Na^+$ for $2\,K^+$ ions across the membrane for each ATP hydrolytic event (Dahl and Hokin, 1974).

As far as ATP hydrolysis can be considered to be the reverse of the ATP synthase reaction, these observations require that H^+ become associated with the membrane sector of the ATPase complex. These are then transported from the intermembrane space back into the mitochondrial matrix during ATP synthesis (Brand *et al.*, 1976). Considerable interest has recently been directed toward studies of the H^+ : site ratio for electron transport (number of H^+ ejected per pair of electrons per energy coupling site). The ratio of 2 required by the original chemiosmotic hypothesis (Mitchell and Moyle, 1965) has been increasingly subject to question. Several laboratories have indicated that a H^+ : site ratio of at least 3 is required to account for coupled synthesis of external ATP (Papa, 1976; Nicholls, 1974; Rottenberg, 1975;

Weichman *et al.*, 1975). Indeed, studies by Brand *et al.* (1976) with mitochondrial particles have yielded experimentally determined H^+ : site ratios of 3 to 4. These numbers appear to be consistent with the H^+ requirements indicated above when other systems involving H^+ transport in mitochondrial particles are considered (Brand and Lehninger, 1977).

Evidence has been obtained to show that H^+ movement linked to ATP hydrolysis and synthesis proceeds via the F_0 proteolipid F_0-3 component of the ATPase complex (Criddle *et al.*, 1977b; see Section X,A below). Reactions 2A and 2B have both been demonstrated with purified F_0-3 incorporated into planar lipid membranes or liposomes. With present evidence, it appears also that this proteolipid may be the only H^+ transporter in the complex, and it must therefore be this site for binding and transporting 2–3H^+ per hydrolysis event (Criddle *et al.*, 1977b). The oligomeric nature of the proteolipid in ATPase (Sierra and Tzagoloff, 1973) suggests a means by which transport of several protons could be accomplished even with only one H^+ binding site per F_0-3 molecule. Also, evidence from changes in oligomycin binding, pH dependence, K^+ transport, and direct size analysis (reviewed below) indicate that the proteolipid can exist in more than one distinct physical and kinetic form dependent upon H^+ binding. This allows formulation of several hypotheses as to how H^+ binding could be energetically linked to ATP synthesis via conformational changes in F_0-3. These will be discussed later.

C. Interconversion of Bound Nucleotides by Transphosphorylation

Moudrianakis and co-workers have described a transphosphorylation reaction similar to that of adenylate kinase, catalyzed by the F_1 coupling factor from chloroplasts (Roy and Moudrianakis, 1971a) and mitochondria (Barnes *et al.*, 1977). As indicated by reaction 3, incubation of ADP with purified chloroplast F_1 resulted in formation of a mixture of AMP, ADP, and ATP. The possibility that this reaction was catalyzed by contaminating adenylate kinase from plant extracts was minimized by pH and inhibition studies. One interesting difference noted in these two enzymes was that the transphosphorylation reaction was independent of Mg^{2+}. As adenylate kinase, ATPase, and ATP synthase reactions do require Mg^{2+}, transphosphorylation may possibly be studied independently of the overall reaction sequence.

A major implication of these studies is the apparent need for at least two nucleotide binding sites on the enzyme to facilitate transphosphorylation. This offers support to kinetic mechanisms of the type proposed by Moudrianakis and Tiefert (1976) and Boyer *et al.* (1977).

D. Oleoyl Phosphate-Dependent ATP Synthesis

Submitochondrial particles, isolated ATPase, and also purified F_1-ATPase have been shown to catalyze an oleoyl phosphate-dependent ATP synthesis, reaction 4 (Hyams *et al.*, 1977; Johnston and Criddle, 1977). This reaction appears to be highly specific for oleoyl phosphate; no other acyl phosphate can serve as a substrate for the reaction. Reaction stoichiometry indicates that one oleoyl phosphate is used to produce one ATP. This reaction is blocked by F_1-ATPase inhibitors, such as efrapeptin, aurovertin, and 1799, but is not inhibited by uncouplers such as CCCP or TTFB or by the inhibitors oligomycin, triethyltin, or venturicidin.

E. Oleoyl Phosphate Phosphatase

Preparations of ATPase from yeast and beef heart catalyze the hydrolysis of oleoyl phosphate (Johnston and Criddle, 1977; Hyams and Griffiths, 1978). This reaction was stimulated by dinitrophenol (DNP) and also by 1799. The level of DNP causing significant stimulation of oleoyl phosphatase is about 12 μM, about the same level required for uncoupling of *in vivo* oxidative phosphorylation. This result together with the studies by Hanstein and Hatefi (1974a,b) and by Hanstein (1976) on uncoupler binding to their ATPase preparation (complex V) suggest that DNP could exert its uncoupling effects by direct interaction with a chemical intermediate of oxidative phosphorylation rather than by effects on membrane permeability (see further discussion of uncoupling below).

F. Thioesterase (Acylthioester Hydrolase)

It has recently been demonstrated that purified preparations of yeast mitochondrial ATPase (but not F_1) also have associated thioesterase activity (Scharf *et al.*, 1979). These preparations catalyze hydrolysis of oleoyl-CoA and, with a slower rate, stearyl- and palmitoyl-CoAs (reaction 6). Specific activity of the thioesterase is about 3–5 % of that for ATP hydrolysis. Hydrolysis of the acylthioester bond is effectively inhibited by two specific inhibitors of ATPase. Oligomycin and venturicidin both block thioesterase activity of enzyme isolated from wild-type (sensitive) yeast strains. Studies with oligomycin-resistant yeast strains that are altered in F_0-3 show no change in sensitivity of thioesterase activity to this antibiotic relative to the wild type. However, studies with mutant strains resistant to venturicidin yield thioesterase preparations that are highly resistant to venturicidin.

The thioesterase studies yield four important conclusions: (1) Some subunit component(s) of ATPase can catalyze thioester hydrolysis; (2) mutations conferring venturicidin resistance to ATPase activity result in venturicidin-resistant thioesterase activity; (3) the oligomycin resistance–venturicidin resistance gene locus on mitochondrial DNA, which has been tentatively linked to F_0-1 of the ATPase (Monk, 1977), has a direct effect on thioesterase activity; (4) since the antibiotic-resistant mutant strains were initially selected as oxidative phosphorylation mutants, the changes noted in ATPase and thioesterase and consequently the reactions that they catalyze, must also be directly related to oxidative phosphorylation. Since F_0-1 has also been implicated as the pantothenic acid-containing protein in the ATPase (Criddle *et al.*, 1977a), it is tempting to propose that this may also be the active site for thioesterase reaction.

G. Transacylation for Formation of Oleoyl-CoA from Oleoyl Phosphate

When oleoyl phosphate is incubated with purified ATPase and CoA in the absence of ADP, the production of oleoyl-CoA can be observed (reaction 7). The reaction requirements have not been thoroughly investigated at this time, but the reaction does appear to require the intact ATPase (Criddle, 1977). The yield of oleoyl-CoA is increased by addition of oligomycin, presumably because of the inhibition of thioesterase activity noted above.

H. Lipoic Acid Involvement in ATP Synthesis

In an important series of papers, Griffiths and co-workers (Griffiths, 1976a,b; Griffiths *et al.*, 1977a,b; Partis *et al.*, 1977) have described a series of reactions involving lipoic acid and a number of its derivatives in ATP synthesis by mitochondria, chloroplasts, and some bacteria. This work has stimulated a great number of studies in other laboratories, but several workers have experienced difficulties in reproducing portions of this work, leaving interpretation of the results open to question.

Griffiths reports observation of lipoic acid-dependent ATP synthesis in studies using either submitochondrial particles or isolated ATPase preparations (Griffiths, 1976a). With submitochondrial particles (with electron transport inhibited), dihydrolipoate was used to drive synthesis of ATP. Oxidized lipoic acid was formed. In reactions with isolated ATPases, addition of oleoyl-CoA and oleate were required in addition to the dihydrolipoate for synthesis to be observed. Questions arise concerning both the specificity of the lipoic acid requirements and the reaction stoichiometry. However, it was demonstrated that no other dithiol reagent tested could substitute for dihydrolipoate in this reaction; oxidized lipoate and lipoamide were ineffective, and 8-methyllipoate was a potent inhibitor both of lipoate-dependent ATP synthesis and of oxidative phosphorylation.

Much of the evidence for lipoic acid involvement has been based on inhibitor studies or work with *E. coli* lipoic acid auxotrophs. Dibutylchloromethyltin (DBCT) was reported to block ATP synthesis by binding covalently to lipoic acid, and indeed evidence was presented that inhibition by DBCT is specifically reversed by addition of excess lipoic acid (Griffiths, 1976b). Membrane-bound ATPase fractions from a lipoic acid auxotroph were shown to require added dihydrolipoate for ATP synthesis. In all cases, the lipoic acid-dependent synthesis was inhibited by the general inhibitors of coupled phosphorylation.

One problem arising in these studies was the lack of stoichiometric relationship between lipoic acid and the amount of ATP formed. It appears that some dihydrolipoate preparations will not work and that "aged" solutions of lipoic acid with some loss of —SH titer work more effectively than fresh solutions (Hyams and Griffiths, 1978). A decrease in ATP yields in parallel with loss of —SH titer was not noted, making an oxidation–reduction reaction of lipoic acid unlikely as the direct link between electron transport and ATP synthesis.

Recently, Haddock and Begg (1977) and Singh and Bragg (1978) have reexamined ATP synthesis in the *E. coli* lipoic acid auxotroph. They could find no requirement for added lipoate in respiration, in ATP-driven proton translocation, or in ATP synthesis. However, DBCT still served as an efficient inhibitor of ATP synthesis in the lipoic acid-deficient samples, suggesting some additional site of DBCT action. No indication was given whether lipoic acid reversed in the block in these studies.

The general lipoic acid-linked reactions proposed by Griffiths to be catalyzed by ATPase are summarized in Table II, reactions 8A, B, and C. Further details of how these reactions proceed through chemical intermediates have been postulated by these authors as an "oleoyl cycle" (Griffiths, 1977). One key prediction of the oleoyl cycle proposal is that the rate of the overall reaction is determined by the individual reaction rate constants and the steady-state concentrations of the intermediates. This suggests that addition of postulated intermediates, such as oleoyl lipoate, oleoyl-CoA, and oleoyl phosphate, should lead to enhanced rates of reactions, such as the ATP–P_i exchange reaction discussed below. Significant stimulation of the exchange rates by these intermediates has been reported (Hyams and Griffiths, 1978).

At present, the most difficult portion of these experiments to document is the direct involvement of reduced lipoic acid. The "aging" of lipoic acid in organic solvent may result in formation of some additional component that is more directly responsible for the reactions observed than is dihydrolipoate. For example, lipoic acid polymers of the type

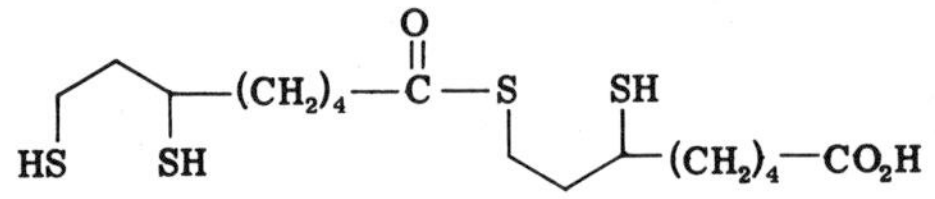

O
||
$-(CH_2)_4-C-S$ SH
HS SH $(CH_2)_4-CO_2H$

or a thiolactone (Boyer *et al.*, 1977)

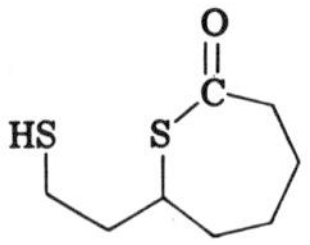

O
||
C
HS S

may be formed. Such thioacyl moieties could exchange with a thiol group in the enzyme to form a high-energy acyl protein intermediate, which could then be used for ATP synthesis.

It seems essential to define clearly whether lipoic acid or some other related small thiol compound is actually involved in ATP synthesis and to outline experimental methods in such detail that workers in all laboratories can obtain similar results. Until then, final decisions cannot be made concerning Griffiths' proposed oleoyl cycle. In the meantime, his studies have greatly stimulated reevaluation of chemical steps in ATP synthesis and have provided a basis for experimental examination of several reactions catalyzed by the ATPase complex.

I. Exchange Reactions

There are four exchange reactions (9A–9D) catalyzed by mitochondrial ATPase that have been described in detail. These are considered separately here, but all must be tied together when considering proposed reaction mechanisms for the ATPase/ATP synthase. Until recently, these exchange reactions were treated largely as catalytic properties of mitochondrial ATPase that could be conveniently used for the characterization of various enzyme preparations. Further use of these reactions as probes for studying the overall mechanism of oxidative phosphorylation has been primarily the work of Boyer and his co-workers (Ernster *et al.*, 1977; Kayalar *et al.*, 1977; Rosing *et al.*, 1977; Hackney and Boyer, 1978). These workers have clearly illustrated the usefulness of the exchange reactions in providing important insights into the chemical events involved in phosphorylation, which in turn place a number of restrictions on any mechanisms proposed from other lines of inquiry. For example, results of their exchange studies are not consistent with one of the earlier-postulated mechanisms proposed by Mitchell (1974) regarding formation of a phosphorylium intermediate in ATP synthesis. It is important to state, however, that studies of exchange reactions do not by themselves lead to a unique model for ATP synthesis.

Analyses of nucleotide binding to various preparations of F_1 and of ATPase are important to studies of the exchange reactions and kinetic investigations. Nucleotide binding to ATPases was the subject of a recent review by Harris (1978) and will not be discussed here except as necessary in the consideration of these other areas.

Reaction A: ATP–P_i exchange

Several mitochondrial ATPase preparations have been described that catalyze a rapic exchange of $^{32}P_i$ with the γ-phosphate of ATP. All seem to require the presence of at least F_0, F_1, and phospholipid for maximal activity. Kagawa *et al.* (1973) were able to define many of the requirements for an enzyme capable of ATP–P_i exchange activity through reconstitution studies. Among the conclusions reached was that vesicular membrane structures were an absolute prerequisite for exchange activity, making this energized reaction process fit into current chemiosmotic theories. Furthermore, the specificity of phospholipids required in these studies suggested that unsaturated fatty acyl groups were essential. Presumably this reflected the need for optimal membrane fluidity or lipid stabilization of the enzyme in a conformation suitable for exchange.

Sadler *et al.* (1974) subsequently used a lysolecithin treatment of beef heart electron transport particles to prepare an enzyme fraction greatly enriched in ATP–P_i exchange activity and showed that this fraction was also enriched in ATPase. The presence of lysolecithin in this preparation and studies of reaction properties suggested that closed vesicles were not needed for the exchange reaction.

Ryrie (1975) has reconsituted a highly purified, active yeast ATP–P_i exchange complex containing the 9 polypeptides of ATPase and added phospholipid. In a more recent study by Ryrie (1977) the enzyme was further purified and an active ATP–P_i exchange complex was reconstituted with the ATPase containing only 8 detectable polypeptide subunits. The exchange reaction was sensitive to uncouplers and other energy-transfer inhibitors (oligomycin and DCCD), but not to electron-transport inhibitors.

Serrano *et al.* (1976) reported the isolation of a purified ATPase from beef heart (with 9 major and 3 minor bands on SDS gels), which again required added phospholipid for maximal rates of both ATP hydrolysis and ATP–P_i exchange. Stiggall *et al.* (1978) have reported an additional method of purifying a beef heart mitochondrial complex capable of catalyzing the ATP–P_i exchange reaction. Their ATPase preparation, referred to as Complex V, exhibits similarity in both size and number of polypeptide subunits to that of Serrano *et al.* (1976); however, an exchange reaction could be catalyzed by this complex without added phospholipid and apparently in the absence of vesicles. While enzyme activity of this preparation was stimulated by added phospholipid, it appears that in the absence of vesicles, the exchange reaction was inhibited by uncouplers [including picrate, which has been shown to be a membrane impermeable uncoupler (Hanstein and Hatefi, 1974a,b)].

It appears from these studies that the only proteins needed for ATP–P_i exchange activity are those in the ATPase complex, that vesicles are not an absolute requirement for exchange, and that transmembrane proton or other gradients are not necessary for uncoupler sensitivity.

By way of comparison, Yoshida *et al.* (1975) have performed careful reconstitution studies with purified ATPase from a thermophilic bacterium and phospholipid vesicles that suggest only 8 protein subunits listed in Table I are required to reconstitute an active ATP–P_i exchange.

The above studies indicate great progress in defining general reaction requirements for ATP–P_i exchange, but they do leave important questions unanswered regarding both the number and identity of polypeptides involved directly in the reaction, the reaction mechanism, and the participation of phospholipids in enzyme structure or catalysis.

A different approach to the study of the ATP–P_i exchange reactions which provides some insight into these questions has been pursued by Hyams and Griffiths (1978). They reported a significant stimulation of the ATP–P_i exchange reaction by addition of dihydrolipoate, oleoyl lipoate, oleoyl-CoA, or oleoyl phosphate to ox heart submitochondrial particles. Studies of inhibitor and uncoupler sensitivities of this stimulation showed differential effects dependent on the oleoyl derivative added. These results are summarized in Table II (reactions 9A, a, b, c, and d). The stimulation of exchange rates was interpreted by the investigators as supportive evidence that the lipoic acid and oleoyl derivatives used are reaction intermediates or are capable of generating intermediates necessary for ATP–P_i exchange reactions. The pattern of inhibition by the uncouplers and inhibitors tested was used to infer an order of reaction intermediates and to define several specific modes of inhibitor and uncoupler action. These results also appear to indicate that vesicles are not required in the exchange reactions and that sites exist on the enzyme itself for uncoupler interaction.

The sensitivity of the ATP–P_i exchange in submitochondrial particles to uncouplers is equal to that of oxidative phosphorylation (Rosing *et al.*, 1977). The loss of ATP–P_i exchange induced by uncouplers seems to parallel the increase in ATPase activity noted. This exchange reaction is also inhibited by arsenate or by removal of ADP from the medium (Kayalar *et al.*, 1977).

Reaction B: ADP–ATP exchange

The exchange reaction illustrated in Table II, reaction 9B, in which an exchange between ADP and ATP is catalyzed by mitochondrial ATPase is the least studied of the exchange reactions. It appears that the enzyme requirements for catalysis of this reaction and the inhibitors blocking this exchange are similar to those seen with the ATP–P_i exchange reaction; i.e., F_1 and F_0 plus phospholipids are required for maximal activity (Rosing *et al.*, 1977; Kayalar *et al.*, 1977).

Reaction C: ATP–HOH exchange

Mitochondrial ATPase is able to catalyze the rapid incorporation of water oxygens into the terminal phosphate of ATP. This reaction was first described in 1954 (Boyer *et al.*, 1954; Cohn and Drysdale, 1955) and has been termed the ATP–HOH exchange. Recent studies in Boyer's laboratory have extended our understanding of this reaction and demonstrated more clearly its involvement with oxidative phosphorylation.

In studies to date, the ATP–HOH exchange requires an intact, membrane-bound ATPase. Studies have usually been performed with submitochondrial particles (Cross and Boyer, 1975; Rosing *et al.*, 1977; Kayalar *et al.*, 1977). No reports of simpler requirements have appeared. It would be interesting to test this reaction with the Complex V preparation of Stiggall *et al.* (1978), which can catalyze ATP–P_i exchange apparently without vesicles.

Two recent papers (Rosing *et al.*, 1977; Kayalar *et al.*, 1977) provide evidence that the ATP–HOH exchange measured is actually the sum of two exchange reactions that give the same net result. These are referred to as the intermediate ATP–HOH exchange and the medium ATP–HOH exchange.

The intermediate ATP–HOH exchange is sensitive to arsenate, which suggests that this exchange is dependent on the binding of P_i to the catalytic site (Kayalar *et al.*, 1977). The medium ATP–HOH exchange is not sensitive to arsenate and is postulated to occur from ATP binding, hydrolysis, and reformation at the same catalytic site without P_i release to the medium (Kayalar *et al.*, 1977).

The total ATP–HOH exchange can be inhibited by uncouplers such as dinitrophenol (Cohn and Drysdale, 1955), CCCP (Cross and Boyer, 1975), and S-13 (Rosing *et al.*, 1977). The sensitivity of this exchange to uncouplers seems to parallel the sensitivity of the ATP–P_i exchange discussed earlier, and inhibition appears to be concomitant with stimulation of ATPase activity.

ATP–HOH exchange can also be inhibited by oligomycin (Cross and

Boyer, 1975). Titration with oligomycin shows parallel sensitivities of the ATP–H_2O exchange, the ATP–P_i exchange, and net oxidative phosphorylation.

Kayalar *et al.* (1977) have also demonstrated that both the intermediate and medium ATP–HOH exchanges are dependent upon the presence of added ADP. Addition of an ATP regenerating system to keep ADP levels low prevents this exchange.

A comparison of the relative rates of the exchange reactions shows that the total ATP–HOH exchange proceeds several times faster than the ATP–P_i exchange, but usually slower than the P_i–HOH exchange (Green, 1974; Cross and Boyer, 1975). A satisfactory explanation of this phenomenon has not been proposed.

Reaction D: P_i–HOH exchange

Mitochondrial ATPase can also catalyze the rapid exchange of oxygen between water and inorganic phosphate. This reaction was also described relatively early (Cohn, 1953), but development of its potential in understanding ATP synthesis has largely been in the recent work of Boyer and his colleagues (Boyer *et al.*, 1973; Cross and Boyer, 1975; Rosing *et al.*, 1977; Kayalar *et al.*, 1977; Hackney and Boyer, 1978).

P_i–HOH exchange can be inhibited by oligomycin with sensitivity closely paralleling that of the other exchange reactions discussed and also of ATP hydrolysis (Cross and Boyer, 1975). Yet, titration of the P_i–HOH exchange with uncouplers shows that significant P_i–HOH exchange remains in the presence of an amount of uncoupler that causes complete inhibition of the ATP–P_i exchange, ATP–HOH exchange, and net oxidative phosphorylation (Boyer *et al.*, 1973).

As the total P_i–HOH exchange is completely blocked by oligomycin but only partially inhibited by uncouplers, it was suggested that an energy source not sensitive to uncouplers was required for P_i–HOH exchange. Cross and Boyer (1975) and Rosing *et al.* (1977) have shown rather conclusively that the energy for P_i–HOH exchange comes from hydrolysis of ATP. The total P_i–HOH exchange is directly proportional to the amount of ATP hydrolysis (Rosing *et al.*, 1977), indicating that oligomycin sensitivity of this exchange is due to inhibition of ATPase activity. Further evidence of the requirement for ATP hydrolysis is shown by glucose-hexokinase inhibition of the total P_i–HOH exchange as well as inhibition of this exchange when formation of ATP by adenylate kinase is blocked (Rosing *et al.*, 1977). These studies enabled Boyer and his co-workers to dissect the total P_i–HOH exchange into two component parts, the medium and the intermediate P_i–H^2O exchanges (see also Russo *et al.*, 1978).

The medium P_i–HOH exchange is that portion of the total P_i–HOH exchange that exhibits uncoupler sensitivity like that of ATP–P_i exchange and net oxidative phosphorylation. This suggests that uncoupler prevents the binding of medium P_i in a mode competent for exchange. Thus, arsenate would also be predicted to block medium P_i–HOH exchange by preventing binding of medium P_i to the catalytic site. This has been observed experimentally (Kayalar *et al.*, 1977). In addition to medium P_i–HOH exchange being dependent on the hydrolysis of ATP, there is also a requirement for ADP (Kayalar *et al.*, 1977).

The intermediate P_i–HOH exchange is defined as the uncoupler-insensitive portion of the total P_i–HOH exchange. It is postulated to occur from exchange of water oxygens into the terminal phosphate of ATP subsequently released as P_i by hydrolysis (Rosing *et al.*, 1977). Thus inhibition of ATP hydrolysis would account for the sensitivity of this exchange to oligomycin as well. Arsenate does not inhibit this exchange, presumably because the intermediate P_i–HOH exchange is not dependent on binding of P_i to the catalytic site (Kayalar *et al.*, 1977).

Ernster *et al.*, (1977) have reconstituted an ATPase from purified F_0 and F_1 components and showed that this complex could catalyze an oligomycin-sensitive uncoupler-insensitive P_i–HOH exchange. Brief mention was also made that this exchange could be catalyzed by F_1 alone in the presence of low concentrations of ATP.

Boyer and his co-workers have proposed an alternating-site mechanism that accounts for all the exchanges discussed above. The mechanism postulates that energy input from reactions of the electron transport chain is utilized to promote productive P_i binding to one site on the ATPase simultaneous with the release of bound ATP from a second site. This mechanism is discussed more fully by Kayalar *et al.* (1976b, 1977) and Hackney and Boyer (1978) and will not be presented here. Studies by Moudrianakis and co-workers have led to similar conclusions (Moudrianakis and Tiefert, 1976).

VII. Kinetics of ATPase Reactions

A. Presteady-State Studies

Only limited information is available on the presteady-state kinetics of mitochondrial ATPase. This is unfortunate, as the studies to date are of considerable interest. For example, Recktenwald and Hess (1977b) have reported a 100-msec lag preceding steady-state hydrolysis of ATP with yeast F_1. It was postulated that this lag represents an ATP-induced conformational transition, which may be an integral part of the catalytic mechanism. While a multiple-conformer postulate seems reasonable (Boyer, 1977), confirmation awaits other corroborating evidence.

In contrast to mitochondrial studies, rather extensive rapid kinetic analysis has been applied to studies of photophosphorylation by intact thylakoids. The results probably pertain quite closely to oxidative phosphorylation. Several workers (Boyer *et al.*, 1975; Smith *et al.*, 1976; Harris and Crofts, 1978) reported rapid formation of [β-^{32}P]ADP in $^{32}P_i$ rapid labeling studies with chloroplast thylakoids. This suggested that perhaps AMP was the initial nucleotide phosphorylated and provided support for the transphosphorylation hypothesis of Roy and Moudrianakis (1971b).

In contrast, Yamamoto and Tonomura (1975) and Vinkler *et al.* (1978) reported that ADP is the initial nucleotide phosphorylated by thylakoid preparations exposed to light or an acid–base transition. More than 4 mol of [^{32}P]ATP per mole of CF_1 can be formed before the detection of a slight amount of [β-^{32}P]ADP (Vinkler *et al.*, 1978).

In apparent agreement with these studies, Hill and Boyer (1967) have observed that ADP is the initial nucleotide acceptor of P_i during oxidative phosphorylation. Other important results that have come from a similar approach are that (1) tightly bound ATP is a likely intermediate in net photophosphorylation (Smith and Boyer, 1976); (2) there exists approximately 1 mol of P_i and ADP per mole of CF_1 "committed" to ATP synthesis during photophosphorylation (Smith and Boyer, 1976); and (3) a lag of 3–7 msec precedes the steady-state rate of ATP synthesis during photophosphorylation (Smith *et al.*, 1976).

B. Steady-State Kinetics of ATPase

1. *General*

As noted, the ATP hydrolysis reaction is the fundamental catalytic property of mitochondrial ATPase that has allowed its subsequent purification and characterization. Consequently, many studies have appeared detailing the substrate specificities, cation requirements, inhibitor effects, and kinetic constants obtained for a wide range of mitochondrial ATPase preparations. These were summarized by Pedersen (1975), and the reader is referred to this review for details of earlier work. In general, most early studies reported that ATP hydrolysis obeyed typical Michaelis–Menten-type kinetics (Pedersen, 1975). However, negative cooperativity with a Hill coefficient of 0.5 has been observed by Ebel and Lardy (1975), and subsequent studies have confirmed and extended these findings (Schuster *et al.*, 1975, 1976; Pedersen, 1976a).

Pedersen (1976a) noted biphasic ATPase kinetics in Tris-Cl buffer catalyzed by both membrane-bound and purified F_1 enzyme. Eadie–Hofstee plots of the kinetic data yield two values of K_m (0.068 mM and 0.21 mM) for ATP hydrolysis in the absence of activation anion. Detailed studies by Schuster *et al.* (1975, 1976) on the kinetics of ATP hydrolysis interpreted by the concepts of Cleveland (1963) lead to the conclusion that the ATPase contains both regulatory and catalytic sites that may bind ATP or other nucleotides.

2. *Anion Effects on ATPase Kinetics*

It is well known that oxy anions, such as bicarbonate, bisulfite, borate, and pyrophosphate, can stimulate the rate of ATP hydrolysis several fold. This effect has been noted for membrane-bound, detergent-solubilized, and purified F_1 ATPases from mitochondrial as well as nonmitochondrial sources (Pedersen, 1975).

The functional significance of anion activation remains obscure, although recent studies have provided some additional insight into a possible mechanism. The suggestion that the activating anion participates directly in catalysis (Mitchell and Moyle, 1971) currently seems less likely than an alternative postulate of anion-induced conformational change in the ATPase enzyme (Ebel and Lardy, 1975). Studies have established that (1) the activation effect is usually greater for F_1 than for the ATPase complex (Ebel and Lardy, 1975); (2) the activation is relatively specific for ATP hydrolysis (Pedersen, 1976a); and (3) the presence of activating anion destroys the negative cooperativity seen for ATP hydrolysis (Ebel and Lardy, 1975).

More extensive studies of HCO_3^- activation compared effects on ATP hydrolysis, ATP–P_i exchange, and ATP-dependent transhydrogenation (Pedersen, 1976a). Only the ATPase activity was significantly enhanced by bicarbonate. Furthermore, bicarbonate-activated F_1 exhibited kinetic constants resembling rather closely those obtained for the membrane-bound ATPase. This prompted the postulate of two possible conformers of F_1; the form with bound HCO_3^- was presumed to resemble F_1 attached to the membrane sector.

Recktenwald and Hess (1977a) have proposed a simple kinetic model based on anion-induced conformational changes to fit their data obtained with yeast F_1. In this model, three units on F_1 (F_1-1 and/or F_1-2) each possess a regulatory site affecting the conformation of a catalytic site with no cooperativity predicted between the individual units. Anions or ATP can bind to the regulatory site(s) and cause changes in enzyme structure, which then yield the more active catalytic site(s). This model predicts a maximum of six nucleotide binding sites on F_1; up to five have been detected experimentally (Harris *et al.*, 1973; Harris, 1978).

Pedersen (1976b) has shown that activation of ATPase can be reversed by sulfhydryl reagents only to the extent of the original activation. Titration of F_1 with *p*-chloromercuribenzoate (CMB) showed no HCO_3^--dependent increase in accessible sulfhydryl groups, prompting the conclusion that a conformational change is induced by HCO_3^- close to a CMB-sensitive site.

The membrane-bound form of ATPase from *R. rubrum* chromatophores is dependent on either Ca^{2+} or Mg^{2+} for ATPase activity, while the F_1 form is strictly Ca^{2+} dependent and even inhibited by Mg^{2+} (Webster *et al.*, 1977). However, F_1 in the presence of activating anion exhibits the same divalent cation specificity as the membrane-bound form.

Additional studies of this type will be of great interest since the question of conformational changes is of central importance to at least one of the current theories of oxidative phosphorylation (Boyer, 1977). Schuster (1977) has observed significant stimulation of ATPase activity by organic solvents and reports a conversion of nonlinear Lineweaver–Burk plots to linear forms in the presence of 20% methanol. Such results may extend the postulate of Pedersen (1976a) to include organic solvents as well as activating anions in a proposed conformational shift of F_1 to a form more closely resembling the native membrane-bound ATPase.

3. *Order of Substrate Binding*

Schuster *et al.* (1977) have performed detailed kinetic studies on the synthesis of ATP with submitochondrial particles testing a possible mandatory binding order of substrates. Evidence for an ordered Bi-Uni reaction mechanism (neglecting the product water) was obtained with P_i binding first, followed by ADP, and subsequent ATP release. Kayalar *et al.* (1977) have disputed these findings and reported a random order of P_i and ADP addition based on exchange reactions catalyzed by submitochondrial particles. This question will have to be settled by further experimentation.

VIII. Relationship of Uncouplers to the Mode of Action of ATPase

It has been establised for many years that a large family of molecules with chemically diverse structures can disrupt coupling between electron flow and ATP synthesis (for reviews, see Hanstein, 1976; Green, 1977). These uncoupler molecules suppress all known (ATP dependent) coupled processes in the mitochondria; active transport of ions, energized transhydrogenation, reverse electron flow, and ATP synthesis. Uncouplers were initially postulated to act upon the terminal steps of oxidative phosphorylation by disruption of high-energy chemical bonds in a manner analogous to the inhibition of glycerol phosphate dehydrogenase by arsenate. However, the ascendancy of the chemiosmotic model for ATP synthesis with its emphasis on proton gradients required an alternative explanation for uncoupler action. This was provided by a large body of information showing that uncouplers in general can act as protonophores and can thereby uncouple by discharging pH gradients. Work with other cation transporters (e.g., valinomycin and nigericin) further indicated a general correlation between the ability of molecules to discharge membrane potentials and their ability also to uncouple ATP synthesis (Montal *et al.*, 1970; Kessler *et al.*, 1976).

Recently this generality has been questioned by observations indicating that some very effective uncouplers are membrane impermeable and do not act as cation transporters. For example, trinitrophenol and tetraphenyl borate are both highly polar molecules that are effective uncouplers only with broken or inverted mitochondria (Hanstein and Hatefi, 1974a; Phelps and Hanstein, 1977). With inverted membranes, these work as classical uncouplers in all energy-linked mitochondrial processes. Hanstein and Hatefi (1974b) and also Kurup and Sanadi (1976, 1977) provided a rationale for this behavior in binding studies using 2-azido-4-nitrophenol (ANP), the photoaffinity labeling analog of dinitrophenol. Their studies demonstrate a specific binding site for ANP on a 29,000 MW protein that copurifies with the beef heart mitochondrial ATPase. Binding of ANP can be competitively inhibited by the addition of a wide variety of protonophoric uncouplers, such as dinitrophenol, S-13, CCCP, and PCP, as well as by sodium azide and the membrane impermeant uncouplers trinitrophenol and tetraphenyl borate. This indicates that there is potentially a common site of action for these uncouplers on ATPase (Phelps and Hanstein, 1977; Hanstein and Hatefi, 1974b).

An additional indication that a site of uncoupler action may be directly on the ATPase comes from studies by Sadler *et al.* (1974) and Stiggall *et al.* (1978) which present evidence that the ATP–P_i exchange reaction can be catalyzed using vesicle-free preparations of purified ATPase. These preparations do contain some bound cholate and phospholipid, which may play a role that is difficult to evaluate, but they do not appear to be part of closed membrane structures. Still, the exchange reaction is inhibited by the uncouplers CCCP and DNP and also by valinomycin plus nigericin. Clearly, these uncouplers can have non-membrane-dependent effects on ATPase. The isolation of an uncoupler-resistant mutant of yeast has been reported (Griffiths *et al.*, 1973). The exact changes resulting from this mutation have not been documented, but it is difficult to rationalize a mutation of this type having a general effect on the ion permeability of mitochondrial membranes. Change in a protein analogous to the uncoupler binding protein would seem most likely to yield this result.

Yet another suggestion of a direct uncoupler action on ATPase is the effects of classical uncouplers on some chemical reactions catalyzed by solubilized ATPases (see Table II). Dinitrophenol and 1799 both stimulate ATPase-catalyzed hydrolysis of oleoylphosphate (Hyams *et al.*, 1977). Oleoyl-CoA stimulation of ATP–P_i exchange is inhibited by valinomycin plus nigericin; oleoyl-*S*-lipoate stimulation is inhibited by these two and also by FCCP, DNP, etc. Returning to a view that uncouplers may react at enzyme sites is also consistent with kinetic studies by Kayalar *et al.* (1976a) showing DNP effects on increasing the K_m for ADP during the reactions of oxidative phosphorylation.

These observations all seem to suggest a mode of uncoupling associated with the ATPase protein. This view leads back to the proposal of uncoupling by chemical "short circuit" as in the early chemical hypotheses. Such studies must not be interpreted to indicate that uncouplers do not also have important direct effects on membrane permeabilities. It is probable that the site or sites of action of each uncoupler molecule may have to be evaluated separately. Phelps and Hanstein (1977) have proposed a mechanism of uncoupler action in which uncoupler binds to the uncoupler binding site on the matrix side of the ATPase membrane complex, undergoes protonation by some H^+ transporting components of the complex, and then discharges the potential gradient driving ATP synthesis by transferring H^+ to water. While this mechanism is written for membrane-bound enzyme, the same rationale could be used to postulate uncoupler-mediated discharge of an H^+-bound intermediate in nonmembranous preparations like those of Stiggall *et al.* (1978).

IX. Ion Translocation by Mitochondrial ATPases

Since the initial observations that hydrolysis of ATP by ATPases bound to mitochondrial particles results in a transmembrane movement of protons, much effort has been directed toward determining the mechanism of the translocation process and identifying what chemical groups are involved (Racker, 1976; Green, 1974; Okamoto *et al.*, 1977; Kozlov and Skulachev, 1977; Criddle *et al.*, 1977b). It was demonstrated that removal of the F_1 unit from the membrane complex made the membranes permeable to H^+ (Ernster, 1975). Permeability could then be reduced by reattachment of F_1 or by addition of inhibitors, such as oligomycin, DCCD, or venturicidin (Hinkle and Horstman, 1971; Racker, 1972; Shchipakin *et al.*, 1976; Okamoto *et al.*, 1977). This was interpreted to mean that a "proton-permeable channel" existed through the F_0 sector that could be blocked by either the F_1 unit or by these inhibitors (Racker, 1972). Experiments with enriched F_0 preparations in reconstituted vesicles confirmed these observations (Okamoto *et al.*, 1977).

The question of whether the entire F_0 unit was required to allow proton permeability or whether some component part of the complex could function in this capacity was approached by preparation of subfractions that could be tested for ionophoric activity. It was found that the chloroform : methanol-soluble proteolipid fractions from chloroplasts and also beef heart mitochondria act as ionophores in increasing conductance through lipid membranes (Racker, 1976; Blondin, 1974; Blondin *et al.*, 1971; Southard *et al.*, 1974). The specificity of this conductance change and its relation to ATP hydrolysis were questionable at this conductance increase was not sensitive to any inhibitors of oxidative phosphorylation, and it was demonstrated that a large number of membrane proteins could also have appreciable effects on membrane conductance (Schubert *et al.*, 1977).

Specificity was, however, clearly demonstrated in studies with the yeast ATPase proteolipid component. Purified preparations of the F_0-3 proteolipid fraction could efficiently discharge photoelectric potentials induced across lipid membranes by bacteriorhodopsin (Criddle *et al.*, 1977b). This ionophoric activity in the lipid membrane was blocked by oligomycin, DCCD, and ethidium bromide, reagents that commonly inhibit ATPase activity and block ATP-driven H^+ translocation. Ionophoric activity was further related to the *in vivo* process of oxidative phosphorylation when it was demonstrated that mutant yeast strains resistant to oligomycin inhibition yielded F_0-3 with ionophoric activity also resistant to oligomycin.

These studies and studies with liposome vesicles (Konoshi *et al.*, 1979) make it clear that the proteolipid F_0-3 component of the mitochondrial ATPase can effect the transport of a net flux of protons through lipid membranes in response to a potential gradient. Such a discharge of potential through mitochondrial membranes, if not linked to some other energy-requiring step, would result in an inefficient loss of energy. H^+ transport must therefore be linked to additional steps of ATP synthesis, making the question of the mechanism of H^+ transport by F_0-3 one of the central problems in energy transduction.

Several diverse observations have provided information about how F_0-3 functions in H^+ transport. First, transport of H^+ across lipid membranes by purified F_0-3 has a pH dependence which appears as a simple titration curve with pK_a near 7.3 (Johnston *et al.*, 1977; Criddle *et al.*, 1979). At high pH the protein is a very poor transporter. Also, oligomycin inhibition of ATPase activity (i.e., binding to F_0-3) is pH dependent. At low pH, ~ 5.5, oligomycin does not inhibit appreciably while at pH 8.5 inhibition by a comparable level of oligomycin is virtually complete (Johnston *et al.*, 1977). Titration curves showing pH dependence of oligomycin inhibition of ionophore activity of purified F_0-3 in liposomes or in planar membranes show dependence on pH nearly identical to that noted for ATPase activity (Johnston and Criddle,

1978). Also, recall that F_0-3 isolated from oligomycin-resistant mutants (Class B) is found as monomers rather than predominantly as oligomers in wild-type enzyme (Tzagoloff *et al.*, 1976; Groot Obbink *et al.*, 1976). When purified F_0-3 is incorporated into liposomes, exposed to various pH levels, and then analyzed by SDS gels, the results show that preparations incubated at pH 6–8 yield predominantly monomer, but above pH 8, oligomeric protein is recovered (Criddle *et al.*, 1979). A correlation thus seems to exist relating oligomer formation and the ability to bind oligomycin as compared to H^+ binding and monomer formation. The suggestion that oligomycin may be interacting most effectively with oligomeric F_0-3 is strengthened by observations that addition of one oligomycin per 4–8 F_0-3 molecules in liposome preparations is sufficient to cause 50% inhibition of ionophoric activity (Criddle *et al.*, 1979). It has previously been shown that addition of 1–2 DCCD per ATPase complex was also sufficient to block ATPase activity *in vivo*, though as many as 6–8 F_0-3 molecules are probably present per enzyme complex.

These observations indicate that direct H^+ binding to the F_0-3 protein is a central step in coupled ATP synthesis. Moreover, the two aggregation states of F_0-3 suggest an equilibrium existing between two enzyme forms, the presence of H^+ favoring formation of what is measured *in vivo* as monomer and oligomycin favoring formation of oligomer.

An additional component influencing oligomycin inhibition is K^+. Addition of K^+ greatly enhanced binding of oligomycin to ATPase (presumably to oligomer) and correspondingly increased inhibition. The following equilibrium reaction has been postulated to explain these results (Johnston *et al.*, 1977).

$$n(\text{H}^+\text{-subunit F}_0\text{-3}) + m\text{K}^+ \xrightleftharpoons{\text{oligomycin}} n\text{H}^+ + (m\text{K}^+\text{-subunit F}_0\text{-3})_n$$

As written, this reaction predicts that K^+ may have a major effect on proton transport by F_0-3 and even suggests that K^+ may be transported by F_0-3 preparations under the appropriate conditions. Potassium ion-transport capability can indeed be noted for F_0-3 or some small contaminant in F_0-3 preparations, but only when F_0-3 protein is present in sufficient amounts to form oligomers (Criddle *et al.*, 1979). This property first became evident when the number of F_0-3 molecules per liposome was varied during studies of H^+ transport. With one F_0-3 per liposome (calculated assuming an average molecular weight of 2×10^6 per liposome) protons could be transported into K^+-loaded vesicles in exchange for K^+ only if valinomycin was added to facilitate transport of the potassium ions (see Fig. 1). No change in the external pH was noted without valinomycin. In contrast, if more F_0-3 was added per liposome, then it was no longer necessary to add valinomycin

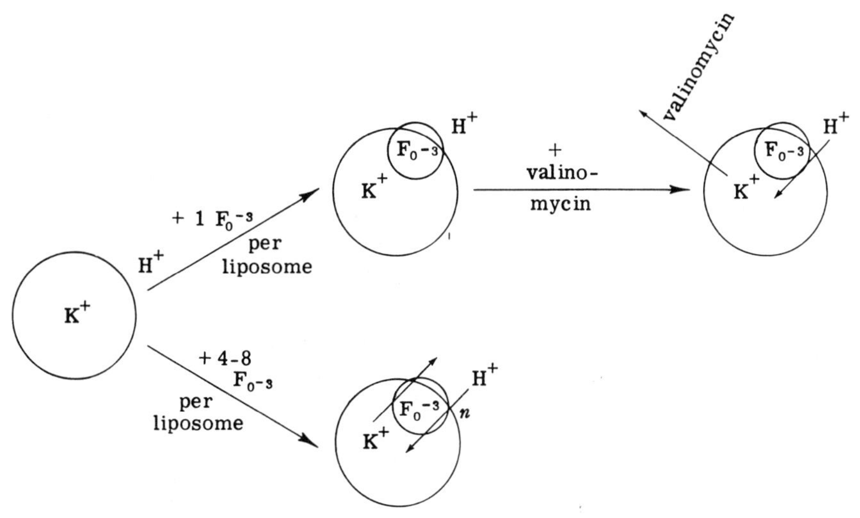

FIG. 1. H^+ and K^+ transport through liposome membranes mediated by F_0-3. Potassium-loaded liposomes prepared from mitochondrial phospholipid or soybean azolecithin are prepared with an average of less than 1 or with 6 F_0-3 protein molecules per liposome. With only one F_0-3 per liposome, the addition of a potassium transporter, such as valinomycin, is required to observe uptake of H^+.

to observe the K^+–H^+ exchange (see Fig. 1). The requirement for K^+ is very specific, as Li^+, Na^+, and Cs^+ are not transported and do not substitute nor compete with K^+, although Rb^+ does. Kinetic studies of this process indicate a carrier-mediated K^+ transport. Again, inhibitors such as oligomycin and DCCD completely block K^+ transport and K^+–H^+ exchange.

When very low levels of F_0-3 are present per vesicle, H^+ must be transferred through the membrane via some type of carrier process. With multiple subunits, such conclusions are not so clear. It appears, from pH dependence studies, oligomycin titration studies, and protein concentration studies, that K^+ transport probably requires some aggregated protein structure. For example, again 50% inhibition of K^+ transport is noted when only one oligomycin molecule is added for every 4–6 F_0-3 molecules. Clearly one oligomycin can block transport activity of more than one F_0-3. These results suggest that a channel formation requiring multiple subunits may be necessary to move K^+ through the membranes. Recently, Scott *et al.* (1978) reported the isolation of a neutral chloroform: methanol-soluble proteolipid from beef heart mitochondrial membranes that is also capable of enhancing K^+ transport across lipid membranes. While this is not a homogeneous preparation, it is likely that their observed activity is due to the F_0-3 proteolipid.

X. Linking Ion Translocation to the Chemical Formation of ATP

Consideration of Table II, reactions that can be catalyzed by ATPase, raises the question of whether these can be put together into a logical sequence that would result in synthesis of ATP. It is evident that components of purified ATPase can carry out both the net transmembrane movement of ions in response to a chemical or electrical potential and also the catalysis of a series of chemical reactions. Inhibitor studies and studies with mutants indicate that both sets of reactions are related to oxidative phosphorylation. Somewhere within this complex the two processes are brought together; i.e., elements of the energy transducing system appear to be contained wholly within the complex. The accumulated evidence to date does not allow explicit interpretation of how this process takes place, but guidelines from ion translocation studies, chemical reactions and kinetic analysis do provide a means for ordering the observed reactions into an outline of consistent steps for ATP production.

A. Proton Movement

The starting point for this consideration will be high-energy protons generated by way of the electron transport process. The form this energy takes can be either a proton gradient as suggested by Mitchell (1977) or a high charge density generated within the lipid bilayer as suggested by Williams (1978).

The "high-energy protons" can be accepted by F_0-3 of the ATPase complex, probably directly, but it is not currently possible to exclude the presence of additional carriers. It is clear that F_0-3 does not need any additional carrier for translocation of protons through artificial membranes (Criddle *et al.*, 1977b; Konishi *et al.*, 1979), and some direct involvement of F_0-3 in the proton translocation steps of oxidative phosphorylation *in vivo* seems well established through the studies comparing properties of the protein from wild-type and inhibitor-resistant mutants of yeasts.

If protons do bind to ATPase F_0-3 as a result of an electrochemical gradient that drives ATP synthesis, then it is clear that this binding must be coupled to some secondary events that can trap the potential energy of those protons. Simple transport of H^+ across the membrane would merely cause a loss of potential. The transport step therefore must be linked directly or indirectly to formation of a high-energy bond (Boyer *et al.*, 1977; Ernster, 1975).

A possible means for maintaining the energy would be via an induced conformational change in the H^+ binding protein, which results in activation of some additional site within the ATPase complex. Such conformational changes in coupled oxidative phosphorylation have been proposed previously (Boyer, 1977), but it has not been possible to evaluate these proposals readily. However, studies of proton binding to F_0-3 now show that protonation does cause major changes in structure of at least this one protein. This was best reflected with *in vitro* studies indicating a change in the state of polymerization by binding H^+. However, changes described earlier in pH dependence of oligomycin binding, potassium transport studies, examination of oligomycin-resistant mutants, etc., also provide evidence for a major conformational change in F_0-3 by proton binding (Johnston *et al.*, 1977; Johnston and Criddle, 1978). These studies do not establish that F_0-3 structural changes can be equated directly to those postulated in the conformational hypothesis, but this hypothesis becomes somewhat more attractive when changes in a specific H^+ acceptor protein can be described.

B. Formation of High-Energy Chemical Bonds

Several models exist by which a change in protein conformation could retain the high-energy state. These have again been discussed in general terms by Boyer *et al.* (1977) in his conformational hypothesis for ATP synthesis. With the knowledge of F_0-3 action it is now possible to make this type of hypothesis somewhat more specific. Binding of protons and a subsequent conformational (possibly oligomer $\leftrightarrow$ monomer) shift could change the environment of a reactive group in the F_0 portion of the enzyme on or near F_0-3. If we postulate the reactive group to be a thiol (based on the chemical reactions of Table II) (Griffiths, 1976b; Criddle *et al.*, 1977a; Scharf *et al.*, 1979), several well defined examples of how high-energy thioacyl bond formation can occur during enzyme catalysis may be cited. For example, Angelides and Fink (1978) have studied ester hydrolysis by the thiol enzyme papain. This enzyme has been demonstrated to exist in two forms at low and high pH. The key steps for our consideration are shown in Fig. 2.

Protonation of papain at a site removed from the catalytic site causes a structural shift, rupturing the hydrogen bond between asparagine and the imidazole side chain of histidine. A proton is then extracted from —SH to establish salt linkages involving $—S^{\ominus}$, $Im^{\oplus}$, and $—CO_2^{\ominus}$, resulting in a partial negative charge on the thiol. This condition has low stability in the nonpolar interior of a protein, so that reaction with a carboxylic acid or ester derivative would proceed rapidly to form a thioacyl intermediate. While this reaction sequence in papain is normally continued via the hydrolysis of the acyl intermediate, a similar reaction in the ATPase complex would require conservation of the high-energy acyl intermediate to drive other

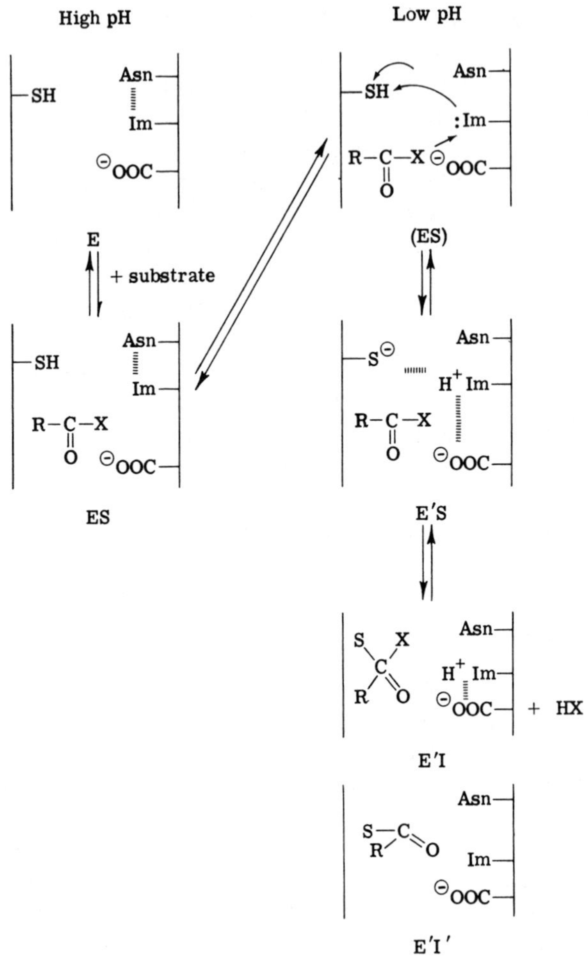

FIG. 2. Key steps in ester hydrolysis by the thiol enzyme papain according to Angelides and Fink (1978). E = enzyme; E′ = protonated enzyme; X = OH or R—CH_2—O; Asn = asparagine; Im = imidazole; $^{\ominus}$COO— = carboxyl side chain of amino acid, ≡ = hydrogen bond or salt linkage; I and I′ = reaction intermediates.

steps in ATP synthesis. This papain analogy provides a simple model for a mechanism of energy transduction linking potential gradient to formation of a high-energy bond.

Is the papain analogy a good one for consideration of ATP synthesis? It does show a well characterized mechanism whereby a link between protonation and formation of a high-energy bond can take place. It is not difficult to envision the ATPase, which catalyzes such reactions as thioester hydrolysis, thioacyl transfer, and coenzyme A-linked ATP synthesis, participating in such a reaction. The activated thiol moiety could involve the pantothenate moiety on subunit F_0-1 (Criddle *et al.*, 1977a), an additional protein-linked thiol (Fluharty and Sanadi, 1960; Shaikh *et al.*, 1976), or other small organic compounds, such as coenzyme A or lipoic acid (Griffiths, 1977).

Many workers have examined mitochondrial particles for protein-bound sulfhydryl groups and thioesters under varying conditions that affect oxidative phosphorylation. Sabadie-Pialoux and Gautheron (1971) reported an oligomycin- and uncoupler-sensitive increase in measured —SH groups upon addition of ADP to mitochondria in state 4, suggesting a change in these groups accompanying oxidative phosphorylation. However, Chude and Boyer (1974) were unable to detect any change in thioester concentration of mitochondrial particles with change in conditions affecting oxidative phosphorylation. This study does not eliminate the possibility that such groups may exist as intermediates. Still it must be emphasized that no direct information is available at this time to define any high energy chemical intermediate of ATP synthesis.

C. Group-Transfer Reactions

The major biochemical role of pantothenic acid derivatives in biological systems is to carry acetyl or fatty acyl groups. An example of this function is seen with the acyl carrier protein involved in fatty acid synthesis (Vagelos *et al.*, 1966). Finding pantothenate as part of the ATPase complex, therefore, suggests that transacylation steps may be important reactions catalyzed by subunits of ATPase. In fact, more than one type of thiol compound may be involved in sequential transacylation reaction steps on the enzyme. The reactions summarized in Table II support the postulate that thioacyl intermediates are formed during ATP synthesis and that oleic acid or an oleic acid ester appears to be the most probable donor acyl group involved. Both thioester hydrolysis and acyl phosphate driven ATP synthesis reactions show specificity for oleate (Scharf *et al.*, 1979; Hyams *et al.*, 1977).

An additional compound of known structure implicated by the reactions of Table II as a possible high-energy intermediate in ATP synthesis is oleoyl phosphate. This high-energy phosphate derivative could presumably be formed by a transacylation reaction between R—C(=O)—S—CoA (or more likely, protein–pantothenine—S—C(=O)—R, where R—CO_2H = oleic acid) and P_i. No direct observation has been made that this reaction is catalyzed by ATPase, though an analogous reverse reaction was shown with oleoyl phosphate plus CoA-SH yielding oleoyl-CoA and P_i (Criddle, 1977). Other methods of formation of oleoyl phosphate, though possible, lack any experimental support at this time. Direct activation of phosphate to form compounds such as thiophosphate has not been shown. Green has presented evidence for phosphate and ADP transporting proteins that may participate in activation by some as yet undertermined mechanism (Green, 1978). Alternatively, such a phosphate carrier protein could be simply a means of moving phosphate to a reaction site in a nonpolar region of the ATPase for reaction with the activated acyl moiety.

D. Formation of ATP

The terminal group-transfer reaction suggested by the chemical reactions outlined in Table II is the formation of ATP from oleoyl phosphate and ADP. Since it is possible to demonstrate this reaction with a number of ATPase preparations, the key question remaining is whether this reaction does in fact represent the *in vivo* terminal chemical step resulting in ATP synthesis (Hyams *et al.*, 1977; Johnston and Criddle, 1977). Oleoyl phosphate has not been identified as a natural component of the mitochondrial ATPase. Some other yet to be identified membrane-bound residue could serve as an acyl phosphate intermediate if such an intermediate indeed exists. The rigid specificity requirements for oleoyl phosphate in the reactions studied and the observed specific inhibitor sensitivities are, however, consistent with this being an intermediate.

XI. Electrochemical Potential-Driven Synthesis of ATP

ATP synthesis driven by large pH gradients has been demonstrated with chloroplasts (Jagendorf and Uribe, 1966) and mitochondrial (Ried *et al.*, 1966) membrane systems. Although these early studies failed to define an exclusive participation of enzyme components of the ATPase complex in this synthesis, subsequent experiments were performed using purified ATPase, reconstituted in bacteriorhodopsin-containing vesicles, to show ATP synthesis dependent upon a light-generated ΔpH (Racker and Stoeckenius, 1974). In addition, purified ATPase from thermophilic bacteria was incorporated into liposomes with no other protein components and a gradient of pH of pH plus K^+ ions was used to generate ATP (Sone *et al.*, 1976). Such experiments leave little doubt that an electrochemical potential and pH gradient can drive ATP synthesis by mitochondrial ATPase in a mechanism that seems independent of many of the reactions outlined in Table II. The means by which ATP is synthesized in these studies is not completely understood but has been widely discussed in terms of various forms of the chemiosmotic hypothesis. The recent review by Kozlov and

Skulachev (1977) summarizes much of the information generated in support of this hypothesis, so it will not be discussed further here.

The existence of the a gradient of "free H^+" in mitochondria or chloroplasts large enough to drive appreciable synthesis *in vivo* has been questioned (Ort *et al.*, 1976), but other mechanisms employing "surface H^+" and/or gradients of other ions have been proposed to account for these differences (Williams, 1978).

XII. Questions Regarding the *in Vivo* Mode of ATP Synthesis

In this review of energy-transducing ATPases, various catalytic properties of these enzymes have been discussed, but there has been no evaluation of how these fit into the *in vivo* mode of enzyme function. It seems clear that ATP can be synthesized via purified ATPase incorporated into phospholipid vesicles using energy supplied as an imposed pH or ion gradient and without evidence of high-energy chemical intermediates. On the other hand, it is also evident that purified ATPase can catalyze a number of defined chemical reactions suggestive of high-energy intermediate participation. A link between a number of these reactions and oxidative phosphorylation has been established through inhibitor and mutuant studies. Finally, kinetic studies in several laboratories and detailed studies of the exchange reactions catalyzed by ATPase have provided strong evidence for a conformationally linked alternating-site mechanism that neither requires nor excludes high-energy chemical intermediates.

It would seem from this summary that we have returned to the long-standing question of whether the chemical, chemiosmotic, or conformational hypothesis for oxidative phosphorylation is correct. Rather than considering these postulates as mutually exclusive, it should be recognized that the experimental results prompting each of these diverse proposals contributes important information about the overall integrated mechanism of ATP synthesis. The major question to be addressed, then, is: How can the total information be integrated into one consistent model?

To approach this question one may ask: Is there more than one mechanism for ATP synthesis by this complex, and if so, does one or the other of these predominate *in vivo*? To what extent do *in vitro* experiments mirror the *in vivo* synthetic process? Can results leading to apparently conflicting theories be reconciled into an integrated mechanism, or do chemiosmotic studies rule out, for example, chemical intermediates? Added to these should be the further questions of whether the *in vivo* pathways for ATP synthesis and hydrolysis are the same and whether some *in vitro* observations of ATP synthesis may reflect the reverse of a hydrolytic pathway rather than the most common *in vivo* synthetic pathway.

The postulate that separate mechanisms operate *in vivo* for the primary modes of ATP synthesis and ATP hydrolysis appears particularly attractive. Separate pathways for synthesis and degradation of key biomolecules in the cell is a well established biochemical principle. Sugars, amino acids, fatty acids, and other components of the cell employ separate mechanisms and enzymes for synthesis and degradation. One major advantage of separate pathways is greater flexibility in regulation. The sensitivity of metabolic pathways to ATP levels or "energy charge" suggests that a secondary level for regulation of ATP concentration may be necessary and would, therefore, seem to require the flexibility provided by separate mechanisms for synthesis and hydrolysis. The ATPase complex may even represent the ultimate in sophistication of this approach, a single integrated enzyme complex catalyzing both the synthesis and hydrolysis of ATP via separate mechanisms. Little direct evidence is available to support this idea, but it is consistent with studies of many kinds and provides a basis for integrating studies relating to each of the major hypotheses for ATP synthesis.

The possible existence of separate primary pathways for synthesis and degradation of ATP is indicated by many diverse findings. For example, ATP hydrolysis can be carried out using isolated F_1 or uncoupler-treated ATPase preparations. It is clear that these reactions are not simply a reversal of the entire ATP synthase mechanism, though they may represent a portion of the same reaction pathway. Also, anion activation with HCO_3^- stimulates ATP hydrolysis but not ATP–P_i exchange (Pedersen, 1976a), and low levels of thiol reagents inhibit ATP synthesis, but have little effect on ATP hydrolysis. The ATPase inhibitor protein exerts a major effect on hydrolysis but does not appear to affect ATP synthesis. Thioesterase activity can be blocked by 10 μM Hg^{2+} with no effects on the rate of ATP hydrolysis (Scharf *et al.*, 1979). Studies of the exchange reactions catalyzed by ATPase have demonstrated a pronounced change in pattern when phosphorylating submitochondrial particles are compared with nonphosphorylating particles (Hackney and Boyer, 1978; Boyer, 1978). Many other observations of this sort could be cited. Each may be explained individually in several ways, but all could be consistent with separate pathways.

It is important to note that the principle of microscopic reversibility requires that, should there be two separate pathways for the primary modes of synthesis and hydrolysis of ATP, both must be reversible. Thus, two paths for synthesis plus two for hydrolysis are defined by this concept. Two of these would represent the primary mechanisms for synthesis and for hydrolysis and the other two are simply consequences of reversal of the primary pathways. Not all components and reactions of the two pathways are necessarily separate. For example, the nucleotide binding sites are probably the same in both cases. Also, both the synthesis and hydrolysis of ATP by intact membrane-bound ATPase complex are tightly coupled to the transport of H^+ through the membrane, probably via subunit F_0-3 in each case. Still, the path of H^+ translocation need not be entirely identical for these two. The synthetic process could require steps for H^+ transport linked through chemical reaction intermediates included in the hydrolytic process that are not. Still, equilibrium considerations dictate that the primary hydrolysis pathway can be reversed directly by imposing a large ΔpH across a membrane-bound ATPase complex in the opposite direction of the proton pump. Net synthesis of ATP would then be noted without involvement of chemical intermediates. The possibility that some studies of ΔpH-induced ATP synthesis may actually be investigations of reversal of the primary hydrolytic reaction should be carefully evaluated. Observations of direct pH-induced ATP synthesis and proposed high-energy chemical intermediates are not mutually exclusive.

This rather simple concept of separation of hydrolytic and synthetic pathways may also be reconciled with exchange reaction data that have led to the alternating-site hypothesis. The water-exchange reactions, ATP–H_2O and P_i–H_2O, may be reflections of the primary hydrolytic pathway dependent on ATP hydrolysis. In contrast, ATP–P_i exchange and ATP–ADP exchange may be linked more closely to the additional steps of a primary ATP synthesis reaction.

This proposal offers many interesting predictions that could be tested in this rapidly progressing field of study.

REFERENCES

Abrams, A. (1965). *J. Biol. Chem.* **240**, 3675–3681.

Agsteribbe, E., Douglas, M., Ebner, E., Koh, T. Y., and Schatz, G. (1976). *In* "Genetics and Biogenesis of Chloroplasts and Mitochondria" (T. Bucher, W. Neupert, W. Sebald, and S. Werner, eds.), pp. 135–141. North-Holland Publ., Amsterdam.

Amzel, L. M., and Pedersen, P. L. (1978). *J. Biol. Chem.* **253**, 2067-2069.

Andreu, J. M., Albendea, J., and Munoz, E. (1973). *Eur. J. Biochem.* **37**, 505–515.

Andreu, J. M., Wrath, R., and Munoz, E. (1978). *FEBS Lett.* **86**, 1–5.

Angelides, K., and Fink, A. L. (1978). *Biochem.* **17**, 2659–2668.

Asami, K., Juntti, K., and Ernster, L. (1970). *Biochim. Biophys. Acta* **205**, 307–311.

Baird, B. A., and Hammes, G. (1976). *J. Biol. Chem.* **251**, 6953–6962.

Barnes, J., Tiefert, M. A., and Moudrianakis, E. N. (1977). *Fed. Proc., Fed. Am. Soc. Exp. Biol.* **36**, 726.

Bertagnolli, B. L., and Hanson, J. B. (1973). *Plant Physiol.* **52**, 431–435.

Blondin, G. A. (1974). *Biochem. Biophys. Res. Commun.* **56**, 97–105.

Blondin, G. A., DeCastro, A. F., and Senior, A. E. (1971). *Biochem. Biophys. Res. Commun.* **43**, 28–35.

Boyer, P. D. (1977). *TIBS* Feb., pp. 38–41.

Boyer, P. D. (1978). Personal communication.

Boyer, P. D., Falcone, A. B., and Harrison, W. H. (1954). *Nature* (*London*) **174**, 401–402.

Boyer, P. D., Cross, R. L., and Momsen, W. (1973). *Proc. Natl. Acad. Sci. U.S.A.* **70**, 2837–2839.

Boyer, P. D., Stokes, B. O., Wolcott, R. G., and Degani, C. (1975). *Fed. Proc. Fed. Am. Soc. Exp. Biol.* **34**, 1711–1717.

Boyer, P., Chance, B., Ernster, L., Mitchell, P., Racker, E., and Slater, E. (1977). *Annu. Rev. Biochem.* **46**, 955–1026.

Bragg, P. D., and Hou, C. (1972). *FEBS Lett.* **28**, 309–312.

Bragg, P. D., and Hou, C. (1975). *Arch. Biochem. Biophys.* **167**, 311–321.

Bragg, P. D., Davies, P. L., and Hou, C. (1973). *Arch. Biochem. Biophys.* **159**, 664–670.

Brand, M. D., and Lehninger, A. L. (1977). *Proc. Natl. Acad. Sci. U.S.A.* **74**, 1955–1959.

Brand, M. D., Reynafarje, B., and Lehninger, A. L. (1976). *J. Biol. Chem.* **251**, 5670–5679.

Capaldi, R. A. (1973). *Biochem. Biophys. Res. Commun.* **53**, 1331–1337.

Catall, K., Lindop, G., Knight, I., and Beechey, R. (1971). *Biochem. J.* **125**, 169–177.

Catterall, W. A., and Pedersen, P. L. (1971). *J. Biol. Chem.* **246**, 4987–4994.

Catterall, W. A., and Pedersen, P. L. (1974). *Biochem. Soc. Spec. Publ.* **4**, 63–88.

Catterall, W. A., Coty, W. A., and Pedersen, P. L. (1973). *J. Biol. Chem.* **248**, 7427–7431.

Chang, T. M., and Penefsky, H. (1973). *J. Biol. Chem.* **248**, 2746–2754.

Chude, O., and Boyer, P. D. (1974). *Arch. Biochem. Biophys.* **160**, 366–371.

Cleland, W. W. (1963). *Biochim. Biophys. Acta* **67**, 173–187.

Cohn, M. (1953). *J. Biol. Chem.* **201**, 735–750.

Cohn, M., and Drysdale, G. R. (1955). *J. Biol. Chem.* **216**, 831–846.

Criddle, R. S. (1977). Unpublished observations.

Criddle, R. S., Edwards, T., Partis, M., and Griffiths, D. E. (1977a). *FEBS Lett.* **84**, 278–282.

Criddle, R. S., Packer, L., and Sheih, P. (1977b). *Proc. Natl. Acad. Sci. U.S.A.* **74**, 4306–4310.

Criddle, R. S., Johnston, R., Packer, L., and Sheih, P. (1979). Submitted for publication.

Cross, R. L., and Boyer, P. D. (1975). *Biochemistry* **14**, 392–398.

Dahl, J. L., and Hokin, L. E. (1974). *Annu. Rev. Biochem.* **43**, 327–351.

Datta, A., and Penefsky, H. S. (1970). *J. Biol. Chem.* **245**, 1537–1544.

Davies, P. L., and Bragg, P. D. (1972). *Biochem. Biophys. Acta* **266**, 273–284.

Deters, D. W., Racker, E., Nelson, N., and Nelson, H. (1975). *J. Biol. Chem.* **250**, 1041–1047.

Doi, R. (1978). *Bacteriol. Rev.* **41**, 568–594.

Douglas, M. G., Koh, Y., Dockter, M. E., and Schatz, G. (1977). *J. Biol. Chem.* **252**, 8333–8335.

Ebel, R. E., and Lardy, H. A. (1975). *J. Biol. Chem.* **250**, 191–196.

Ebner, E., and Maier, K. (1977). *J. Biol. Chem.* **252**, 671–676.

Edwards, D. L., and Unger, B. (1978). *Fed. Proc., Fed. Am. Soc. Exp. Biol.* **37**, 1520.

Enns, R., and Criddle, R. S. (1977a). *Arch. Biochem. Biophys.* **182**, 587–600.

Enns, R., and Criddle, R. S. (1977b). *Arch. Biochem. Biophys.* **183**, 742–752.

Ernster, L. (1975). *In* "Enzymes: Electron Transport Systems" (P. Desnuelle and A. M. Michelson, eds.), pp. 253–276. Elsevier, Amsterdam.

Ernster, L., Carlsson, C., and Boyer, P. D. (1977). *FEBS Lett.* **84**, 283–286.

Esch, F., and Allison, W. S. (1978). *Fed. Proc., Fed. Am. Soc. Exp. Biol.* **37**, 1519.

Farron, F. (1970). *Biochemistry* **9**, 3823–3828.

Farron, J., and Racker, E. (1970). *Biochemistry* **9**, 3829–3836.

Ferguson, S. J., Lloyd, W. S., and Radda, G. K. (1975). *Eur. J. Biochem.* **54**, 127–133.

Ferguson, S. J., Lloyd, W. S., and Radda, G. K. (1976). *Biochem. J.* **159**, 347–353.

Fernandez-Moran, H. (1962). *Circulation* **26**, 1039–1065.

Fluharty, A., and Sanadi, D. R. (1960). *Proc. Natl. Acad. Sci. U.S.A.* **46**, 608–615.

Foster, D. L., and Fillingame, R. H. (1978). *Fed. Proc., Fed. Am. Soc. Exp. Biol.* **37**, 1520.

Frigeri, L., Galante, Y., Hanstein, W., and Hatefi, Y. (1977). *J. Biol. Chem.* **252**, 3147–3152.

Futai, M. (1977). *Biochem. Biophys. Res. Commun.* **79**, 1231–1237.
Goffeau, A., Landry, Y., Foury, F., Briquet, M., and Colson, A. (1973). *J. Biol. Chem.* **248**, 7097–7105.
Green, D. E. (1974). *Biochim. Biophys. Acta* **346**, 27–78.
Green, D. E. (1977). *TIBS* May, pp. 113–116.
Green, D. E. (1978). Personal communication.
Griffiths, D. E. (1976a). *Biochim. J.* **160**, 809–812.
Griffiths, D. E. (1976b). *In* "Genetics and Biogenesis of Chloroplasts and Mitochondria" (T. Bucher, W. Neupert, W. Sebald, and S. Werner, eds.), pp. 175–185. North-Holland Publ., Amsterdam.
Griffiths, D. E. (1977). *Biochem. Soc. Trans.* **5**, 1283–1285.
Griffiths, D. E., Houghton, R. L., and Lancashire, W. E. (1973). *In* "Biogenesis of Mitochondria" (A. M. Kroon and C. Saccone, eds.), pp. 215–224. Academic Press, New York.
Griffiths, D. E., Hyams, R. L., Bertoli, E., and Carver, M. (1977a). *Biochem. Biophys. Res. Commun.* **75**, 449–456.
Griffiths, D. E., Cain, K., and Hyams, R. L. (1977b). *Biochem. Soc. Trans.* **5**, 205–207.
Groot Obbink, D., Hall, R., Linnane, A. W., Lukins, H. B., Monk, B. G., Spithill, T., and Trembath, M. K. (1976). *In* "The Genetic Function of Mitochondrial DNA" (C. Saccone and A. Kroon, eds.), pp. 363–173. North-Holland Publ., Amsterdam.
Hachimori, A., Muramatsu, N., and Nosoh, Y. (1970). *Biochim. Biophys. Acta* **206**, 426–437.
Hackney, D. D., and Boyer, P. D. (1978). *J. Biol. Chem.* **253**, 3164–3170.
Haddock, B., and Begg. Y. A. (1977). *Biochem. Biophys. Res. Commun.* **79**, 1150–1154.
Hanstein, W. (1976). *Biochim. Biophys. Acta* **456**, 129–149.
Hanstein, W., and Hatefi, Y. (1974a). *Proc. Natl. Acad. Sci. U.S.A.* **71**, 288–292.
Hanstein, W., and Hatefi, Y. (1974b). *J. Biol. Chem.* **249**, 1356–1362.
Harris, D. A. (1978). *Biochim. Biophys. Acta* **463**, 245–273.
Harris, D. A., and Crofts, A. R. (1978). *Biochim. Biophys. Acta* **502**, 87–101.
Harris, D. A., Rosing, J., Van de Stadt, R. J., and Slater, E. C. (1973). *Biochim. Biophys. Acta* **314**, 149–153.
Hill, R. D., and Boyer, P. D. (1967). *J. Biol. Chem.* **242**, 4320–4323.
Hinkle, D. C., and Horstman, L. L. (1971). *J. Biol. Chem.* **246**, 6024–6029.
Hyams, R. L., and Griffiths, D. E. (1978). *Biochem. Biophys. Res. Commun.* **80**, 104–111.
Hyams, R. L., Carver, M. A., Partis, M. D., and Griffiths, D. E. (1977). *FEBS Lett.* **82**, 307–314.
Jagendorf, A., and Uribe, E. (1966). *Proc. Natl. Acad. Sci. U.S.A.* **55**, 170–177.
Johnston, R., and Criddle, R. S. (1977). *Proc. Natl. Acad. Sci. U.S.A.* **74**, 4919–4923.
Johnston, R., and Criddle, R. S. (1978). *Fed. Proc., Fed. Am. Soc. Exp. Biol.* **37**, 1519.
Johnston, R., Scharf, S., and Criddle, R. S. (1977). *Biochem. Biophys. Res. Commun.* **77**, 1361–1368.
Johnston, R., Criddle, R. S., Sheih, P., and Packer, L. (1979a). In preparation.
Johnston, R., Lagier, S., and Criddle, R. S. (1979b). Personal observations.
Kagawa, Y., and Racker, E. (1966). *J. Biol. Chem.* **241**, 2461–2466.
Kagawa, Y., Kandrach, A., and Racker, E. (1973). *J. Biol. Chem.* **248**, 676–684.
Kayalar, C., Rosing, J., and Boyer, P. (1976a). *Biochem. Biophys. Res. Commun.* **72**, 1153–1159.
Kayalar, C., Rosing, J., and Boyer, P. D. (1976b). *Fed. Proc. Fed. Am. Soc. Exp. Biol.* **35**, 1601.
Kayalar, C., Rosing, J., and Boyer, P. D. (1977). *J. Biol. Chem.* **252**, 2486–2491.
Kemp, A., and Out, T. A. (1975). *Proc. K. Ned. Akad. Wet., Ser. C* **73**, 143–166.
Kessler, R., Van de Zande, H., Tyson, L., and Blondin, G. (1976). *Proc. Natl. Acad. Sci. U.S.A.* **74**, 2241–2245.
Klein, G., Satre, M., and Vignais, P. (1977). *FEBS Lett.* **84**, 129–134.
Knowles, A. F., Guillory, R. J., and Racker, E. (1971). *J. Biol. Chem.* **246**, 2672–2679.

Konishi, T., Packer, L., and Criddle, R. S. (1979). *Methods in Enzymology* (in press).
Kovác, L., and Weissová, K. (1968). *Biochim. Biophys. Acta* **153**, 55–59.
Kozlov, I. A., and Mikelsaar, H. N. (1974). *FEBS Lett.* **43**, 212–214.
Kozlov, I. A., and Skulachev, V. P. (1977). *Biochim. Biophys. Acta* **463**, 29–90.
Kurup, C. K. R., and Sanadi, D. R. (1976). *J. Bioenerg.* **8**, 218–224.
Kurup, C. K. R., and Sanadi, D. R. (1977). *Indian J. Biochem. Biophys.* **14**, 188–191.
Lambeth, D. O., and Lardy, H. A. (1971). *Eur. J. Biochem.* **22**, 355–362.
Lang, R., and Racker, E. (1974). *Biochim. Biophys. Acta* **333**, 180–186.
Lardy, H., Reed, P., and Chiu-Lin, C. H. (1975). *Fed. Proc., Fed. Am. Soc. Exp. Biol.* **34**, 1707–1710.
Lunardi, J., Lauquin, G., and Vignais, P. (1977). *FEBS Lett.* **86**, 317–323.
Machleidt, W., Michel, R., Neupert, W., and Wachter, E. (1976). *In* "Genetics and Biogenesis of Chloroplasts and Mitochondria" (T. Bucher, W. Neupert, and S. Werner, eds.), pp. 195–198. North-Holland Publ., Amsterdam.
MacLennan, D. H., and Tzagoloff, A. (1968). *Biochemistry* **7**, 1603–1610.
Mahler, H. R. (1973). *Crit. Rev. Biochem.* **1**, 381–460.
Mahler, H. R., and Bastos, R. N. (1974). *Proc. Natl. Acad. Sci. U.S.A.* **71**, 2241–2245.
Marcus, F., Schuster, S., and Lardy, H. (1976). *J. Biol. Chem.* **251**, 1775–1780.
Mitchell, P. (1961). *Nature* (*London*) **191**, 144–148.
Mitchell, P. (1974). *FEBS Lett.* **43**, 189–194.
Mitchell, P. (1977). *FEBS Lett.* **78**, 1–19.
Mitchell, P., and Moyle, J. (1965). *Nature* (*London*) **208**, 147–151.
Mitchell, P., and Moyle, J. (1967). *Biochem. J.* **105**, 1147–1162.
Mitchell, P., and Moyle, J. (1971). *Bioenergetics* **2**, 1–11.
Monk, B. C. (1977). Ph.D. Thesis, Monash University, Melbourne.
Monk, B. C., Trembath, M. K., Kellerman, G. M., and Linnane, A. W. (1979). Submitted for publication.
Montal, M., Lee, B., and Chance, C.-P. (1970). *J. Membr. Biol.* **2**, 201–234.
Moudrianakis, E. N., and Tiefert, M. A. (1976). *J. Biol. Chem.* **251**, 7796–7801.
Moyle, J., and Mitchell, P. (1973). *FEBS Lett.* **30**, 317–320.
Muller, J. L. M., Rosing, J., and Slater, E. C. (1977). *Biochim. Biophys. Acta* **462**, 422–437.
Nelson, N. (1976). *Biochim. Biophys. Acta* **456**, 314–328.
Nelson, N., Nelson, H., and Racker, E. (1972). *J. Biol. Chem.* **247**, 7657–7662.
Nelson, N., Deters, D., Nelson, H., and Racker, E. (1973). *J. Biol. Chem.* **248**, 2049–2055.
Nelson, N., Kanner, B., and Gutnik, D. (1974). *Proc. Natl. Acad. Sci. U.S.A.* **71**, 2720–2724.
Nicholls, D. G. (1974). *Eur. J. Biochem.* **50**, 305–315.
Nieuwenhuis, F., van der Drift, J., Voet, A. B., and van Dam, K. (1974). *Biochim. Biophys. Acta* **368**, 461–463.
Okamato, H., Sone, N., Hirata, H., Yoshida, M., and Kagawa, Y. (1977). *J. Biol. Chem.* **252**, 6125–6131.
Ort, D. R., Dilley, R. A., and Good, N. E. (1976). *Biochim. Biophys. Acta* **449**, 108–124.
Out, T. A., and Kemp, A. (1974). *Biochem. Soc. Trans.* **2**, 516–517.
Packer, L. (1978). Personal communication.
Papa, S. (1976). *Biochim. Biophys. Acta* **456**, 39–84.
Partis, M. D., Hyams, R. L., and Griffiths, D. E. (1977). *FEBS Lett.* **75**, 47–51.
Pedersen, P. L. (1975). *Bioenergetics* **6**, 243–275.
Pedersen, P. L. (1976a). *J. Biol. Chem.* **251**, 934–944.
Pedersen, P. L. (1976b). *Biochem. Biophys. Res. Commun.* **71**, 1182–1188.
Penefsky, H. (1974). *In* "The Enzymes" (P. Boyer, ed.), 3rd Ed., Vol. 10, pp. 375–394. Academic Press, New York.

Perlman, P. S., and Mahler, H. R. (1970). *Bioenergetics* **1**, 113–138.
Phelps, D., and Hanstein, W. (1977). *Biochem. Biophys. Res. Commun.* **79**, 1245–1254.
Pullman, M. E., and Monroy, G. (1963). *J. Biol. Chem.* **238**, 3762–3769.
Pullman, M. E., Penefsky, H., Datta, A., and Racker, E. (1960). *J. Biol. Chem.* **235**, 3322–3329.
Racker, E. (1970). "Membranes of Mitochondria, and Chloroplasts," pp. 127–168. Van Nostrand-Reinhold, Princeton, New Jersey.
Racker, E. (1972). *J. Membr. Biol.* **10**, 221–235.
Racker, E. (1976). "A New Look at Mechanisms in Bioenergetics," pp. 128–135. Academic Press, New York.
Racker, E., and Stoeckenius, W. (1974). *J. Biol. Chem.* **249**, 662–663.
Racker, E., Tyler, D. E., Estabrook, R. W., Conover, R. W., Parsons, T. E., and Chance, B. (1965). *In* "Oxidases and Related Redox Systems" (T. E. King, H. S. Mason, and M. Morrison, eds.), pp. 1077–1094. Wiley, New York.
Ragan, C. I., and Racker, E. (1973). *J. Biol. Chem.* **248**, 2563–2569.
Recktenwald, D., and Hess, B. (1977a). *FEBS Lett.* **76**, 25–28.
Recktenwald, D., and Hess, B. (1977b). *FEBS Lett.* **80**, 187–189.
Ried, R. A., Moyle, J., and Mitchell, P. (1966). *Nature (London)* **212**, 257–258.
Rosing, J., Kayalar, C., and Boyer, P. D. (1977). *J. Biol. Chem.* **252**, 2478–2485.
Rottenberg, H. (1975). *Bioenergetics* **7**, 61–74.
Roy, H., and Moudrianakis, E. N. (1971a). *Proc. Natl. Acad. Sci. U.S.A.* **68**, 464–468.
Roy, H., and Moudrianakis, E. N. (1971b). *Proc. Natl. Acad. Sci. U.S.A.* **68**, 2720–2724.
Russell, J., Jeng, S., and Guillory, R. J. (1976). *Biochem. Biophys. Res. Commun.* **70**, 1225–1234.
Russell, L. K., Kirkley, S. A., Klegman, T. R., and Chan, S. H. P. (1976). *Biochem. Biophys. Res. Commun.* **73**, 434–443.
Russo, J. A., Lamos, C. M., and Mitchell, R. A. (1978). *Biochemistry* **17**, 473–480.
Ryrie, I. J. (1975). *Arch. Biochem. Biophys.* **168**, 704–711.
Ryrie, I. J. (1977). *Arch. Biochem. Biophys.* **184**, 464–475.
Sabadie-Pialoux, N., and Gautheron, D. (1971). *Biochim. Biophys. Acta* **234**, 5–19.
Sadler, M. H., Hunter, D. R., and Haworth, R. A. (1974). *Biochem. Biophys. Res. Commun.* **59**, 804–812.
Satav, J., and Criddle, R. S. (1979). In preparation.
Satre, M., Klein, G., and Vignais, P. V. (1976). *Biochim. Biophys. Acta* **453**, 111–120.
Scharf, S., Ohlrogge, J., and Criddle, R. (1979). Submitted for publication.
Schatz, G. (1968). *J. Biol. Chem.* **243**, 2192–2199.
Schatz, G., and Mason, T. (1974). *Annu. Rev. Biochem.* **43**, 51–87.
Schatz, G., Penefsky, H., and Racker, E. (1967). *J. Biol. Chem.* **242**, 2552–2560.
Schuster, S. M. (1977). *Fed. Proc., Fed. Am. Soc. Exp. Biol.* **36**, 902.
Schubert, D., Bend, H., Domning, B., and Wiedner, G. (1977). *FEBS Lett.* **74**, 47–49.
Schuster, S. M., Ebel, R. E., and Lardy, H. A. (1975). *J. Biol. Chem.* **250**, 7848–7853.
Schuster, S. M., Gertschen, R. J., and Lardy, H. A. (1976). *J. Biol. Chem.* **251**, 6705–6710.
Schuster, S. M., Reinhart, G. D., and Lardy, H. A. (1977). *J. Biol. Chem.* **252**, 427–432.
Scott, K., Shi, G., and Brierly, G. P. (1978). *Fed. Proc., Fed. Am. Soc. Exp. Biol.* **37**, 1567.
Sebald, W. (1977). *Biochim. Biophys. Acta* **463**, 433–440.
Sebald, W., and Wachter, E. (1978). *In* "Energy Conservation in Biological Membranes" (G. Schafer and M. Klingenberg, eds.), pp. 228–236. Springer-Verlag, Berlin and New York.
Sebald, W., Graf, T., and Wild, G. (1976). *In* "Genetics and Biogenesis of Chloroplasts and Mitochondria" (T. Bucher, W. Neupert, and S. Werner, eds.), pp. 167–174. North-Holland Publ., Amsterdam.
Sebald, W., Sebald-Althaus, M., and Wachter, E. (1977). *In* "Genetics and Biogenesis of

Mitochondria" (W. Bandlow, R. J. Schweyen, K. Wolf, and E. Kaudewitz, eds.), pp. 433–440. de Gruyter, Berlin.
Senior, A. E. (1973). *Biochim. Biophys. Acta* **301**, 249–277.
Senior, A. E. (1975). *Biochemistry* **14**, 660–664.
Serrano, R., Kanner, B., and Racker, E. (1976). *J. Biol. Chem.* **251**, 2453–2461.
Shaikh, F., Joshi, S., and Sanadi, D. R. (1976). *Fed. Proc., Fed. Am. Soc. Exp. Biol.* **35**, 1557.
Shannon, C., Enns, R., Wheelis, L., Burchiel, K., and Criddle, R. S. (1973). *J. Biol. Chem.* **248**, 3004–3011.
Shchipakin, V., Chuchlova, E., and Evtodienko, Y. (1976). *Biochem. Biophys. Res. Commun.* **69**, 123–127.
Sierra, M., and Tzagoloff, A. (1973). *Proc. Natl. Acad. Sci. U.S.A.* **70**, 3155–3159.
Singh, A. P., and Bragg, P. D. (1978). *Biochem. Biophys. Res. Commun.* **80**, 161–167.
Smith, D. J., and Boyer, P. D. (1976). *Proc. Natl. Acad. Sci. U.S.A.* **73**, 4314–4318.
Smith, D. J., Stokes, B. O., and Boyer, P. D. (1976). *J. Biol. Chem.* **251**, 4165–4171.
Sone, N., Yoshida, M., Hirata, H., and Kagawa, Y. (1975). *J. Biol. Chem.* **250**, 7917–7923.
Sone, N., Yoshida, M., Hirata, H., and Kagawa, Y. (1976). *J. Biol. Chem.* **252**, 2956–2960.
Southard, J. H., Blondin, G. A., and Green, D. E. (1974). *J. Biol. Chem.* **249**, 678–681.
Stekhoven, F. S., Waitkus, R. F., and van Moerkerk, H. (1972). *Biochemistry* **11**, 1145–1150.
Sternweiss, P. C. (1977). Ph.D. Thesis, Cornell Univ., Ithaca, New York.
Sternweiss, P. C. (1978). *J. Biol. Chem.* **253**, 3123–3128.
Stiggall, D., Galante, Y., and Hatefi, Y. (1978). *J. Biol. Chem.* **253**, 956–964.
Trembath, M. K., Monk, B., Kellerman, G., and Linnane, A. W. (1975a). *Mol. Gen. Genet* **140**, 333–337.
Trembath, M. K., Monk, B. C., Kellerman, G., and Linnane, A. W. (1975b). *Mol. Gen. Genet.* **141**, 9–22.
Trembath, M. K., Malloy, P., Sriprakash, K. S., Cutting, G., Linnane, A. W., and Lukins, H. B. (1976). *Mol. Gen. Genet.* **145**, 43–52.
Tzagoloff, A. (1970). *J. Biol. Chem.* **245**, 1545–1551.
Tzagoloff, A. (1976). *In* "The Enzymes of Biological Membranes" (A. Martinosi, ed.), pp. 103–124. Plenum, New York.
Tzagoloff, A., and Meagher, P. (1971). *J. Biol. Chem.* **246**, 7328–7336.
Tzagoloff, A., MacLennan, D., and Byington, K. (1968). *Biochemistry* **7**, 1596–1602.
Tzagoloff, A., Rubin, M., and Sierra, M. (1973). *Biochim. Biophys. Acta* **301**, 71–104.
Tzagoloff, A., Akai, A., and Foury, F. (1976). *FEBS Lett.* **65**, 391–395.
Vagelos, P. R., Majerus, P. W., Alberts, A. W., Larrabee, A. R., and Sihaud, G. P. (1976). *Fed. Proc., Fed. Am. Soc. Exp. Biol.* **25**, 1485–1494.
Vambutas, V. K., and Racker, E. (1965). *J. Biol. Chem.* **240**, 2660–2667.
Van de Stadt, R. J., DeBoer, B., and Van Dam, K. (1973). *Biochim. Biophys. Acta* **292**, 338–349.
Verschour, G. J., Van Der Sluis, P. R., and Slater, E. C. (1977). *Biochim. Biophys. Acta* **462**, 438–449.
Vignais, P. V. (1976). *Biochim. Biophys. Acta* **456**, 1–38.
Vinkler, C., Rosen, G., and Boyer, P. D. (1978). *J. Biol. Chem.* **253**, 2507–2510.
Vogel, G., and Steinhart, R. (1976). *Biochemistry* **15**, 208–216.
Wachter, E., and Sebald, W. (1977). *In* "Genetics and Biogenesis of Mitochondria" (W. Bandlow, R. J. Schweyen, K. Wolf, and F. Kaudewitz, eds.), pp. 441–449. de Gruyter, Berlin.
Wagenvoord, R. J., Vanderkran, K., and Kemp, A. (1977). *Biochim. Biophys. Acta* **460**, 17–24.
Wakabayashi, K. (1972). *J. Antibiot.* **25**, 476–480.
Wakabayashi, K., Kubota, M., Yoshida, M., and Kagawa, Y. (1977). *J. Mol. Biol.* **117**, 515–519.
Webster, G. D., Edwards, P. A., and Jackson, J. B. (1977). *FEBS Lett.* **76**, 29–35.

Weichman, A., Beem, E., and Van Dam, K. (1975). *In* "Electron Transfer Chains and Oxidative Phosphorylation" (E. Quagliariello, S. Papa, F. Palmieri, E. Slater, and N. Siliprandi, eds.), pp. 335–342. North-Holland Publ., Amsterdam.
Williams, R. J. P. (1978). *FEBS Lett.* **85**, 9–18.
Wingfield, P. T., and Boxer, D. H. (1975). *Biochem. Soc. Trans.* **3**, 763–765.
Yamamoto, T., and Tonomura, Y. (1975). *J. Biochem.* (*Tokyo*) **77**, 137–146.
Yoshida, M., Sone, N., Hirata, H., and Kagawa, Y. (1975). *J. Biol. Chem.* **240**, 2660–2667.

Ionophores and Ion Transport Across Natural Membranes

ADIL E. SHAMOO AND THOMAS J. MURPHY
Department of Radiation Biology and Biophysics
University of Rochester School of Medicine and Dentistry
Rochester, New York

I. Definitions

The term ionophoric or ionophorous activity has been used to mean various particular properties of a substance depending on the method employed to assay for the ionophoric material. The common characteristic of all methods employed may be used to construct a general definition of ionophores. An ionophore is therefore defined as a substance that enhances the movement or incorporation of an ion from an aqueous phase into an hydrophobic phase (Shamoo and Goldstein, 1977). Note that this definition is more general than that of Pressman (1976).

A carrier is a lipid-soluble molecule that binds substrate on one side of the membrane, crosses the membrane, and dissociates on the other side, releasing free substrate. A channel is a hydrophilic or ionophilic pathway that spans the membrane. While a large, nonselective, water-filled hole in the membrane fits this definition rather well, it does not fit the above definition for an ionophore since an ion in a water-filled pore never leaves

ISBN 0-12-152509-0

the aqueous medium. In spite of this distinction, we will treat all channels as equals in this review.

Ionic dependency for an ionophore implies that the establishment of an ionophore-mediated function is dependent on the presence of a certain ion. Ionic selectivity for an ionophore implies that the ionophore mediates the transport of a certain ion more than other ions.

II. Introduction

With the breakthroughs of Mueller *et al.* (1962) and Bangham *et al.* (1965) in the formation of artificial, lipid-bilayer membranes, two facts became immediately apparent.

1. The ion permeability of natural membranes must be mediated by nonlipid membrane components. This fact was demonstrated by the large (several orders of magnitude) difference between the ion permeability of natural membranes and the permeability of purely lipid membranes.
2. These artificial membranes have great potential as tools for the study of transmembrane ion transport mediated by nonlipid substances. As a matter of a fact, Mueller and his colleagues demonstrated in their original bilayer paper that the addition of a proteinaceous substance, later called excitability-inducing material (EIM), to the bilayer matrix dramatically increased membrane ion permeability and conferred on the membrane a voltage-dependent conductance similar to that measured in nerve and muscle membranes.

Encouraged by this early success, a great number of investigators proceeded to add a wide variety of natural proteins to artificial membranes. It was found that it was very easy to induce nonspecific increases in ion permeability—too easy, perhaps. Substances that have no known ion transport function *in vivo* were shown to dramatically increase the ion permeability of artificial membranes. It proved more difficult to reconstruct in artificial membranes ion transport systems that behave in the same way as they do *in vivo*. Much significant progress, however, has been made on this front in this decade.

Two related but distinct uses have been successfully made of artificial membranes in the study of ion transport systems: reconstitution of entire ion transport systems, and the study of ionophores isolated from ion transport systems. In the former approach, a protein or protein complex responsible for some ion transport function is isolated from the natural membrane and inserted as an intact unit into an artificial membrane. The primary advantage of this approach is that, provided the protein under study is sufficiently purified, the transport activity that is reconstituted can be unambiguously ascribed to the protein. The approach is similar in principle to removing the clockwork from its case, winding the mainspring, and watching it go in order to verify that the ticking and hand movement emanate from the complex of springs and gears. There are other advantages of studying the reconstituted system over the native system. The contributory role of specific membrane lipids can be studied by varying the lipids used to form the membrane. Details of the transport process can be more easily studied in the absence of other parallel transport processes. Perhaps the most important piece of information that can be gained from reconstitution studies is the minimum number of components required for full transport activity. This can be achieved by removing system components until the transport activity ceases, like removing pieces from a clockwork until ticking and/or hand movement ceases.

In the study of ionophores isolated from ion transport systems, to continue the analogy, one continues to disassemble the clock, undaunted by the fact that the hands are still and the ticking has stopped, and examines those properties of the springs, wheels, and gears isolated from the clock that can be related to the operation of the intact system. A protein or protein complex is physically broken down to smaller pieces and the ion transport or ionophoric properties of the pieces are investigated. Characteristics of the ion transport induced in artificial membranes by the pieces are correlated with the properties of the intact system (e.g., transported ion selectivity, common inhibitors of ionophoric activity, and transport function of the intact system). An attempt is made to keep track of the spatial relationship of the pieces in the intact system so that the importance of interactions between the various pieces is not missed.

For clocks or ion transport systems, it would be difficult to gain any useful information about the intact system from the study of small pieces of the system without some conceptual framework or general idea of how the intact system must work. A simple conceptual framework for the study of ion transport systems has been proposed by Shamoo and Goldstein (1977). In this model(s), ion transport systems are made up of one, two, or all three basic parts: a nonselective channel, a selective gate, and an energy transducer. The ionophoric properties of these pieces can be inferred from their function. The nonselective channel should behave as such in a membrane. The selective gate should have an affinity for the transported ion and may or may not be hydrophobic. If it is hydrophobic, it will probably behave as an ion-selective ionophore. The energy transducer will probably have no ionophoric activity but should contain the binding site for the energy source substrate (e.g., ATP). Thus, the ionophoric properties of various pieces can give clues as to what functional roles the pieces play in the intact system. This conceptual framework is of course highly speculative, so that inferences drawn from this approach must be tentative. Once this functional assignment to the various pieces of a transport system is made, tentative though this assignment may be, new experimental approaches to the problem of energy transduction are opened up. We will see that for one ion transport system, Ca^{2+} + Mg^{2}-ATPase, it has proved possible to design experiments to check the tentative functional assignments obtained from preliminary study of the ionophoric activity of the individual pieces and to proceed to the study of the interaction between the pieces.

The distinction between the two experimental approaches just outlined, reconstitution of the intact ion transport system and study of ionophores isolated from an ion transport system, should be stressed. In the latter approach, direct and detailed comparison of the properties of a piece of a system with those of the intact system must be tempered with the realization that the coupling between various parts of a complex system is extremely important in the function of the whole. To return to the previous analogy, one can open a clock and remove the mechanical oscillator that controls speed and study the uncontrolled whir of gears and spinning of hands, thereby gaining insight into the drive mechanism of the clock. While it is true that this rapid burst of movement bears little resemblance to the even movement of an intact timepiece, its relevance to the keeping of time cannot be denied.

There is a large body of literature on the ionophoric properties of various antibiotics and the use of these ionophores in the study of the function of biological membranes, both as experimental models and as experimental tools. This literature has been extensively reviewed (McLaughlin and Eisenberg, 1975; Pressman, 1976; Gomez-Puyou and Gomez-Lojero, 1977).

In this review, we will focus on the study of biological ion transport systems by reconstitution and the study of ionophores isolated from the transport systems. These are rapidly expanding fields of study, and a great deal of truly significant progress has been made and is being made in this decade. We will see that experimental techniques are being refined; new and more versatile artificial membrane assay systems are being developed. We will speculate on the role of ionophores in transport and energy transduction, and on the future direction to be taken in order to achieve an understanding of the mechanism of transport across biological membranes. Reviews on related subjects with various scopes have recently been published (Montal, 1976; Shamoo and Goldstein, 1977; Blumenthal and Shamoo, 1979; Criddle *et al.*, this volume).

III. Methods for Measuring Ionophoric Activity

Black lipid membranes. Mueller *et al.* (1962) first described the formation of thin lipid films or black lipid membranes (BLMs) in aqueous media. These workers proposed a molecular structure for the films that is very similar to the bimolecular leaflet suggested by Davson and Danielli (1952) for the structure of membranes of cells. Electron micrographs of these films (Mueller *et al.*, 1962; Henn *et al.*, 1967) and thermodynamic considerations Danielli, 1966; Tien and Dawidowicz, 1966; Ohki and Fukuda, 1967; Good, 1969) make this proposed structure a very probable one, and it has been widely accepted as accurate.

It has been demonstrated that these membranes contain a significant amount of organic solvent, not only in the boundary region but also in the hydrophobic core of the bilayer itself (Henn and Thompson, 1967; Pagano *et al.*, 1972; Fettiplace *et al.*, 1971). The presence of the organic solvent may cause chemical alteration of the proteins incorporated into the bilayer or change the bilayer structure sufficiently to inactivate the transport system. For example, Haydon *et al.* (1977) have demonstrated a correlation between the anesthetic (nerve-impulse blocking) and BLM hydrophobic core thickening properties of *n*-alkanes of various lengths, commonly used solvents for BLM formation. BLMs have the additional disadvantage of relatively short lifetime. This consideration is important because the achievement of an incorporated protein's equilibrium configuration in the membrane is not always instantaneous. Murphy (1977) found that, at least for the case of BLMs doped from a vesicular preparation, doped bilayer electrical properties continued to change throughout the lifetime of the membrane even after the doping material had been removed from the membrane bathing solution.

In spite of these difficulties, BLMs have proved to be extremely useful in the study of ionophores. The ion selectivity of an ionophore can be quantitated as the ratio of the bilayer permeability to various ions when the ionophore is incorporated into the bilayer. These permeability ratios can be calculated from the transmembrane voltage for zero transmembrane current when there is a gradient of the concentration of ion species across the membrane. When all permeable ions present are of single valance (all ± 1 or all ± 2 or all ...), the Goldman (1943) equation can be used to calculate the permeability ratio from the zero current membrane voltage and the ion species concentrations on either side of the membrane. The Goldman equation has recently been generalized for the case of mixed univalent and divalent permeable ion species (Shamoo and Goldstein, 1977; Pickard, 1976). When the permeability ratio for two ion species is 1, the ionophore is not selective between the two ion species. When the permeability ratio is far from 1, the ionophore is selective to one of two ion species.

In this approach for calculating the ionic selectivity of an ionophore, it is assumed that the major current pathway through the bilayer is via the ionophore molecule only. For this assumption to be true, two conditions must be met; the conductance of the bilayer must be significantly greater in the presence of the ionophore (preferably several hundredfold) than in the absence of the ionophore, and there must not be additional parallel current pathways induced in the bilayer when the ionophore is incorporated. While the former condition is usually met with little difficulty, it is often difficult to ensure that nonselective ionic leaks (e.g., nonspecific leaks around the protein) are not present in the system, especially in the case of channels. Consequently, estimates of ion selectivity tend to err on the side of nonselectivity.

Bilayers formed from monolayers. Montal and Mueller (1972) developed a bilayer system formed by folding together two lipid monolayers. In this system the concentration of organic solvent in the bilayer hydrophobic core is considerably reduced from that in BLMs. This technique also has the advantage that bilayers with asymmetric lipid compositions, mimicking the asymmetric lipid composition of natural membranes, can easily be formed.

Vesicles. Bangham *et al.* (1965) described the spontaneous formation of lipid vesicles, mostly multilamellar, spherical structures of outside diameter in the several micron range, upon exposure of dried phospholipid to salt solutions. Upon prolonged sonication at temperatures above the liquid crystalline transition temperature of the phospholipid, smaller (250–500 Å diameter) but unilamellar vesicles can be produced in abundance (Papahadjopoulos and Miller, 1967; Papahadjopoulos and Watkins, 1967; Huang, 1969). Kasahara and Hinkle (1977) have recently described a freeze–thaw procedure for the production of vesicles with increased volume to surface ratio. The freeze–thaw procedure has the additional advantage of reducing the required sonication time (which may disrupt the incorporated protein).

Vesicles have two major advantages over BLMs, long lifetime and the absence of organic solvent. Extensive use, pioneered by Racker and his colleagues (see other sections of this review), has been made of these structures in the reconstitution of intact ion transport systems. They have also been useful in the study of ionophores. Their major disadvantage is that transmembrane voltage cannot be directly monitored and controlled during the course of an experiment.

Giant vesicles. Mueller and Rudin (1968) described the formation of large (diameters in the 100-μm range) vesicles by emulsifying phospholipids in organic solvent with aqueous solutions of water-soluble protein and salt. The electrical properties of these membranes were investigated using microelectrodes and found to be quantitatively similar to those of BLMs. Concentration of organic solvent and protein in the final membranes is uncertain.

Reeves and Dowben (1968) reported on the spontaneous formation of large vesicles (diameters in the range 0.5 to 10 μm). Their method was essentially similar to the method of Bangham *et al.* (1965) except that the salt solution was replaced by distilled water or nonelectrolyte solution. A thin layer of phospholipid was spread on the walls of a glass vessel, the organic spreading solvent was evaporated away under nitrogen, and water or nonelectrolyte solution was carefully added to the vessel. After an incubation period of 2 hours at room temperature, during which mechanical vibration was carefully avoided, the aqueous phase contained the vesicles. Electron micrographs of the vesicles showed them to be bounded by one or two trilaminar membranes (Reeves and Dowben, 1968).

Mueller (personal communication) has found that if salt is rigorously excluded from the system and the incubation is carried out overnight at a temperature near 5°C, the diameters of the resulting vesicles are increased by more than an order of magnitude. Salt can be added to vesicles formed in nonionic solutions by dilution of the vesicles into isosmotic or hypoosmotic solutions (Murphy and Shamoo, unpublished observations). These vesicles are called giant vesicles.

The electrical properties of giant vesicles can be studied using microelectrodes. Because of the high resistance of the membrane, however, the small current leak around the electrode becomes significant in these measurements. Mueller's group (Antanavege *et al.*, 1978; Murphy and Shamoo, unpublished observations) has recently shown that giant vesicles can be fused to a circular hole in a thin plastic partition separating two aqueous phases. By applying a short voltage pulse across the vesicle, the vesicle membrane on one side of the plastic partition can be broken, leaving a single bilayer across the hole. The resulting single bilayers are stable for hours and have the obvious advantage that ion concentration on both sides of the membrane can be varied and controlled.

Giant vesicles appear to offer the best of the BLM and vesicle worlds. Because they are stable for weeks at 5°C, protein can be incorporated into the membrane well before experimental measurements are taken. There is no organic solvent present in the final membrane. Transmembrane voltage can be directly monitored and controlled, and ion concentrations on either side of the membrane controlled during an experiment. Because of their extremely large volume-to-surface ratio, these structures may also prove to be useful in nonelectrical studies of ion fluxes.

Equilibrium extraction and bulk phase transport. These methods measure directly the ability of an ionophore to incorporate an ion into a hydrophobic medium. The interphase of the aqueous and hydrophobic media, however, is not a particularly good model for the membrane. The advantages and disadvantages of these systems have been discussed elsewhere (Shamoo and Goldstein, 1977).

IV. Natural Membrane Ion Transport Systems Containing Ionophores

A. $Ca^{2+} + Mg^{2+}$-ATPase from Sarcoplasmic Reticulum

It is well established that intracellular free calcium concentration determines the state of contraction of the actomyosin complex and the actomyosin ATPase activity in muscle. In skeletal muscle, the free calcium concentration is regulated by the sarcoplasmic reticulum (SR) (MacLennan and Holland, 1975). While the mechanism for calcium release from SR is not well understood, the role of $Ca^{2+} + Mg^{2+}$-ATPase in the active uptake of calcium by the SR is clear. This enzyme accounts for most of the protein in the SR membrane and has been purified in active form by several procedures (MacLennan, 1970; MacLennan and Holland, 1975; Ikemoto *et al.*, 1971; Meissner *et al.*, 1973; Warren *et al.*, 1974a; Deamer, 1973; Racker, 1979). For this and other reasons this enzyme has been an ideal system for studying the active transport process. It was the first active transport system to be successfully reconstituted in an artificial membrane system (Racker, 1972) and also the most extensively characterized one using the approach of isolating ionophores from an ion transport system.

1. *Reconstitution of the Active* Ca^{2+}*-Transport System*

Active calcium transport by $Ca^{2+} + Mg^{2+}$-ATPase from skeletal muscle sarcoplasmic reticulum was first reconstituted in soybean phospholipid vesicles by Racker (1972). Since this initial finding, Racker's group (Racker, 1973; Racker and Eytan, 1973; Racket *et al.*, 1975a,b) and Metcalf's group (Warren *et al.*, 1974a,b) have published several reports of reconstitution of this enzyme in vesicles using a variety of lipids. Data from these studies indicate that no specific phospholipid is absolutely required for ATP hydrolysis or active Ca^{2+}-transport activity. Some lipids are more effective than others in restoring active Ca^{2+} accumulation by the enzyme, especially in the absence of a calcium complexer such as oxalate inside the vesicles, but these differences may be accounted for by differences in the tightness of the lipid vesicles to passive calcium leakage (Warren *et al.*, 1974a). On the other hand, modification of phosphatidylethanolamine (PE) by acetylation (Knowles *et al.*, 1975) or by fluorescamine (Hidalgo and Tong, 1978) completely inhibits active Ca^{2+} transport. Hidalgo and Tong (1978) reported that fluorescamine-modified PE did not affect the ATP hydrolysis activity of the enzyme. These data suggest an intimate involvement of phospholipid in the coupling of ATP hydrolysis with active Ca^{2+} transport.

Dean and Tanford (1977, 1978) showed that extensive delipidation of the enzyme by treatment with deoxycholate in the presence of 20% glycerol (residual phospholipid was reduced to as low as 1 mol of phospholipid per mole of enzyme) resulted in loss of ATP hydrolysis activity, which could be restored by the addition of a wide variety of ionic and nonionic detergents, including sodium dodecyl sulfate. Among the detergents tested, only deoxycholate, cholate, and the cationic detergent trimethyl tetradecylammonium chloride failed to produce some activation. The authors interpreted the failure of the cationic detergent as an indication that a positive charge on the polar head group could not be tolerated.

a. Electrogenicity. Because of the difficulties associated with the morphology of the sarcoplasmic reticulum, it has been impossible to directly monitor the electrical events associated with active Ca^{2+} transport *in situ*. Moreover, active Ca^{2+} transport has not been reconstituted up to now in an artificial membrane system that is amenable to direct electrical measurement. Consequently, there are few data on the question of whether Ca^{2+} transport is electrogenic (involves the net movement of charge) or some other ion or ions are actively cotransported with calcium to balance the charge movement. There is some indirect evidence that dephosphorylation of the enzyme during active pumping of Ca^{2+} may require the movement of Mg^{2+} through the enzyme in the direction opposite to Ca^{2+} movement (Inesi *et al.*, 1970; Kanazawa *et al.*, 1971; Yamamoto, 1972; MacLennan and Holland, 1975). The stoichiometry and driving force for Mg^{2+}, however, have not been defined experimentally. Using a transmembrane potential-sensitive cyanine dye in SR vesicles, Madeira (1978) obtained evidence that a potential change across the membrane does occur with a time course paralleling Ca^{2+} uptake. Unfortunately, the observed fluorescence changes were not calibrated against known changes in membrane potential, so the amplitude of the voltage change cannot be compared with expected values based on membrane impedance and measured active calcium flux.

b. Monomer as the Functional Unit. Malan *et al.* (1975) compared the density of intramembranous particles observed in freeze-fracture electron microscopy of purified sarcoplasmic reticulum membranes with the calculated density of ATPase polypeptides and found that each particle must account for about three ATPase molecules. Tanford's group (LeMaire *et al.*, 1976) found that there was a high degree of self-association for the enzyme. Moreover, they found that, when the detergent-solubilized enzyme was chromatographed on Tween 80 Sepharose 4B, monomers and oligomers could be separated. The specific ATPase activity of the monomer was far below that of the oligomer. These authors suggested that "some form of denaturation occurs when the degree of association falls below a critical minimum." These data seemed to suggest that the functional calcium pumping unit was an oligomer (trimer or tetramer). The issue was partially resolved when Dean and Tanford (1978) showed that when the enzyme is almost completely delipidated and reactivated with detergent, tendency for self-association by the enzyme is dramatically reduced and the resulting monomeric enzyme was completely active with respect to ATP hydrolysis.

This does not necessarily mean the ATPase monomer is competent to actively transport Ca^{2+}. For a system in which active transport and hydrolytic function are tightly coupled with a defined stoichiometry, however, it is difficult to rationalize the need for more than one independently competent energy source. The easiest interpretation of the data now available is that, in the membrane, the enzyme does indeed have a strong tendency for self-association (which is not surprising for an enzyme that makes up most of the protein in the membrane in which it is embedded), but that the functional unit is composed of a single, monomeric peptide chain.

2. *Study of Ionophores Isolated from the Active* Ca^{2+}*-Transport System*

a. The Proteolipid. Racker and co-workers (Racker *et al.*, 1975a,b; Racker and Eytan, 1975) prepared a heat-stable water-soluble proteolipid, called coupling factor, from sarcoplasmic reticulum during final stages of $Ca^{2+} + Mg^{2+}$-ATPase purification. This proteolipid was shown to increase the calcium transported per ATP hydrolyzed in the reconstituted system and cause the release of ^{45}Ca from inside vesicles. These data were first taken as evidence that the proteolipid was an ionophore similar to A23187 (Racker and Eytan, 1975). It was subsequently found, however, that the proteolipid causes nonspecific increases in BLM conductance (Shamoo and Goldstein, 1977) and nonspecific leakage of ions from vesicles (Racker, personal communication). On the other hand, Laggner and Graham (1976) reported that the proteolipid reduced both the nonspecific ion and water permeabilities in BLMs formed from sarcoplasmic reticulum phospholipids or egg yolk phosphatidylcholine. The exact role of the proteolipid in the overall sarcoplasmic reticulum function is yet to be determined.

b. Ionophoric Properties of the ATPase Molecule. Shamoo and co-workers have made extensive use of BLMs as an assay for the calcium ionophoric activity of the intact enzyme and several fragments obtained from controlled tryptic digestion of the enzyme. Conductance increments were shown with the $Ca^{2+} + Mg^{2+}$-ATPase by either (1) succinylation of the intact enzyme in order to solubilize the enzyme, (2) dispersion of the insoluble ATPase in the bilayer bathing solution by prolonged sonication (about 15 minutes), or (3) tryptic digestion of ATPase molecule (Shamoo and MacLennan, 1974; Shamoo and Ryan, 1975; Shamoo, *et al.*, 1976; Shamoo, MacLennan and Eldefrawi, 1976; Shamoo and MacLennan, 1975; Shamoo *et al.*, 1975, 1977; Shamoo and Abramson, 1977; Abramson and Shamoo, 1979). The tryptic digest of the ATPase molecule was water soluble

TABLE I

RELATIVE DIVALENT CATION PERMEABILITY AND CATION VS. ANION PERMEABILITY ($P_{Ca^{2+}}:P_{Cl^-}$) FOR THE UNDIGESTED SUCCINYLATED $Ca^{2+} + Mg^{2+}$ –ATPASE AND THE VARIOUS TRYPTIC FRAGMENTS[a,b]

	Undigested succinylated $Ca^{2+} + Mg^{2+}$-ATPase	45,000-Dalton fragment	55,000-Dalton fragment	20,000-Dalton fragment
$P_{Ca^{2+}}/P_{Ba^{2+}}$	0.55	1.01	0.50	0.69
$P_{Ca^{2+}}/P_{Sr^{2+}}$	1.49	1.13	1.50	1.10
$P_{Ca^{2+}}/P_{Mg^{2+}}$	1.89	1.29	1.80	1.34
$P_{Ca^{2+}}/P_{Mn^{2+}}$	2.04	1.16	2.00	1.39
$P_{Ca^{2+}}/P_{Cl^-}$	4.30	0.92	4.30	2.30

[a] (From Abramson and Shamoo, 1979.)

[b] All bilayers are made with oxidized cholesterol, and the bathing solution is buffered with 5 m*M* HEPES, Tris, pH 7.3.

whether succinylated or not, but only displayed ionophoric properties without succinylation (Shamoo and MacLennan, 1974). The relative permeabilities exhibited by this material are shown in Table I.

Using the same SR $Ca^{2+} + Mg^{2+}$-ATPase, Tashmukhamedor (1976) and Levitsky (personal cummunication) have confirmed the ionorphoric properties of the protein in BLMs. Moreover, Carafoli's group (Chiesi *et al.*, 1977) has recently shown that the intact cardiac $Ca^{2+} + Mg^{2+}$-ATPase prepared by a similar method has nearly the same ionophoric activity and the same selectivity in BLMs.

The observed selectivity is consistent with the selectivity of active calcium transport. It is not necessarily required that this ionophoric activity have exactly the same selectivity exhibited in active transport. The specificity for Ca^{2+} may be due to the inherent specificity of the biochemical reaction (i.e., the specificity for Ca^{2+} of the Ca^{2+} binding site involved in the regulation of ATP hydrolysis) or the ion selective gate or both in series. Thus, the measured passive ionophoric selectivity is presumably in series with the biochemical selectivity. It should be emphasized that under conditions used to measure ionophoric activity of the transport enzyme, active transport activity is not present. The hydrolytic function and the transport function have been uncoupled by succinylation, prolonged sonication, tryptic digestion, organic solvent present in the BLM, or some other method. The ionophoric properties of the transport protein measured in BLMs need not be the same as the passive ionophoric properties of the actively pumping protein, but only consistent with it. Similarly, the specificity of the hydrolytic function need not be the same as that of the active transport, but rather consistent with it. Passive ionophoric activity by the native enzyme would be inconsistent with sarcoplasmic reticulum function. The native sarcoplasmic reticulum cannot be freely permeable to Ca^{2+} except during Ca^{2+} release at the beginning of contraction. Moreover, there is no definitive evidence or necessity for the transport ATPase being the pathway for Ca^{2+} release. The fact that $Ca^{2+} + Mg^{2+}$-ATPase does not act as a free calcium ionophore *in situ* is consistent with the idea that the hydrolytic function of the pump is tightly coupled to transport. Thus, in order to observe Ca^{2+} permeability within the ATPase molecule, an uncoupling of the two functions may be necessary.

Martonosi and co-workers (DeBoland *et al.*, 1975; Jilka *et al.*, 1975) used vesicles to study the passive ionophoric properties of the ATPase enzyme. They found a relatively nonselective increase in the membrane permeability to calcium, sucrose, sodium, choline, and sulfate. The observed nonspecific increase in vesicular permeability due to transport ATPase may represent leakage around the protein molecule in the region where the hydrophobic protein is associated with the hydrophobic core of the lipid membrane.

The ionophoric activity of the "intact" enzyme in BLMs is inhibited competitively by several monovalent and divalent cations. In addition, this activity is inhibited by several known inhibitors of active calcium transport in SR (see Shamoo and Goldstein, 1977, for a comprehensive review).

c. Isolation of Enzyme Fragments and Functional Assignments. Exposure of sarcoplasmic reticulum to trypsin in the presence of 1 *M* sucrose results in degradation of the 102,000-dalton enzyme to two fragments of 55,000 and 45,000 daltons with subsequent appearance of fragments of 30,000 and 20,000 daltons (Shamoo *et al.*, 1976a,b; Thorley-Lawson and Green, 1975; Stewart *et al.*, 1976). The various fragments were purified either by sodium dodecyl sulfate (SDS) column chromatography or by SDS-preparative gel electrophoresis (Shamoo *et al.*, 1976b; Ryan *et al.*, 1976; Abramson and Shamoo, 1979). Attempts to purify the fragments without SDS have failed both in this laboratory and that of Green. Yu *et al.* (1978) have recently reported the isolation of the 55,000-dalton and 45,000-dalton fragments with deoxycholic acid and thiol-Sepharose 4B batch-affinity techniques. The 55,000-dalton fragment became attached to the beads, and the 45,000-dalton fragment remained in the supernatant. However, no ATPase activity was detected after the isolation (Yu, personal communication). Figure 1 summarizes the details of the digestion pattern.

i. The large nonspecific channel (45,000-*dalton fragment*). The use of antibodies against the various fragments indicated that the 55,000-dalton fragment was in large part exposed to the cytoplasm *in vivo* whereas the 45,000-dalton fragment was only partially exposed (Shamoo *et al.*, 1976b;

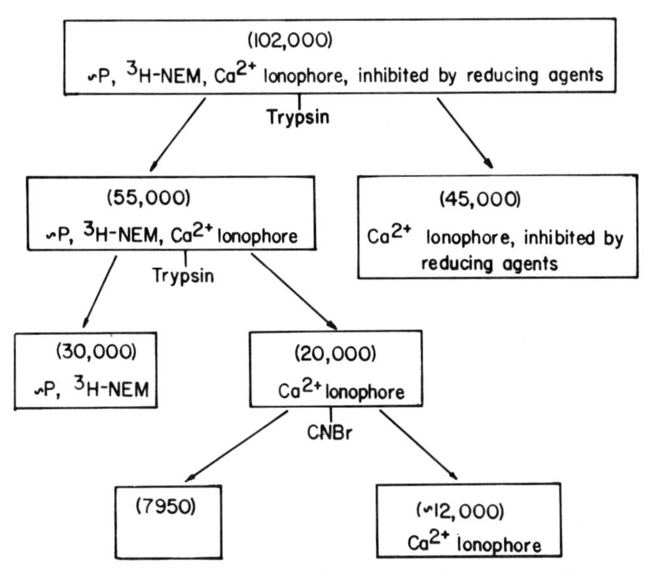

FIG. 1. Digestion pattern for Ca^{2+} + Mg^{2+} -ATPase. ^{3}H-NEM, *N*-[^{3}H]ethylmaleimide.

Ryan *et al.*, 1976). The 45,000-dalton fragment was found to be a relatively nonselective, divalent cation-dependent ionophore when incorporated into BLMs. Ionophoric activity of the fragment was inhibited by low concentration of $LaCl_3$ and $HgCl_2$ (Abramson and Shamoo, 1979). Thorley-Lawson and Green (1977) have shown that there are three disulfide bonds in the 45,000-dalton fragment that are not accessible to reduction without delipidation of the SR by detergents. Preincubation with excess reducing agent completely inhibits the ionophoric properties of the intact enzyme and the 45,000-dalton fragment, but not of the 55,000-dalton fragment. Moreover, the selectivity ratios for Ca^{2+} vs. Cl^- and for Ca^{2+} vs. other divalent cations is almost identical for the intact enzyme and the 55,000-dalton fragment (see Table I). From these data, Abramson and Shamoo (1979) concluded that the pathway for Ca^{2+} transport inside the enzyme consists of the 55,000-dalton and 45,000-dalton fragments in a series arrangement. All these data are consistent with the role of the 45,000-dalton fragment being that of the nonselective channel part of the transport system (Shamoo and Goldstein, 1977).

ii. The energy transducer (30,000-*dalton fragment*). The 55,000-dalton fragment and the 30,000-dalton fragment (which is derived from the 55,000-dalton piece, see Fig. 1) contain the site of phosphorylation and of *N*-[^{3}H]ethylmaleimide binding, indicating the presence of the hydrolytic

Ca^{2+} + Mg^{2+}-ATPase site. Cleavage of the bond between the 30,000-dalton fragment and the 20,000-dalton fragment does not affect the hydrolytic activity of the enzyme (Scott and Shamoo, 1977; see section on coupling of ATP hydrolysis with Ca^{2+} transport in this review). As expected, the 30,000 dalton fragment shows no ionophorous activity in the BLM assay.

iii. The ion-selective gate (20,000-*dalton fragment*). The 20,000-dalton fragment contains the site of Ca^{2+}-dependent and -selective ionophoric activity as shown in Table I and Fig. 1. However, the Ca^{2+} dependency and selectivity was not as pronounced as that in the intact enzyme or in the 55,000-dalton fragment (Shamoo, 1978) (see Table I). The ionophoric properties of the 20,000-dalton fragment have been confirmed in experiments involving vesicles (Shamoo, unpublished observations). Further digestion of the 20,000-dalton fragment with cyanogen bromide indicated that the Ca^{2+} ionophoric activity is associated with a smaller fragment (Shamoo *et al.*, 1976b). The cyanogen bromide fragments have been recently isolated using gel filtration and hydroxylapatite chromatography (Klip and MacLennan, 1978). Preliminary BLM assay experiments indicate that the Ca^{2+} selective and dependent ionophoric activity resides in the CNBr fragment of about 12,000 daltons (Klip and MacLennan, 1978; Shamoo, unpublished observations).

iv. Arrangement of the three pieces in situ. Shamoo and Ryan (1975) have proposed a model for the ATPase molecule that appears to be consistent with the tryptic digestion data, the interaction of SR membrane with antibodies to the various fragments, and the overall function of the enzyme (Shamoo and Goldstein, 1977). This model is depicted in Fig. 2. It must be noted that the structural data on this molecule are not complete and that the relative positions of the pieces, especially the 30,000- and 20,000-dalton pieces, with respect to the membrane are at best tentative.

d. Interaction between the Three Pieces in the Working Enzyme. In order to study the interaction of the three functional parts of the Ca^{2+}-transport enzyme, Scott and Shamoo (Scott and Shamoo, 1977 and unpublished observations; Shamoo *et al.*, 1977) correlated the appearance of the ATP hydrolytic activity and the loss of Ca^{2+} uptake activity with the appearance and disappearance of the various fragments upon exposure of SR vesicles to low levels of trypsin. These data indicate that the cleavage of the intact enzyme to 55,000-dalton and 45,000-dalton fragments has no effect on either Ca^{2+} uptake or ATP hydrolysis, which is consistent with the finding that the intact enzyme and the 55,000-dalton fragment have almost identical properties in ionophore selectivity. However, digestion of the 55,000-dalton fragment to 30,000- and 20,000-dalton peptides uncouples the two functions, Ca^{2+} uptake being abolished while ATP hydrolysis remains unaffected. The uncoupling indicates that the bond between the 30,000-

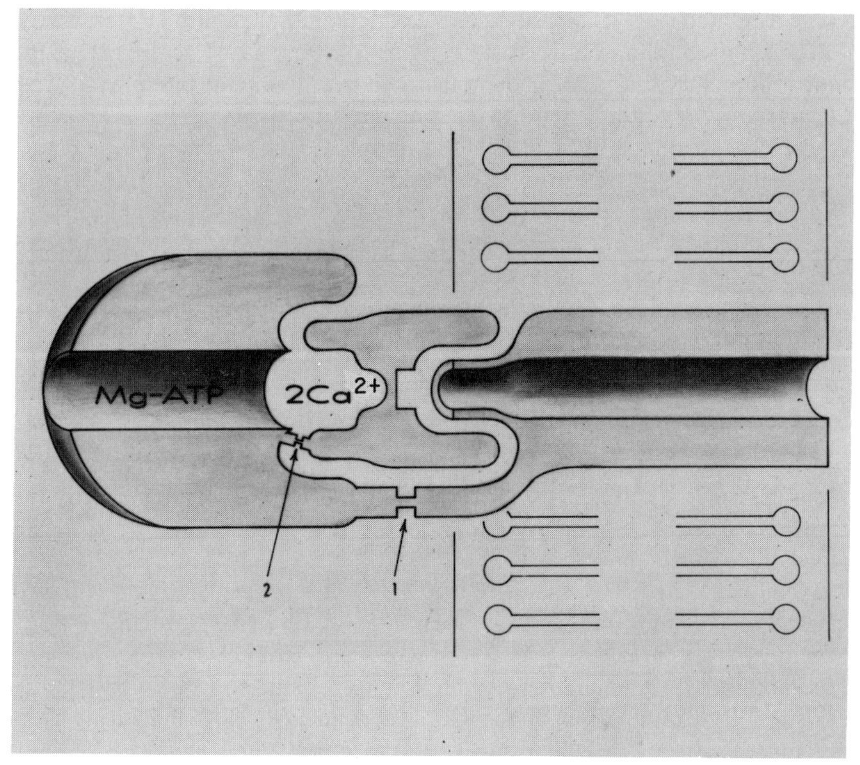

FIG. 2. Model for active transport of calcium in sarcoplasmic reticulum. (Adapted from Shamoo *et al.*, 1977.)

and 20,000-dalton fragments is essential for energy transduction between the scalar chemical hydrolysis and vectorial ion transport. It should be remembered that cleavage of this bond also affects the selectivity and dependency of the Ca^{2+} ionophore site. Passive permeability studies show that loss of Ca^{2+} uptake by digestion is not the result of increased Ca^{2+} permeability of the sarcoplasmic reticulum membrane. Also, the digested enzyme has the same steady-state level of phosphoenzyme as control SR, consistent with the lack of effect of digestion on the overall steady-state rate of ATP hydrolysis.

Berman *et al.* (1977) have shown similar data on uncoupling of ATPase activity from Ca^{2+} transport, utilizing the pretreatment of SR at low pH for a short time. They showed that such a treatment caused a decrease in Ca^{2+} transport without any effect on ATPase activity. Similar uncoupling was observed by Shamoo *et al.* (1976a), who found that a low concentration of

Hg^{2+} inhibited Ca^{2+} transport without concomitant loss of ATPase activity.

The tight coupling between ATP hydrolysis and Ca^{2+} transport in the intact SR membrane has been demonstrated by running the pump backward, i.e., using a Ca^{2+} gradient to drive a reaction producing ATP (Kanazawa *et al.*, 1970; Yamada *et al.*, 1972; Makinose and Hasselbach, 1971; Panet and Selinger, 1972; Deamer and Baskin, 1972). Racker and colleagues (Racker, 1977) have demonstrated that it is possible to use purified Ca^{2+} + Mg^{2+}-ATPase to form ATP in the absence of a Ca^{2+} gradient or a membranous structure. The high-energy phosphorylated enzyme is first formed in the absence of Ca^{2+}, then Ca^{2+} and ADP are simultaneously added. The amount of ATP formed is stoichiometric with the amount of phosphorylated enzyme present on the addition of Ca^{2+} and ADP. Ca^{2+} can be removed using EGTA, and the cycle repeated. On the basis of this result and calorimetric studies indicating a large enthalpy change associated with the binding of Mg^{2+} to the enzyme, Racker has proposed that the backward reaction in the membrane involves displacement of Mg^{2+} from its binding site by the transported Ca^{2+}. Access by Ca^{2+} to the site is controlled by the gate (Racker, 1977).

Ikemoto (1974, 1975) found that at 22°C, the purified enzyme had two classes of calcium-binding sites: two high-affinity (α) sites and 3 low-affinity (β) sites. However, at 0°C, there were three classes of sites: one high-affinity (α) site, one intermediate-affinity (γ) site, and three low-affinity (β) sites. By examination of the Ca^{2+} dependence of ATP hydrolysis at 0°C, he was able to show that Ca^{2+} binding to the α site stimulated ATPase, binding to the β site had no effect, and binding to the γ site was inhibitory. These data are consistent with the observation of Sumida and Tonomura (1974) that at 0°, the Ca^{2+} transported per ATP hydrolyzed was 1:1 whereas at higher temperatures the ratio is 2:1 (Hasselbach and Makinose, 1963; Weber *et al.*, 1966).

Scott and Shamoo (1978 and unpublished observations) have studied the effect of tryptic cleavage between the 20,000- and 30,000-dalton pieces of the enzyme on Ca^{2+} binding. This cleavage leaves the total number of binding sites unchanged, but reduces the affinity of one of the two high-affinity sites to intermediate affinity, reminiscent of the temperature effect studied by Ikemoto. When Ca^{2+} binding to the digested enzyme is measured at 4°, a large portion of the remaining high-affinity sites is converted to intermediate affinity, suggesting that reduced temperature and tryptic cleavage are affecting separate high-affinity sites. Further studies of calcium uptake and hydrolytic activity as a function of temperature and tryptic digestion should be valuable in sorting out the roles of these two high-affinity Ca^{2+} binding sites.

B. $Na^+ + K^+$-ATPASE

1. *Reconstitution Studies*

One of the ubiquitous properties of animal cells is the asymmetric distribution of various ions, particularly sodium and potassium, across the plasma membrane. After the discovery by Skou (1957, 1971) of the $Na^+ + K^+$-ATPase, a large amount of indirect evidence was collected linking this enzyme to the establishment and maintenance of the sodium/potassium gradient across the membrane. It was not until Goldin and Tong (1974) reconstituted ouabain-inhibitable ATP-dependent Na^+ transport by the purified $Na^+ + K^+$-ATPase from canine renal medulla that a direct linkage was experimentally demonstrated. Subsequently, reconstitution of active transport of sodium was demonstrated for $Na^+ + K^+$-ATPase isolated from *Squalus acanthias* rectal gland (Hilden *et al.*, 1974; Hilden and Hokin, 1975), from canine brain (Goldin and Sweadner, 1975; Sweadner and Goldin, 1975), and from the electric organ of the eel (Racker and Fisher, 1975).

Originally it was thought, both from results obtained from the study of $Na^+ + K^+$-ATPase *in situ* and in reconstitution experiments, that the ion selectivity and stoichiometry of transport varied with the functional role of the enzyme in the various membranes. The stoichiometry for the antiport movement of Na^+ and K^+ obtained in *in situ* studies of nerve and red blood cell ($3\,Na^+ : 2\,K^+$) was also found in reconstitution studies for rectal gland ATPase (Hilden and Hokin, 1975) and brain ATPase (Goldin and Sweadner, 1975; Sweadner and Goldin, 1975). On the other hand, results from *in situ* studies on the ascending limb of the loop of Henle indicated a ouabain-sensitive active transport of Cl^- along with the transport of Na^+ ($1\,Na^+ : 1\,Cl^-$) (Burg and Stoner, 1974; Morel and deRouffignac, 1973; Windhager and Giebisch, 1961; Malnic *et al.*, 1966). This finding was supported by the original reconstitution work of Goldin and Tong (1974). These authors reported that the ratio of $K^+ : Na^+$ transported was far below the 2 : 3 ratio measured in nerve and red blood cells and that Cl^- was actively cotransported with Na^+ in the reconstituted system. Anner *et al.* (1977), however, showed that $Na^+ + K^+$-ATPase from kidney medulla did transport K^+ but the $Na^+ : K^+$ ratio was 5 : 1. They suggested that the K^+ transport mechanism was more labile than the Na^+ mechanism and that it might be possible to obtain the same 3 : 2 ratio observed in the enzyme from other tissues. Using more refined techniques resulting in a more efficient *in vitro* transport system, Goldin (1977) obtained the same $3\,Na^+ : 2\,K^+$ antiport stoichiometry for $Na^+ + K^+$-ATPase from the loop of Henle that was obtained for the enzyme isolated from other membranes. Moreover, he obtained no evidence for the involvement of the enzyme in the active transport of Cl^-. Goldin concluded that $Na^+ + K^+$-ATPase from this source behaved exactly the same as $Na^+ + K^+$-ATPase from other membranes and that the active transport of Cl^- in the loop of Henle must be accounted for by some other differences in that membrane.

Jain *et al.* (1972, 1973) reported the incorporation of several membrane fragment fractions rich is ATPase and acetylcholinesterase activity isolated from rat brain cortical tissues into oxidized cholesterol membranes. Fragments were in the form of vesicles. Upon addition of ATP to the solution on the side of the membrane on which the vesicles were added, they observed development of an open-circuit membrane potential and a short-circuit membrane current that was dependent on the presence of sodium, potassium (or rubidium), and magnesium in the bathing solution. These phenomena were abolished by addition of ouabain to the bathing solution on the side of the membrane opposite that on which the membrane fragments were added. The amplitude of the short-circuit current and open-circuit voltage and the reproducibility in these experiments, however, were extremely low. Moreover, Hyman (1977) demonstrated that short-circuit currents and open-circuit voltages of similar amplitude could be obtained from BLMs containing acid phospholipids (no ATPase present). He attributed these phenomena to a surface effect resulting from the binding of ATP to the membrane.

2. *Ionophores Isolated from the* $Na^+ + K^+$-*ATPase*

Shamoo and co-workers (Shamoo and Albers, 1973; Shamoo and Ryan, 1975) used BLMs to study the ionophoric properties of the tryptic digest of electroplax microsomes rich in $Na^+ + K^+$-ATPase and of the various fragments of the ATPase molecule. They found that a highly Na^+-dependent and specific ionophore resides in a small fragment of the smaller of the two subunits that make up the enzyme. Cyanogen bromide digestion of the molecule released a fraction displaying K^+-dependent ionophoric activity.

Despite the fact that the Na^+-ionophore from $Na^+ + K^+$-ATPase was not isolated owing to its presence in various heterogeneous peptidic fragments, it provided the first evidence that ionophores could be covalently bound to a transport protein and that cleavage of chemical bonds is required to release ionophoric activity in some systems.

C. Electrically Excitable Membranes

The report on the formation of BLMs by Mueller *et al.* (1962) sparked a great deal of hope among those working on electrically excitable (nerve and muscle) membranes that BLMs could immediately be used to study the molecular basis for the voltage- and time-dependent membrane conductance changes associated with excitability. Mueller and Rudin (1963, 1968; Mueller *et al.*, 1962) published studies of substances that induced properties in BLMs that mimicked electrical excitability. While these studies served as useful models, the excitability-inducing material (EIM) and antibiotics, have no known membrane ion transport function *in vivo*. Despite initial hopes, the sodium channel has not been isolated and reconstituted to date.

Barnola and Villegas (1976) found that transmembrane Na^+ flux could be stimulated in lobster nerve plasma membrane vesicles with veratrine or bactrachotoxin, toxins known to hold sodium channels open *in situ*. These effects were reversed by tetrodotoxin, a known blocker of sodium channels. Villegas *et al.* (1977) found that addition of lipids to the plasma membrane vesicles by the method of Kasahara and Hinkle (1977) increased the flux per milligram of protein, particularly if the added lipids contained acid phospholipids. Murphy (1977) studied current noise in BLMs doped with plasma membranes from frog ventricle and observed both 1/f noise and Lorentzian noise. The Lorentzian noise components indicated the presence of gated channels (Stevens, 1972). From the dependence of Lorentzian noise variance on voltage, it was determined that the channels were selective for potassium over sodium and thus were potassium channels. The voltage dependence of the Lorentzian corner frequency was qualitatively consistent with this conclusion. Lorentzian noise was not, however, affected by tetraethylammonium, a known blocker of potassium channels.

While these studies provide some hope that the expectations of the last two decades may eventually be fulfilled, it is clear that significant progress is required before reconstitution experiments will shed any light on the molecular basis for electrical excitability.

D. ACh Receptor

1. *The Cholinergic Response* in Situ

The acetylcholine (ACh) receptor is an extremely diversified ion transport system appearing in a large variety of animal cells. The cholinergic response is important *in vivo* in neuromuscular junction (skeletal muscle), cardiac muscle, and smooth muscle, as well as the electric organ of electric fish. While ACh receptors in all these tissues are almost certainly closely related, there are some differences in the details of the cholinergic response in the various tissues. In skeletal muscle and electric organ, for example, the cholinergic response is excitatory and the cholinergic induced conductance pathway is relatively nonselective among cations (has an equilibrium potential near zero) (Magleby and Stevens, 1972; Lester *et al.*, 1975). In cardiac atrium, the cholinergic induced conductance causes hyperpolarization (DelCastillo and Katz, 1955; Hutter and Trautwein, 1956) consistent with the pathway being highly potassium selective. Clark (1926) found that ACh in the 10^{-8} to 10^{-4} M concentration range produced a graded reduction in contraction strength. Cholinergic induced changes in ventricle plasma membrane electrical properties, however, have not been observed except at ACh concentrations close to 1 mM (Lueken and Shütz, 1938; Bammer, 1952; Hoffman and Cranefield, 1960). Edman and Schild (1962) and Schatzmann (1961, 1964) have postulated that ACh acts in smooth muscle by releasing calcium from an intracellular store, probably the cytoplasmic reticulum. The work of Parisi *et al.* (1975) shows that choline-containing compounds can cross negatively charged membranes rather easily.

Most detailed electrophysiological studies on ACh receptors *in situ* and most attempts to reconstitute the ACh receptor have been carried out on ACh receptors in the neuromuscular junction and in the electric organ of electrical fish. Studies on the effects of antibodies raised to electric organ ACh receptors on neuromuscular junction ACh receptors have demonstrated the close relationship between the receptor molecules from these two sources (Patrick and Lindstrom, 1973; Sugiyama *et al.*, 1973; Bevan *et al.*, 1976).

Agents have been found that modify the properties of the cholinergic response. The reducing agent dithiothreitol reversibly reduces the single-channel conductance and shortens the mean open times (Ben-Haim *et al.*, 1975). Aliphatic alcohols increase mean channel open times up to 5-fold (Gage *et al.*, 1975). This effect may involve changes in the membrane matrix (Gage *et al.*, 1975; Neher and Stevens, 1977) and thus be an important consideration in reconstitution studies.

Different cholinergic agonists induce ACh receptor responses that differ in detail (i.e., single-channel conductance and mean channel open times). These effects were best studied *in situ* by fluctuation analysis. In fact, the study of ACh receptors represents the most successful application of fluctuation analysis techniques to date. The reader is referred to an excellent review and discussion by Neher and Stevens (1977). Little use has been made of fluctuation analysis in ACh receptor reconstitution studies.

Eldefrawi *et al.* (1975a) demonstrated an ACh-dependent binding of calcium by isolated ACh receptors and suggested that calcium may play an important role in the cholinergic response. Chang and Neumann (1976; Neumann and Chang, 1976) found that the binding of 2 ACh molecules caused the release of 4–6 Ca^{2+} ions from the receptor molecule. In addition, they showed that cholinergic inhibitors caused re-uptake of the Ca^{2+} ions by the receptor. Because of the many-faceted and profound action of calcium on transmitter release and on ion transport in general at the neuromuscular junction, investigation of the role of calcium in the cholinergic response in intact membrane preparations has met with some rather formidable difficulties. Alburquerque *et al.* (1973) have suggested that the cholinergic response is mediated by two distinct protein components, a receptor molecule and a channel whose conductance is modulated via the receptor protein by ACh. Again, investigation into this possibility cannot be carried out particularly well using intact membrane preparations. The answers to these questions therefore most probably await a reasonably successful reconstitution of the purified ACh receptor in an artificial membrane system.

2. *Attempts to Reconstitute the ACh Receptor*

Numerous attempts have been reported for the reconstitution of purified or partially purified ACh receptor in artificial lipid membrane systems (Shamoo and Goldstein, 1977). To date none of the reconstituted systems had excitability that mimics in detail the physiological events (Stevens, 1974; Katz and Miledi, 1972). For example, Kemp *et al.* (1973) reported a 4-fold acceleration in the rate of conductance increase of bilayers doped with an ACh receptor preparation from rat diaphragm in response to the addition of ACh on the side of the membrane opposite that on which ACh receptor was added. Shamoo and Eldefrawi (1975) found it necessary to mildly treat the protein with trypsin in order to obtain a consistent cholinergic induced conductance increase. Once the conductance increased, it could not be reversed by addition of curare or α-bungarotoxin. Schlieper and DeRobertis (1977) have recently reported an increase in membrane conductance and current noise upon application of ACh to BLMs doped with ACh receptors. These phenomena were reversed upon addition of *d*-tubocurarine. Spectral analysis of the cholinergic-induced current noise has not been reported by these workers. Murphy (1977) found that bilayers doped with membrane vesicles from frog ventricle responded to 10^{-4} *M* ACh or carbamylcholine with a nonspecific, voltage-dependent increase in conductance. The opening/closing kinetics of the cholinergic-induced channels were studied by spectral analysis of the membrane current noise. The voltage dependence of the characteristic time constant for change in channel conductance was found to be qualitatively similar to that observed for neuromuscular junction ACh receptors in intact membranes, but significantly slower. Hazelbauer and Changeux (1974) and Michaelson and Raftery (1974; Michaelson *et al.*, 1976) reported the cholinergic induced release of ^{22}Na efflux from ACh receptor-doped vesicles.

In virtually all these studies, reproducibility was rather low and the kinetics of the response, when studied, proved to be significantly slower than has been observed in *in situ* studies. For reconstitution studies to have any advantage over *in situ* measurements in gaining meaningful information about an ion transport system, it must be established that the experimenter has more control of the reconstituted system than he would have of the transport system *in situ*. While the above-cited studies show much progress toward the ahievement of this situation for the case of ACh receptors, it is clear that the stage has not yet been reached where meaningful conclusions about the mechanism of the cholinergic response can be drawn with confidence from reconstitution studies.

One of the problems faced by an experimenter attempting to reconstitute the ACh receptor is that of desentization. *In vivo*, the concentration of ACh at the ACh receptor is high enough to elicit a response only transiently because of the presence of acetylcholine esterase. When the concentration of ACh or other cholinergic agonists is artificially maintained for longer periods (seconds), the ACh receptor "desensitizes" or becomes less sensitive to cholinergic agents. This effect has been observed both in muscle end plates (Katz and Thesleff, 1957) and in electroplaque (Lester *et al.*, 1975). In reconstitution studies, the investigator is faced with obtaining a response comparable to *in situ* response without comparable means of obtaining *in situ*-like jumps in agonist concentration. Studying the cholinergic response in electroplaque, Lester and Chang (1977) have described the use of an intriguing cholinergic agonist, 3,3-bis-α-(trimethylammonium)methylazobenzene (bis-Q). This compound exists in two isomeric forms, one of which is an active agonist for the ACh receptor. Moreover, the compound can be converted to predominantly one isomer or the other by exposure to light of the appropriate wavelength. Thus, *in vivo*-like jumps in agonist concentration at the membrane can be achieved with short flashes of light. This compound may well be useful in future reconstitution studies.

3. *Isolation of Ionophores from the ACh Receptor*

To date, no detailed studies have been published on the isolation of ionophores from this ion transport system. Several features of the receptor have been demonstrated, however, that suggest that this approach would be useful in elucidating the mechanism of the cholinergic response. The receptor protein is made up of several subunits (Lindstrom and Patrick, 1974; Meunier *et al.*, 1974; Chang, 1974; Michaelson *et al.*, 1974; Eldefrawi *et al.*, 1975a; Carroll *et al.*, 1973; Eldefrawi *et al.*, 1975b; Hucho *et al.*, 1978). On the basis of the differential effects of α-bungarotoxin and perhydrohistrionicotoxin on the cholinergic response *in situ*, Albuquerque *et al.* (1973) proposed that this ion transport system is composed of two types of sites, a cholinergic agonist binding site and a conductance modulator (ionophorous) site.

The cholinergic agonist binding site and the ionophore site must be tightly coupled as evidenced by the fact that single-channel conductance and channel kinetics depend on the agonist (see Section IV,D,1 above). The role of Ca^{2+} in this coupling, if any, has yet to be elucidated.

E. Mitochondrial Membrane

Mitochondria have long been recognized as the primary source of cellular ATP. The production of ATP is carried out in the mitochondrial membrane and involves transmembrane proton and voltage gradients. In addition, it has become apparent in the last few years that this organelle is also involved in the buffering of intracellular ion concentrations, especially Ca^{2+} (see Bygrave, 1977, 1978, for recent and comprehensive reviews). Consequently, the mitochondrial membrane is an extremely rich source of proton and ion transport systems. For several reasons, detailed study of mitochondrial ion transport systems *in situ* are met with rather formidable experimental difficulties. The electrophysiological techniques that have been so useful in characterizing ion transport systems in cell plasm membranes are very difficult (data are difficult both to obtain and to interpret) because of the small size and complex structure of the mitochondrion. Other types of studies are confounded by the abundance of ion-binding molecules (proteins and lipids) in the organelle, which render the free concentration of various ions, especially of Ca^{2+}, only a small fraction of the total ion content of the structure.

These factors would appear to make the use of artificial membrane in reconstitution studies; the method of choice for the assay of ionophores in studying ion transport systems from the mitochondrial membrane. However, because of the high incidence of artifacts in these studies, *in situ* results must be relied upon as a guide in the interpretation of the results. Thus, the eventual unraveling of mechanism of ion transport in mitochondria will require a close interplay between *in situ* studies and reconstitution studies.

Ionophores Isolated from Mitochondrial Membrane.

There is a compelling body of evidence from *in situ* studies that calcium transport in mitochondria is mediated by a carrier embedded in the inner mitochondrial membrane (see Bygrave, 1977, 1978, for a comprehensive review). Numerous attempts have been made to isolate this calcium carrier. Carafoli and co-workers (Carafoli, 1975; Carafoli and Sottocasa, 1974;

Prestipino *et al.*, 1974) reported that the mitochondrial Ca^{2+} binding glycoprotein displayed divalent cation-dependent but nonselective ionophoric activity in BLMs. This protein, however, may be an extrinsic membrane binding protein rather than a Ca^{2+} carrier (Carafoli, 1976; Prestipino *et al.*, 1974) and its exact role in Ca^{2+}-transport has yet to be determined.

Blondin (1974a,b, 1975) found that several nonpeptidic organic compounds isolated from beef heart mitochondria displayed Ca^{2+} ionophoric activity in equilibrium extraction and bulk-phase transport experiments. The same laboratory (Blondin *et al.*, 1977) has isolated a K^{+}/Ca^{2+} ionophore from mitochondria membrane using tryptic digestion and diazomethane treatment (to eliminate the charged groups generated by the trypsin cleavage) to release ionophoric activity. The ionophoric activity was assayed for using the aminonaphthalensulfonic acid fluorescence method of Feinstein and Felsenfeld (1971) and the ability of the ionophore to induce uncoupler-sensitive cation uptake in mitochondria. The ionophore had a molecular weight of 1600 and was part of larger protein with molecular weight greater than 10,000.

Shamoo and Jeng (1978) and Jeng *et al.* (1978) isolated a Ca^{2+} ionophore from calf heart inner mitochondrial membrane using an electron paramagnetic resonance assay based on the relative binding properties of the protein to Ca^{2+}, Mn^{2+}, and Mg^{2+}. The molecular weight of this protein was estimated to be about 3000 by urea–SDS gel electrophoresis and amino acid analysis. Experiments involving extraction with organic solvents showed selectivity toward divalent cations over monovalent cations. The relative selectivity sequence for divalent cations was $Ca^{2+}, Sr^{2+} > Mn^{2+} > Mg^{2+}$. Ruthenium red and La^{3+} inhibited the protein-mediated extraction of Ca^{2+} into the organic solvent. The calcium translocation, determined in a Pressman cell, was selectively driven by a hydrogen ion gradient. The protein as isolated contained a large amount of tightly bound phospholipid. Extensive delipidation of the protein, however, left its ionophoric activity intact (Jeng and Shamoo, unpublished observations). To reflect its believed function in the mitochondrial membrane, the protein was named calciphorin.

Schein *et al.* (1976) isolated a voltage-dependent, anion-selective ionophore from *Paramecium* mitochondria and used bilayers formed from monolayers to study its properties. The ionophore formed a channel in BLMs as evidenced by the occurrence of uniformly sized discrete jumps in conductance under conditions where only a few ionophores were present in the membrane. The ionophore showed a steep dependence on membrane voltage that peaked at zero transmembrane voltage and was symmetric about that value. The voltage dependence of conductance correlated well with the mean time during which the channel is open with values for a few other ionophores.

F. Other Ion Transport Systems Containing Ionophores

An impressively detailed array of data has been collected by Rothstein's group and Passow's group on anion transport in red blood cells (see review in Shamoo and Goldstein, 1977). Many of these data were collected without recourse to the use of artificial membranes. Isolation and characterization of ionophores from gastric mucosal membrane and various other sources has recently been reviewed in detail by Shamoo and Goldstein (1977) and Blumenthal and Shamoo (1978). Montal (1976) has extensively reviewed the use of artificial membranes in the study of rhodopsin. The reader is referred to these review articles for expansion beyond the scope of this review.

V. General Properties of Ion Transport Systems

Most natural ion transport systems probably contain ionophores. Some of the simpler systems, in fact, may be nothing more than an ionophore embedded in the membrane. For example, the passive potassium conductance system responsible for the membrane resting potential in nerve, muscle, and other cells appears to be "turned on" all the time. Regulation of such a system needs to be no more elaborate than perhaps controlling the concentration of a K^+-selective ionophore (carrier or channel) in the membrane.

For many other ion transport systems, however, fine control transport on a very short time scale necessitates a more direct regulation of the ionophoric activity *in vivo*. Consequently, the ionophore is only part of the protein or lipoprotein complex that makes up the ion transport system. This is the case for all the transport systems discussed in this review. For active transport systems, the energy source or energy transducer also serves as a focus for fine control of ion movement.

In order to elucidate the mechanism of ion transport, it is necessary not only to study the ionophoric properties of membrane proteins and the details of the control system, but also to study the interaction between these two parts of the overall system. By dissecting the complex that makes up the total system and identifying the function of the various parts, the study of this interaction can be made possible.

The study of ionophores isolated from ion transport systems can contribute to the unraveling of the detailed mechanism of ion transport in two ways. (1) Once the (usually small) part of the system that is the ionophore is isolated and identified, the structural details that account for ion selectivity can be studied. (2) Once the ionophore and ionophoric activity regulating parts of the system have been identified, the interaction between the two can be more readily studied and elucidated.

VI. Summary and Concluding Remarks

It should be clear from our discussion that neither the construction of a model ion transport system in an artificial membrane system nor the isolation of a selective ionophore from the membrane for the model is an end in itself. There should be at all times a strong interplay between *in situ* studies of ionophore in order to resolve what originally this field started to resolve—namely, the elucidation of the molecular mechanism of transport systems and, more important, the mechanism of energy transduction. We should not lose sight for a moment of what is our essential goal and of the fact that all these systems are subservient to our original objective.

In order to achieve our goal in the description of the mechanism of ion transport, unfortunately, we must disassemble the machinery for such transport. Once this is accomplished, we are faced with the problem of ensuring that the disassembled pieces have in fact the function assigned to them. This can partially be achieved by comparing characteristics in common between the pieces and the parent protein, such as common specificity and selectivity and common inhibitors. We should be vigilant in using multiple methods and criteria to make the connection between the isolated pieces and the parent protein. Furthermore, we should continue to develop better and more efficient reconstituted systems in order to avoid collection of large volumes of data where protein-induced BLM conductance increases have nothing or little to do with the transport systems *in situ*, and to avoid misinterpretation of data, as happened when it was erroneously concluded that Cl^- was actively transported by Na + K^+-ATPase (Goldin and Tong, 1974). In the case of Ca^{2+} + Mg^{2+}-ATPase sufficient data are now available to correlate well the properties of the reconstituted system with the *in situ* function, and therefore it could serve as a model for future elucidation of transport mechanism.

The identification of ion transport site (ionophore) and the ATP hydrolytic enzyme provide the best lead yet to the study of the mechanism of energy transduction. It could provide the means by which one can study the spatial relationship between the ATP hydrolytic site and the ion transport site and correlate this with the functional state of the enzyme. It would be very useful if a system were developed whereby one could control the state of phosphorylation and dephosphorylation of the enzyme to study the relationship between the ionophoric and ATP hydrolytic sites under such controlled states.

ACKNOWLEDGMENTS

This paper is based on work performed under contract with the U.S. Department of Energy (DOE) at the University of Rochester Biomedical and Environmental Research Project and has been assigned Report No. UR-3490-1414. This paper was also supported in part by NIH Grant 1 RO1 AM 18892; Program Project Grant ES-10248; the Muscular Dystrophy Association (USA); and the Upjohn Company. A.E.S. is an established Investigator of the American Heart Association.

REFERENCES

Abramson, J. J., and Shamoo, A. E. (1979). *J. Membr. Biol.* **44**, 233–258.

Albuquerque, E. X., Barnard, E. A., Chiu, T. H., Lapa, A. J., Dolly, O. J., Jansson, S.-E., Daly, J., and Witkop, B. (1973). *Proc. Natl. Acad. Sci. U.S.A.* **70**, 949–953.

Anner, B. M., Lane, L. K., Schwartz, A., and Pitts, B. J. R. (1977). *Biochim. Biophys. Acta* **467**, 340–345.

Antanavege, J., Chien, T. F., Ching, Y. C., Dunlap, L., Mueller, P., and Rudy, B. (1978). *Biophys. J.* **21**, 122a.

Bammer, H. (1952). *Pfluegers Arch. Gesampte Physiol. Menschen Tiere* **255**, 476–484.

Bangham, A. D., Standish, M. M., and Watkins, J. C. (1965). *J. Mol. Biol.* **13**, 238–252.

Barnola, F. V., and Villegas, R. (1976). *J. Gen. Physiol.* **67**, 81–90.

Ben-Haim, D., Dreyer, F., and Peper, K. (1975). *Pfluegers Arch.* **355**, 19–26.

Berman, M. C., McIntosh, D. B., and Kench, J. E. (1977). *J. Biol. Chem.* **252**, 994–1001.

Bevan, S., Heinemann, S., Lennon, V. A., and Lindstrom, J. (1976). *Nature (London)* **260**, 438.

Blondin, G. A. (1974a). *Ann. N.Y. Acad. Sci.* **227**, 392–397.

Blondin, G. A. (1974b). *Biochem. Biophys. Res. Commun.* **56**, 97–105.

Blondin, G. A. (1975). *Ann. N.Y. Acad. Sci.* **264**, 98–111.

Blondin, G. A., Kessler, R. J., and Green, D. E. (1977). *Proc. Natl. Acad. Sci. U.S.A.* **74**, 3667–3671.

Blumenthal, R., and Shamoo, A. E. (1979). *In* "The Receptors: A Comprehensive Treatise" (R. D. O'Brien, ed.). Plenum, New York.

Burg, M., and Stoner, L. (1974). *Fed. Proc., Fed. Am. Soc. Exp. Biol.* **33**, 31–36.

Bygrave, F. L. (1977). *Curr. Top. Bioenerg.* **6**, 260–318.

Bygrave, F. L. (1978). *Biol. Rev. Cambridge Philos. Soc.* **53**, 43–79.

Carafoli, E. (1975). *Mol. Cell. Biochem.* **8**, 133–140.

Carafoli, E. (1976). *In* "Mitochondria" (L. Packer and A. Gomez-Puyou, eds.), pp. 47–60. Academic Press, New York.

Carafoli, E., and Sottocasa, G. (1974). *In* "Dynamics of Energy-Transducing Membranes" (L. Ernster, R. W. Estabrook, and E. C. Slater, eds.), pp. 455–469. Elsevier, Amsterdam.

Carroll, R. C., Eldefrawi, M. E., and Edelstein, S. J. (1973). *Biochem. Biophys. Res. Commun.* **58**, 864.

Chang, H. W. (1974). *Proc. Natl. Acad. Sci. U.S.A.* **71**, 2133.

Chang, H., and Neumann, E. (1976). *Proc. Natl. Acad. Sci. U.S.A.* **73**, 3364.

Chiesi, M., Prestipino, G. F., Wanke, E., and Carafoli, E. (1977). *In* "Calcium-Binding Proteins and Calcium Function" (R. H. Wasserman *et al.*, eds.), pp. 188–191. North-Holland Publ., Amsterdam.

Clark, A. J. (1926). *J. Physiol. (London).* **61**, 530–556.

Danielli, J. F. (1966). *J. Theor. Biol.* **12**, 439–441.
Davson, H., and Danielli, J. F. (1952). "The Permeability of Natural Membranes," 2nd ed. Cambridge Univ. Press, London and New York.
Deamer, D. W. (1973). *J. Biol. Chem.* **248**, 5477–5485.
Deamer, D. W., and Baskin, R. J. (1972). *Arch. Biochem. Biophys.* **153**, 47–54.
Dean, W. L., and Tanford, C. (1977). *J. Biol. Chem.* **250**, 2013.
Dean, W. L., and Tanford, C. (1978). *Biochemistry* **17**, 1683–1690.
DeBoland, A. R., Jilka, R. L., and Martonosi, A. N. (1975). *J. Biol. Chem.* **250**, 7501–7510.
Del Castillo, J., and Katz. B. (1955). *Nature* (*London*) **175**, 1035.
Edman, K. A. P., and Schild, H. O. (1962). *J. Physiol.* (*London*) **161**, 424–441.
Eldefrawi, M. E., Eldefrawi, A. T., Penfield, L. A., O'Brien, R. D., and Van Campen, D. (1975a). *Life Sci.* **16**, 925.
Eldefrawi, M. E., Eldefrawi, A. T., and Shamoo, A. E. (1975b). *Ann. N.Y. Acad. Sci.* **264**, 183–202.
Eldefrawi, M. E., Eldefrawi, A. T., and Wilson, D. B. (1975c). *Biochemistry* **14**, 4304.
Feinstein, M. B., and Felsenfeld, H. (1971). *Proc. Natl. Acad. Sci. U.S.A.* **68**, 2037–2041.
Fettiplace, R., Andrews, D. M. and Haydon, D. A. (1971). *J. Membr. Biol.* **5**, 277–296.
Gage, P. W., McBurney, R. N., and Schneider, G. T. (1975). *J. Physiol.* (*London*) **244**, 409–429.
Goldin, S. M. (1977). *J. Biol. Chem.* **252**, 5630–5642.
Goldin, S. M., and Sweadner, K. J. (1975). *Ann. N.Y. Acad. Sci.* **264**, 387–396.
Goldin, S. M., and Tong, S. W. (1974). *J. Biol. Chem.* **249**, 5907–5915.
Goldman, D. E. (1943). *J. Gen. Physiol.* **27**, 37–61.
Gomez-Puyou, A., and Gomez-Lojero, C. (1977). *Curr. Top. Bioenerg.* **6**, 222–259.
Good, R. J. (1969). *J. Colloid. Interface Sci.* **31**, 540–544.
Hasselbach, W., and Makinose, M. (1963). *Biochem. Z.* **339**, 94.
Haydon, D. A., Hendry, B. M., Levinson, S. R., and Requena, J. (1977). *Biochim. Biophys. Acta* **470**, 17–34.
Hazelbauer, G. L., and Changeux, J.-P. (1974). *Proc. Natl. Acad. Sci. U.S.A.* **71**, 1479.
Henn, F. A., and Thompson, T. E. (1967). *J. Mol. Biol.* **31**, 227–235.
Henn, F. A., Decker, G. L., Greenawalt, J. W., and Thompson, T. E. (1967). *J. Mol. Biol.* **24**, 51–58.
Hidalgo, C., and Tong, S. (1978). *Biophys. J.* **21** ; 46a.
Hilden, S., and Hokin, L. E. (1975). *J. Biol. Chem.* **250**, 6296–6303.
Hilden, S., Rhee, H. M., and Hokin, L. E. (1974). *J. Biol. Chem.* **249**, 7432–7440.
Hoffman, B. F., and Cranefield, P. E. (1960). "Electrophysiology of the Heart." McGraw-Hill, New York.
Huang, C. H. (1969). *Biochemistry* **8**, 344–352.
Hucho, F., Bandini, G., and Suarez-Isla, B. A. (1978). *Eur. J. Biochem.* **83**, 335–340.
Hutter, O. F., and Trautwein, W. (1956). *J. Gen. Physiol.* **39**, 715–733.
Hyman, E. S. (1977). *J. Membr. Biol.* **37**, 263–275.
Ikemoto, N. (1974). *J. Biol. Chem.* **249**, 649–651.
Ikemoto, N. (1975). *J. Biol. Chem.* **250**, 7219–7224.
Ikemoto, N., Bhatnager, G. M., and Gergely, J. (1971). *Biochem. Biophys. Res. Commun.* **44**, 1510–1517.
Inesi, G., Maring, E., Murphy, A. J., and MacFarland, B. H. (1970). *Arch. Biochem. Biophys.* **138**, 285–294.
Jain, M. K., White, F. P., Strickholm, A., Williams, E., and Cordes, E. H. (1972). *J. Membr. Biol.* **8**, 363–388.
Jain, M. K., White, F. P., Strickholm, A., Williams, E., and Cordes, E. H. (1973). *J. Membr. Biol.* **11**, 195–196.

Jeng, A. Y., Ryan, T. E., and Shamoo, A. E. (1978). *Proc. Natl. Acad. Sci. U.S.A.* **75**, 2125–2129.
Jilka, R. L., Martonosi, A. N., and Tillack, J. W. (1975). *J. Biol. Chem.* **250**, 7511–7524.
Kanazawa, T., Yamada, S., and Tonomura, Y. (1970). *J. Biochem.* (*Tokyo*) **68**, 593–595.
Kanazawa, T., Yamada, S., Yamamoto, T., and Tonomura, Y. (1971). *J. Biochem.* (*Tokyo*) **70**, 95–123.
Kasahara, M., and Hinkle, P. C. (1977). *J. Biol. Chem.* **252**, 7384–7390.
Katz, B., and Miledi, R. (1972). *J. Physiol.* (*London*) **224**, 665.
Katz, B., and Thesleff, S. (1957). *J. Physiol.* (*London*) **138**, 63–80.
Kemp, G., Dolly, J. A., Barnard, E. A., and Wenner, C. E. (1973). *Biochem. Biophys. Res. Commun.* **54**, 607–613.
Klip, A., and MacLennan, D. H. (1978). "Frontiers of Biological Energetics: Electrons to Tissues" (A. Scarpa, F. Dutton, and J. Leigh, eds.), Vol. 2, 1137–1141. Academic Press, New York.
Knowles, A. F., Kandrach, A., Racker, E., and Khorana, H. E. (1975). *J. Biol. Chem.* **250**, 1809–1813.
Laggner, P., and Graham, D. E. (1976). *Biochim. Biophys. Acta* **433**, 311–317.
LeMaire, M., Møller, J. V., and Tanford, C. (1976). *Biochemistry* **15**, 2336–2342.
Lester, H. A., and Chang, H. W. (1977). *Nature* (*London*) **266**, 373–374.
Lester, H. A., Changeux, J.-P., and Sheridan, R. E. (1975). *J. Gen. Physiol.* **65**, 797–816.
Lindstrom, J., and Patrick, J. (1974). *In* "Synaptic Transmission and Neuronal Interaction" (M. V. L. Bennett, ed.), p. 191. Raven, New York.
Lueken, B., and Schütz, E. (1938). *Z. Biol.* (*Munich*) **99**, 186–197.
McLaughlin, S., and Eisenberg, M. (1975). *Annu. Rev. Biophys. Bioeng.* **4**, 335–366.
MacLennan, D. H. (1970). *J. Biol. Chem.* **245**, 4508–4518.
MacLennan, D. H., and Holland, P. C. (1975). *Annu. Rev. Biophys. Bioeng.* **4**, 377–404.
Madeira, V. M. C. (1978). *Arch. Biochem. Biophys.* **185**, 316–325.
Magleby, K. L., and Stevens, C. F. (1972). *J. Physiol.* (*London*) **223**, 173–197.
Makinose, M., and Hasselbach, W. (1971). *FEBS Lett.* **12**, 271–272.
Malan, N. T., Sabbadini, R., Scales, D., and Inesi, G. (1975). *FEBS Lett.* **60**, 122–125.
Malnic, G., Klose, R. M., and Giebisch, G. (1966). *Am. J. Physiol.* **211**, 529–540.
Meissner, G., Conner, G. E., and Fleischer, S. (1973). *Biochim. Biophys. Acta* **298**, 246–269.
Meunier, J. C., Sealock, R., Olsen, R., and Changeux, J.-P. (1974). *Eur. J. Biochem.* **45**, 371.
Michaelson, D. M., and Raftery, M. A. (1974). *Proc. Natl. Acad. Sci. U.S.A.* **71**, 4768–4772.
Michaelson, D., Vandlen, R., Bode, J. Moody, T., Schmidt, J., and Raftery, M. A. (1974). *Arch. Biochem. Biophys.* **165**, 796.
Michaelson, D. M., Duguid, J. R., Miller, D. L., and Raftery, M. A. (1976). *J. Supramol. Struct.* **4**, 419–425.
Montal, M. (1976). *Annu. Rev. Biophys. Bioeng.* **5**, 19–175.
Montal, M., and Mueller, P. (1972). *Proc. Natl. Acad. Sci. U.S.A.* **69**, 3561.
Morel, F., and deRouffignac, C. (1973). *Annu. Rev. Physiol.* **35**, 17–54.
Mueller, P., and Rudin, D. O. (1963). *J. Theor. Biol.* **4**, 268–280.
Mueller, P., and Rudin, D. O. (1968). *J. Theor. Biol.* **18**, 222–258.
Mueller, P., Rudin, D. O., Tien, H. T., and Wescott, W. C. (1962). *Circulation* **26**, 1167–1171.
Murphy, T. J. (1977). Ph.D. Thesis, Univ. of Illinois, Urbana-Campaign.
Neher, E., and Stevens, C. F. (1977). *Annu. Rev. Biophys. Bioeng.* **6**, 345–381.
Neumann, E., and Chang, H. (1976). *Proc. Natl. Acad. Sci. U.S.A.* **73**, 3794.
Ohki, S., and Fukuda, N. (1967). *J. Theor. Biol.* **15**, 362–375.
Pagano, R. E., Ruysschaert, J. M., and Miller, I. R. (1972). *J. Membr. Biol.* **10**, 11–30.
Panet, R., and Selinger, Z. (1972). *Biochim. Biophys. Acta* **255**, 34–42.
Papahadjopoulos, D., and Miller, I. R. (1967). *Biochim. Biophys. Acta* **135**, 624–638.

Papahadjopoulos, D., and Watkins, J. C. (1967). *Biochim. Biophys. Acta* **135**, 639–652.
Parisi, M., Adragna, C., and Salas, P. J. I. (1975). *Nature* (*London*) **258**, 245–247.
Patrick, J., and Lindstrom, J. (1973). *Science* **180**, 871.
Pickard, W. F. (1976). *Math. Biosci.* **30**, 99–111.
Pressman, B. C. (1976). *Annu. Rev. Biochem.* **45**, 501–530.
Prestipino, G., Ceccarelli, D., Conti, F., and Carafoli, E. (1974). *FEBS Lett.* **45**, 99–103.
Racker, E. (1972). *J. Biol. Chem.* **247**, 8198–8200.
Racker, E. (1933). *Biochem. Biophys. Res. Commun.* **55**, 224–230.
Racker, E. (1977). *In* "Calcium-Binding Proteins and Calcium Function" (R. H. Wasserman *et al.* eds.), pp. 155–163. North-Holland Publ., Amsterdam.
Racker, E. (1979). *Membr. Biochem.* (in press).
Racker, E., and Eytan, E. (1973). *Biochem. Biophys. Res. Commun.* **55**, 174–178.
Racker, E., and Eytan, E. (1975). *J. Biol. Chem.* **250**, 7533–7534.
Racker, E., and Fisher, L. W. (1975). *Biochem. Biophys. Res. Commun.* **67**, 1144–1150.
Racker, E., Chien, T. F., and Kandrach, A. (1975a). *FEBS Lett.* **57**, 14–18.
Racker, E., Knowles, A. F., and Eytan, E. (1975b). *Ann. N.Y. Acad. Sci.* **264**, 17–33.
Reeves, J. P., and Dowben, R. M. (1968). *J. Cell. Physiol.* **73**, 49–60.
Ryan, T. E., Woods, G. M., Kirkpatrick, F. H., and Shamoo, A. E. (1976). *Anal. Biochem.* **72**, 359–365.
Schatzmann, H. J. (1961). *Pfluegers Arch. Menschen Tiere* **274**, 295–310.
Schatzmann, H. J. (1964). *In* "Pharmacology of Smooth Muscle" (E. Bulbring, ed.), pp. 57–69. Pergamon, New York.
Schein, S. J., Colombini, M., and Finkelstein, A. (1976). *J. Membr. Biol.* **30**, 99–120.
Schlieper, P., and DeRobertis, E. (1977). *Biochem. Biophys. Res. Commun.* **75**, 886–894.
Scott, T., and Shamoo, A. E. (1977). *Biophys. J.* **17**, 185a.
Scott, T. L., and Shamoo, A. E. (1978). *Biophys. J.* **21**, 47a.
Shamoo, A. E. (1978). *J. Membr. Biol.* **43**, 227–242.
Shamoo, A. E., and Abramson, J. J. (1977). *In* "International Symposium on Calcium-Binding Proteins and Calcium Function in Health and Disease" (R. H. Wasserman *et al.*, eds.), pp. 173–180. Elsevier/North Holland, Amsterdam.
Shamoo, A. E., and Albers, R. W. (1973). *Proc. Natl. Acad. Sci. U.S.A.* **70**, 1191–1194.
Shamoo, A. E., and Eldefrawi, M. E. (1975). *J. Membr. Biol.* **25**, 47–63.
Shamoo, A. E., and Goldstein, D. A. (1977). *Biochim. Biophys. Acta* **472**, 13–53.
Shamoo, A. E., and Jeng, A. Y. (1978). *Ann. N.Y. Acad. Sci.* **307**, 235–237.
Shamoo, A. E., and MacLennan, D. H. (1974). *Proc. Natl. Acad. Sci. U.S.A.* **71**, 3522–3526.
Shamoo, A. E., and MacLennan, D. H. (1975). *J. Membr. Biol.* **25**, 65–74.
Shamoo, A. E., and Ryan, T. E. (1975). *Ann. N.Y. Acad. Sci.* **264**, 83–97.
Shamoo, A. E., Thompson, T. E., Campbell, K. P., Scott, T. L., and Goldstein, D. A. (1975). *J. Biol. Chem.* **250**, 8289–8291.
Shamoo, A. E., MacLennan, D. H., and Eldefrawi, M. E. (1976a). *Chem.-Biol. Interact.* **12**, 41–52.
Shamoo, A. E., Ryan, T. E., Stewart, P. S., and MacLennan, D. H. (1976b). *J. Biol. Chem.* **251**, 4147–4154.
Shamoo, A. E., Scott, T. L., and Ryan, T. E. (1977). *J. Supramol. Struct.* **6**, 345–353.
Skou, J. C. (1957). *Biochim. Biophys. Acta* **23**, 394–401.
Skou, J. C. (1971). *Curr. Top. Bioenerg.* **4**, 357–398.
Stevens, C. F. (1972). *Biophys. J.* **12**, 1028–1047.
Stevens, C. F. (1974). *In* "Synaptic Transmission and Neuronal Interactions" (M. V. L. Bennet, ed.), p. 45. Raven, New York.
Stewart, P. S., MacLennan, D. H. and Shamoo, A. E. (1976). *J. Biol. Chem.* **251**, 712–719.

Sugiyama, H., Benda, P., Meunier, J.-C., and Changeux, J.-P. (1973). *FEBS Lett.* **35**, 124.
Sumida, M., and Tonomura, Y. (1974). *J. Biochem.* (*Tokyo*) **75**, 283–297.
Sweader, K. J., and Goldin, S. M. (1975). *J. Biol. Chem.* **250**, 4022–4024.
Tashmukhamedor, B. A. (1976). *IUPAB Proc., Copenhagen*
Thorley-Lawson, D. A., and Green, N. M. (1975). *Eur. J. Biochem.* **59**, 193–200.
Thorley-Lawson, D. A., and Green, N. M. (1977). *Biochem. J.* **167**, 739.
Tien, H. T., and Dawidowicz, E. A. (1966). *J. Colloid Interface Sci.* **22**, 438–453.
Villegas, R., Villegas, G. M., Barnola, F. V., and Racker, E. (1977). *Biochem. Biophys. Res. Commun.* **79**, 210–217.
Warren, G. B., Toon, P. A., Birdsall, N. J. M., Lee, A. G., and Metcalfe, J. C. (1974a). *Proc. Natl. Acad. Sci. U.S.A.* **71**, 622–626.
Warren, G. B., Toon, P. A., Birdsall, N. J. M., Lee, A. G., and Metcalfe, J. C. (1974b). *FEBS Lett.* **41**, 122–124.
Weber, A., Herz, R., and Reiss, I. (1966). *Biochem. Z.* **345**, 329.
Windhager, E. E., and Giebisch, G. (1961). *Am. J. Physiol.* **200**, 581.
Yamada, S., Sumida, M., and Tonomura, Y. (1972). *J. Biochem.* (*Tokyo*) **72**, 1537–1548.
Yamamoto, T. (1972). *In* "Muscle Proteins, Muscle Contraction and Cation Transport" (Y. Tonamura, ed.), pp. 305–356. Univ. Park Press, Baltimore, Maryland.
Yu, B. P., Masoro, E. J., and Downs, J. (1978). *Mol. Cell. Biochem.* **19**, 3–6.

Reaction Mechanisms for ATP Hydrolysis and Synthesis in the Sarcoplasmic Reticulum

TAIBO YAMAMOTO, HARUHIKO TAKISAWA, and YUJI TONOMURA
Department of Biology
Faculty of Science
Osaka University
Toyonaka, Osaka, Japan

ISBN 0-12-152509-0

I. Introduction

A. General Aspects

The membrane system in muscle cells, which has a role in controlling muscle contraction, consists of the plasma membrane with its tubular infoldings (the T-system) running transversely to the fiber axis and a reticular structure (the sarcoplasmic reticulum, SR) that forms a network surrounding the myofibrils (Bennett and Porter, 1953; Porter and Palade, 1957; Huxley, 1964; Peachey, 1965).

When the T-system communicates the depolarization of the plasma membrane to the interior of the muscle fiber (Huxley and Taylor, 1958), Ca^{2+} is released from SR to bind with troponin located on thin filaments and induces muscle contraction. When Ca^{2+} is reabsorbed by SR, the muscle relaxes (cf. Ebashi, 1976). The T-system and SR are interconnected in a unique structure called the triad (Porter and Palade, 1957). The mechanism by which the stimulation is transmitted through the triad is still unknown. Early studies on the structure and function of SR have been widely reviewed by such scholars as Porter (1961), Sandow (1965), Ebashi and Endo (1968), Huxley and Simmons (1971), and Costantin (1975).

Biochemical studies on the mechanisms for ATP hydrolysis and synthesis in the SR membrane system have been done mostly using fragmented SR (FSR) isolated from muscle homogenates as a microsomal fraction. Hasselbach and Makinose (1961) and Ebashi and Lipmann (1962) first observed that FSR can remove a significant amount of Ca^{2+} from the medium in the presence of ATP and Mg^{2+}. Hasselbach and Makinose (1961, 1962, 1963) also reported that FSR shows Ca^{2+}, Mg^{2+}-dependent ATPase activity. Furthermore, they demonstrated that 2 mol of Ca^{2+} were transported into the vesicles when 1 mol of ATP was hydrolyzed and that FSR could produce a Ca^{2+} concentration gradient of 1000 to 5000 across the membrane. These findings strongly suggest that Ca^{2+} is actively transported into FSR by the Ca^{2+}-dependent ATPase reaction.

FSR isolated from skeletal muscle shows evidence of empty vesicles about 7 nm wide (Hasselbach and Elfvin, 1967). The surface of the vesicle membrane is covered with particles about 4 nm in diameter (Inesi and Asai, 1968; Martonosi, 1968; Ikemoto *et al.*, 1968). As mentioned in Section II, these particles contain headpieces and stalks, and the catalytic site for the ATPase reaction is located in the headpiece. MacLennan (1970) and MacLennan *et al.* (1971) purified the ATPase of FSR solubilized with deoxycholate (DOC) by stepwise fractionation with ammonium acetate. The purified ATPase preparation consists of a major protein of the ATPase with a molecular weight of about 1×10^5 and various kinds of phospholipids. This preparation re-forms membrane when detergent is removed. The requirements of divalent cations for the purified ATPase are the same as those for the intact SR. However, the re-formed membrane from the purified ATPase is incapable of accumulating Ca^{2+}. Recently, Racker (1972, 1973) and Knowles and Racker (1975b) succeeded in reconstituting ATP-supported Ca^{2+} uptake by mixing purified ATPase and excess amounts of soybean phospholipids. Warren *et al.* (1974a,b) also reported that the activity of Ca^{2+} uptake was reconstituted even when the intrinsic lipid of the purified ATPase was completely substituted by a synthetic lipid. These results indicate that the active transport of Ca^{2+} through the SR membrane requires only two components, the ATPase protein and phospholipids.

B. Historical Survey of Studies on the Reaction Mechanisms of ATPase and Ca^{2+} Transport

FSR offers many advantages for studying molecular mechanism of active transport. (1) FSR can be easily isolated in large quantities from muscle homogenates. ATPase protein accounts for more than two-thirds of the total protein in the FSR membrane. (2) The ATPase reaction is tightly coupled to Ca^{2+} transport; 2 mol of Ca^{2+} are transported for each mole of ATP hydrolyzed under a variety of conditions. (3) The sidedness of the FSR membrane where the enzyme reacts with the substrate, products, and cation can be clearly distinguished. In addition, the sidedness of the membrane can be easily destroyed without loss of the ATPase activity by treating the membrane with some detergents or glycolethylenediaminetetraacetic acid (EGTA) in alkaline pH. (4) The concentrations of Ca^{2+} and Mg^{2+} outside the membrane can be controlled with chelators like EGTA and ethylenediaminetetraacetic acid (EDTA), and the concentration of Ca^{2+} inside the vesicle can be controlled with oxalate and P_i. (5) Furthermore, the process of Ca^{2+} transport is completely reversible; the outward movement of Ca^{2+} is coupled to ATP synthesis from ADP and P_i. These advantages of FSR have led in recent years to rapid and detailed developments in studies on the molecular mechanism of coupling between Ca^{2+} transport and the ATPase reaction.

Hasselbach and Makinose (1962, 1963) and Ebashi and Lipmann (1962) found that FSR catalyzed phosphate exchange between ADP and ATP in the presence of Ca^{2+} and Mg^{2+}. Hasselbach (1964b) and Makinose (1966) postulated that a phosphoprotein was formed as an intermediate of the reaction during the active transport of Ca^{2+}. Thereafter, the phosphorylated intermediate (EP) was isolated as an acid-stable phosphoprotein by Yamamoto and Tonomura (1967, 1968) and Makinose (1969). This finding facilitated studies on the coupling mechanism between Ca^{2+} transport and the ATPase reaction in SR membrane, and transient kinetic studies on the formation and decomposition of EP have been done by Kanazawa *et al.* (1971), Froehlich and Taylor (1975, 1976), and Kurzmack and Inesi (1977). Direct evidence for Ca^{2+} translocation across the SR membrane during EP formation has been obtained by Sumida and Tonomura (1974) and Kurzmack *et al.* (1977). Yamada and Tonomura (1972b) and Ikemoto (1975, 1976) demonstrated that the apparent affinity of the enzyme for Ca^{2+} in the presence of Mg^{2+} decreased drastically when it was phosphorylated. Recently, it has been shown that the ATPase reaction of SR is reversible. Makinose and Hasselbach (1971) made the discovery that Ca^{2+} efflux from FSR is coupled to ATP synthesis from ADP and P_i, i.e., the ATPase reaction of SR is reversible. The mechanism of the reverse reaction of the Ca^{2+}-dependent ATPase in the presence of Ca^{2+} gradient across the membrane was investigated in detail later by Makinose (1972) and Yamada *et al.* (1972). Kanazawa (1975) and Knowles and Racker (1975a) found that, even in the absence of Ca^{2+} gradient, ATPase could produce EP by reacting with P_i, and that the EP produced could react with ADP to form ATP in the presence of a large amount of Ca^{2+}.

This review principally covers the reaction mechanisms for energy transductions in the active transport of Ca^{2+} and considers the following major topics: (1) the reaction mechanism of Ca^{2+}-dependent ATPase, (2) the coupling mechanism of the ATPase reaction to Ca^{2+} transport, and (3) the mechanism of ATP synthesis in the reverse reaction of ATPase. This article is intended not to give an overall view of cation transport by SR, but to review recent important investigations concerning the molecular bioenergetics of the active transport of Ca^{2+} in the SR membrane system. The reader is referred to the reviews by Hasselbach (1964a, 1974), Martonosi (1972), Tonomura (1972), and Tada *et al.* (1978) for general studies on SR, and to those by Inesi (1972) and MacLennan and Holland (1976) for the structure of the SR membrane.

II. Structure of the Sarcoplasmic Reticulum (SR) Membrane

In this section, we document briefly recent studies on the molecular structure of SR membrane, focusing our attention on the problems related to the reaction of ATPase and its coupling to Ca^{2+} transport. Detailed knowledge about the structure of the membrane, particularly the membrane components and the structural features of the ATPase molecule is necessary to analyze the molecular mechanism of active Ca^{2+} transport across the membrane (Section VI). Furthermore, as described in Section IV, the kinetic properties of the Ca^{2+}-dependent ATPase change dramatically when the SR membrane is solubilized.

A. Ultrastructure of the SR Membrane

A number of studies on the ultrastructure of the SR membrane using electron microscopy, X-ray diffraction, electron paramagnetic resonance (EPR), and chemical modifications have demonstrated that the SR membrane is organized in a manner essentially compatible with the fluid mosaic model of Singer and Nicolson (1972).

The majority of the lipids of the SR membrane are several kinds of phospholipids: 65–73% phosphatidylcholine, 12–19% phosphatidylethanolamine, 6–12% phosphatidylserine, 8–9% phosphatidylinositol, and small amounts of other lipids (Martonosi *et al.*, 1968; Meissner and Fleischer, 1971; MacLennan *et al.*, 1971; Owens *et al.*, 1972). Phosphatidylcholine is known to be essential for the activity of SR ATPase (Martonosi, 1963; Martonosi *et al.*, 1968; Fiehn and Hasselbach, 1970; Nakamura and Ohnishi, 1975; Knowles *et al.*, 1976). However, whether other phospholipids directly participate in the ATPase reaction and Ca^{2+} transport by SR has not been ascertained. Phospholipid constituents of FSR can be divided into two classes: one is bilayer lipids, and the other is the so-called "annulus lipids," which bind tightly to the ATPase protein (Warren *et al.*, 1974a,b, 1975). When the purified ATPase of SR was delipidated by increasing amounts of detergent, the full ATPase activity was maintained only at a ratio above 30 mol of lipid per mole. At lower ratios, the activity was irreversibly reduced to a negligible level at about 15 lipid molecules for each ATPase molecule (Warren *et al.*, 1974b, 1975). These results suggest that the annulus lipids are essential for maintaining the structure of the ATPase protein necessary for its full activity (Warren *et al.*, 1975).

The composition of protein in the SR membrane has been extensively investigated, and four kinds have been identified (cf. MacLennan and Holland, 1975, 1976; Tada *et al.*, 1978). The ATPase protein with a molecular weight of about 1×10^5, which participates directly in active Ca^{2+} transport, accounts for 60–80% of the total protein of the SR membrane (Inesi, 1972; Yamada *et al.*, 1972; Meissner *et al.*, 1973). Recently, Meissner (1975) separated FSR into light and heavy fractions using a sucrose density gradient, and reported that in the light fraction about 90% of the total protein was the ATPase protein whereas the amount was 55–60% in the heavy fraction. The SR membrane contains several other kinds of protein; 5–19% calsequestrin with a molecular weight of 4.4×10^4 (MacLennan and Wong,

1971; MacLennan, 1974) and a small amount of the high-affinity Ca^{2+}-binding protein with a molecular weight of about 5.5×10^4 (Ostwald and MacLennan, 1974; Meissner *et al.*, 1973). These are proteins with strong negative charges and contain many Ca^{2+}-binding sites, although the affinities of these sites for Ca^{2+} are much lower than that of the ATPase protein. In addition, these proteins exist on the interior surface of FSR as extrinsic proteins. Therefore, they are assumed to be related to storage of Ca^{2+} (MacLennan and Holland, 1975, 1976). MacLennan *et al.* (1972) noted that proteolipid represents another constituent of intrinsic proteins. Proteolipid was extracted from the purified ATPase preparation with acidified chloroform methanol. Racker and Eytan (1975) reported that it markedly improved the coupling efficiency of ATP hydrolysis to Ca^{2+} transport in reconstituted vesicles from phospholipids and the purified ATPase. However, the mechanism is still unknown.

One of the most important questions related to the active transport of Ca^{2+} in FSR is whether the ATPase molecule is exposed to both the inner and outer surfaces of the membrane. Electron microscopic observation of negatively stained membrane of SR shows that the membrane is covered with particles about 4 nm in diameter, which are connected with stalks about 2 nm in diameter (Section I). The same particles were observed on vesicles formed from the purified ATPase preparation. These particles can be removed by exhaustive digestion by trypsin with accompanying loss of the ATPase activity and Ca^{2+} transport (Migala *et al.*, 1973; Stewart and MacLennan, 1974, 1975). Furthermore, the antibody against the purified ATPase binds to SR on the outer surface of the membrane (Stewart *et al.*, 1976). Hg-phenylazoferritin also modifies SH groups in ATPase on the outer surface of the membrane (Hasselbach and Elfvin, 1967). These findings suggest that at least part of the ATPase molecule is exposed to the outer surface of the SR membrane. However, there is not enough evidence to show that the ATPase molecule spans the membrane and is exposed also to the internal surface. Deamer and Baskin (1969), using the freeze fracture method, demonstrated that within the membrane were many particles 8–9 nm in diameter, which were considered to be ATPase molecules. The majority of these particles were distributed in the outer leaflet, but a few were in the inner leaflet of the intact SR membrane. However, Hidalgo and Ikemoto (1977) recently reported that 30% of the amino groups of ATPase protein were labeled by fluorescamine in intact FSR, whereas 72% were labeled when leaky vesicles were formed from the purified ATPase preparation. In order to explain these results, they suggest that the ATPase molecule is exposed also to the internal surface of the SR membrane.

Another important problem related to active Ca^{2+} transport is whether the functional unit of the transport system is formed from a monomer or oligomer of the ATPase molecule within the FSR membrane. Possible formation of an oligomer of the ATPase molecule has been indicated using techniques of cross-linking (Murphy, 1976a; Chyn and Martonosi, 1977), sedimentation in detergents (Rizzolo *et al.*, 1976; Le Maire *et al.*, 1976), fluorescence energy transfer, and electron microscopy (Vanderkooi *et al.*, 1977). Furthermore, Inesi (1972) concurred with the above possibility by pointing out the discrepancy between the densities of the surface and intramembranous particles of the ATPase. However, Jørgensen *et al.* (1978) clearly showed that the monomeric ATPase retains most of the catalytic features of vesicular ATPase. It is premature to conclude that ATPase molecules form an oligomer as a functional unit within the intact SR membrane. This problem will be discussed again in Section IV in connection with the analysis of the presteady-state kinetics of the ATPase reaction.

B. Molecular Structure of ATPase

The purified ATPase prepared by MacLennan's procedures contained phospholipid in an amount equal to that found in FSR. Its phospholipid composition was identical to that of the intact SR membrane (MacLennan, 1970; Owens *et al.*, 1972). The sodium dodecyl sulfate (SDS) gel electrophoretic pattern of the purified preparation showed a single band of the ATPase protein with a molecular weight of 10^5. The substrate specificity and requirement for divalent cations of this preparation were identical to those of intact SR. Furthermore, 0.7–0.9 mol of EP could be formed per mole of ATPase (MacLennan *et al.*, 1971; Meissner and Fleischer, 1973). These findings indicate that the ATPase protein with a molecular weight of 10^5 contains one catalytic site for ATP hydrolysis.

The ATPase protein was degraded into two fragments with molecular weights of 5.2×10^4 to 6.0×10^4 and 4.5×10^4 to 5.5×10^4 when FSR was exposed briefly to trypsin (Thorley-Lawson and Green, 1973; Stewart and MacLennan, 1974; Inesi and Scales, 1974). Even after this digestion, the activities of both ATP hydrolysis and EP formation were retained. ^{32}P was incorporated from [γ-^{32}P] ATP into the larger subfragment as $E^{32}P$. This indicates that the larger subfragment contains the site for ATP hydrolysis (Thorley-Lawson and Green, 1973; Stewart *et al.*, 1976). Further digestion of SR with trypsin degraded the larger subfragment into two fragments with molecular weights of about 3×10^4 and 2×10^4, but did not affect the smaller subfragment. When the active site of ATP hydrolysis was labeled with [γ-^{32}P] ATP prior to the tryptic digestion, ^{32}P was recovered in the subfragment with a molecular weight of 3×10^4 (Thorley-Lawson and Green, 1973; Stewart *et al.*, 1976). Recently, the primary structure of the phosphorylation site was investigated in detail by several workers;

the results will be described in Section III. Each of these tryptic subfragments of the ATPase protein were isolated by Thorley-Lawson and Green (1975), Stewart and MacLennan (1975), and Stewart *et al.* (1976) using gel filtration in SDS. Stewart *et al.* (1976) studied the localization of these subfragments on the ATPase molecule using immunological techniques. They reported that the subfragment of 5.2×10^4 to 6.0×10^4 molecular weight was mostly exposed on the outer surface of the membrane, whereas subfragments with molecular weights of 2×10^4 and 4.5×10^4 to 5.5×10^4 were mostly buried within the membrane.

De Boland *et al.* (1975) and Jilka *et al.* (1975) reported that the passive movement of Ca^{2+} across the liposome membrane made of lecithin or phospholipids increased when the purified ATPase was added. Shamoo and MacLennan (1974) also suggested that the ATPase has ionophoric properties. Stewart *et al.* (1976) and Shamoo *et al.* (1976) reported that the tryptic subfragment with a molecular weight of 2×10^4 exhibited Ca^{2+} ionophore activity and supposed that this subfragment functions as an ionophore with the Ca^{2+}-binding site during Ca^{2+} transport in SR. However, the increase in the ionic current due to the addition of the "ionophoric" subfragment to a lipid bilayer was extremely low, and the exact nature of the ionophoric property of ATPase remains to be extensively explored.

III. Properties of ATP Hydrolysis by SR in the Steady State

FSR accumulates Ca^{2+} inside the vesicle coupled with Ca^{2+}-dependent ATP hydrolysis (see Section I), and the Ca^{2+}-dependent ATPase molecule itself is the carrier protein for Ca^{2+} transport across the FSR membrane.

Two problems remain on the reaction mechanism of FSR-ATPase. One is that the ATPase reaction is complex and vectorial when the ATPase is located in the SR membrane (see Section IV). In order to avoid this problem, disruption of the membrane structure with diethyl ether (Inesi *et al.*, 1967), alkali-EGTA (Duggan and Martonosi, 1970), or Triton X-100 (Yamada *et al.*, 1971) is usually employed. The other problem is that the membrane of FSR exhibits two types of ATP hydrolysis, Ca^{2+}-dependent and Ca^{2+}-independent ATPase activities (Hasselbach and Makinose, 1961, 1962). It is uncertain whether the two activities are attributable to different enzymes or whether they represent different manifestations of the same enzyme. Furthermore, the Ca^{2+}-independent ATPase activity varies with the FSR preparation (Weber *et al.*, 1966). This problem can be avoided by performing the experiments using either FSR preparations that show a negligibly small Ca^{2+}-independent ATPase activity or conditions under which the Ca^{2+}-independent ATPase activity is negligible. However, the Ca^{2+}-dependent ATPase activity is usually obtained as the difference between the total ATPase activity and the Ca^{2+}-independent ATPase activity (cf.

Hasselbach and Makinose, 1961; Yamamoto and Tonomura, 1967). This is discussed in more detail in a review by Tada *et al.* (1978) and also in Section V,B.

In this section, we describe the properties of the overall reaction of ATP hydrolysis in FSR in the steady state. They provide a basic framework for understanding the molecular mechanism of the coupling of the ATPase reaction with Ca^{2+} transport, which will be covered in Sections IV, V, and VI.

A. Phosphoprotein (EP) as a Reaction Intermediate

The existence of a high-energy phosphorylated intermediate as a component of the reaction mechanism was proposed on the basis of the observation that FSR catalyzed a rapid ATP–ADP exchange that exhibited the same dependence on Ca^{2+} concentration as did ATP hydrolysis and Ca^{2+} uptake by FSR (see Section I,B). Yamamoto and Tonomura (1967, 1968) and Makinose (1969) found that a protein of FSR was phosphorylated when the Ca^{2+}-dependent ATPase reaction of this membrane with [γ-^{32}P]ATP was quenched by trichloroacetic acid (TCA). This finding was confirmed by Martonosi (1967, 1969), Inesi and Almendares (1968), and Inesi *et al.* (1970). The maximum amount of phosphate incorporated was found to correspond to about 1 mol per mole of the ATPase protein (see Section II,B). Phosphoprotein levels, at the steady state, depended on the concentrations of Ca^{2+} and ATP and paralleled the Ca^{2+}-dependent ATPase activity (Yamamoto and Tonomura, 1967; Martonosi, 1969; Inesi *et al.*, 1970). These results strongly suggested that the phosphorylated protein is a true intermediate of the Ca^{2+}-dependent ATPase reaction. This assumption was later proved, when EP formation and its decomposition into $E + P_i$ were analyzed extensively (Sections IV and V).

The phosphoprotein isolated after quenching by TCA is stable at acidic pH, but unstable at alkaline pH (Yamamoto and Tonomura, 1967, 1968; Makinose, 1969). It is decomposed by hydroxylamine (Yamamoto and Tonomura, 1967; Makinose, 1969) to produce hydroxamate (Yamamoto *et al.*, 1971). These stability characteristics indicate that this phosphoprotein is similar to the acylphosphoprotein intermediate of Na^{+}, K^{+}-dependent ATPase, reported by Nagano *et al.* (1965) and Hokin *et al.* (1965). Bastide *et al.* (1973) examined the chemical and electrophoretic properties of ^{32}P-labeled phosphoryl peptide produced after proteolysis of phosphorylated intermediates of Ca^{2+}-dependent ATPase of FSR and Na^{+}, K^{+}-dependent ATPase of kidney microsomes. They found that the aspartyl residue was phosphorylated and that the probable tripeptide sequence at the active site of both ATPases was (Ser or Thr)-Asp-Lys. Degani and Boyer (1973) reported that the reductive cleavage of the acyl phosphate bond of the phosphoprotein by sodium [^{3}H]borohydride yielded homoserine after acid hydrolysis of the protein, confirming that the phosphoryl group was covalently bound to the β-carboxyl group of aspartate. Recently, Allen and Green (1976) isolated a tryptic peptide containing 31 residues from the active site of the Ca^{2+}-dependent ATPase, and they partially determined its sequence.

Since the formation and decomposition of the EP intermediate can be measured kinetically, the analyses of these two partial reactions have given us valuable information for clarifying the reaction mechanism of ATPase (see Section V). However, it should be noted that studies on EP were done using the phosphoprotein isolated after quenching by TCA. It is uncertain whether the properties obtained for EP represent the inherent nature of the intermediate in its intact form.

B. Substrate Specificity

The response of Ca^{2+}-dependent ATPase activity to ATP concentration is of the Michaelis–Menten type, at least at neutral pH and low ATP concentrations. But steady-state kinetics cannot tell us what kind of reaction intermediates exist in the ATP-hydrolyzing process. Therefore, the Michaelis constant (K_m) is the reciprocal of the "apparent" affinity constant of ATP for ATPase and is several micromolar (Weber *et al.*, 1966; Yamamoto and Tonomura, 1967; Vianna, 1975). ATPase also catalyzes hydrolysis of other natural nucleoside triphosphates (NTP), and the velocity of hydrolysis is in the following sequence: ATP (1.0) > ITP (0.8) > GTP (0.7) > CTP (0.55) > UTP (0.25) [numbers in parentheses represent velocities of hydrolysis relative to that of ATP (Makinose and The, 1965)]. Furthermore, other phosphate compounds, such as acetyl phosphate (de Meis, 1969a,b; de Meis and Hasselbach, 1971; Pucell and Martonosi, 1971), carbamyl phosphate (Pucell and Martonosi, 1971), and *p*-nitrophenyl phosphate (Inesi, 1971), were shown to serve as substrates, although the rates of their hydrolysis are extremely low. Active Ca^{2+} transport is supported by all these NTPs and phosphate compounds, and the coupling of 2 mol of Ca^{2+} for each mole of P_i liberated is strictly maintained (Makinose and The, 1965; Friedman and Makinose, 1970; Inesi, 1971).

C. Divalent Cation Requirement

The Ca^{2+}-dependent hydrolysis of ATP by FSR requires both Ca^{2+} and Mg^{2+} for its full activity. In the presence of saturating concentrations of Mg^{2+} (equimolar to ATP; see below), the rate of ATP hydrolysis rises with increasing Ca^{2+} concentrations, reaching maximal velocity at about 1 μM Ca^{2+} (Hasselbach and Makinose, 1963; Makinose and Hasselbach, 1965; Makinose and The, 1965). At concentrations above 0.1 mM, Ca^{2+} inhibits ATP hydrolysis (see Sections V,B and VI). Half-maximal activation occurs at 0.3–0.5 μM Ca^{2+} (Weber *et al.*, 1966; Yamamoto and Tonomura, 1967). The Hill coefficient for activation by Ca^{2+} is nearly 2 (The and Hasselbach, 1972; Vianna, 1975), which is consistent with the observation that the molar ratio of coupling between ATP hydrolysis and Ca^{2+} transport is 2 (see Section VI).

Mg^{2+} is considered to have at least two important roles in ATP hydrolysis by SR. One is to accelerate the decomposition of EP formed during the reaction (see Section V,A,4), and the other is to form an equimolar complex with ATP and serve as the true substrate for the Ca^{2+}-dependent ATPase. The latter role of Mg^{2+} was implied by the findings that Mg^{2+} concentration in excess of that of ATP gave optimal activity, and that the Ca^{2+}-dependency profile of the enzyme activity was not altered over a broad range of Mg^{2+} concentrations (Weber *et al.*, 1966; Yamamoto and Tonomura, 1967). Furthermore, Vianna (1975) demonstrated that the Lineweaver–Burk plot of the Ca^{2+}-dependent ATPase activity, in the presence of equimolar concentrations of Mg^{2+} and ATP, was linear only when plotted against the reciprocal of the concentration of the Mg-ATP complex. Froehlich and Taylor (1975) also observed that the initial velocity of EP formation was proportional to the amount of the Mg-ATP complex. Thus, the Mg-ATP complex is generally accepted as the true substrate for ATPase.

Souza and de Meis (1976) and Garrahan *et al.* (1976) observed EP formation without adding Mg^{2+} to SR and suggested that the main effect of the added Mg^{2+} was to increase the rate of phosphorylation. These results do not contradict the possibility that the Mg-ATP complex is the true substrate of the reaction, because the contaminating Mg^{2+} had not been completely removed from their experimental systems. EP might be formed in the absence of added Mg^{2+} in spite of the low Mg-ATP concentration, because EP decomposition was inhibited at low Mg^{2+} concentrations (see Section V,A,4).

ATP can bind the ATPase without forming the Mg-ATP complex. Pang and Briggs (1977) measured the binding of adenosine 5′-(β,γ-methylene) triphosphate (AMP-PCP) to FSR by centrifugation and studied the effect of Mg^{-2+} and Ca^{2+} on the AMP-PCP binding. They found that FSR bound about 1 mol of AMP-PCP per mole of ATPase and that ATP and ADP competitively inhibited the AMP-PCP binding. The dissociation constants for ATP and ADP were 3.5×10^{-5} and 3.3×10^{-6} M, respectively. Since these data were obtained in the absence of Mg^{2+}, i.e., in the presence of EDTA, FSR has a high affinity for the metal-free forms of ATP, ADP, and AMP-PCP. Mg^{2+} concentrations in excess of 1×10^{-4} *M* inhibited AMP-PCP binding. The effect of Ca^{2+} on AMP-PCP binding was biphasic. At concentrations between 1×10^{-6} and 1×10^{-4} *M*, Ca^{2+} inhibited analog binding, but above 1×10^{-4} *M*, Ca^{2+} facilitated analog binding.

D. Effects of pH and Temperature

The effect of pH on the Ca^{2+}-dependent ATPase activity exhibits a bell-shaped profile with optimal activity at pH 6.5–7.5. The enzyme activity decreases at alkaline pH, but the EP level increases because EP decomposition is inhibited at alkaline pH (Yamamoto and Tonomura, 1968).

The transition temperature at about 20°C derived from the temperature–activity profile of ATPase activity (Johnson and Inesi, 1969) is interesting in relation to the phase transition of the phospholipid bilayer of FSR (Tada *et al.*, 1978). The break in the Arrhenius plot at 20°C has been considered to be largely due to the change in the fluidity of the bilayer lipid (Inesi *et al.*, 1973) or to the phase transition of the protein-bound annular phospholipids (Hesketh *et al.*, 1976). However, Kirino *et al.* (1977) recently studied the fluctuation in the secondary structure of the Ca^{2+}-dependent ATPase molecule in the membrane using the hydrogen-deuterium exchange reaction, and they showed a correlation between the fluctuation of the protein structure and the ATPase activity. Furthermore, all the ATPase activities of FSR and purified ATPase, regardless of whether the lipids have or have not been replaced with dioleoyllecithin or egg yolk lecithin, showed a break at about 18°C, and the correlation between the secondary structure of the protein and the activity was observed with all these preparations (Anzai *et al.*, 1978). They attributed the break to fluctuations in the secondary structure of the ATPase molecule.

Phospholipids are generally accepted as being essential, particularly for the step of EP decomposition in the transport ATPase reaction. This was first suggested by Martonosi *et al.* (1971) for Ca^{2+}-dependent ATPase of SR, and also by Taniguchi and Tonomura (1971) for Na^{+}, K^{+}-dependent ATPase. This suggestion was confirmed by Nakamura *et al.* (1976), who measured the rate constant of EP decomposition using SR preparations in which the phospholipids had been substituted with synthetic phospholipids. For a more complete discussion of the interactions between lipids and ATPase, the reader is referred to the review by Tada *et al.* (1978). Recently, Dean and Tanford (1977) found that some nonionic detergents could be substituted for phospholipids with retention of the ATPase activity.

Therefore, it should be interesting to clarify which groups in phospholipids and detergents are required to retain the ATPase activity.

IV. EP Formation and P_i Liberation in the Presteady State

The preceding section dealt with the general properties of the Ca^{2+}-dependent ATPase reaction, that is, substrate specificity, the K_m value for ATP and Ca^{2+}, requirement for Mg^{2+}, and pH and temperature dependence. As ATPase molecules are distributed anisotropically in the membrane, and transport Ca^{2+} unidirectionally across the membrane coupled with ATP hydrolysis, the detailed reaction mechanism cannot be determined using enzyme kinetics of the steady state which has usually been applied to the reaction of soluble enzymes. Therefore, to clarify the reaction mechanism of Ca^{2+}-dependent ATPase, kinetic analyses in the presteady state (in this section) and of the partial reaction (Section V) must be done, taking into consideration the vectorial properties of the elementary steps.

It should be noted that the ionic composition of the inner environment of FSR is continuously changing as ATP is hydrolyzed and Ca^{2+} is transported, and that a Ca^{2+} concentration gradient or membrane potential may affect the kinetics of the ATP hydrolysis reaction. When the membrane structure is destroyed by detergent, anisotropic events do not play a significant role and the reaction itself can be analyzed. In this section, we first describe the presteady state reaction of solubilized ATPase, which can be easily analyzed, then the presteady state reaction of FSR. Finally, we summarize the regulation of the ATPase activity of FSR.

A. Solubilized ATPase

Figure 1A shows the initial phase of the reaction of SR ATPase solubilized with Triton X-100 (Yamada *et al.*, 1971). The amount of EP increased rapidly with time and reached the steady-state level. P_i liberation showed a definite presteady state, which closely corresponded to the period when the amount of EP was increasing.

EP was obtained as an acid-stable phosphoprotein when the reaction was stopped by TCA (see Section III,A). Its true chemical structure during the reaction was not known. However, the presteady state reaction could be quantitatively analyzed by assuming EP to be an intermediate: the observed time course of P_i liberation agreed well with that calculated from the observed time course of EP formation, assuming that EP was an intermediate

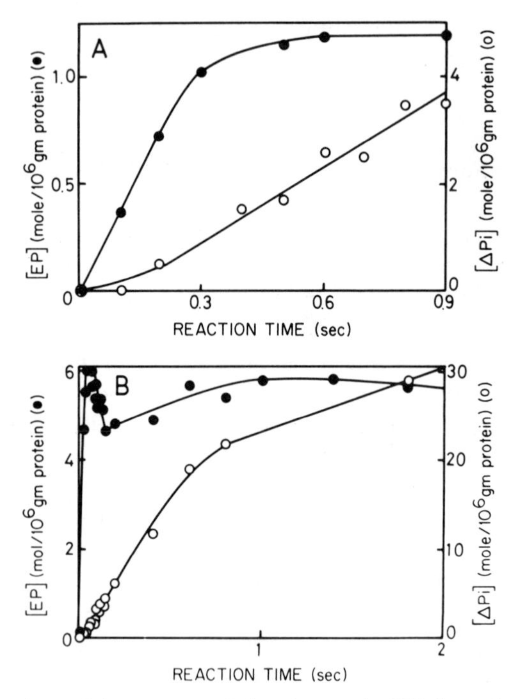

FIG. 1. Comparison of time courses of phosphoprotein (EP) formation and P_i liberation during the initial phase of the reaction. (A) Sarcoplasmic reticulum (SR) ATPase solubilized with Triton X-100. (From Yamada *et al.*, 1971.) (B) Fragmented SR. (From Takisawa and Tonomura, 1978a.)

in the reaction with a specific turnover rate (see Fig. 1A). In case of FSR, the same result was obtained only when the reaction was performed with low ATP concentrations (Kanazawa *et al.*, 1971; cf. Section IV,B for the reaction in high ATP concentrations).

Double-reciprocal plots of the ATPase activity and the amount of EP of the solubilized ATPase in the steady state against the ATP concentration gave straight lines, and the Michaelis constants both for ATPase activity and EP formation were equal (Yamada *et al.*, 1971; Jørgensen *et al.*, 1978). Therefore, the reaction of solubilized ATPase can be explained quantitatively by assuming that the reaction proceeds in a sequential route consisting of at least one enzyme–substrate complex (ES) and one EP without bound ADP:

$$\mathrm{E + S} \rightleftharpoons \mathrm{ES} \rightleftharpoons \mathrm{EP + ADP} \longrightarrow \mathrm{E + ADP + P_i}$$

When FSR was treated with a detergent, such as Triton X-100 or DOC, the ATPase activity was very labile (Ikemoto *et al.*, 1971; MacLennan and

Holland, 1976). Le Maire *et al.* (1976) have recently succeeded in preparing solubilized ATPase with stable activity, using a nonionic detergent, dodecyl octaoxyethyleneglycol monoether. This ATPase has the same kinetic properties as those described above (Takisawa and Tonomura, 1978a,b), and the ATPase reaction of solubilized enzyme will be analyzed in more detail in the near future.

B. Fragmented SR (FSR)

As described above, the reaction mechanism of solubilized ATPase can be explained by a simple scheme. However, to clarify the reaction mechanism of the coupling of ATPase with active Ca^{2+} transport, we must investigate the mechanism of ATP hydrolysis by FSR. As shown in Fig. 1, the time courses of EP formation and P_i liberation by the ATPase reaction of FSR were found to be much more complicated than those of solubilized ATPase. Thus, Kanazawa *et al.* (1971) measured the reaction of FSR in the presteady state using a simple rapid-mixing apparatus and showed that the time course of P_i liberation consisted of a lag phase, a burst phase, and a steady phase, whereas EP was formed without a lag phase and its amount rapidly reached the steady-state level after the start of the reaction. More recently, Froehlich and Taylor (1975) measured the time course of EP formation and P_i liberation in detail, using a rapid-quenching method. They observed an overshoot in the phosphorylation reaction and a P_i burst that coincided with the transient decay of EP when the reaction was started by adding ATP (cf. Figs. 1B and 2A). They proposed the following reaction scheme to explain these observations:

$$\mathrm{E + S} \rightleftharpoons \mathrm{ES} \rightleftharpoons \mathrm{EP + ADP} \rightleftharpoons \mathrm{E \cdot P + ADP} \longrightarrow \mathrm{E + ADP} + \mathrm{P_i}$$

EP is rapidly hydrolyzed to an acid-labile phosphate intermediate (E · P), which is in equilibrium with EP. The EP overshoot and P_i burst can be explained by assuming the existence of E · P.

They thought that the P_i burst reported by Kanazawa *et al.* (1971) was the same phenomenon they had observed. However, the P_i burst that coincided with the transient decay of EP occurred about 50 msec after the start of the reaction, whereas the burst phase of P_i liberation reported by Kanazawa *et al.* (1971) continued for longer than 1 second (see Fig. 1B and also Fig. 2A). The size of P_i liberated during the burst phase of the latter type was about four times greater than the number of available enzymic sites. Therefore, we can reasonably assume that the reaction mechanisms

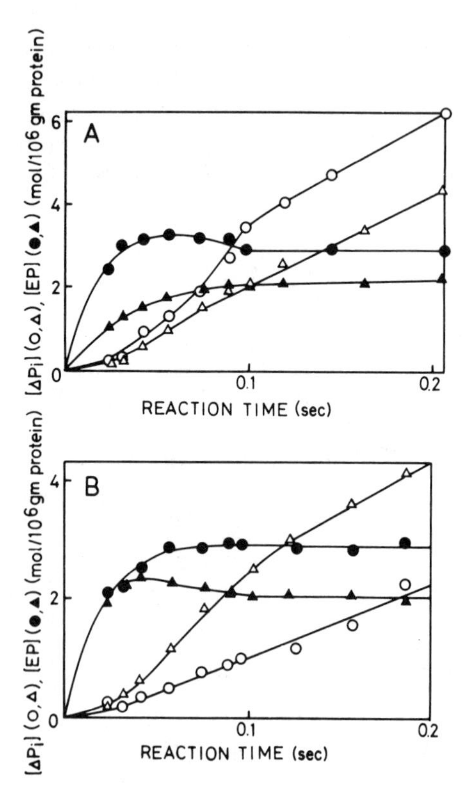

FIG. 2. Absence of phosphoprotein (EP) overshoot and P_i burst after starting the reaction by adding [γ-^{32}P]ATP with Ca^{2+} to FSR or [γ-^{32}P]ATP to FSR preloaded with Ca^{2+}. (A) The reaction was started by adding [γ-^{32}P]ATP [○, ●] or [γ-^{32}P]ATP + Ca^{2+} [△, ▲]. (B) ○, ●, Ca^{2+}-preloaded FSR; △, ▲, Ca^{2+}-unloaded FSR. (From Takisawa and Tonomura, 1978a.)

for these P_i bursts are different. Hereafter, we will call the P_i burst observed by Froehlich and Taylor (1975) the P_i burst of the first type and that observed by Kanazawa *et al.* (1971) the P_i burst of the second type. Properties of the P_i burst of the first type will be discussed in this subsection and those of the P_i burst of the second type in Section IV,C.

More recently, Froehlich and Taylor (1976) observed the initial burst of P_i liberation by FSR even in the absence of Ca^{2+}. Since its time course was very similar to that of the P_i burst of the first type, they assumed the Ca^{2+}-independent ATPase reaction was an alternate pathway of the enzyme. They proposed a flip-flop model in which the enzyme functions as a dimer for coupled transport of Ca^{2+} and Mg^{2+}. Kurzmack and Inesi (1977) confirmed the existence of the initial P_i burst of the first type both in the presence and in the absence of Ca^{2+}. However, they found that the burst size of the Ca^{2+}-independent ATPase reaction at saturating substrate concentrations was about three times greater than the number of available enzymic sites. This finding contradicts the assumption of Froehlich and Taylor (1976) that the P_i burst of the Ca^{2+}-independent ATPase reaction is caused by the formation of $E \cdot P$ (see Section V,A,3 for further evidence against $E \cdot P$ formation as a cause). At present, the relation of the Ca^{2+}-independent ATPase to the Ca^{2+}-dependent one is not clear (see Section III).

The mechanism of the Ca^{2+}-dependent ATPase reaction, which is coupled with active Ca^{2+} transport, should be studied under conditions where the Ca^{2+}-independent ATPase activity is negligibly small. Takisawa and Tonomura (1978a) studied the Ca^{2+}-dependent ATPase reaction in the presteady state using FSR preparations showing negligibly small activity in the absence of Ca^{2+}. When the reaction was started by adding ATP, EP overshoot and P_i burst of the first type were observed (see Fig. 2A). However, when the reaction was started by adding Ca^{2+} and ATP to FSR in the presence of EGTA (Fig. 2A), both the EP overshoot and the P_i burst of the first type did not occur, although the rate of P_i liberation and the amount of EP in the steady state were almost equal to those of the reaction started with ATP. The differences between the reaction profile in the initial phase of the ATP start and that of the Ca^{2+} + ATP start cannot be due to a slow binding of Ca^{2+} to the ATPase, because Ca^{2+} binds very rapidly with the SR ATPase (see Section VI,A,1), and Ca^{2+} and ATP bind with the ATPase in a random sequence (see Section V,A,2). The EP overshoot and the P_i burst of the first type were also eliminated when SR was preloaded with a small amount of Ca^{2+} ions (Fig. 2B), although the rate of P_i liberation and the amount of EP in the steady state were affected only slightly. Takisawa and Tonomura (1978a) also observed that the time course of P_i liberation showed a relatively long lag phase and a burst phase, and the apparent rate constant of EP decomposition, i.e., $v/[EP]$ (see Section V,A,4), showed a very complex pattern during the initial phase of the reaction. It increased from the initial value, reached a maximum, then decreased to the steady-state level. Neither the mechanism proposed in Section IV,A to explain the kinetic properties of solubilized SR ATPase nor that by Froehlich and Taylor (1975) mentioned above can explain the complicated time course of the $v/[EP]$ value. The long lag phase of P_i liberation may be explicable if we assume that P_i or acid-labile $E \cdot P$ is formed through at least two TCA-stable EP intermediates. This assumption is consistent with Ikemoto's proposal (Ikemoto, 1976) of the sequential formation of two acid-stable EP intermediates differing in Ca^{2+} affinity.

There are several ways to explain these complicated kinetic properties of FSR ATPase without assuming the existence of an acid-labile intermediate (E · P). First is a complicated reaction scheme proposed by increasing the number of intermediates. For example, the possibility exists that P_i is liberated not only from the EP intermediate, but also from other intermediates such as EATP, as suggested by Lowe and Smart (1977) based on analysis of the Na^+,K^+-dependent ATPase reaction in the presteady state. Another possibility is that two EP intermediates exist and have different rate constants for decomposition, which was proposed by Nakamura and Tonomura (1978) to explain the presteady-state and steady-state kinetic behavior of the SR in the *p*-nitrophenylphosphatase reaction (see Section V,C).

The second way is to assume that "apparent" rate constants of several steps change with the reaction time owing to the conformational change of the membrane structure during the reaction. Sumida *et al.* (1976) have already shown that the P_i burst of the second type is due to the change in the "apparent" rate constant of EP decomposition with time (see Section IV,C).

The third way is to assume interactions between ATPase molecules in the membrane. Vanderkooi *et al.* (1977) observed that energy transfer between ATPase molecules modified covalently with different fluorescent reagents was not affected by up to 10-fold dilution of the lipid phase of the vesicles. They suggested the existence of ATPase oligomers in the membrane (see Section II,A). Therefore, the possibility exists that a flip-flop reaction such as described by Froehlich and Taylor (1976) occurs when the ATPase molecules form the oligomer in the membrane. However, Le Maire *et al.* (1976) indicated that the solubilized ATPase prepared by their method existed as oligomers (mainly tetramers), and we found that the ATPase reaction of this solubilized preparation followed a simple reaction scheme (see Section IV,A). Furthermore, the maximum amount of EP formed was 1 mol per mole of ATPase. This conflicts with the flip-flop mechanism.

In the above discussion, we tacitly assumed that the physicochemical state of ATPase is homogeneous at the start of the reaction. However, Yamamoto and Tonomura (1977) showed that the number of lysine residues in various subfragments of SR ATPase, which are modified by 2,4,6-trinitrobenzenesulfonate, is not integral (cf. Table II). Therefore, the possibility exists that the state of the ATPase in the membrane is somewhat heterogeneous. Furthermore, the complicated reaction in the initial phase did not occur when the membrane of FSR was solubilized (Section IV,A). Therefore, it seems likely that interactions between the ATPase molecules and phospholipids are heterogeneous in the initial state, and that this affects the kinetic properties of the ATPase reaction in the presteady state.

C. Regulation of the ATPase Activity by ATP and Ca^{2+}

The ATPase activity of FSR from skeletal muscle is known to be regulated by three factors. First, the substrate ATP itself accelerates the ATPase reaction. Yamamoto and Tonomura (1967) and Kanazawa *et al.* (1971) found that the double-reciprocal plot of the ATPase activity against ATP concentration consisted of two straight lines, and both V_{max} and apparent K_m values of the steady-state ATPase at higher ATP concentrations were greater than those at lower concentrations. Kanazawa *et al.* (1971) also observed that the amount of EP in the steady state was proportional to the ATPase activity over a wide range of ATP concentrations. [Acceleration of the ATPase reaction by ATP at high concentrations was not observed on solubilized ATPase (Yamada *et al.*, 1971; Takisawa and Tonomura, 1978a).] That high ATP concentration does not accelerate EP decomposition was shown by Kanazawa *et al.* (1971), Martonosi *et al.* (1974), and Sumida *et al.* (1976). The remarkable stimulation of ATPase activity by high ATP concentration resulted from ATP-induced acceleration of the velocity of EP formation. Actually, the double-reciprocal plot of the initial rate of EP formation at high ATP concentration exhibited a downward deviation from linearity (Kanazawa *et al.*, 1971; Froehlich and Taylor, 1975). Recently, Dupont (1977) measured the binding of Mg-ATP to FSR in the absence of Ca^{2+} using a Millipore filtration method, and he observed two kinds of binding sites with different affinity for ATP. The high-affinity site showed high specificities for ATP. Its amount was 1 mol per mole of ATPase, and the K_d value was 2–3 μM. His results indicate that the high-affinity site is located at the active center of ATPase (cf. Section III,A). On the other hand, the low-affinity site was of low specificity, and ATP could easily be displaced by other polyphosphates, such as adenosine 5′-(α,β-methylene) triphosphate (AMP-CPP) or PP_i. The K_d value for ATP was about 500 μM. Dupont also observed that PP_i and AMP-CPP, which is a very poor substrate for the ATPase, remarkably increased the ATPase activity. He suggested that this was due to the binding of PP_i or AMP-CPP to the low-affinity binding site. The accelerating effect of ATP was also found with Na^+,K^+-dependent ATPase (Kanazawa *et al.*, 1970).

The second possible regulatory process involves the Ca^{2+}-induced inhibition of EP decomposition. The Mg^{2+}-dependent decomposition of EP is inhibited competitively by Ca^{2+} inside the vesicle (see Section VI,A). Therefore, accumulation of Ca^{2+} within the vesicles decreases the ATPase activity. Actually, Weber (1971) found that the Ca^{2+}-dependent hydrolysis of ATP by FSR decreased during the initial phase of the reaction when there was considerable uptake of Ca^{2+} into the vesicle.

The third possible regulatory process involves the large transition of

v/[EP] during the initial phase of the reaction. Kanazawa *et al.* (1971) and Yamada *et al.* (1971) found a pronounced transition of the v/[EP] value of FSR ATPase from a high value in the initial phase to a small one at the steady state. This transition induces the P_i burst of the second type, as mentioned in Section IV,B, which occurs after completion of the EP overshoot and the P_i burst of the first type. The transition might be an important factor determining the time course of relaxation of muscle, which is induced by removal of Ca^{2+} ions from myofibrils into SR. Sumida *et al.* (1976) directly showed that the P_i burst of the second type is caused by the transition of the k_d value (the apparent rate constant of EP decomposition); they measured the value of k_d directly after the EP overshoot had occurred, and showed that the transition takes place according to the following equation:

$$k_d = (k_{d,\,initial} - k_{d,\,steady}) \times \exp(-k_{tr} \cdot t) + k_{d,\,steady}$$

where $k_{d,\,initial}$ and $k_{d,\,steady}$ are the values of k_d in the initial and steady phases of the reaction, respectively, and k_{tr} is the rate constant for the transition in the k_d value during the initial phase. Sumida *et al.* (1976) reported that in a typical experiment the ratio of $k_{d,\,initial}$ to $k_{d,\,steady}$ was 7.3 and the value of k_{tr} was 0.5 sec^{-1}. They also showed that the time course of P_i liberation calculated from the observed rates of EP formation and the value of k_d obtained from the above equation agreed well with experimentally observed time courses of P_i liberation. It should be added that the P_i burst of the first type occurred during only a very short period, and this contributes insignificantly to the time course of P_i liberation during a few seconds after the start of the reaction. The mechanism of the P_i burst of the second type has not been clarified yet, although Yamada *et al.* (1971) suggested that the P_i burst is caused by an ATP-induced cooperative change in the conformation of the vesicular structure of FSR on the basis of the following observations: (a) the transition in v/[EP] is independent of the concentration of divalent cations inside the vesicles; (b) the P_i burst occurs even in the presence of a concentration of ATP too low to saturate the phosphorylation sites; and (c) it does not occur when the membranes are treated with Triton X-100. However, the possibility that the P_i burst of the second type is caused by the inhibition of ATPase activity due to Ca^{2+} inside the vesicle cannot be excluded, although Yamada *et al.* (1971) showed that the P_i burst was not affected by the presence of oxalate when the Ca^{2+} concentration inside was assumed to be low.

These three regulatory processes of ATPase activity disappear when the membrane is dispersed. It can be easily understood that the inhibition of ATPase activity by Ca^{2+} inside the vesicle disappears when the membrane is dispersed. However, it is not clear why the stimulation of ATPase activity by ATP and the P_i burst of the second type disappear.

Finally, we should add that the ATPase activity of cardiac SR is known to be regulated by cyclic AMP-dependent protein kinase. More recently, the ATPase activity of FSR from skeletal muscle was reported to be regulated by Ca^{2+}-dependent protein kinase (Hörl *et al.*, 1978; Hörl and Heilmeyer, 1978). Since this is beyond the scope of the present article, the reader is referred to the review by Tada *et al.* (1978).

V. Reaction Mechanisms of ATP and *p*-Nitrophenyl Phosphate Hydrolysis by SR

As mentioned in Sections IV,B and C, many complicated and difficult problems remain unsolved in the kinetics of the reaction catalyzed by ATPase molecules in the membrane. However, to clarify the molecular mechanism of Ca^{2+} transport, the relation between the elementary steps of the ATPase reaction and Ca^{2+} transport should be investigated using FSR. In this section, we review the present knowledge of the reaction mechanism of FSR ATPase based on kinetic studies of partial reactions (Kanazawa *et al.*, 1971). The coupling mechanism between the ATPase reaction and Ca^{2+} transport is described in Section VI. Kanazawa *et al.* (1971) studied the ATPase reaction under physiological conditions where sufficient amounts of alkali metal salts such as K^+ were present. Shigekawa *et al.* (1978) and Shigekawa and Dougherty (1978a,b) recently made very interesting discoveries about the kinetic properties of EP in the absence of alkali metal salts. Therefore, the reaction mechanisms of ATPase in the presence and in the absence of added alkali metal salts are described separately in Sections V,A and V,B, respectively. Finally, Section V,C describes the reaction mechanism of *p*-nitrophenyl phosphate (pNPP) hydrolysis by SR ATPase.

A. ATP Hydrolysis in the Presence of Alkali Metal Salts

1. *Elementary Steps of the ATPase Reaction*

Figure 3 is a probable reaction mechanism of FSR ATPase, where ^{i} and ^{o} indicate the inside and outside of the membrane vesicles, respectively (cf. Section VI). Two moles of Ca^{2+} and 1 mol of ATP bind in a random sequence to 1 mol of the ATPase active site, E, on the outer surface of the membrane, forming the first Michaelis complex $^{o}CaE^{oATP}$. This is converted into a second complex $Ca{*}E^{oATP}$. The conversion is accelerated by high concentrations of ATP. EP without bound ADP is formed from $Ca{*}E^{oATP}$ via a

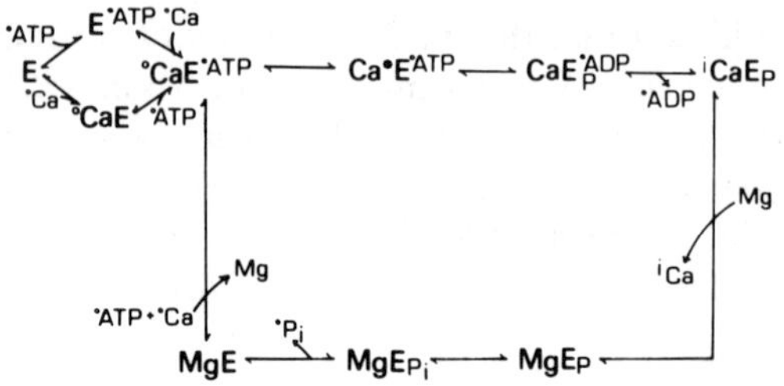

FIG. 3. Reaction scheme for sarcoplasmic reticulum (SR) ATPase. For explanation, see text.

transient complex CaE_P^{ADP}. EP formation induces translocation of Ca^{2+} from outside to inside the membrane. ADP is released from the enzyme on the outside of the vesicle, and Ca^{2+} is released from the enzyme into the interior of the vesicle. In the presence of a sufficient amount of alkali metal salt, MgE_P formed from $^iCaE_P$ instantaneously decomposes to E + P_i, and P_i is liberated from the enzyme to the exterior of the vesicle.

To discuss whether a ligand reacts with ATPase on the outside or inside of the membrane vesicles, we have to define operationally the inside and outside of the membrane. The distinction is easy when a ligand is impermeable to the membrane. Radioactive substances and cations are considered to exist outside the membrane if they are readily removed by washing with a neutral solution containing unlabeled substances and a chelating agent. On the other hand, they are considered to exist within the membrane vesicle if they cannot be removed by washing, if necessary with unlabeled substances and/or a chelating agent, but can be removed after treatment of the membrane with detergents or at alkaline pH to make the membrane leaky (see Section III).

2. *EP Formation from E and ATP*

The initial step in Ca^{2+} transport is the recognition of Ca^{2+}, which can be analyzed kinetically as formation of the first Michaelis complex, $^oCaE^{oATP}$ (see Section VI). Immediately after the addition of ATP to FSR in the presence of Ca^{2+}, EP rapidly forms without a lag phase and reaches the steady-state level within 1 second at 15°C (see Sections IV,A and B). The absence of a lag phase in EP formation, even after addition of an extremely low concentration of ATP that is membrane impermeable (Weber *et al.*, 1966), suggests that the binding of ATP to the ATPase site to form EP occurs on the outer surface of the membrane.

Kanazawa *et al.* (1971) measured the velocity of EP formation, v_f, under relatively low ATP concentrations and various Ca^{2+} concentrations. Figure

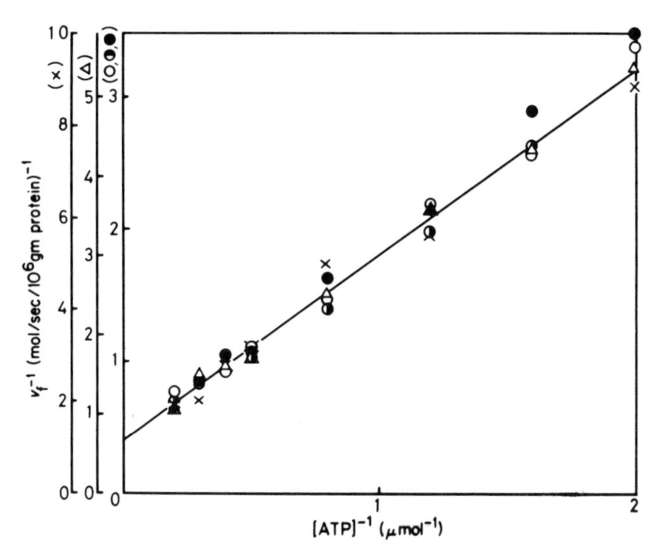

FIG. 4. Relationship between the initial rate of phosphoprotein (EP) formation (v_f) in fragmented sarcoplasmic reticulum (FSR) and the ATP concentration, [ATP], at various concentrations of Ca^{2+} in the low ATP concentration range. $CaCl_2$, 0.5 m*M*; $MgCl_2$, 5 m*M*; KCl, 0.1 *M*. Concentration of EGTA: ◑, 0.40; ○, 0.45; ●, 0.51; △, 0.54; ×, 0.57 m*M*. (From Kanazawa *et al.*, 1971.)

4 shows that a linear relationship exists between the reciprocal of the rate of EP formation, v_f^{-1}, and the reciprocal of the ATP concentration $[ATP]^{-1}$. When the Ca^{2+} concentration is reduced, the maximum value (V_f') of v_f at sufficiently high ATP concentrations decreases while the Michaelis constant, K_f, remains constant. This indicates that EP is formed via an enzyme–ATP–Ca^{2+} complex produced by binding ATP and Ca^{2+} to ATPase in a random manner. Furthermore, the double-reciprocal plot of V_f', obtained from Fig. 4, against the square of the free Ca^{2+} concentration gives a straight line. Coffey *et al.* (1975) also measured the dependence of v_f on Ca^{2+} concentrations and found the Hill coefficient to be about 2. These results indicate that the reaction occurs when one ATPase binds with two Ca^{2+}. The following equation was obtained (Kanazawa *et al.*, 1971):

$$v_f = \frac{V_f}{\{1 + (K_{Ca}/[Ca])^2\}\{1 + K_m/[ATP]\}}$$

where V_f, K_{Ca}, and K_m represent the maximal velocity of EP formation and concentrations of Ca^{2+} and ATP that give the half-maximal velocity of EP formation, respectively. When we use 0.2 μM as the apparent dissociation constant of the Ca-EGTA complex (Schwarzenbach, 1960), the value of K_{Ca} becomes about 4 μM.

Yamamoto and Tonomura (1967) measured the dependence of the ATPase activity in the steady state on various concentrations of Ca^{2+} and ATP and suggested that the reaction occurs in an ordered sequence; first, ATP binds ATPase; then Ca^{2+} binds the EATP complex. This conflicts with the above conclusion by Kanazawa *et al.*(1971). The disagreement may originate from the fact that Kanazawa *et al.* measured the initial rate of EP formation whereas Yamamoto and Tonomura measured the overall reaction at steady state. As indicated by MacLennan and Holland (1976), this disagreement can be easily explained by assuming that Mg^{2+} dissociates from the enzyme and permits Ca^{2+} binding only after the enzyme has bound ATP.

3. *Formation of ATP from EP and ADP*

The reactivity of EP needs to be known because EP formation accompanies Ca^{2+} translocation from the outside to the inside of the vesicle and the change in affinities for divalent cations (see Section VI). As in the case of Na^+,K^+-dependent ATPase, the reactivities of EP to ADP and divalent cations have been studied by many researchers, and this subsection describes the reaction of EP with ADP. The effects of divalent cations on EP formation and decomposition are described in Section VI.

The existence of a high-energy phosphorylated intermediate that forms ATP when allowed to react with ADP was first suggested on the basis of the observation that FSR catalyzes a rapid ATP–ADP exchange in the presence of Ca^{2+} (see Section I). Kanazawa *et al.* (1971) later demonstrated that ATP is formed from EP and ADP. Their typical experiment is shown in Fig. 5. When a large amount of EGTA was added to remove Ca^{2+} from the medium after EP formation, further formation was instantaneously terminated. This was accompanied by the liberation of P_i from EP following first-order kinetics. However, when ADP was added simultaneously with EGTA, EP decreased rapidly without a lag phase not accompanying the liberation of P_i, and the amount of ATP formed was equal to that of the decrease in EP. We can reasonably assume that ADP does not permeate the membrane like ATP (cf. Weber *et al.*, 1966) and ATP is formed from EP and ADP without a lag phase. Therefore, ADP reacts with EP on the outside of the vesicle. Because the amount of ATP formed from EP and the added ADP are equal to the decrease in EP, the possibility that an appreciable amount of $E \cdot P$ (cf. Section IV,B) exists in equilibrium with EP can be ruled out. This conclusion is supported by the result that almost all the [γ-^{32}P]ATP added was converted to $E^{32}P$ when the enzyme concentration was sufficiently higher than that of ATP (Kanazawa *et al.*, 1971; Sumida *et al.*, 1976).

At alkaline pH, where EP decomposes very slowly (see Section IV,D),

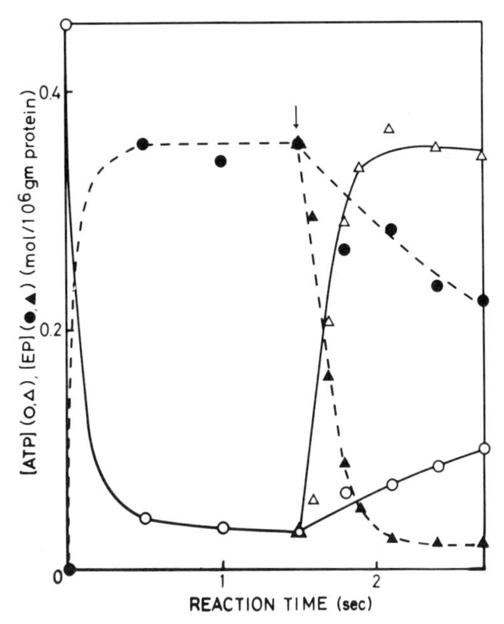

FIG. 5. Formation of ATP from phosphoprotein (EP) and ADP. SR protein, 10 mg/ml; $CaCl_2$, 50 μM; $MgCl_2$, 1 mM; KCl, 0.15 M; pH 8.8. At 1.5 seconds after the addition of 4.6 μM [γ-^{32}P]ATP (↓), EGTA was added to stop EP formation with (●,○) or without (▲, △) ADP. (From Kanazawa *et al.*, 1971.)

addition of ADP with EGTA to EP resulted in an almost completely stoichiometric reaction: EP + ADP → E + ATP (Fig. 5). This finding suggested that EP did not contain enzyme-bound ADP (Kanazawa *et al.*, 1971). Takisawa and Tonomura (1978a) estimated the amount of ADP bound to the enzyme during the ATPase reaction by measuring the amount of ADP remaining in the SR ATP–ATPase system coupled with sufficient amounts of creatine kinase and creatine phosphate, and found it to be negligibly small.

The rate of EP formation was accelerated by high concentrations of ATP, as mentioned in Section IV,C. The step of the ES complex formation from the initial reactants is considered to be a very rapid process that is in quasi-equilibrium, since EP formation occurs without a lag phase even in extremely low concentrations of ATP and the formation of $E^{32}P$ from [γ-^{32}P]ATP stops immediately when a large amount of unlabeled ATP is added. It is also evident that the rate of EP formation from its direct precursor was not accelerated by ATP, because the reverse reaction, i.e., the formation of ATP from EP and ADP, was not stimulated by ATP. In order to explain the acceleration of the forward reaction by high concentrations of ATP, Kanazawa *et al.*(1971) postulated the existence of another type of Michaelis complex, $Ca{*}E^{\circ ATP}$, after the first Michaelis complex, $^{\circ}CaE^{\circ ATP}$, and proposed that the rate of formation of the second complex from the first is accelerated by high concentrations of ATP.

4. *Decomposition of EP*

When Ca^{2+} is translocated from outside to inside the membrane, it has to be released from the enzyme into the vesicular lumen. This process is related to EP decomposition, which is the main topic of this subsection.

The first-order rate constant of EP decomposition can be measured by two different methods, if only one species of EP exists in the reaction system (or if several EP species exist but are in rapid equilibrium). One is to measure the ratio v_o/[EP], i.e., the rate of ATPase reaction per unit of EP concentration at steady state. The other is to directly measure the decomposition of $E^{32}P$ after terminating its formation by the addition of EGTA to remove Ca^{2+} or unlabeled ATP to dilute the radioactive ATP and thus halt ^{32}P incorporation into the enzyme (Kanazawa *et al.*, 1971). The time course of the EP decomposition was found to follow first-order reaction kinetics without a lag phase, and the rate constant thus obtained was named k_d (Section IV,C). Since the FSR membrane is impermeable to EGTA and ATP (Weber *et al.*, 1966), these findings indicate that Ca^{2+} and ATP reacted with the ATPase on the outer surface of the membrane (see also Section V,A,2).

Inesi *et al.* (1970), Martonosi (1969), Kanazawa *et al.* (1971), and Panet *et al.* (1971) found that the decomposition of EP requires Mg^{2+}. It is important to know whether Mg^{2+} reacts with EP on the outside or inside of the membrane vesicle, in relation to the problem of whether or not Mg^{2+} is transported as a counterion to Ca^{2+} (Section VI). When the ATPase reaction of FSR was stopped by adding EGTA to remove Ca^{2+}, EP decomposition started instantaneously (the time course of the EP decomposition did not follow first-order kinetics, because the experiment was performed at alkaline pH with a low concentration of Mg^{2+}) (Fig. 6). When Mg^{2+} was removed by EDTA after the EP formation, the decay of EP gradually ceased, but readdition of a large amount of Mg^{2+} restored it. However, when the membrane had been solubilized with detergents, addition of EDTA instantaneously terminated EP decomposition (Kanazawa *et al.*, 1971). These observations suggested that the FSR membrane is more permeable to Mg^{2+} than Ca^{2+}, and that Mg^{2+} participates in EP decomposition inside the vesicle. Unfortunately, the permeability of the membrane to Mg^{2+} has not been measured precisely. Miyamoto and Kasai (1976) tentatively reported that Mg^{2+} did not permeate the FSR membrane. It

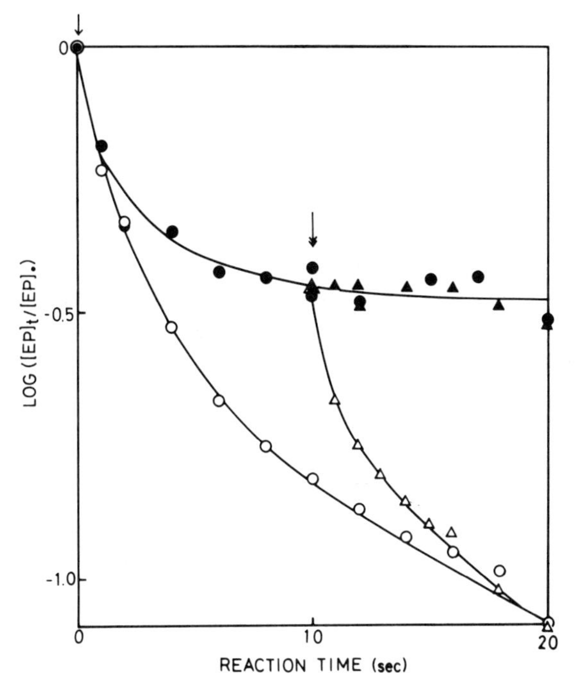

FIG. 6. Inhibition of phosphoprotein (EP) decay by removing free Mg^{2+} with EDTA. FSR was phosphorylated with [γ-^{32}P]ATP in 0.2 mM $MgCl_2$ and 0.1 M KCl at pH 8.5. The amount of Ca^{2+} contaminating the reaction mixture was usually sufficient for the phosphorylation reaction to occur. ○, Ca^{2+} was removed from the system with EGTA at ↓ ; ●, EDTA was added at ↓ to remove Ca^{2+} and Mg^{2+} ; △, EDTA was added at ↓, and $MgCl_2$ and EGTA was added at ↡; ▲, EDTA was added at ↓, and EGTA was added at ↡. (From Kanazawa *et al.*, 1971.)

seems premature at present to decide whether Mg^{2+} participates in EP decomposition on the interior of the membrane or in the pocket of the ATPase molecule, although the latter possibility seems more reasonable (cf. Sections VI and VII).

Employing deoxycholate-treated membrane of FSR, in which the inside and outside of the membrane are kinetically indistinguishable, Yamada and Tonomura (1972b) measured the dependence of v_o/[EP] on the concentration of both Mg^{2+} and Ca^{2+} and found that Ca^{2+} inhibited the decomposition of EP by competing at the site where Mg^{2+} accelerated it. They found that the ratio of affinity for Ca^{2+} to that for Mg^{2+} at the EP-decomposition step was very small compared to that at the EP-formation step (cf. Section VI). Thus, they concluded that the EP formation that coincides with the translocation of Ca^{2+} across the membrane is accompanied by a dramatic change in the affinities of the enzyme for Ca^{2+} and Mg^{2+}. This conclusion was confirmed by Ikemoto (1975), who observed the release of Ca^{2+} from ATPase accompanying EP formation.

Employing SR vesicles whose exchangeable Mg^{2+} was thoroughly removed by washing with *trans*-1,2-diaminocyclohexanetetraacetic acid (CDTA), Garrahan *et al.* (1976) recently examined the effects of Mg^{2+} on EP decomposition. They showed that EP formed in such membranes in the absence of added Mg^{2+} did not decompose even when a large amount of Mg^{2+} was added subsequently, whereas decomposition of the EP formed in the presence of added Mg^{2+} was not prevented when Mg^{2+} was removed by adding CDTA. They also obtained essentially similar findings with the SR solubilized with Triton X-100. Based on these observations, Garrahan *et al.* (1976) suggested that the Ca^{2+}-binding sites required for EP formation are distinct from the Mg^{2+}-binding site(s) required for its decomposition, and that the Mg^{2+}-binding site is occluded during EP formation, thus preventing the chelating agent from removing Mg^{2+} bound to EP. Their observation, that EP formed in the absence of added Mg^{2+} did not decompose when a large amount of Mg^{2+} was added, was not consistent with the results obtained by Kanazawa *et al.* (1971) (see above). Since Mg-ATP is the true substrate for the ATPase reaction (see Section III), its concentration increases when Mg^{2+} is added to the system in the absence of other added Mg^{2+}. Therefore, the absence of Mg^{2+} acceleration of decomposition of the EP formed without added Mg^{2+} observed by Garrahan *et al.* (1976) may be due to a counterbalance of incomplete termination of the reaction by unlabeled ATP and EGTA in the presence of a large amount of added Mg^{2+}, and an increase in the rate of EP formation by Mg^{2+}. Furthermore, Klodos and Skou (1977) showed that CDTA chelates Mg^{2+} very slowly. Another possibility is that CDTA treatment of the membranes altered the kinetic properties of EP, as Garrahan *et al.* (1976) suggested. Bound, but not free, Mg^{2+} was also reported to be required for the normal decomposition of EP in Na^+,K^+-dependent ATPase (Fukushima and Post, 1978). More precise kinetic analyses on the time course of EP decomposition and P_i production should be done on FSR to clarify the role of the so-called tightly bound Mg^{2+}.

B. ATP Hydrolysis in the Absence of Alkali Metal Salts

The rate of Ca^{2+} uptake by FSR is accelerated by addition of a low concentration of alkali metal salts, such as K^+ (Katz and Repke, 1967; Yamada *et al.*, 1970; Shigekawa and Pearl, 1976; Duggan, 1977). Shigekawa and Pearl (1976) observed that in the absence of added alkali metal salts, the rate of Ca^{2+}-dependent ATP hydrolysis is low, concomitant with slow Ca^{2+} accumulation, and the rate of EP decomposition is much lower than that in the presence of optimal salt concentrations.

Recently, Shigekawa *et al.* (1978) and Shigekawa and Dougherty (1978a) measured the rate constant of EP decomposition by two different methods (v_o/[EP] and k_d; cf. Section V,A,4) to characterize the kinetic properties of EP formed in the absence of added alkali metal salts. EP decomposition was accelerated by alkali metal salts, free divalent cations (Mg^{2+} or Ca^{2+}), or the Mg-ATP complex. Shigekawa and Dougherty (1978a) also reported that EP formed in the absence of added alkali metal salts did not contain bound ADP as well as EP in the presence of alkali metal salts (Section V,A,3). However, the former EP rarely reacted with ADP under the condition where the Ca^{2+} concentration was low (Shigekawa and Dougherty, 1977, 1978b), whereas the latter EP did (Section V,A,3).

They found that the apparent rate constant of K^+-dependent decomposition of the EP formed in the absence of K^+ measured as k_d (cf. Section V,A,4) was approximately 4-fold higher than the value of v_o/[EP] in the steady state and in the presence of K^+. Therefore, they assumed a sequential formation of EP, i.e., ADP-sensitive E_1P and ADP-insensitive E_2P: $E_1P \rightarrow E_2P \rightarrow E + P_i$, and acceleration of the decomposition of E_2P by alkali metal salts. Thus, in the presence of added alkali metal salts, the main EP intermediate was E_1P, while in the absence of salts, it was E_2P.

The transition from E_1P to E_2P was more clearly indicated by examining the reactivity of EP with ADP during the transient phase of the reaction (Shigekawa and Dougherty, 1978b). E_1P (ADP-sensitive EP) was formed first and then converted to E_2P (ADP-insensitive EP) with low concentrations of free Mg^{2+} and Ca^{2+} at 0° and pH 6.8. This conversion ($E_1P \rightarrow E_2P$) was accelerated by free Mg^{2+}, and this action of Mg^{2+} was antagonized by free Ca^{2+}. Ca^{2+} accelerated the reaction of the $E_2P \rightarrow E_1P$ conversion. The free Ca^{2+} concentration that produced half-maximal acceleration of the $E_2P \rightarrow E_1P$ conversion was 0.6 to 1.0 m*M* in 1 to 2 m*M* Mg^{2+} and was affected only slightly by changing the free Mg^{2+} concentration (see Fig. 7). These findings suggested that the ADP-insensitive EP had a much lower Ca^{2+} affinity than the enzyme prior to phosphorylation (Section V,A,2).

The reaction scheme proposed by Shigekawa and Dougherty (1978b) is consistent with the reaction scheme shown in Fig. 3, which is proposed on the basis of experiments in the presence of added alkali metal salts. This scheme shows that CaE_P (ADP-sensitive) is converted to MgE_P (ADP-insensitive). De Meis and de Mello (1973) reported that ATP accelerated the decomposition of EP, whereas Kanazawa *et al.* (1971) reported otherwise (see Sections IV,C and V,A,3). This is because the former performed

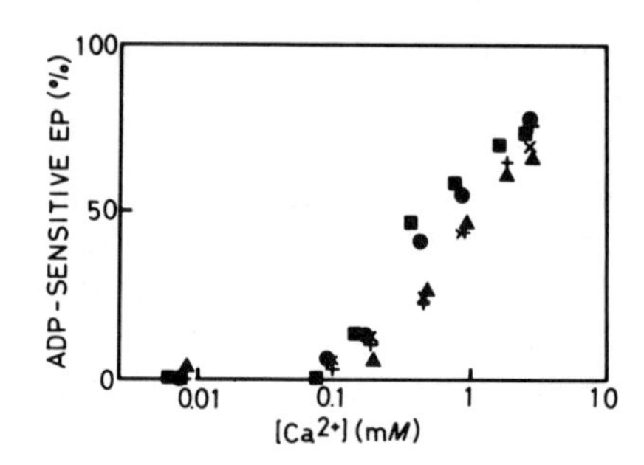

FIG. 7. Effect of Ca^{2+} on the conversion of ADP-insensitive phosphoprotein (EP) to ADP-sensitive EP. ADP-insensitive EP was formed by allowing purified ATPase to react with ATP in the absence of added alkali metal salts. The ADP-insensitive EP was converted into ADP-sensitive EP as the amount of $CaCl_2$ added was increased. Final free Mg^{2+} concentration: ■, 68.3–105.1 μM; ●, 0.592–0.729 mM; ×, 1.699–1.878 mM; +, 2.230–2.396 mM; ▲, 3.820–3.963 mM. (From Shigekawa and Dougherty, 1978b.)

their experiments in the absence of added alkali metal salts, where the rate-limiting step, $E_2P \rightarrow E + P_i$, is accelerated by ATP. In contrast, the acceleration of EP decomposition by ATP does not occur in the presence of an optimal amount of K^+, since the conversion of E_1P to E_2P is rate-limiting in the presence of K^+.

Although the results obtained by Shigekawa *et al.* (1978) and Shigekawa and Dougherty (1978a,b) can be easily explained by the Fig. 3 reaction scheme, the possibility exists that E_1P and E_2P are formed via different routes, and the decomposition of E_1P into $E + P_i$ is the main pathway under physiological conditions where a sufficient amount of K^+ is present. Two-route mechanisms have also been proposed for Na^+,K^+-dependent ATPase (Yamaguchi and Tonomura, 1978) and SR-pNPPase (Section V,C). At present, it is uncertain whether the sequential mechanism (Fig. 3) or the two-route mechanism is correct. More detailed investigations must be done to establish the relation between the reaction mechanisms in the presence and in the absence of alkali metal salts.

It should be added that detailed kinetic analyses of EP formation from E and ATP have not yet been done in the absence of added alkali metal salts. The rate of ATP hydrolysis and the amount of EP at steady state increased with increasing Ca^{2+} concentrations both in the presence and in the absence of added alkali metal salts, but at very low Ca^{2+} concentrations both the amount of EP and the ATPase activity at steady state were significantly lower in the presence of added K^+ than in its absence (Shigekawa and Pearl, 1976; Shigekawa *et al.*, 1978; Ribeiro and Vianna, 1978). This indicates that the rate of EP formation at very low Ca^{2+} concentrations is lowered by adding K^+.

C. Hydrolysis of *p*-Nitrophenyl Phosphate

Many high-energy phosphate compounds can serve as substrates for SR ATPase, and active Ca^{2+} transport is coupled with hydrolysis of these substrates (see Section III,B). Inesi (1971) demonstrated that a low-energy phosphate compound, *p*-nitrophenyl phosphate (pNPP), was hydrolyzed by SR ATPase and could induce active Ca^{2+} transport with the coupling of 2 mol of Ca^{2+} for each mole of *p*-nitrophenol (pNP) liberated. Furthermore, he estimated the free energy of the steady-state concentration gradient ($\Delta G = RT \ln[^{i}Ca]/[^{o}Ca]$) both in the presence of pNPP or of ATP. These values agreed well with the ΔG values calculated for hydrolysis of pNPP and ATP. Na^{+},K^{+}-dependent ATPase also hydrolyzes pNPP (Rega and Garrahan, 1976), although pNPP cannot support active transport of Na^{+} and K^{+} (Garrahan and Rega, 1972). At present, the SR-pNPP system is the only example known where a low-energy phosphate compound serves as a substrate in a biological energy-transducing process.

When pNPP was used as a substrate for SR ATPase, a phosphoprotein was found in the presence of a high concentration of Ca^{2+} at alkaline pH, where its decomposition was extremely inhibited (cf. Sections III,C and III,D), although the intermediate could not be detected under physiological conditions (Nakamura and Tonomura, 1978). The pH stability and hydroxylamine sensitivity of TCA-denatured phosphoprotein were similar to those of EP in the ATPase reaction. In addition, ATP was formed stoichiometrically when ADP reacted with the phosphoprotein produced from pNPP, indicating that it is kinetically indistinguishable from EP in the ATPase reaction, which had been assumed to be a high-energy phosphate compound. The phosphoprotein intermediate formed during the hydrolysis of pNPP thus was referred to as EP.

Nakamura and Tonomura (1978) measured the decay of $E^{32}P$ in the pNPPase reaction after addition of excess unlabeled pNPP and found that the time course of EP decay was biphasic (having fast and slow phases) and the ratio between the components of the fast and slow phases was not affected by addition of pNP. Although the fast phase was almost independent of divalent cations, the slow one was accelerated by Mg^{2+} and inhibited competitively by Ca^{2+} (see Section V,A,4).

Based on these findings, Nakamura and Tonomura (1978) assumed that two types of EP intermediates with and without bound pNP existed and decayed with different rate constants (see Fig. 8). Since the rate of pNP liberation at the steady state was much lower than the sum of $k_d'[E_P^{pNP}]$ and $k_d[EP]$, the fast-decomposing intermediate was interpreted to have decomposed as follows:

$$E_P^{pNP} \xrightarrow{k_d'} E^{pNP} + P_i \longrightarrow E + pNP + P_i$$

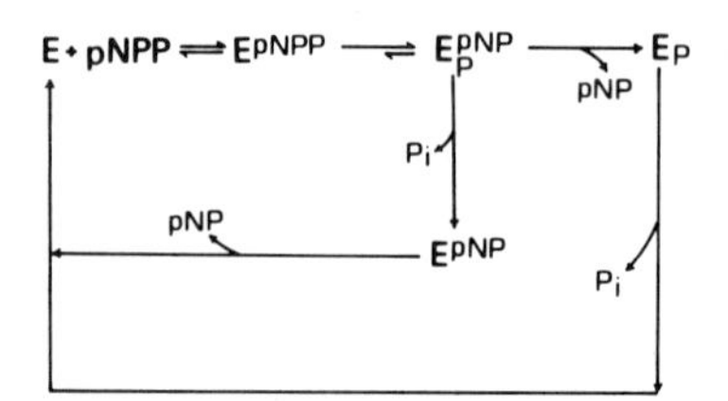

FIG. 8. Reaction scheme for hydrolysis of *p*-nitrophenyl phosphate (pNPP) catalyzed by sarcoplasmic reticulum (SR) ATPase. For explanation, see text.

Such an assumption was supported by the finding that pNP was liberated rapidly during the initial phase and the amount of initial burst of pNP was higher than the amount of EP. They showed that all the results obtained could be explained quantitatively by the reaction scheme in Fig. 8. In the reaction of Na^+,K^+-dependent ATPase, Yamaguchi and Tonomura (1978) suggested that EATP was hydrolyzed via two routes, i.e., one through E_P^{ADP} and EP and the other through E_P^{ADP} and E^{ADP}. Their mechanism is essentially the same as that given in Fig. 8. However, it was proposed to explain the experimental results under the extreme conditions of high Ca^{2+} concentration and alkaline pH. When ATP is used as a substrate for SR ATPase under physiological conditions (low Ca^{2+} concentration and neutral pH), no EP with bound ADP is detectable (see Section V,A,3), and the main route for the ATPase reaction is considered to be the outer route of the scheme.

VI. Kinetic Analysis of the Coupling of Ca^{2+} Transport with ATP Hydrolysis

The transport reaction of a solute by a carrier protein in a biological membrane consists of three elementary processes (Pardee, 1968a,b) in which (1) the solute is recognized by a carrier, (2) translocated by the carrier across the membrane, then (3) released from the carrier. This section aims to clarify the relationship between the three elementary steps of Ca^{2+} transport and various steps in the Ca^{2+}-dependent ATPase reaction in the SR membrane, where the Ca^{2+}-dependent ATPase itself serves as a carrier protein in the Ca^{2+}-transporting system. Almost all studies described in this section were performed in the presence of sufficient amounts of K^+ where the amount of ADP-insensitive EP (E_2P) was negligibly small and only ADP-sensitive EP (E_1P) was detected as EP. Refer to the reaction scheme in Fig. 3, which shows the kinetic mechanism for the coupling between ATP hydrolysis and Ca^{2+} transport.

A. Change of Affinity for Divalent Cations

1. *Recognition of Divalent Cations*

The step of EP formation by the reaction of ATPase with ATP is closely related to the recognition step of the ATPase for Ca^{2+}. As described in Section V,A,2, 1 mol of ATP and 2 mol of Ca^{2+} bind rapidly to 1 mol of the enzyme in a random sequence to form $^{o}CaE^{oATP}$ on the outer surface of the SR membrane. In the case of FSR, the apparent dissociation constant of the binding of Ca^{2+} to E was estimated to be about 4 μM from the dependence of the rate of EP formation on the Ca^{2+} concentration (Section V,A,2). On the other hand, Yamada and Tonomura (1972b) measured the rate of EP formation over broad concentration ranges of Ca^{2+} and Mg^{2+} using a purified ATPase preparation. They showed that the Ca^{2+}-dependent EP formation was competitively inhibited by Mg^{2+}. They obtained 0.35 μM and 10.6 mM as the apparent dissociation constants of CaE and MgE (K_{Ca} and K_{Mg}), respectively. Namely, the affinity of the purified enzyme for Ca^{2+} was about 30,000 times stronger than for Mg^{2+} in the enzyme state, E. It is well known that Sr^{2+} is also transported actively by SR (Van der Kloot, 1965; Weber *et al.*, 1966; Mermier and Hasselbach, 1976). Yamada and Tonomura (1972b) obtained 27.5 μM as the value of K_{Sr}.

Since the binding of Ca^{2+} to the enzyme occurs independently of ATP binding, it was measured directly by many investigators using nonkinetic methods. These results are summarized in Table I for comparison with those obtained by the kinetic studies described above. The values for the Ca^{2+} binding to the high affinity-site of SR agree with those obtained by kinetic studies, except the results of Chiu and Haynes (1977). All the other measurements indicated that 2 mol of Ca^{2+} bind tightly to 1 mol of ATPase, independently of ATP. Chiu and Haynes (1977) measured the time course of Ca^{2+} binding to FSR by means of a stopped-flow technique employing a Ca^{2+} indicator dye, Arsenazo III; they found that the binding of Ca^{2+} to ATPase reached the maximum level within the dead time of the apparatus (3 msec). This result is consistent with the conclusion drawn from kinetic analysis of the EP formation that the reaction step, $E + Ca^{2+} \rightleftharpoons CaE$, is in rapid equilibrium (Sections V,A,2 and 3). The studies by Ikemoto (1974) using an equilibrium-dialysis technique suggested that the ATPase contains three classes of Ca^{2+}-binding sites, α, β, and γ, with dissociation constants of 0.3, 20, and 1000 μM, respectively, each of which can bind about 1 mol of Ca^{2+} at 0°C. Ikemoto (1975) found that there were two α and no β sites per mole of ATPase at 22°C. The temperature-induced alteration in the number of Ca^{2+}-binding sites of ATPase is described again in Section VI,B in relation to the effect of temperature on the molar ratio of Ca^{2+} transport and and ATP hydrolysis.

TABLE I

AFFINITIES AND AMOUNTS OF Ca^{2+} BINDING

Preparation	Method	Conditions	K_{Ca} (μM)	Amount of bound Ca^{2+} (nmol/mg)	Reference[a]
FSR	Equilibrium dialysis	pH 6.5, 4°			
		0.6 *M* KCl, 1 m*M* $MgCl_2$	1.3	10	(1)
		No KCl or $MgCl_2$	0.4	10–20	(1)
	Millipore filtration	pH 6.5, 4°			
		5 m*M* $MgCl_2$	—	13	(2)
	Centrifugation	pH 6.5, 4°			
		0.1 *M* KCl, 2 m*M* $MgCl_2$	0.94	12	(3)
	Fluorescence of calcein	pH 7.4, 25°			
		0.1 *M* KCl	17.5	35	(4)
Purified ATPase	Centrifugation	pH 7.4, 0°			
		0.1 *M* KCl, 5 m*M* $MgCl_2$	5.5	13	(5)
		0.1 *M* KCl, 1 m*M* $MgCl_2$	3.8	14	(5)
	Equilibrium dialysis	pH 7.4, 0°–4°			
		0.1 *M* KCl, 1 m*M* $MgCl_2$	0.5–1.5	12	(6)
		pH 7.0, 0°	α 0.3	7	(7)
		0.1 *M* KCl, 5 m*M* $MgCl_2$	β 20	10	(7)
			γ 1000	13	(7)
	Flow dialysis	pH 7.4, 20°			
		0.1 *M* KCl, 5 m*M* $MgCl_2$	1.2–1.3	~20	(8)

[a] (1) Chevallier and Butow (1971); (2) Fiehn and Migala (1971); (3) Miyamoto and Kasai (1976); (4) Chiu and Haynes (1977); (5) Meissner (1973); (6) Meissner *et al.* (1973); (7) Ikemoto (1974); (8) Yates and Duance (1976).

2. Ca^{2+} *Release from ATPase*

When EP is formed by the reaction of the ATPase of FSR with ATP, the product ADP is released into the external medium (Section V,A,3). The Ca^{2+} on EP is replaced by Mg^{2+}, then released from the enzyme into the internal lumen of FSR. When EP decomposes, the product P_i is released into the external medium (Section VIII,A,2). Yamada and Tonomura (1972b) measured the ratio of v_o/[EP] over a broad range of Mg^{2+} and Ca^{2+} concentrations employing ATPase purified from SR treated with DOC. They found that the Mg^{2+}-dependent EP decomposition was competitively inhibited by Ca^{2+} and the ratio of K_{Mg}/K_{Ca} was about 2.5. This was far less than that of K_{Mg}/K_{Ca} (=30,000), which was obtained from kinetic analysis of the EP formation as described in the preceding subsection. In other words, the enzyme can distinguish Ca^{2+} from Mg^{2+} only slightly in the state EP, but has a much stronger affinity for Ca^{2+} in the state E.

The above conclusion was supported by Ikemoto (1975, 1976), who directly measured the time courses of Ca^{2+} release and rebinding to the enzyme during the formation and decomposition of EP in the presence of 5 m*M* $MgCl_2$. Using the stopped-flow technique and employing Ca^{2+} indicator, Arsenazo III, Ikemoto (1976) found that the release of Ca^{2+} occurred only after a significant delay and reached the maximum level at a considerably slower rate than the EP formation, whereas the latter began immediately after the addition of ATP to purified ATPase in the presence of Mg^{2+} (Fig. 9). He explained this phenomenon as the sequential formation

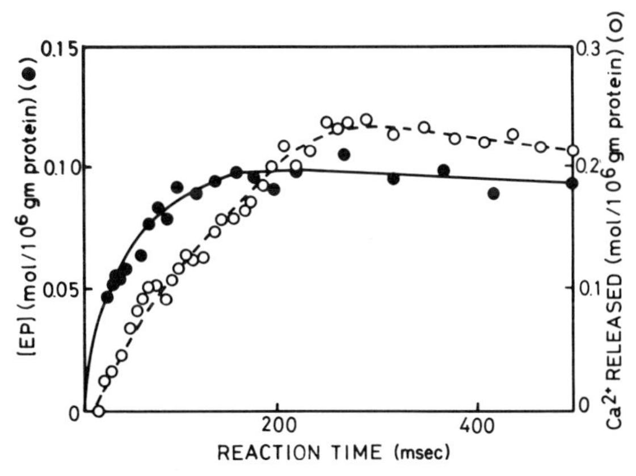

FIG. 9. Relation between Ca^{2+} release and EP formation of purified ATPase. The time course of the Ca^{2+} release was followed using Arsenazo III as a Ca^{2+} indicator. (From Ikemoto, 1976.)

of two kinds of EP which have different affinities for Ca^{2+}. However, an alternative explanation may be that the delay is due to a change in the secondary structure of the enzyme which causes lower affinity for Ca^{2+}. The existence of the two kinds of EP which have different affinities for Ca^{2+} will be discussed again in Section VIII.

3. *Remaining Problems*

Thus, kinetic studies on the Ca^{2+}-dependent ATPase led us to the conclusion that the formation of EP results in a large change in affinity of the enzyme for Ca^{2+} and Mg^{2+}. We still do not know how the structure of the Ca^{2+}-binding site on the ATPase protein changes when EP is formed. Even the chemical structure itself is entirely obscure at the moment, but Shamoo *et al.* (1976) and Stewart *et al.* (1976) reported that the tryptic subfragment of the ATPase protein with a molecular weight of 2×10^4 exhibited Ca^{2+} ionophore activity (see Section II). Furthermore, recent chemical modifications of FSR have provided evidence for conformational changes in ATPase induced by EP formation. Yamamoto and Tonomura (1976, 1977) measured the maximum number of modified lysine residues of the ATPase protein after the reaction of FSR with an impermeable reagent, 2,4,6-trinitrobenzenesulfonate (TBS), in the enzymic states of E, E_{Ca}, E^{ATP}, and E_p. They found that the number of modified lysine residues changed markedly with a change in the state (cf. Table II). Murphy (1976b, 1978) found that the reactivity of the SH groups of the ATPase protein with dinitrofluorobenzene varied significantly with the enzymic state.

As described in Sections III and V, EP formation from E and ATP requires the presence of Ca^{2+}, whereas EP decomposition into E and P_i requires Mg^{2+}. In addition, kinetic studies indicated that these divalent cations compete for the binding site in the E and EP states. There is circumstantial evidence that Mg^{2+} is located in the interior of the FSR membrane or in the pocket of the ATPase molecule during participation in EP decomposition (Section V,A,4). However, at present we have no convincing information on the distribution of Mg^{2+} in the SR membrane. Furthermore, we do not know what kind of cation is transported as a counterion during the Ca^{2+} transport through the SR membrane. The problem was first studied by Carvalho and Leo (1967), who measured the amounts of various kinds of cation released from FSR during the Ca^{2+} uptake. Based on the finding that Mg^{2+} was released most extensively among the cations, they suggested that it could be transported as a counterion of Ca^{2+}. From their results, we can reasonably conclude that Mg^{2+} enhances EP decomposition by substituting for Ca^{2+} bound to EP on the interior surface of the SR membrane. Then we must assume that Mg^{2+} is easily taken up into the vesicle by passive transport to continue the active transport of Ca^{2+}. However, the permeability of the SR membrane to divalent cations, including Mg^{2+}, is generally considered to be very low (cf. Miyamoto and Kasai, 1976). Therefore, the possibility remains that some anions (such as phosphate) are cotransported with Ca^{2+}, or that the transported Ca^{2+} binds to anionic sites located in the SR membrane.

B. Translocation of Ca^{2+} across the SR Membrane

1. *Kinetic Evidence for* Ca^{2+} *Translocation Coupled with EP Formation*

The formation of EP from E and ATP requires external Ca^{2+} (Section V,A,2). Furthermore, when ADP is added to EP in the presence of K^+, ATP is formed by the reverse reaction of EP formation (Section V,A,3). Therefore, if the reverse reaction can be shown experimentally to require the presence of Ca^{2+} inside the SR vesicle, we can conclude that Ca^{2+} is translocated from outside to inside the SR vesicle coupled with EP formation. In fact, Kanazawa *et al.* (1971) found that [γ-^{32}P]ATP was formed from ADP and $E^{32}P$ produced by the reaction of E with [γ-^{32}P]ATP even when the outer Ca^{2+} had been removed completely by adding excess EGTA or when [γ-^{32}P]ATP was diluted with nonradioactive ATP. Therefore, they concluded that the reaction step of ATP formation from EP and ADP was independent of the external Ca^{2+}.

Kanazawa *et al.* (1971) phosphorylated FSR by [γ-^{32}P]ATP in the presence of Ca^{2+} at alkaline pH when the membrane was leaky. At intervals after termination of further $E^{32}P$ formation by adding EGTA, ADP was added, and the amount of $E^{32}P$ decay accompanied by [γ-^{32}P]ATP formation was measured. The fraction of $E^{32}P$ which decayed in response to ADP decreased with time after the addition of EGTA. On the contrary, when $E^{32}P$ formation was stopped by adding nonradioactive ATP, the amount of the ADP-sensitive fraction of $E^{32}P$ did not decrease with time. Furthermore, when FSR was solubilized with Triton X-100, addition of EGTA and ADP after E^{32}-P formation did not induce the formation of [γ-^{32}P]ATP, whereas simultaneous addition of ADP and unlabeled ATP induced the reverse reaction (Kanazaw *et al.*, 1971). These findings showed that the reverse reaction requires the presence of Ca^{2+} inside the SR vesicle. It was thus suggested that Ca^{2+} is translocated from the outer medium to the inside of the vesicle membrane during EP formation:

$$ {}^{o}CaE^{ATP} \rightleftharpoons {}^{i}CaE_{P} + ADP $$

The same conclusion was drawn by Makinose (1973), who showed that when ADP was added to FSR that had been partially loaded with Ca^{2+}, a rapid exchange reaction between ITP and ADP occurred with concomitant rapid exchange of Ca^{2+} between the inside and outside of the SR vesicle.

2. *Direct Evidence for Coupling of* Ca^{2+} *Translocation with EP Formation*

As described in the preceding subsection, kinetic studies on the partial reaction of the Ca^{2+}-dependent ATPase, i.e., EATP $\rightleftharpoons$ EP + ADP, suggested that EP formation was coupled to Ca^{2+} translocation across the SR membrane. Direct evidence for this suggestion was obtained by Sumida and Tonomura (1974). They measured the time course of EP formation and compared it with that of Ca^{2+} transport in the presence of a low Ca^{2+} concentration and the absence of added Mg^{2+} at 0°C. Under these conditions, EP was rapidly formed, while P_i liberation was very slow, as shown in Fig. 10A. This slow rate was due to restraint of EP decomposition caused by the absence of added Mg^{2+} (Section V,A,4). The time course of Ca^{2+} uptake measured by the EGTA-stop method showed a fast initial phase corresponding to the time course of EP formation, followed by a slower steady phase corresponding to the time course of slow P_i liberation (Fig. 10B).

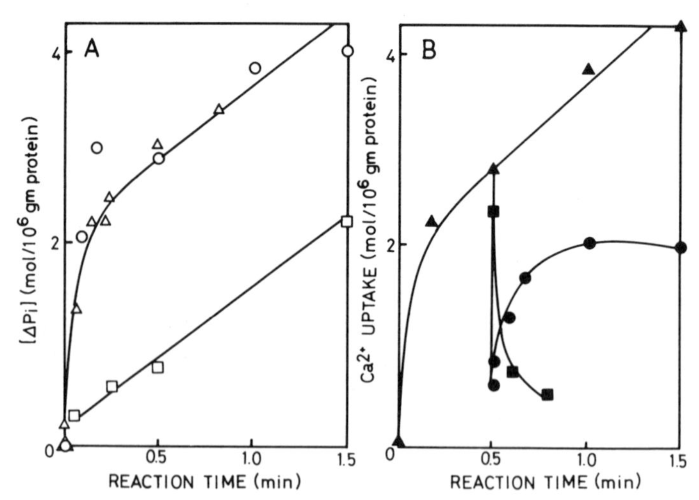

FIG. 10. Coupling between ATPase reaction and Ca^{2+} translocation. (A) Time course of P_i liberation. At intervals, EGTA (○), EGTA + $MgCl_2$ (△), or EGTA + ADP (□) was added to the reaction mixture to stop EP formation. (B) Time course of Ca^{2+} uptake. The reaction was stopped by the addition of EGTA + $MgCl_2$. Simultaneous addition of EGTA and ADP (■) caused a rapid release of Ca^{2+} from SR. When the time interval between the additions of EGTA + $MgCl_2$ and ADP was increased (●), the amount of ADP-sensitive Ca^{2+} leakage decreased. (From Sumida and Tonomura, 1974.)

The molar ratio of the Ca^{2+} uptake and EP formation was 1 under these conditions. This is consistent with the observation by Ikemoto (1974) that there was 1 mol of high-affinity sites for Ca^{2+} binding per mole of ATPase at 0°C, while there were 2 mol of the site at room temperature (Section VI,A,1). In fact, Kurzmack *et al.* (1977) reported more recently that 2 mol of Ca^{2+} were taken up by SR when 1 mol of EP was formed at room temperature. Furthermore, when the reaction of Ca^{2+} uptake by SR was terminated by the simultaneous addition of EGTA and ADP, its time course exhibited only the slow phase that corresponded to the slow liberation of P_i from EP. This is because the addition of ADP induced the rapid release from SR of Ca^{2+} that had been taken up during EP formation.

Sumida and Tonomura (1974) measured the time course of the Ca^{2+} release from SR by adding ADP at intervals after terminating EP formation by adding EGTA. As shown in Fig. 10B, the amount of Ca^{2+} released by the addition of ADP decreased with an increase in the interval between the addition of EGTA and that of ADP. The time course of the decrease corresponded well to that of EP decrease after addition of EGTA. These findings indicate that Ca^{2+} is translocated from the outside to the inside of the SR membrane by E_P formation, and that the Ca^{2+} bound to E_P is liberated into the lumen of the vesicle when E_P decomposes to $E + P_i$, while the Ca^{2+} is exposed on the outer surface of the membrane when E_P is converted to E^{ATP} by adding ADP (Sumida and Tonomura, 1974).

VII. Molecular Models for Ca^{2+} Transport

In preceding sections, we described kinetic studies on the coupling mechanism between ATP hydrolysis and Ca^{2+} transport in SR which have advanced extensively in recent years. In this section, we discuss the functional movements of the ATPase molecule within the SR membrane which induce the ATP-dependent transport of Ca^{2+}.

The "circulating carrier model" proposed by Shaw (1954) is now widely accepted as a framework for constructing a more advanced model for active cation transport. This model was originally proposed to explain the active transport of Na^+ and K^+ across the cell membrane. It was assumed that the cation could be translocated from one side to the other of the membrane in combination with a carrier whose affinity for the cation changes reversibly in association with the cycle of the transport reaction.

In the case of Ca^{2+} transport in SR, where the ATPase molecule serves as a carrier (Section VI), Ca^{2+} is translocated from outside to inside the membrane during the EP formation. This process accompanies a large decrease in affinity of the carrier for Ca^{2+} in the presence of Mg^{2+}. At

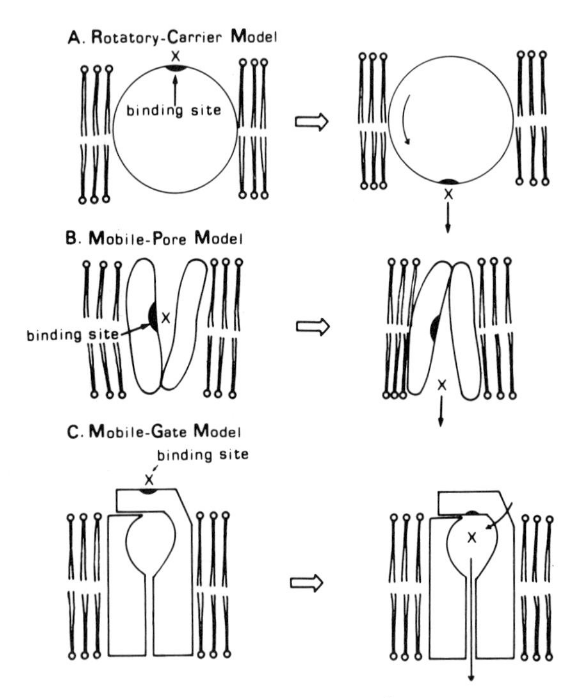

FIG. 11. Molecular models for active transport of Ca^{2+}. In each model, × represents the Ca^{2+} to be translocated. (A) Rotatory-carrier model. [From Tonomura and Morales (1974) and Yamamoto and Tonomura (1976, 1977).] (B) Mobile-pore model. [From Singer (1974) and Singer and Nicolson (1972).] (C) Mobile-gate model. For explanation, see text.

present, a number of models have been proposed in order to understand the molecular mechanism of Ca^{2+} transport across the SR membrane on the structural basis of the ATPase molecule. These models can be divided into two categories, i.e., "rotatory model" and "mobile-pore model" (Fig. 11). The former assumes that Ca^{2+} bound to the ATPase is translocated across the membrane when the enzyme molecule rotates within the membrane. In the latter, it assumed that the ATPase protein(s) forms a pore through which Ca^{2+} is transported and the translocation of Ca^{2+} is coupled to the energy-linked conformational change of the protein-lined pore.

A. ROTATORY MODEL

The rotatory model was originally proposed by Yariv *et al.* (1972) as the simplest model to explain the movement of a cation carrier, and substantial evidence for the model was first provided by Tonomura and Morales (1974) in the Ca^{2+} transport system of FSR. They prepared the spin-labeled *N*-ethylmaleimide (NEM)-bound SR, and measured the time course of the decrease in EPR spectra due to quenching of the spin by externally added ascorbate. The time course of the quenching was shown to be composed of two different components with high and low rate constants. They indicated that the component with the high quenching rate constant corresponded to labels located on the exterior of the SR membrane, while that with the low rate constant corresponded to labels located on the interior of the membrane. The relative amount of the two components depended largely on the conditions, where SR was fixed on one of the enzymic states of CaE, E^{ATP}, or E_P. Especially, the ratio of the fast component to the slow one decreased significantly in the enzymic state of E_P where Ca^{2+} was translocated across the membrane. Almost all their results could be explained by assuming rotation of the ATPase molecule within the membrane during ATP hydrolysis.

Yamamoto and Tonomura (1976) measured the number of exposed lysine residues of SR ATPase protein after the exposure of FSR to 2,4,6-trinitrobenzenesulfonate (TBS), which did not permeate the membrane and did not affect the ATPase activity significantly. They found that 2, 3, 1, and 3–4 mol of lysine residues per mole of ATPase were modified by TBS when FSR was fixed in the enzymic states of E, CaE, E^{ATP}, and E_P, respectively. When the enzymic state was changed during the reaction of FSR with TBS, the number of modified residues increased but did not exceed the sum of those at each state. These results indicated that the exposed lysine residues of the ATPase molecule varied with the enzymic state, and at least some of them were common to different enzymic states. Although these results could be explained easily by the rotatory model, the amount of modified lysine residues with TBS was far less than the total amount of lysine residues, i.e., 50 mol per mole of ATPase (Meissner *et al.*, 1973). Therefore, the above results suggest that a large rotational movement occurs in a restricted area on the ATPase molecule.

Yamamoto and Tonomura (1977) examined the change in localization of exposed lysine residues among the three tryptic subfragments (see Section II,B) with a change in the enzymic state of SR ATPase. As shown in Table II, the number of modified residues in each mole of the subfragments varied with the enzyme state. Further, the number of modified residues in each subfragment sometimes was not an integral number, while their sum per mole of ATPase molecule usually was. This suggests heterogeneity in the states of ATPase proteins within the SR membrane, as discussed in Section IV,B.

Thus, the rotatory model allows us to visualize the continuous changes in structure and function of the enzyme molecule during the Ca^{2+} transport

TABLE II

AMOUNTS OF EXPOSED LYSINE RESIDUES IN THREE TRYPTIC SUBFRAGMENTS OF ATPASE PROTEIN

Enzyme state	TNP/subfragment (mole/mole)[a,b]			
	10^5 MW	4.9×10^4 MW	3.2×10^4 MW	2.1×10^4 MW
E	2.0	1.0	0.5	0.5
E^{ATP}	1.0	0.5	0.5	0
E_p	3.5	1.5	1.0	1.0

[a] For simplicity, the numbers of exposed lysine residues were approximated to 0, 0.5, 1, 1.5, 2, or 3.5.

[b] The amounts of bound TNP were determined by assuming that $\varepsilon = 1 \times 10^4$ mol^{-1} cm^{-1} (Pull, 1970). From Yamamoto and Tonomura (1977).

by FSR. However, as pointed out by Singer (1974), it is thermodynamically unlikely that a protein as large as ATPase would rotate as a whole across the hydrophobic region of a lipid bilayer. Furthermore, the ATPase protein is surrounded by tightly bound phospholipids (cf. Hesketh *et al.*, 1976). Dutton *et al.* (1976) recently found that the antibody of the anti-2,4-dinitrophenol (DNP) did not prevent either Ca^{2+} transport or ATPase reaction of FSR that had been previously labeled with [^{3}H]dinitrophenol. Therefore, the rotatory movement described above probably occurs only in a restricted area of the ATPase.

B. MOBILE-PORE MODEL

This model was first proposed by Singer and Nicolson (1972) in relation to their "fluid mosaic" model of biomembrane. It assumes that cation is transported through a channel lined by a protein or protein complex. A more refined form of this kind of model was proposed by Kyte (1974, 1975) to explain the transport of Na^+ and K^+ across cell membrane. Dutton *et al.* (1976) supported the mobile-pore model, based on the observation that the antibody against the anti-DNP inhibited neither the ATPase activity nor the Ca^{2+} transport of DNP-SR, as mentioned above. It is tacitly assumed in the mobile-pore model of the Singer and Nicolson type that the Ca^{2+}-dependent ATPase forms an oligomer within the membrane as a functional unit for Ca^{2+} transport, but this has not been proved (see Section II,A). A model of this type was also proposed by Racker (1976) to explain the coupling mechanism between proton transport across mitochondrial membrane and the synthesis or hydrolysis of ATP by the coupling factor 1 (F_1). He assumed that F_1, located on the matrix side of the inner mitochondrial membrane, is connected to another factor (F_0) through which a proton passes easily. A variant of this model was proposed by Racker (1976) for the Ca^{2+} transport by SR membrane. He assumed that Ca^{2+} is transported through an intramembranous channel, formed by the proteolipid, which is connected with the ATPase on the outer surface of the membrane. However, the structural and functional significance of the proteolipid in the SR membrane is still unknown (Section II). On the basis of observations that the tryptic subfragment of ATPase with a molecular weight of about 2×10^4 exhibited ionophoretic activity (Section II), Shamoo and Goldstein (1977) thought that the subfragment might form a channel or pore through which Ca^{2+} could be transported.

In the mobile-pore model, a cation can be transported across the membrane unaccompanied by large movements of the ATPase molecule as a whole in which the polar groups of the enzyme remain always outside the membrane. However, in this type of model, it is difficult to explain the continuous transition in molecular structure of the carrier protein during cation transport, which is indicated by chemical modification and ascorbate quenching techniques. Indeed, much more complicated molecular motions must be assumed, if this model is to be employed for interpreting the results obtained by the techniques described in Section VII,A.

Both the rotatory-carrier and mobile-pore models are constructed on fragmentary information obtained by indirect measurements and are not satisfactory for interpreting the molecular mechanism of the active transport of Ca^{2+} in SR membrane. Considering their merits and demerits, we propose another model, a "mobile-gate model" (Fig. 11), on the following assumptions. The ATPase monomer or oligomer constructs a gate, a pocket, and a pore in the SR membrane. Ca^{2+} binds to a specific site on the outer surface of the gate, then is translocated from the outer medium to the interior of the pocket in the ATPase by rotational movement(s) of the gate, which is coupled with EP formation. In the pocket, the Ca^{2+} bound to the divalent cation site is replaced by Mg^{2+} owing to the pronounced change in affinity that accompanies EP formation. The Ca^{2+} released is preferentially transported through the pore because (1) the divalent cation is transported through the pore as a hydrated ion, and the radius of the hydrated Ca^{2+} (6 Å) is smaller than that of Mg^{2+} (8 Å), or (2) the divalent cation is transported through the pore in the dehydrated form, and the energy of dehydration for Ca^{2+} is far less than that of Mg^{2+}. Furthermore, the tight coupling between Ca^{2+} transport and ATP hydrolysis in the SR membrane can be understood by assuming that the permeability of the part of the pocket for Ca^{2+} is low, while that for Mg^{2+} is not low. This model is also consistent with the complicated functions of Mg^{2+} in the ATPase reaction described in Sections V,A,4 and VI,A,3,

VIII. Reaction Mechanism of ATP Synthesis

In the preceding sections, we described mechanisms of the forward reaction of the Ca^{2+}-dependent ATPase and the coupling mechanism between ATP hydrolysis and Ca^{2+} transport in the SR membrane. Another useful approach for clarifying the mechanism of active Ca^{2+} transport is to investigate the ATP synthesis coupled with the reversal of Ca^{2+} transport. Many investigators have attempted to synthesize ATP from ADP and P_i by reversing the ion pump since Jagendorf and Uribe (1966) demonstrated ATP formation caused by acid–base transition of chloroplasts. The ATP formation linked with the chemiosmotic coupling of the Na^+,K^+ pumping system in erythrocyte membrane was observed later by Garrahan and Glynn (1967), Lant and Whittam (1968), and Glynn and Lew (1970).

In spite of exhaustive studies, the mechanisms of ATP synthesis in chloroplasts and mitochondria are still unknown, and it is uncertain whether they are essentially the same as those in Na^+ and Ca^{2+} pumps (Racker, 1976). There are many difficulties in studying the reaction mechanism of ATP synthesis by the reversal of Na^+,K^+-dependent ATPase because of complicated properties in the structure and function of this system. In contrast, studies of the ATP synthesis in SR membrane have rapidly advanced in recent years since Makinose and Hasselbach (1971) demonstrated ATP formation coupled to the outflow of Ca^{2+} from FSR, because the system of Ca^{2+} pumping is relatively simple, the coupling between ATP hydrolysis and Ca^{2+} transport is tight, and the reaction mechanism of the Ca^{2+}-dependent ATPase has been studied in detail (see Section V). More recently, Kanazawa (1975), Knowles and Racker (1975a), and de Meis and Tume (1977) found that EP and ATP were formed from P_i and ADP even in the absence of the Ca^{2+} concentration gradient across the membrane. Taniguchi and Post (1975) also found that EP and ATP were formed by the reaction of Na^+,K^+-dependent ATPase with P_i and ADP. The reader is referred to an excellent review by Hasselbach (1978) on the reversibility of the Ca^{2+} pump.

A. ATP Synthesis by Ca^{2+}-Loaded FSR

1. *Overall Reaction*

As already described in Section I, the Ca^{2+} transport coupled with ATP hydrolysis results in formation of a steep gradient of Ca^{2+} concentration across the FSR membrane. Makinose (1971) found that when the net uptake of Ca^{2+} had been completed, a rapid exchange of ATP $\rightleftharpoons$ P_i occurred associated with the exchange of Ca^{2+} between the inside and outside of the vesicles. Both exchange reactions were prevented by the treatment of SR with phospholipase C or ether. Makinose (1971) suggested the existence of a reverse reaction of ATPase that is coupled with the efflux of Ca^{2+} from FSR. Similar results were obtained by Racker (1972) and Racker *et al.* (1975) with reconstituted SR vesicles and by de Meis and Carvalho (1974), de Meis and Sorenson (1975), and de Meis *et al.* (1975) with intact FSR.

Barlogie *et al.* (1971) observed that the efflux of Ca^{2+} from SR vesicles, which was very low even when the external concentration of Ca^{2+} was reduced by adding EGTA, was enhanced about 100 times by the simultaneous addition of ADP and P_i. Makinose and Hasselbach (1971) demonstrated that 1 mol of ATP was formed when 2 mol of Ca^{2+} were released from the vesicles in the presence of ADP and P_i outside the membrane. Similar results were also obtained by Panet and Selinger (1972). These findings suggested that the osmotic energy produced by the concentration gradient of Ca^{2+} across the SR membrane can be converted into chemical energy for ATP formation. Detailed studies on the reaction mechanism of the energy conversion between ATP synthesis and the Ca^{2+} gradient were initiated when Yamada *et al.* (1972) and Makinose (1972) found the formation of EP by reaction of P_i with ATPase in Ca^{2+}-loaded FSR, as described below.

2. *EP Formation by Reaction of* Ca^{2+}*-Loaded FSR with* P_i

Makinose (1972) prepared Ca^{2+}-loaded FSR vesicles by utilizing acetylphosphate as an energy source for the Ca^{2+} uptake. On the other hand, Yamada and Tonomura (1972a) and Yamada *et al.* (1972) incubated the vesicles without an energy supply, in a medium containing a large excess of Ca^{2+}, for several hours until the electrochemical activity of Ca^{2+} inside the vesicle became equal to that of the outer medium. Yamada and co-workers (1972), Yamada and Tonomura (1972a), and Makinose (1972) found independently that the Ca^{2+}-loaded FSR showed a rapid incorporation of P_i immediately after the Ca^{2+} concentration of the outer medium had been reduced by adding EGTA. The radioactivity of $^{32}P_i$ was incorporated into the SR ATPase protein with a molecular weight of 10^5 (Yamada *et al.*, 1972). The pH stability and hydroxylamine sensitivity of the phosphoprotein isolated by adding TCA were similar to those of acylphosphate and indistinguishable from those of EP formed by the forward reaction of the SR ATPase (Yamada *et al.*, 1972; Makinose, 1972). Furthermore, as described below, the phosphorylated enzyme formed by the backward reaction could phosphorylate ADP to form ATP, like EP formed by the forward reaction. Both Yamada *et al.* (1972) and Makinose (1972) concluded that the phosphorylated enzyme formed in the backward reaction was identical to EP formed in the forward reaction.

The following two findings indicated that P_i interacts with the ATPase on the exterior of the SR membrane: (a) the formation of EP from E and P_i was inhibited competitively by ATP added outside the membrane (Yamada *et al.*, 1972), and (b) the intravesicular $^{32}P_i$, which had been incorporated into the reconstituted vesicle of SR as a precipitating anion for Ca^{2+}, did not serve as a substrate for $E^{32}P$ formation or the ATP $\rightleftharpoons$ $^{32}P_i$ exchange reaction, while the external $^{32}P_i$ served as a substrate for both (de Meis and Carvalho, 1976).

Yamada *et al.* (1972) measured the dependence of the amount of EP formed from P_i on the concentration gradient of Ca^{2+} in the presence of 20 m*M* $MgCl_2$ and 5 m*M* P_i (Fig. 12). The steady-state level of EP, [EP], increased with the electrochemical potential of Ca^{2+} across the SR membrane. The double reciprocal plot of [EP] against the square of the Ca^{2+} gradient was linear, indicating that 2 mol of Ca^{2+} were bound to 1 mol of

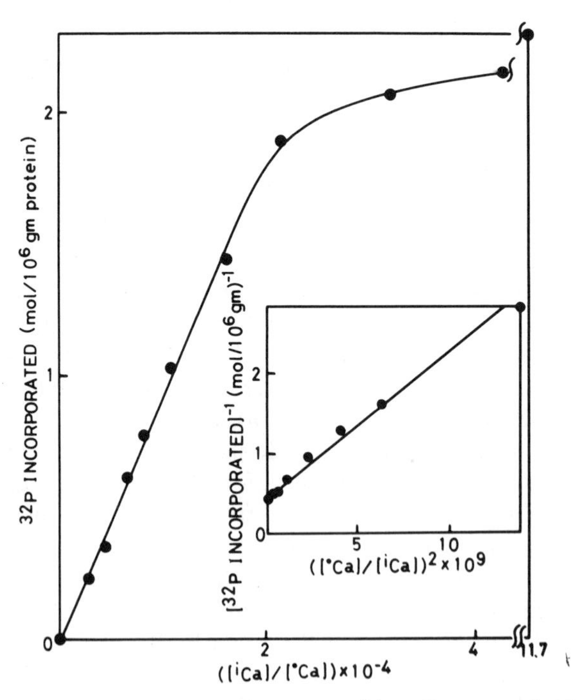

FIG. 12. Dependence of the amount of EP on the Ca^{2+} gradient across the FSR membrane. Inset: The reciprocal of [EP] was replotted against the square of the Ca^{2+} gradient, [ⁱCa]/[°Ca]. (From Yamada *et al.*, 1972.)

ATPase in the reaction of EP formation. The dependence of [EP] on the concentrations of P_i and Mg^{2+} showed a saturation curve of the Michaelis–Menten type and was given by the following equation:

$$\frac{[EP]}{\varepsilon} = 1 \Bigg/ \left\{ 1 + \phi \left(\frac{[^{o}Ca]}{[^{i}Ca]} \right)^2 \cdot \frac{1}{[Mg][P_i]} \right\}.$$

$[^{i}Ca]$ and $[^{o}Ca]$ are the electrochemical activities of Ca^{2+} inside and outside the SR vesicle, respectively. This equation indicates that all the active sites, ε, can be phosphorylated under optimum reaction conditions. The value of the constant, ϕ, was calculated to be 2×10^5 M^2. Thus, in the presence of 20 m*M* $MgCl_2$ and 5 m*M* P_i, the half-saturation value of the Ca^{2+} gradient was estimated to be about 4×10^4. The free-energy change during the EP formation by chemiosmotic coupling was about -12 kcal under the conditions described above.

The amount of EP formed by the reaction of the Ca^{2+}-loaded SR with P_i rapidly decreased immediately after the addition of ADP to EP, and 1 mol of ATP was formed for each 2 mol of Ca^{2+} released from the vesicle (Yamada and Tonomura, 1972a; Yamada *et al.*, 1972; Makinose, 1972). Thus, by these experiments all the processes of the Ca^{2+}-dependent ATPase reaction were found to be reversible.

B. ATP Synthesis by Ca^{2+}-Free FSR

In the preceding subsection, we described the mechanism of ATP formation in the presence of the Ca^{2+} gradient across the SR membrane. However, Kanazawa and Boyer (1973) and Vale *et al.* (1976) recently showed that the formation of EP and probably the single cycle of ATP synthesis from E, P_i, and ADP can occur even in the absence of the Ca^{2+} gradient.

1. *Phosphorylation of* Ca^{2+}*-Free FSR by* P_i

Kanazawa and Boyer (1973) found that the Ca^{2+}-dependent ATPase catalyzed the ^{18}O exchange reaction between P_i and H_2O when FSR was incubated with P_i in the presence of Mg^{2+} and a large excess of EGTA. The $P_i \rightharpoonup HOH$ exchange reaction was enhanced by Mg^{2+}, but inhibited by a small amount of external Ca^{2+}. The concentration of Ca^{2+} required for the half inhibition of the $P_i \rightharpoonup HOH$ exchange rate was equal to that for EP formation from E and ATP (Sections V and VI). The Hill coefficient of the Ca^{2+} inhibition was about 2. Since the Mg^{2+}-dependent $P_i \rightharpoonup HOH$

exchange was inhibited competitively by external Ca^{2+}, Kanazawa and Boyer suggested that Mg^{2+} activated the exchange reaction on the exterior of the SR membrane. They also showed that EP was formed from E and P_i under the conditions for the $P_i \rightleftharpoons$ HOH exchange reaction. Furthermore, they found that both the phosphorylation reaction with P_i and the $P_i \rightleftharpoons$ HOH exchange could be catalyzed by FSR that had been treated with EGTA at high alkaline pH to remove Ca^{2+} completely. These findings suggested that the EP formation from E and P_i did not require the Ca^{2+} gradient across the membrane.

2. *ATP Synthesis during the Release of Membrane-Bound* Ca^{2+} *from FRS*

Recently, Vale *et al.* (1976) proposed another type of reaction for the ATP formation in FSR. Vale and Carvalho (1975) had previously distinguished membrane-bound Ca^{2+} from the intravesicular free Ca^{2+} by using the ionophore, X 537 A. Vale *et al.* (1976) recently observed that, when ADP and P_i together with EGTA were added to FSR that had been pretreated with X 537 A to equilibrate the Ca^{2+} inside and outside the vesicle, a significant amount (about 14 nmol) of ATP was formed during the release of about 40 nmol of Ca^{2+} from 1 mg of SR protein. They also observed that simultaneous addition of ADP, P_i, and EGTA resulted in ATP formation (about 6 nmol) by FSR that had been exposed to Ca^{2+} for a short time. Under these conditions, almost all the Ca^{2+} could exist in the outer medium because the permeability of the membrane for Ca^{2+} was extremely low (Section VIII,A). Therefore, they suggested that the free energy produced by the release of the membrane-bound Ca^{2+} might be converted into energy for the ATP synthesis through conformational changes in the ATPase molecule. However, it is uncertain from which site the Ca^{2+} is released by the addition of EGTA. Furthermore, their experiment does not exclude the possibility of the existence of a short-lived Ca^{2+} gradient after addition of EGTA, even though they used X 537 A-treated vesicles.

C. ATP Synthesis by Solubilized SR ATPase

1. *EP Formation by Solubilized Membrane*

As mentioned in Section VIII,B,1, Kanazawa and Boyer (1973) demonstrated that the leaky FSR catalyzed the $P_i \rightleftharpoons$ HOH exchange reaction and EP formation with P_i. Further studies by Kanazawa (1975) showed clearly that the activity of EP formation from P_i remained even after solubilization of the SR membrane by Triton X-100. Similar results were obtained by de Meis and Carvalho (1976) employing SR membrane treated with diethyl ether. Kanazawa (1975) showed that the phosphorylation of the solubilized SR by P_i required the presence of a high concentration of Mg^{2+}, and the Mg^{2+}-dependent EP formation was inhibited competitively by a small amount of Ca^{2+} with a Hill coefficient of about 2. He measured the dependence of the amount of EP formed on the concentrations of P_i and Mg^{2+} and proposed the following mechanism:

$$E + P_i + Mg^{2+} \rightleftharpoons MgE_{P_i} \rightleftharpoons MgE_P$$

A similar mechanism was also proposed by Boyer *et al.* (1977) using the FSR ATPase. Kanazawa (1975) observed that the phosphorylation of the solubilized enzyme by P_i was enhanced extensively by raising the temperature, as shown in Fig. 13. He calculated the values of the standard free energy, enthalpy, and entropy of the step, $MgE_{P_i} \rightleftharpoons MgE_P$, to be +0.4 kcal/mol, +16 kcal/mol, and +50 e.u., respectively, and suggested that the phosphorylation of the enzyme by P_i was induced by the marked increase in entropy. More recently, Kanazawa and Katabami (1977) observed, using the leaky membrane treated with EGTA at high alkaline pH, that the Arrhenius plot of the equilibrium constant between $E + P_i + Mg^{2+}$ and MgE_P had a break at around 18°C, as did the rate of EP decomposition of the forward reaction (Section III,D).

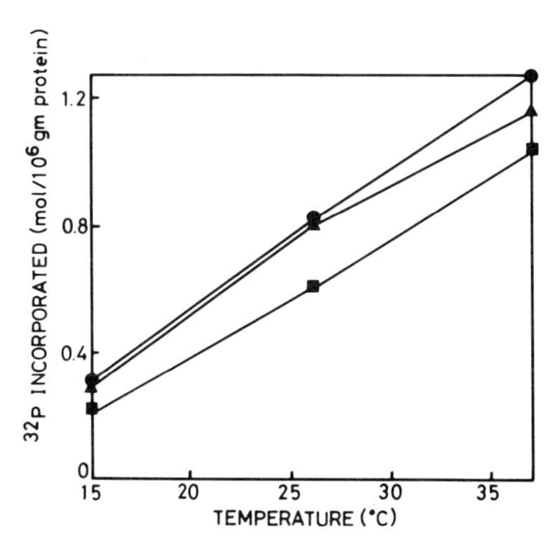

FIG. 13. Temperature dependence of the phosphorylation of the SR ATPase solubilized with Triton X-100 by P_i. Phosphorylation was started with 10 (●), 15 (▲), and 20 m*M* (■) P_i in the presence of high concentrations of $MgCl_2$ and EGTA. (From Kanazawa, 1975.)

Recently, Rauch *et al.* (1977) measured the temperature dependence of the apparent second-order rate constant of EP formation from E and P_i using FSR in the absence of Ca^{2+} gradient. They observed the break in the Arrhenius plot at 25°C and the activation enthalpy for EP formation was +26.3 and +17.7 kcal/mol below and above 25°C, respectively. However, the kinetic properties of the reaction that they measured were rather complicated. For example, the net phosphorylation and the phosphate exchange with medium P_i exhibited biphasic time courses when the reaction had been started by adding P_i or Mg^{2+} to Ca^{2+}-free ATPase above 25°C, whereas when the reaction had been started by Ca^{2+} removal from ATPase, the time course was monophasic at all temperatures between 20°C and 37°C (Rauch *et al.*, 1977). Therefore, at present the kinetic and mechanistic meaning of the activation enthalpies given above are not clear.

2. *ATP Synthesis from ADP and* P_i *by Purified ATPase*

In both cases of solubilized and leaky SR, EP is produced from the enzyme and P_i by reversal of the ATPase reaction, as mentioned before. But conclusive evidence for ATP formation from ADP and the EP thus formed has not yet been obtained. However, Knowles and Racker (1975a) recently reported ATP formation using purified ATPase preparation. The ATPase was first incubated with P_i in the presence of Mg^{2+} and EGTA to form EP. The subsequent addition of ADP with a large amount of Ca^{2+} caused ATP to be produced in an amount equal to that of the decrease in EP (Fig. 14). They added large amounts of hexokinase and glucose to the reaction mixture and assayed ATP synthesis as the amount of glucose 6-phosphate produced in order to avoid the ATP $\rightleftharpoons$ P_i exchange reaction, which may be caused by contaminating ATP in the ADP preparation, and also to eliminate a side reaction catalyzed by adenylate kinase.

Two possibilities may be considered as the energy source for the ATP synthesis in the purified ATPase system; one is that the ATP produced exists in a tightly bound form, and the other is that a conformational change(s) in the enzyme occurs that induces the formation of ATP from EP and ADP. In other words, the energy level of the enzyme state after the conformational change(s) is lower than that of the original one. Knowles and Racker (1975a) eliminated the former possibility, since the synthesized ATP rose into the supernatant as free ATP when the reaction mixture was centrifuged without terminating the reaction with an acid. If so, energy is required to revert the enzyme state to the original one in order to repeat the cycle of ATP synthesis. This energy might be supplied by the Ca^{2+} efflux across the membrane of Ca^{2+}-loaded SR or by the release of Ca^{2+} from the binding site on the membrane on addition of EGTA (cf. Sections VIII,A,3 and B,2).

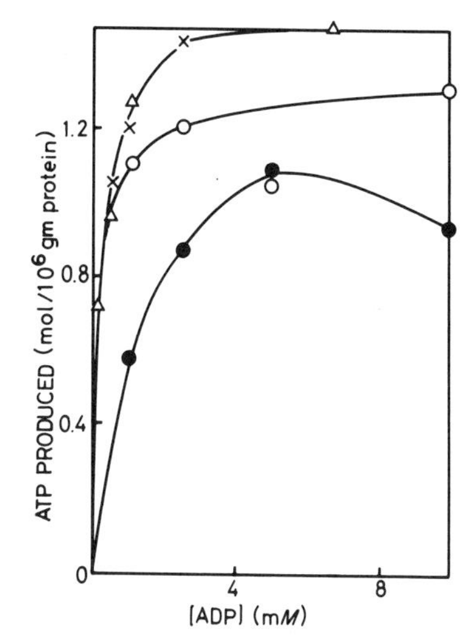

FIG. 14. ATP formation from ADP and EP produced by the reverse reaction. EP was produced by the reaction of P_i with purified ATPase in 10 mM $MgCl_2$ and 1 mM EGTA. After incubation for an appropriate time, 3.3 (●), 5 (○), 6.7, or 10 mM (△ or ×) $CaCl_2$ and the various concentrations of ADP indicated were added. (From Knowles and Racker, 1975a.)

As described in Section VI,B, Kanazawa *et al.* (1971) and Sumida and Tonomura (1974) showed the existence of two kinds of EP: $^{i}CaE_P$, which can react with ADP to produce ATP; and ADP-insensitive EP (MgE_P) formed in the absence of Ca^{2+} inside FSR. The existence of two types of EP differing in reactivity with ADP was also observed by Beil *et al.* (1977), who studied the kinetic properties of EP formed from E and P_i, both in the presence and in the absence of Ca^{2+} gradient, using FSR. On the other hand, Knowles and Racker (1975a) showed that the EP formed by reaction of the purified ATPase with P_i in the presence of Mg^{2+} could react with ADP to form ATP, only when a large amount of Ca^{2+} was added together with ADP. Very recently, Shigekawa and Dougherty (1977, 1978b) reported that most of the EP formed by the forward reaction of the purified ATPase with ATP in the absence of alkali metal salts did not react with ADP, and almost all was converted into ADP-sensitive EP by the addition of a high concentration of Ca^{2+} (Fig. 7).

At present, we cannot say whether the two kinds of EP (ADP-sensitive EP and ADP-insensitive EP) found by Kanazawa *et al.* (1971) and Sumida and Tonomura (1974) using FSR are the same as those found by Knowles and Racker (1975a) and Shigekawa and Dougherty (1978b) using the purified ATPase preparation. Furthermore, we do not know whether the

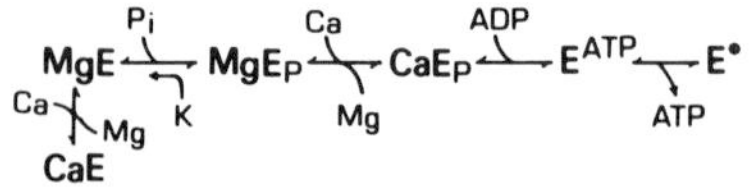

$$\begin{array}{c} \mathrm{MgE} \xrightleftharpoons[\mathrm{K}]{\mathrm{P_i}} \mathrm{MgE_P} \xrightleftharpoons[\mathrm{Mg}]{\mathrm{Ca}} \mathrm{CaE_P} \xrightleftharpoons{\mathrm{ADP}} \mathrm{E^{ATP}} \xrightleftharpoons[\mathrm{ATP}]{} \mathrm{E^{*}} \\ \mathrm{Ca} \updownarrow \mathrm{Mg} \\ \mathrm{CaE} \end{array}$$

FIG. 15. Reaction scheme for ATP synthesis from ADP and P_i catalyzed by SR ATPase. For explanation, see text.

EP formed from E and P_i in the presence of Ca^{2+} gradient across the membrane is the same as that formed in the absence of the gradient. To solve these problems, we must measure in detail the kinetic and thermodynamic properties of phosphoenzymes produced under various conditions. Recently, Beil *et al.* (1977) showed that the phosphoenzymes formed in the presence and in the absence of Ca^{2+} gradient differed in the affinity of the ATPase for P_i, the enthalpy of formation, and the kinetics of P_i incorporation. However, the possibility remains that the apparently different properties of the phosphoenzymes are due to artifacts induced by destruction of the membrane structure, not to their intrinsic properties.

Thus, many problems remain to be solved. However, if we assume that the ATP hydrolysis or synthesis occurs through a single route, and the ADP-insensitive and -sensitive EP are produced sequentially, then many experimental results on the two kinds of EP can be explained by assuming the reaction scheme shown in Fig. 15, where E* indicates the conformational state of the enzyme with an energy level lower than that of E. MgE_P is formed by the reaction of MgE with P_i, i.e., by the reverse reaction of decomposition of MgE_P shown in Fig. 3. The reaction is inhibited by a small amount of external Ca^{2+} because of formation of the CaE complex, i.e., ^{o}CaE in Fig. 3. The ADP-insensitive MgE_P complex is converted into ADP-sensitive CaE_P when a sufficient amount of Ca^{2+} is added to MgE_P, while the decomposition of MgE_P is accelerated by K^+. This scheme is consistent with the observations by de Meis *et al.* (1975) that the leaky SR membrane catalyzed the ATP $\rightleftharpoons$ P_i exchange reaction when the concentration of K^+ was very low, and the activity was enhanced by increasing the concentration of Ca^{2+} in the reaction mixture.

IX. Concluding Remarks

This review outlines characteristics of the reaction mechanisms of ATP hydrolysis and ATP synthesis, coupled and uncoupled to Ca^{2+} transport in the SR membrane. As described here, kinetic studies on the reaction mechanisms of SR ATPase have markedly advanced in the last ten years. However, many problems remain. Although almost all the kinetic properties of the solubilized SR ATPase reaction can be explained quantitatively by a simple mechanism assuming that the enzyme is homogeneous and behaves independently, several complicated phenomena have been observed in the presteady state kinetics of the ATPase reaction of the enzyme located in SR membrane (see Section IV). Two kinds of P_i burst, overshoot in the time course of EP formation and acceleration of EP formation by ATP, were observed when FSR was used, but none of them occurred in the case of solubilized ATPase. These phenomena are considered to be caused by either the interaction between ATPase molecules or heterogeneous interactions between ATPase and phospholipids within the SR membrane. Approaches to answering the question of whether the Ca^{2+}-dependent ATPase molecule exists within the FSR membrane as a monomer or an oligomer have just begun (see Sections IV,B and VII,B). On the other hand, it is well known that phospholipid plays an essential role in the step of EP decomposition (Section III,D). However, the detailed mechanism for the interaction between phospholipids and ATPase in accelerating EP decomposition is still unknown. Furthermore, we do not know the exact relation between EP formed by the reaction of Ca^{2+}-loaded SR and EP formed by the reaction of the purified ATPase with P_i (Section VIII).

In kinetic analyses, we tacitly assume that after the turnover of the reaction the Ca^{2+}-dependent ATPase returns to its original state. However, as mentioned in Section V, kinetic studies on the steady state of FSR ATPase indicated that ATP and Ca^{2+} bind to the enzyme in an ordered sequence, and presteady-state kinetics provided evidence for the binding of ATP and Ca^{2+} in a random sequence. Thus, it is not clear in the case of the SR membrane system whether we can adopt the rate constants obtained by presteady-state kinetic techniques for the analysis of the steady-state reaction without modification.

In Section VI, we described findings to show that Ca^{2+} is translocated from outside to inside the membrane during EP formation, and that the Ca^{2+} is released inside the vesicle which is induced by a decrease in the affinity of the ATPase for Ca^{2+} in the presence of Mg^{2+}. However, many problems remain unsolved with regard to the mechanism of the function of Mg^{2+} in the ATPase reaction. It is important to know whether Mg^{2+} required for EP decomposition functions on the external or internal surface of the membrane. As mentioned in Sections V,A,4 and VII,B, it seems likely that in EP decomposition, Mg^{2+} acts within a pocket of the ATPase molecule.

To propose a realistic molecular model of active Ca^{2+} transport across the SR membrane, we must solve these difficult kinetic problems. Various kinds of models for active cation transport have been proposed without exact knowledge about these problems. Therefore, they are oversimplified models that can explain only fragmentary facts, as mentioned in Section VII,B.

ACKNOWLEDGMENTS

Our sincere thanks are due to Dr. Hiroshi Nakamura of the Faculty of Engineering Science, Osaka University, and Dr. Yoshihiro Fukushima and Mr. Yoichi Nakamura of our laboratory for their valuable advice in the writing of this review.

REFERENCES

Allen, G., and Green, N. M. (1976). *FEBS Lett.* **63**, 188–192.
Anzai, K., Kirino, Y., and Shimizu, H. (1978). *J. Biochem.* (*Tokyo*) **84**, 815–821.
Barlogie, B., Hasselbach, W., and Makinose, M. (1971). *FEBS Lett.* **12**, 267–268.
Bastide, F., Meissner, G., Fleischer, S., and Post, R. L. (1973). *J. Biol. Chem.* **248**, 8385–8391.
Beil, F. U., von Chak, D., and Hasselbach, W. (1977). *Eur. J. Biochem.* **81**, 151–164.
Bennett, H. S., and Porter, K. R. (1953). *Am. J. Anat.* **93**, 61–105.
Boyer, P. D., de Meis, L., Carvalho, M. G. C., and Hackney, D. D. (1977). *Biochemistry* **16**, 136–140.
Carvalho, A. P., and Leo, B. (1967). *J. Gen. Physiol.* **50**, 1327–1352.
Chevallier, J., and Butow, R. A. (1971). *Biochemistry* **10**, 2733–2737.
Chiu, V. C. K., and Haynes, D. H. (1977). *Biophys. J.* **18**, 3–22.
Chyn, T., and Martonosi, A. (1977). *Biochim. Biophys. Acta* **468**, 114–126.
Coffey, R. L., Lagwinska, E., Oliver, M., and Martonosi, A. (1975). *Arch. Biochem. Biophys.* **170**, 37–48.
Costantin, L. L. (1975). *Prog. Biophys. Mol. Biol.* **29**, 197–224.
Deamer, D. W., and Baskin, R. J. (1969). *J. Cell Biol.* **42**, 296–307.
Dean, W. L., and Tanford, C. (1977). *J. Biol. Chem.* **252**, 3551–3553.
De Boland, A. R., Jilka, R. L., and Martonosi, A. N. (1975). *J. Biol. Chem.* **250**, 7501–7510.
Degani, C., and Boyer, P. D. (1973). *J. Biol. Chem.* **248**, 8222–8226.
de Meis, L. (1969a). *J. Biol. Chem.* **244**, 3733–3739.
de Meis, L. (1969b). *Biochim. Biophys. Acta* **172**, 343–344.
de Meis, L., and Carvalho, M. G. C. (1974). *Biochemistry* **13**, 5032–5038.
de Meis, L., and Carvalho, M. G. C. (1976). *J. Biol. Chem.* **251**, 1413–1417.
de Meis, L., and de Mello, M. C. F. (1973). *J. Biol. Chem.* **248**, 3691–3701.
de Meis, L., and Hasselbach, W. (1971). *J. Biol. Chem.* **246**, 4759–4763.
de Meis, L., and Sorenson, M. M. (1975). *Biochemistry* **14**, 2739–2744.
de Meis, L., and Tume, R. K. (1977). *Biochemistry* **16**, 4455–4463.
de Meis, L., Carvalho, M. G. C., and Sorenson, M. M. (1975). *In* "Concepts of Membranes in Regulation and Excitation" (M. R. Silva and G. Suarez-Kurtz, eds.), pp. 7–19. Raven, New York.
Duggan, P. F. (1977). *J. Biol. Chem.* **252**, 1620–1627.
Duggan, P. F., and Martonosi, A. (1970). *J. Gen. Physiol.* **56**, 147–167.
Dupont, Y. (1977). *Eur. J. Biochem.* **72**, 185–190.
Dutton, A., Rees, E. D., and Singer, S. J. (1976). *Proc. Natl. Acad. Sci. U.S.A.* **73**, 1532–1536.
Ebashi, S. (1976). *Annu. Rev. Physiol.* **38**, 293–313.
Ebashi, S., and Endo, M. (1968). *Prog. Biophys. Mol. Biol.* **18**, 123–183.

Ebashi, S., and Lipmann, F. (1962). *J. Cell Biol.* **14**, 389–400.
Fiehn, W., and Hasselbach, W. (1970). *Eur. J. Biochem.* **13**, 510–518.
Fiehn, W., and Migala, A. (1971). *Eur. J. Biochem.* **20**, 245–248.
Friedman, Z., and Makinose, M. (1970). *FEBS Lett.* **11**, 69–72.
Froehlich, J. P., and Taylor, E. W. (1975). *J. Biol. Chem.* **250**, 2013–2021.
Froehlich, J. P., and Taylor, E. W. (1976). *J. Biol. Chem.* **251**, 2307–2315.
Fukushima, Y., and Post, R. L. (1978). *J. Biol. Chem.* **253**, 6853–6862.
Garrahan, P. J., and Glynn, I. M. (1967). *J. Physiol. (London)* **192**, 237–256.
Garrahan, P. J., and Rega, A. F. (1972). *J. Physiol. (London)* **223**, 595–617.
Garrahan, P. J., Rega, A. F., and Alonso, G. L. (1976). *Biochim. Biophys. Acta* **448**, 121–132.
Glynn, I. M., and Lew, V. L. (1970). *J. Physiol. (London)* **207**, 393–402.
Hasselbach, W. (1964a). *Prog. Biophys. Mol. Biol.* **14**, 167–222.
Hasselbach, W. (1964b). *Fed. Proc. Fed. Am. Soc. Exp. Biol.* **23**, 909–912.
Hasselbach, W. (1974). *In* "The Enzymes" (P. D. Boyer, ed.), 3rd Ed., Vol. 10, pp. 431–467. Academic Press, New York.
Hasselbach, W. (1978). *Biochim. Biophys. Acta* **515**, 23–53.
Hasselbach, W., and Elfvin, L.-G. (1967). *J. Ultrastruct. Res.* **17**, 598–622.
Hasselbach, W., and Makinose, M. (1961). *Biochem. Z.* **333**, 518–528.
Hasselbach, W., and Makinose, M. (1962). *Biochem. Biophys. Res. Commun.* **7**, 132–136.
Hasselbach, W., and Makinose, M. (1963). *Biochem. Z.* **339**, 94–111.
Hesketh, T. R., Smith, G. A., Houslay, M. D., McGill, K. A., Birdsall, N. J. M., Metcalfe, J. C., and Warren, G. B. (1976). *Biochemistry* **15**, 4145–4151.
Hidalgo, C., and Ikemoto, N. (1977). *J. Biol. Chem.* **252**, 8446–8454.
Hörl, W. H., and Heilmeyer, L. M. G., Jr. (1978). *Biochemistry* **17**, 766–772.
Hörl, W. H., Jennissen, H. P., and Heilmeyer, L. M. G., Jr. (1978). *Biochemistry* **17**, 759–766.
Hokin, L. E., Sastry, P. S., Galsworthy, P. R., and Yoda, A. (1965). *Proc. Natl. Acad. Sci. U.S.A.* **54**, 177–184.
Huxley, A. F., and Simmons, R. M. (1971). *Nature (London)* **233**, 533–538.
Huxley, A. F., and Taylor, R. E. (1958). *J. Physiol. (London)* **144**, 426–441.
Huxley, H. E. (1964). *Nature (London)* **202**, 1067–1071.
Ikemoto, N. (1974). *J. Biol. Chem.* **249**, 649–651.
Ikemoto, N. (1975). *J. Biol. Chem.* **250**, 7219–7224.
Ikemoto, N. (1976). *J. Biol. Chem.* **251**, 7275–7277.
Ikemoto, N., Sreter, F. A., Nakamura, A., and Gergely, J. (1968). *J. Ultrastruct. Res.* **23**, 216–232.
Ikemoto, N., Bhatnagar, G. M., and Gergely, J. (1971). *Biochem. Biophys. Res. Commun.* **44**, 1510–1517.
Inesi, G. (1971). *Science* **171**, 901–903.
Inesi, G. (1972). *Annu. Rev. Biophys. Bioeng.* **1**, 191–210.
Inesi, G., and Almendares, J. (1968). *Arch. Biochem. Biophys.* **126**, 733–735.
Inesi, G., and Asai, H. (1968). *Arch. Biochem. Biophys.* **126**, 469–477.
Inesi, G., and Scales, D. (1974). *Biochemistry* **13**, 3298–3306.
Inesi, G., Goodman, J. J., and Watanabe, S. (1967). *J. Biol. Chem.* **242**, 4637–4643.
Inesi, G., Maring, E., Murphy, A. J., and McFarland, B. H. (1970). *Arch. Biochem. Biophys.* **138**, 285–294.
Inesi, G., Millman, M., and Eletr, S. (1973). *J. Mol. Biol.* **81**, 483–504.
Jagendorf, A. T., and Uribe, E. (1966). *Proc. Natl. Acad. Sci. U.S.A.* **55**, 170–177.
Jilka, R. L., Martonosi, A. N., and Tillack, T. W. (1975). *J. Biol. Chem.* **250**, 7511–7524.
Jørgensen, K. E., Lind, K. E., Røigaard-Petersen, H., and Møller, J. V. (1978). *Biochem. J.* **169**, 489–498.
Johnson, P., and Inesi, G. (1969). *J. Pharmacol. Exp. Ther.* **169**, 308–314.

Kanazawa, T. (1975). *J. Biol. Chem.* **250**, 113–119.

Kanazawa, T., and Boyer, P. D. (1973). *J. Biol. Chem.* **248**, 3163–3172.

Kanazawa, T., and Katabami, F. (1977). *Proc. Congr. Bioenerg., 3rd, Tokyo* **3**, 35–37.

Kanazawa, T., Saito, M., and Tonomura, Y. (1970). *J. Biochem. (Tokyo)* **67**, 693–711.

Kanazawa, T., Yamada, S., Yamamoto, T., and Tonomura, Y. (1971). *J. Biochem. (Tokyo)* **70**, 95–123.

Katz, A. M., and Repke, D. I. (1967). *Circ. Res.* **21**, 767–775.

Kirino, Y., Anzai, K., Shimizu, H., Ohta, S., Nakanishi, M., and Tsuboi, M. (1977). *J. Biochem. (Tokyo)* **82**, 1181–1184.

Klodos, I., and Skou, J. C. (1977). *Biochim. Biophys. Acta* **481**, 667–679.

Knowles, A. F., and Racker, E. (1975a). *J. Biol. Chem.* **250**, 1949–1951.

Knowles, A. F., and Racker, E. (1975b). *J. Biol. Chem.* **250**, 3538–3544.

Knowles, A. F., Eytan, E., and Racker, E. (1976). *J. Biol. Chem.* **251**, 5161–5165.

Kurzmack, M., and Inesi, G. (1977). *FEBS Lett.* **74**, 35–37.

Kurzmack, M., Verjovski-Almedia, S., and Inesi, G. (1977). *Biochem. Biophys. Res. Commun.* **78**, 772–776.

Kyte, J. (1974). *J. Biol. Chem.* **249**, 3652–3660.

Kyte, J. (1975). *J. Biol. Chem.* **250**, 7443–7449.

Lant, A. F., and Whittam, R. (1968). *J. Physiol. (London)* **199**, 457–484.

Le Maire, M., Møller, J. V., and Tanford, C. (1976). *Biochemistry* **15**, 2336–2342.

Lowe, A. G., and Smart, J. W. (1977). *Biochim. Biophys. Acta* **481**, 695–705.

MacLennan, D. H. (1970). *J. Biol. Chem.* **245**, 4508–4518.

MacLennan, D. H. (1974). *J. Biol. Chem.* **249**, 980–984.

MacLennan, D. H., and Holland, P. C. (1975). *Annu. Rev. Biophys. Bioeng.* **4**, 377–404.

MacLennan, D. H., and Holland, P. C. (1976). *In* "The Enzymes of Biological Membranes" (A. Martonosi, ed.), Vol. 3, pp. 221–259. Plenum, New York.

MacLennan, D. H., and Wong, P. T. S. (1971). *Proc. Natl. Acad. Sci. U.S.A.* **68**, 1231–1235.

MacLennan, D. H., Seeman, P., Iles, G. H., and Yip, C. C. (1971). *J. Biol. Chem.* **246**, 2702–2710.

MacLennan, D. H., Yip, C. C., Iles, G. H., and Seeman, P. (1972). *Cold Spring Harbor Symp. Quant. Biol.* **37**, 469–477.

Makinose, M. (1966). *Biochem. Z.* **345**, 80–86.

Makinose, M. (1969). *Eur. J. Biochem.* **10**, 74–82.

Makinose, M. (1971). *FEBS Lett.* **12**, 269–270.

Makinose, M. (1972). *FEBS Lett.* **25**, 113–115.

Makinose, M. (1973). *FEBS Lett.* **37**, 140–143.

Makinose, M., and Hasselbach, W. (1965). *Biochem. Z.* **343**, 360–382.

Makinose, M., and Hasselbach, W. (1971). *FEBS Lett.* **12**, 271–272.

Makinose, M., and The, R. (1965). *Biochem. Z.* **343**, 383–393.

Martonosi, A. (1963). *Biochem. Biophys. Res. Commun.* **13**, 273–278.

Martonosi, A. (1967). *Biochem. Biophys. Res. Commun.* **29**, 753–757.

Martonosi, A. (1968). *Biochim. Biophys. Acta* **150**, 694–704.

Martonosi, A. (1969). *J. Biol. Chem.* **244**, 613–620.

Martonosi, A. (1972). *In* "Metabolic Pathways" (L. E. Hokin, ed.), Vol. 6, pp. 317–349. Academic Press, New York.

Martonosi, A., Donley, J., and Halpin, R. A. (1968). *J. Biol. Chem.* **243**, 61–70.

Martonosi, A., Donley, J. R., Pucell, A. G., and Halpin, R. A. (1971). *Arch. Biochem. Biophys.* **144**, 529–540.

Martonosi, A., Lagwinska, E., and Oliver, M. (1974). *Ann. N.Y. Acad. Sci.* **227**, 549–567.

Meissner, G. (1973). *Biochim. Biophys. Acta* **298**, 906–926.

Meissner, G. (1975). *Biochim. Biophys. Acta* **389**, 51–68.
Meissner, G., and Fleischer, S. (1971). *Biochim. Biophys. Acta* **241**, 356–378.
Meissner, G., and Fleischer, S. (1973). *Biochem. Biophys. Res. Commun.* **52**, 913–920.
Meissner, G., Conner, G. E., and Fleischer, S. (1973). *Biochim. Biophys. Acta* **298**, 246–269.
Mermier, P., and Hasselbach, W. (1976). *Eur. J. Biochem.* **69**, 79–86.
Migala, A., Agostini, B., and Hasselbach, W. (1973). *Z. Naturforsch., Teil C* **28**, 178–182.
Miyamoto, H., and Kasai, M. (1976). *Proc. Jpn. Biophys. Soc., 15th, Hiroshima* p. 180.
Murphy, A. J. (1976a). *Biochem. Biophys. Res. Commun.* **70**, 160–166.
Murphy, A. J. (1976b). *Biochemistry* **15**, 4492–4496.
Murphy, A. J. (1978). *J. Biol. Chem.* **253**, 385–389.
Nagano, K., Kanazawa, T., Mizuno, N., Tashima, Y., Nakao, T., and Nakao, M. (1965). *Biochem. Biophys. Res. Commun.* **19**, 759–764.
Nakamura, H., Jilka, R. L., Boland, R., and Martonosi, A. N. (1976). *J. Biol. Chem.* **251**, 5414–5423.
Nakamura, M., and Ohnishi, S. (1975). *J. Biochem.* (*Tokyo*) **78**, 1039–1045.
Nakamura, Y., and Tonomura, Y. (1978). *J. Biochem.* (*Tokyo*) **83**, 571–583.
Ostwald, T. J., and MacLennan, D. H. (1974). *J. Biol. Chem.* **249**, 974–979.
Owens, K., Ruth, R. C., and Weglicki, W. B. (1972). *Biochim. Biophys. Acta* **288**, 479–481.
Panet, R., and Selinger, Z. (1972). *Biochim. Biophys. Acta* **255**, 34–42.
Panet, R., Pick, U., and Selinger, Z. (1971). *J. Biol. Chem.* **246**, 7349–7356.
Pang, D. C., and Briggs, F. N. (1977). *J. Biol. Chem.* **252**, 3262–3266.
Pardee, A. B. (1968a). *Science* **162**, 632–637.
Pardee, A. B. (1968b). *J. Gen. Physiol.* **52**, 279s–288s.
Peachey, L. D. (1965). *J. Cell Biol.* **25**, 209–231.
Porter, K. R. (1961). *J. Biophys. Biochem. Cytol.* **10**, Suppl., 219–226.
Porter, K. R., and Palade, G. E. (1957). *J. Biophys. Biochem. Cytol.* **3**, 269–300.
Pucell, A., and Martonosi, A. (1971). *J. Biol. Chem.* **246**, 3389–3397.
Pull, I. (1970). *Biochem. J.* **119**, 377–385.
Racker, E. (1972). *J. Biol. Chem.* **247**, 8198–8200.
Racker, E. (1973). *Biochem. Biophys. Res. Commun.* **55**, 224–230.
Racker, E. (1976). "A New Look at Mechanisms in Bioenergetics." Academic Press, New York.
Racker, E., and Eytan, E. (1975). *J. Biol. Chem.* **250**, 7533–7534.
Racker, E., Chien, T.-F., and Kandrach, A. (1975). *FEBS Lett.* **57**, 14–18.
Rauch, B., von Chak, D., and Hasselbach, W. (1977). *Z. Naturforsch., Teil C* **32**, 828–834.
Rega, A. F., and Garrahan, P. J. (1976). *In* "The Enzymes of Biological Membranes" (A. Martonosi, ed.), Vol. 3, pp. 303–314. Plenum, New York.
Ribeiro, J. M. C., and Vianna, A. L. (1978). *J. Biol. Chem.* **253**, 3153–3157.
Rizzolo, L. J., Le Maire, M., Reynolds, J. A., and Tanford, C. (1976). *Biochemistry* **15**, 3433–3437.
Sandow, A. (1965). *Pharmacol. Rev.* **17**, 265–320.
Schwarzenbach, G. (1960). "Die Komplexometrische Titration." Enke, Stuttgart.
Shamoo, A. E., and Goldstein, D. A. (1977). *Biochim. Biophys. Acta* **472**, 13–53.
Shamoo, A. E., and MacLennan, D. H. (1974). *Proc. Natl. Acad. Sci. U.S.A.* **71**, 3522–3526.
Shamoo, A. E., Ryan, T. E., Stewart, P. S., and MacLennan, D. H. (1976). *J. Biol. Chem.* **251** 4147–4154.
Shaw, T. I. (1954). Ph.D. Thesis, Cambridge Univ., Cambridge, England.
Shigekawa, M., and Dougherty, J. P. (1977). *Biochem. Biophys. Res. Commun.* **76**, 784–789.
Shigekawa, M., and Dougherty, J. P. (1978a). *J. Biol. Chem.* **253**, 1451–1457.
Shigekawa, M., and Dougherty, J. P. (1978b). *J. Biol. Chem.* **253**, 1458–1464.
Shigekawa, M., and Pearl, L. J. (1976). *J. Biol. Chem.* **251**, 6947–6952.

Shigekawa, M., Dougherty, J. P., and Katz, A. M. (1978). *J. Biol. Chem.* **253**, 1442–1450.
Singer, S. J. (1974). *Annu. Rev. Biochem.* **43**, 805–833.
Singer, S. J., and Nicolson, G. L. (1972). *Science* **175**, 720–731.
Souza, D. O. G., and de Meis, L. (1976). *J. Biol. Chem.* **251**, 6355–6359.
Stewart, P. S., and MacLennan, D. H. (1974). *J. Biol. Chem.* **249**, 985–993.
Stewart, P. S., and MacLennan, D. H. (1975). *Ann. N.Y. Acad. Sci.* **264**, 326–334.
Stewart, P. S., MacLennan, D. H., and Shamoo, A. E. (1976). *J. Biol. Chem.* **251**, 712–719.
Sumida, M., and Tonomura, Y. (1974). *J. Biochem.* (*Tokyo*) **75**, 283–297.
Sumida, M., Kanazawa, T., and Tonomura, Y. (1976). *J. Biochem.* (*Tokyo*) **79**, 259–264.
Tada, M., Yamamoto, T., and Tonomura, Y. (1978). *Physiol. Rev.* **58**, 1–79.
Takisawa, H., and Tonomura, Y. (1978a). *J. Biochem.* (*Tokyo*) **83**, 1275–1284.
Takisawa, H., and Tonomura, Y. (1978b). Unpublished observations.
Taniguchi, K., and Post, R. L. (1975). *J. Biol. Chem.* **250**, 3010–3018.
Taniguchi, K., and Tonomura, Y. (1971). *J. Biochem.* (*Tokyo*) **69**, 543–557.
The, R., and Hasselbach, W. (1972). *Eur. J. Biochem.* **28**, 357–363.
Thorley-Lawson, D. A., and Green, N. M. (1973). *Eur. J. Biochem.* **40**, 403–413.
Thorley-Lawson, D. A., and Green, N. M. (1975). *Eur. J. Biochem.* **59**, 193–200.
Tonomura, Y. (1972). "Muscle Proteins, Muscle Contraction and Cation Transport." Univ. of Tokyo Press, Tokyo and Univ. Park Press, Baltimore, Maryland.
Tonomura, Y., and Morales, M. F. (1974). *Proc. Natl. Acad. Sci. U.S.A.* **71**, 3687–3691.
Vale, M. G. P., and Carvalho, A. P. (1975). *Biochim. Biophys. Acta* **413**, 202–212.
Vale, M. G. P., Osorio e Castro, V. R., and Carvalho, A. P. (1976). *Biochem. J.* **156**, 239–244.
Van der Kloot, W. G. (1965). *Comp. Biochem. Physiol.* **15**, 547–565.
Vanderkooi, J. M., Ierokomas, A., Nakamura, H., and Martonosi, A. (1977). *Biochemistry* **16**, 1262–1267.
Vianna, A. L. (1975). *Biochim. Biophys. Acta* **410**, 389–406.
Warren, G. B., Toon, P. A., Birdsall, N. J. M., Lee, A. G., and Metcalfe, J. C. (1974a). *Proc. Natl. Acad. Sci. U.S.A.* **71**, 622–626.
Warren, G. B., Toon, P. A., Birdsall, N. J. M., Lee, A. G., and Metcalfe, J. C. (1974b). *Biochemistry* **13**, 5501–5507.
Warren, G. B., Houslay, M. D., Metcalfe, J. C., and Birdsall, N. J. M. (1975). *Nature* (*London*) **255**, 684–687.
Weber, A. (1971). *J. Gen. Physiol.* **57**, 50–63.
Weber, A., Herz, R., and Reiss, I. (1966). *Biochem. Z.* **345**, 329–369.
Yamada, S., and Tonomura, Y. (1972a). *J. Biochem.* (*Tokyo*) **71**, 1101–1104.
Yamada, S., and Tonomura, Y. (1972b). *J. Biochem.* (*Tokyo*) **72**, 417–425.
Yamada, S., Yamamoto, T., and Tonomura, Y. (1970). *J. Biochem.* (*Tokyo*) **67**, 789–794.
Yamada, S., Yamamoto, T., Kanazawa, T., and Tonomura, Y. (1971). *J. Biochem.* (*Tokyo*) **70**, 279–291.
Yamada, S., Sumida, M., and Tonomura, Y. (1972). *J. Biochem.* (*Tokyo*) **72**, 1537–1548.
Yamaguchi, M., and Tonomura, Y. (1978). *J. Biochem.* (*Tokyo*) **83**, 977–987.
Yamamoto, T., and Tonomura, Y. (1967). *J. Biochem.* (*Tokyo*) **62**, 558–575.
Yamamoto, T., and Tonomura, Y. (1968). *J. Biochem.* (*Tokyo*) **64**, 137–145.
Yamamoto, T., and Tonomura, Y. (1976). *J. Biochem.* (*Tokyo*) **79**, 693–707.
Yamamoto, T., and Tonomura, Y. (1977). *J. Biochem.* (*Tokyo*) **82**, 653–660.
Yamamoto, T., Yoda, A., and Tonomura, Y. (1971). *J. Biochem.* (*Tokyo*) **69**, 807–809.
Yariv, J., Steinberg, I. Z., Kalb, A. J., Goldman, R., and Katchalski, E. (1972). *J. Theor. Biol.* **35**, 459–465.
Yates, D. W., and Duance, V. C. (1976). *Biochem. J.* **159**, 719–728.

CURRENT TOPICS IN BIOENERGETICS, VOLUME 9

Flavoproteins, Iron Proteins, and Hemoproteins as Electron-Transfer Components of the Sulfate-Reducing Bacteria

JEAN LEGALL
Laboratoire de Chimie Bacterienne, CNRS
Marseille, France

DANIEL V. DERVARTANIAN and HARRY D. PECK, JR.
Department of Biochemistry
University of Georgia
Athens, Georgia

ISBN 0-12-152509-0

I. Introduction

The intricate electron-transfer system characteristic of sulfate-reducing bacteria belonging to the genus *Desulfovibrio* has prompted considerable research in the field of structure/function relationships of the different proteins that constitute the oxidation–reduction chains. These proteins belong to certain classes of redox carriers and have been discussed in recent specialized reviews: flavoproteins (Mayhew and Ludwig, 1975), cytochromes (Dickerson and Timkovich, 1975), nonheme iron proteins (Hall, 1977). The goal of the present review is to discuss the most recent publications that deal with the discovery of new redox proteins and propose explanations for their physical and biological functions. Since the last review (LeGall and Postgate, 1973) dealing with sulfate-reducing bacteria was published, little progress has been made in understanding the biological functions of these electron-transfer proteins. Although numerous redox carriers are known, the mechanism of electron transfer is still far from being completely worked out. This is in part due to the fact that the pathway for sulfate reduction is still very controversial, as discussed in Section IX, which deals with the terminal siroheme reductases.

Sulfate-reducing bacteria essentially obtain their energy for growth from the oxidation of a limited number of organic acids, ethanol, or H_2. The electrons derived from these substrates are used to reduce various sulfur compounds to the level of H_2S. Sulfate-reducing bacteria are classified into two genera (Postgate and Campbell, 1966; Campbell and Postgate, 1965), *Desulfotomaculum*[1] and *Desulfovibrio.* The former genus contains the spore-forming sulfate reducing bacteria as well as thermophilic species; a recent addition to this genus *Dm. acetoxidas* has the unique ability of oxidizing acetate to CO_2 with sulfate as electron acceptor (Widdel and Pfennig, 1977). Very intriguing differences between the two types of sulfate reducing bacteria are the absence of *c*-type cytochromes in *Desulfotomaculum* and the presence of a soluble hydrogenase of high specific activity in *Desulfovibrio.*

A very important discovery has been made by Biebl and Pfennig (1977). They found that not only are some sulfate reducers able to utilize colloidal sulfur as "respiratory" substrates, but bacteria belonging to the genus *Desulfuromonas* (Pfennig and Biebl, 1976) are unable to utilize sulfate and are restricted to the reduction of elemental sulfur. Then, to draw a parallel between the sulfate "respirers" and the aerobic world, there exist obligate

[1] Abbreviations used: *D.*, *Desulfovibrio*; *Drm.*, *Desulfuromonas*; *C.*, *Clostridium*; *Dm.*, *Desulfotomaculum*; HIPIP, high-potential iron sulfur protein; M_r, molecular weight; APS, adenylyl sulfate; FMN, flavinadenine mononucleotide; NMR, nuclear magnetic resonance; NADH + H^+, reduced nicolinamide adenine dinucleotide.

as well as facultative sulfate- or sulfur-reducing bacteria. Since the characterization of some redox proteins from *Drm. acetoxidans* has recently been reported, a discussion of their properties will be included in this review.

The fact that sulfate-reducing bacteria are probably present-day representatives of very ancient organisms is reflected by the extreme divergence that characterizes the different species. This renders extremely attractive the study of the redox proteins, such as flavodoxins, rubredoxins, ferredoxins, cytochromes *c*, that constitute families of homologous proteins. Important variations exist in the primary structure of these proteins; some are such that a complete change in the function of the protein may be anticipated (Bruschi *et al.*, 1977).

II. Flavodoxin

A. Biosynthesis

The biosynthesis of flavodoxin poses a very interesting problem in regulation. It was first noted by Knight and Hardy (1966) that clostridial flavodoxin is synthesized only in iron-depleted media. In sulfate-reducing bacteria, the picture is more complicated, since the relative proportion of flavodoxin versus ferredoxin varies for each species and also according to the growth conditions. It has to be pointed out that the normal metabolism of these bacteria leads to the formation of almost insoluble ferrous sulfide, so that the organisms are most probably always growing under iron-limited conditions. Thus, iron does not seem to be the sole factor in the regulation of the biosynthesis of this protein.

Unless the properties of the protein have tremendously changed, rendering its normal purification difficult, flavodoxin appears to be absent from *Drm. acetoxidans* and from *D. desulfuricans*, strain Norway 4. On the other hand, *D. vulgaris*, strain Hildenborough, synthesizes large amounts of flavodoxin and only very little ferredoxin, even in media containing high concentrations of iron, a peculiarity shared with *D. vulgaris*, strain Miyazaki (Irie *et al.*, 1973). The physiological significance of such differences is not understood, since, at least in desulfovibriones, it seems that flavodoxin replaces ferredoxin in all the known reactions in which the latter has been implicated. A redox reaction that would be specific for flavodoxin is yet to be found. However, Barton and Peck (1970) reported that, although electron transfer between hydrogenase and sulfite reductase was restored by its addition, flavodoxin was not effective in the coupling of oxidative phosphorylations in particulate preparations of *D. gigas* and thus could not replace ferredoxin.

The fact that both organisms that have been found to be devoid of flavodoxin are sulfur reducers is not sufficient to imply that ferredoxin can be utilized only in the reduction of colloidal sulfur.

B. Structure

Both primary (Dubourdieu *et al.*, 1973) and tertiary structures (Watenpaugh *et al.*, 1972, 1973a) of *D. vulgaris*, strain Hildenborough, flavodoxin are known. They have been discussed in detail, in comparison with *Clostridium* MP flavodoxin by Mayhew and Ludwig (1975). Since it cannot be hoped that structural data derived from X-ray crystallography can be utilized routinely to study all redox proteins, nuclear magnetic resonance (NMR) has proved to be an excellent method for such comparative studies, in addition to bringing new structural and dynamic data on the protein in solution. Favaudon *et al.* (1979) have studied flavodoxins from three different sulfate reducers, namely: *D. vulgaris*, *D. gigas*, and *D. salexigens*, by ^{1}H- and ^{31}P-NMR in the three redox states of the holoproteins and the apoproteins. According to this study, the configuration of the FMN ribityl 5′-phosphate linkage with the apoprotein is gauche–gauche and does not change markedly upon reduction. The apoproteins have a stable tertiary structure; however, larger differences are observed upon removal of the FMN than in the three redox states of the protein.

The careful comparison of both *Clostridium* MP and *D. vulgaris* flavodoxins X-ray structures by Mayhew and Ludwig (1975) lead to the rather surprising conclusion that many of the variations between the two proteins occur at the vicinity of the FMN-isoalloxazine ring. Although more conservatism has been expected in flavodoxins belonging to the same genus, the NMR results as well as the X-ray crystallography data indicate several important differences between the three *Desulfovibrio* flavodoxins. The resolution of both FMN methyl resonances in the *D. vulgaris* protein indicates that the dimethylbenzene end, in both oxidized and reduced states, is exposed to the solvent, but only one of these methyl resonances is observed in reduced *D. gigas* flavodoxin, suggesting that FMN is more deeply buried within the protein. Even more different is the flavodoxin from *D. salexigens* as the methyl resonances do not even appear in its spectra; the disappearance of these resonances is due to hindrance of the free rotation of the methyl groups.

An aromatic residue, probably a tryptophan in *D. gigas* and a tyrosine in *D. salexigens* flavodoxin, is present in the near vicinity of the flavin group as observed crystallographically for Trp^{60} and Tyr^{98} in *D. vulgaris* flavodoxin. Large differences in the low-field resonances in light water are consistent with different amino acids in the vicinity of the FMN. Finally, *D. salexigens* is much less stable than the other two with respect to proton exchange and to denaturation.

III. Ferredoxin

A. Structure/Function

D. gigas ferredoxin is active in the reduction of sulfite to sulfide from molecular hydrogen (LeGall and Dragoni, 1966) as well as in the phosphoroclastic reaction leading to the formation of hydrogen and acetyl phosphate from pyruvate (Hatchikian and LeGall, 1970). As shown on Fig. 1, this dual function poses a problem since the two reactions imply that the carrier that is active in the system possesses two redox potentials. The mechanism of action of flavodoxin which has the same function as ferrodoxin (LeGall and Hatchikian, 1967; Hatchikian and LeGall, 1970) can readily be interpreted, since the particular structure of this protein allows the stabilization of the FMN semiquinone. As a consequence, the protein can have two redox potentials (Dubourdieu *et al.*, 1975): E'_{o1} (semiquinone/hydroquinone) $= -440$ mV and E'_{o2} (semiquinone/oxidized form) $= -150$ mV.

The three-states theory, as formulated by Carter *et al.* (1972), forbids such an explanation for this ferredoxin since the molecule contains only one 4Fe–4S* cluster (Laishley *et al.*, 1969). This theory postulates that a given nonheme iron protein containing 4Fe–4S* clusters can exist in only one of two redox states, which can be formally written as follows:

$$3Fe^{3+} \cdot 1Fe^{2+} \underset{}{\overset{1e}{\rightleftarrows}} 2Fe^{3+} \cdot 2Fe^{2+} \text{ "HIPIP" state}$$

$$2Fe^{3+} \cdot 2Fe^{2+} \underset{}{\overset{1e}{\rightleftarrows}} 1Fe^{3+} \cdot 3Fe^{2+} \text{ "Ferredoxin" state}$$

The high-potential iron–sulfur protein (HIPIP) state is paramagnetic in its oxidized form and has a high redox potential, whereas the "ferredoxin"

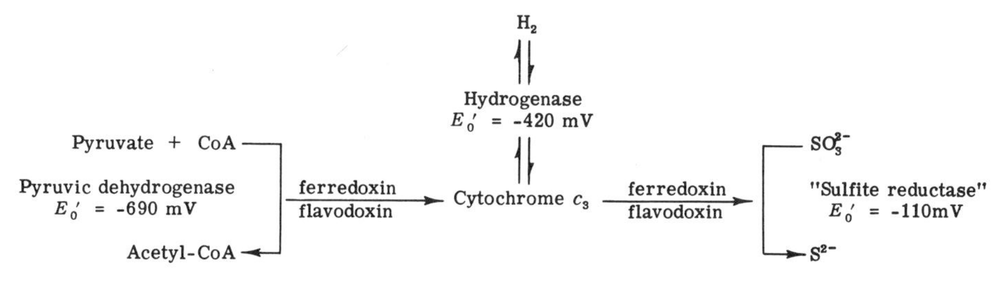

FIG. 1. Electron-transfer pathway from pyruvate to sulfite in *Desulfovibrio* sp.

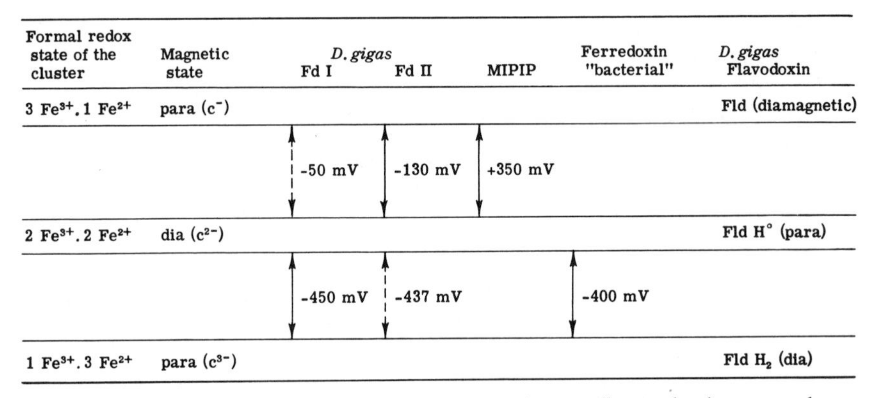

FIG. 2. Redox states of *Desulfovibrio gigas* ferredoxin according to the three-states hypothesis. An analogus type of situation may exist on a single molecule in the case of ferredoxin I from *Azotobacter vinelandii*. This ferredoxin has two 4Fe–4S* clusters, but one cluster is low potential (E_M = −424 mV) and the other high potential (E_M = 320 mV) (Yoch and Caritheus, 1978).

state is diamagnetic in its oxidized form and has a low potential. HIPIP and ferredoxins have different apoproteins although the structure of their redox centers (the 4Fe–4S* cluster) is the same. The difference in the redox potential is thought to be due to the different hydrogen-bonding systems between the cluster and the apoproteins. A close examination of the properties of *D. gigas* ferredoxin by Bruschi *et al.* (1976) allowed the conclusion that the same monomer of molecular weight ratio (M_r) 6000 containing one 4Fe–4S* cluster could exist in a trimeric (Fd I) or a tetrameric form (Fd II) that could be separated using classical chromatographic methods. These two forms have different magnetic properties that are related to different redox potentials (Moura *et al.*, 1977b; Cammack *et al.*, 1977).

Fd I behaves like a typical ferredoxin; it is diamagnetic in its oxidized form and has an E'_0 of −440 mV. Fd II, like "HIPIP" proteins, is paramagnetic in its oxidized form, and has a much higher potential, −130 mV. Another trimeric form (Fd I′), present in small amounts, seems to represent an intermediary state that is able to function indifferently in the two redox states. A summary of these results is given in Fig. 2.

B. PHYSIOLOGICAL ACTIVITY

The physiological activities of Fd I and Fd II confirm the physicochemical data (Bruschi *et al.*, 1976; Moura *et al.*, 1978b): Fd II is more active than Fd I in the reduction of sulfite, whereas only Fd I is active in the phosphoroclastic reaction, a low-potential reaction. In fact, Fd II inhibits the endogenous production of hydrogen, but after a lag phase of several minutes the reaction starts. This is interpreted as an interconversion of Fd II (more oxidized) to Fd I.

Bianco and Haladjian (1977) have studied both Fd I and Fd II by differential pulse polarography. The results are somewhat different from the one given by titration using the platinum electrode in conjunction with electron paramagnetic resonance (EPR) (Cammack *et al.*, 1977), since E'_0 values were found with values of -300 mV and -320 mV for Fd I and Fd II, respectively. Such differences have already been noted by Weitzman *et al.* (1971) in the case of *C. pastorianum* ferredoxin; they can be explained by absorption phenomena occurring during the electrolytic reduction.

IV. Cytochromes

A. Cytochromes *c*

Several *c*-type cytochromes are present in *Desulfovibrio* and *Desulfuromonas* species (LeGall *et al.*, 1975; Probst *et al.*, 1977). Some are monoheme proteins, such as *D. vulgaris* c_{553}, but the typical cytochrome of the genus *Desulfovibrio* is the multiple heme cytochrome c_3. Cytochrome c_7, which is found in *Desulfuromonas*, has such a high degree of structural homology with cytochrome c_3 that it has to be considered as belonging to this family of cytochromes

In the following discussion, the term cytochrome *c* will be restricted to the monoheme cytochromes *c*. As already noted, the sulfate-reducing bacteria that belong to the genus *Desulfotomaculum* are devoid of cytochrome *c*. This is not due to the incapability of these organisms to synthesize protoheme since a *b*-type cytochrome is present (Campbell and Postgate, 1965).

1. *Cytochrome* c_{553}

The primary structure of cytochrome c_{553} from *D. vulgaris* is known (Bruschi and LeGall, 1972). Although it has a rather low redox potential, it cannot be related to the cytochrome c_3 family since it has a single heme per molecule. The presence of an absorption band at 695 nm favors methionine as being the sixth coordination ligand of the heme iron (LeGall *et al.*, 1971b); c_{553} can then be related to the cytochrome *c* family. Almassy and Dickerson (1978) have built a phylogenetic tree of this family, based on structural information. It shows cytochrome c_{553} diverging very early from all other cytochromes *c*, at a time when no oxygen was present in the atmosphere. It is interesting to note that cytochrome c_{553} reacts quite rapidly with oxygen. This property could be related to its primitivity, since the modern mitochondrial cytochrome *c* has "learned" not to react with a gas that has become very common in our atmosphere.

An NMR study of cytochromes *c* by Cookson *et al.*, (1978) shows that *D. vulgaris* c_{553}, together with another low-potential cytochrome c_{554} from a halotolerant *Micrococcus*, have heme environments that are not very similar to those of the other cytochromes *c*.

2. *Multiheme c-Type Cytochromes*

a. Cytochromes c_3. Cytochromes c_3 are a unique class of hemoproteins that contain either four hemes (M_r = 13,000) or eight hemes (M_r = 26,000) per molecule, are autoxidable, and bear a negative oxidation–reduction potential.

The amino acid sequences of c_3 (M_r = 13,000) from three strains are known [*D. desulfuricans*, strain El Agheila Z; *D. vulgaris* and *D. gigas* (Dobson *et al.*, 1974)]. The amino acid sequence analysis of the four-heme cytochrome c_3 from *D. desulfuricans*, strain Norway 4, is almost completed. This latter cytochrome c_3 has a molecular weight of 16,000, and its N-terminal end is approximately eight amino acids longer than any other c_3. In each case the amino acid sequence of cytochrome c_3 shows extensive differences, which is quite unlike the situation for the *c*-type cytochromes from mitochondria.

The physiological electron donor for cytochrome c_3 (M_r = 13,000) is hydrogenase, since reduced hydrogenase readily reduces cytochrome c_3. Little information is available as to the amino acid sequence, natural electron donor specificity, or physiochemical properties of cytochrome c_3 (M_r = 26,000; moreover, it is markedly different in amino acid composition from cytochrome c_3 (M_r = 13,000). Consequently, the studies to be described will deal exclusively with cytochrome c_3 (M_r = 13,000 or 16,000).

Frey *et al.* (1976) have conducted X-ray crystallographic studies at 3.2 Å resolution on cytochrome c_3 from *D. desulfuricans*, strain Norway 4, of M_r = 16,000. These studies have shown definitely that the molecule accommodates four hemes that appear to be closely packed. The iron–iron distances range from 11 Å to 17 Å and three heme planes are roughly parallel to each other while the fourth heme is tilted. A 2.5 Å resolution electron-density map currently in preparation should reveal important structural features such as the hydrogen bond network and the heme environment.

Bianco and Haladjian (1979) have studied stopped-flow kinetics for the reduction of cytochrome c_3 from *D. vulgaris* by sodium dithionite or carboxy radicals. With each reductant, biphasic kinetics were observed and no transient intermediates were detected in the 440–610 nm region.

The rate of reduction for the four protein-bond hemes was not identical, as two heme groups were reduced more rapidly than the other two hemes. The data were consistent with the absence of heme–heme electron exchange within a single protein molecule.

Extensive EPR and NMR studies have been carried out by DerVartanian and LeGall (1974), Dobson *et al.* (1974), and McDonald *et al.* (1974) on cytochromes c_3 from *D. vulgaris* and *D. gigas*. Unless otherwise specified, cytochrome c_3 from *D. vulgaris* refers to the Hildenborough strain. In the case of cytochrome c_3 from *D. gigas*, NMR studies by Moura *et al* (1977c) and EPR redox potential titrations (Xavier, Moura, LeGall and DerVartanian, unpublished observations) have recently been carried out in the presence and in the absence of ferredoxin II, the tetrameric form, which in the reduced state can reduce cytochrome c_3. The latter results suggest a specific interaction between ferredoxin II and cytochrome c_3.

EPR and NMR studies on cytochrome c_3 from *D. vulgaris*, extensively discussed by DerVartanian and LeGall (1974), have shown that the four hemes of cytochrome c_3 are in nonequivalent sites and more recently have been demonstrated by DerVartanian *et al.* (1978) to exhibit four different oxidation–reduction midpoint potentials of -284 mV, -310 mV, -319 mV, and -324 mV with n values near 1. In this regard, Niki *et al.* (1977) employing polarographic and cyclic voltametry reported that cytochrome c_3 from *D. vulgaris*, strain Miyazaki, has a midpoint potential of 270 mV $\pm$ 1 mV, with an n value less than unity. They also proposed that the four-heme cytochrome c_3 probably accepted four electrons stepwise as a single unit and that the four hemes were composed of an interlocking heme cluster.

EPR studies by DerVartanian and LeGall (1974) have shown that cytochrome c_3 from *D. vulgaris* is quite unreactive toward exogenous ligands but, on reduction to the ferro form, becomes highly reactive and exhibits characteristic EPR signals for heme-ligands such as imidazole, cyanide, and nitric oxide.

EPR and NMR studies have shown that histidine is the likely ligand at the 5th and 6th coordinative heme-ligand positions. Studies by DerVartanian and LeGall (1978) show that substitution of the more hydrophobic imidazole for histidine at this 6th ligand positions of 2 hemes of ferrocytochrome c_3 results in a spectrally distinct low-spin c_3–imidazole complex characterized by a blue shift (from a more polar to a less polar state) in the α-band at 77°K from 550.5 to 547.4 nm. When ferrocytochrome c_3 in the presence of bound imidazole is reoxidized to the ferric state, there appears a slow decrease in optical absorption and EPR signal intensities, followed by the appearance of highly distorted high-spin ferric heme resonances. The substitution of imidazole for histidine in cytochrome c_3 is a suicide reaction, since cytochrome c_3 is then highly labile and disintegrates irreversibly.

NMR studies by Dobson *et al.* (1974) led them to propose an outline structure of *D. vulgaris* cytochrome c_3. Histidines were located at heme-ligand positions 5 and 6, and each heme was also bound to two thioether links. The distance between hemes was about 10 Å, but, in spite of this close proximity of hemes, there is a relatively slow electron exchange between the hemes. In general, the structure proposed by Dobson *et al.* (1974) is in basic agreement with the 3.2 Å X-ray crystallographic data of Frey *et al.* (1976) for the cytochrome c_3 from *D. desulfuricans*, strain Norway 4.

In contrast to *D. vulgaris* cytochrome c_3, EPR spectroscopy on ferrocytochrome c_3 from *D. gigas* revealed only two low-spin ferric heme resonances with g_2 values at 2.959 and 2.853, supportive for histidine at the 6th ligand position of heme (Xavier *et al.*, 1979). The g_y value was 2.341 for each heme. The g_x value was at 1.653 and represented (as for g_y) a superimposition of g_x values from all four hemes. It was found that no significant changes in either line shape or signal width occurred at either heme signal of cytochrome c_3 from *D. gigas* during the oxidation–reduction EPR potentiometric titrations either in the absence or in the presence of ferredoxin II. Computer fits of the Nernst equation for cytochrome c_3 in the absence or in the presence of ferredoxin II clearly indicated that, even though only two g_z heme signals were discernible by EPR spectroscopy, the potentiometric EPR titrations showed two distinct slopes for each heme signal. The midpoints determined in the absence of ferredoxin II were $E_{MIA} = -235$ mV and $E_{MIB} = -315$ mV for the first heme signal (at $g_z = 2.959$) and $E_{MIIA} = -235$ mV and $E_{MIIB} = -306$ mV for the second heme signal at $g_z = 2.853$. In the presence of ferredoxin II, the first midpoint potential of both heme signals was unaltered but the second midpoint potential became more negative, changing from -315 mV to -329 mV for heme signal IB and from -306 mV to -315 mV for heme signal IIB. There is thus a small but measurable effect of ferredoxin II on the midpoint potential of two of the EPR-detectable hemes of cytochrome c_3. The second experimental EPR support for an interaction of cytochrome c_3 with ferredoxin II derives from the observation that the g_x signal intensity of ferricytochrome c_3 is decreased 60% in the presence of ferredoxin II. NMR reoxidation patterns of *D. gigas* ferrocytochrome c_3 in the absence and in the presence of ferredoxin II showed definite shifts and broadening in NMR heme resonances and suggested that electron exchange rates between the 4 hemes were altered in the presence of ferredoxin. This interaction is likely to be electrostatic in character, since the interaction between ferredoxin II and cytochrome c_3 is dependent on a competition with phosphate. The EPR and NMR results support an interaction between ferredoxin and cytochrome c_3.

Cytochrome c_3 (*D. vulgaris*) has been found to differ in midpoint potentials among the 4 hemes by 40 mV (DerVartanian *et al.*, 1978) and by 80 mV in the case of *D. gigas* c_3 (Xavier *et al.*, 1979). The midpotential of the most negative heme is −329 mV, which makes cytochrome c_3 considerably more negative in redox potential than previously reported. The biological significance for the E_M values of the various hemes is not well understood. It is interesting to note, however, that when cytochrome c_3 for *D. vulgaris* was reduced with sodium dithionite in the presence of hydroxylamine (resulting in ammonia formation on subsequent reoxidation of cytochrome c_3) the four hemes of cytochrome c_3 could not be completely reduced. Hemes I, II, III, and IV (referring to the four distinct g_z values of ferricytochrome c_3 indicating four nonequivalently located hemes) were in fact 75%, 64%, 43%, and 59% reduced, respectively. The reduction state of these hemes is related to the determined E_M values found in Table I of DerVartanian and LeGall (1974). This observation in turn suggests that the reduction level of the hemes is critical to the nature of the biological reaction(s) in which cytochrome c_3 is involved.

b. Cytochrome c_7 (c-551.5) from Desulfuromonas acetoxidans. Cytochrome c_7 is a low-potential, three heme-containing protein of molecular weight 9800 (Probst *et al.*, 1977). Cytochrome *c*-551.5 from *Drm. acetoxidans* was established to be identical to cytochrome c_7 from the mixed culture of the phototrophic green-sulfur bacterium *Chlorobium* with the newly discovered anaerobic sulfur-reducing genus, *Drm. acetoxidans* (Pfennig and Biebl, 1976).

Ferricytochrome c_7 shows a characteristic very low absorbance at 280 nm which is confirmed by the very low levels of aromatic amino acids found in the protein. The ferric form has absorption maxima at 352 and 408 nm while ferrocytochrome c_7 has maxima at 418, 522.5, and 551.5 nm. High concentrations of ascorbate (100 m*M*) reduce the cytochrome while CO has no effect on the absorption spectrum of ferrocytochrome c_7. The amino acid sequence is known (Dobson *et al.*, 1974). In contrast to the four-heme c_3-type cytochrome, cytochrome c_7 lacks residues 55–90, which include the cystein sequences 58–64 and histidine 47. In other words, the binding group for the fourth heme is absent, explaining why only three hemes are present in cytochrome c_7. Xavier and Moura (1978) have published the 270 MHz NMR spectrum of ferricytochrome c_7 showing the well resolved resonances in the methyl region due to spin-decoupling from three-heme shifting probes on a single, small polypeptide chain. Cytochrome c_3 (M_r = 13,000) clearly functions as a cofactor for hydrogenase; however, the absence of hydrogenase in *Drm. acetoxidans* (Pfennig and Biebl, 1976) suggests that the physiological role of cyctochrome c_7 is quite different from that of cytochrome c_3.

B. Cytochrome *b*

The presence of cytochrome *b* was reported in *D. africanus* by Jones (1971) and more recently by Hatchikian and LeGall (1972) in *D. gigas*.

In the latter case, the *b*-type cytochrome was found associated with a particulate hydrogen-fumarate reductase activity. The presence of the *b*-type cytochrome was suggested by the inhibition of this system with 2-heptyl-4-hydroxyquinoline *N*-oxide and antimycin A. The presence of cytochrome *b* was established by low-temperature spectroscopy of reduced particulate fractions revealing α and β peaks at 550 and 522 nm and by extraction of particles with acid–acetone. The protoheme structure of cytochrome *b* was identified by its reduced pyridine hemochromogen peaks at 419, 523, and 556 nm in the acid–acetone fraction. It should be noted that cytochrome *b* is the one cytochrome in common between the two genera of sulfate-reducing bacteria. *Desulfovibrio* and *Desulfotomaculum.* Cytochrome *b* has also been reported in *Drm. acetoxidans*, but only when the bacteriom is grown on fumarate of malate (Pfennig and Biebl, 1976).

V. Molybdenum-Containing Iron-Sulfur Protein from *Desulfovibrio gigas*

Moura *et al.* (1976) reported on an acidic protein purified from *D. gigas* that had a molecular weight of 120,000 and contained twenty atoms of iron and labile sulfide and one atom of molybdenum per molecule. Only one N-terminal amino acid was found, and it was glycine. The absorption spectrum of the protein revealed maxima at 425 and 467 and a shoulder at 545 nm and was most similar to the 2Fe–2S* type iron–sulfur cluster. Activities of this protein that were tested included electron transport between hydrogenase and bisulfite reductase, and formate dehydrogenase. In each case the test was negative, and the biological role of the protein is thus presently unknown. There was a close similarity in amino acid composition between this protein and plant-type 2Fe–2S*-containing ferredoxins. No evidence for the presence of subunits was found.

Moura *et al.* (1978c) carried out EPR potentiometric titrations of this molybdo-iron-sulfur protein. At least three different types of two iron–two sulfur centers could be distinguished by EPR measurements in the range of 12°–77°K. These centers had the following E_M values: type IA, $E_M = 260$ mV; type IB, $E_M = 440$ mV; type II, $E_M = -285$ mV. The molybdenum signal of the protein required long equilibration times (about 30 minutes) in the presence of dye mediators because of the very slowly equilibrating molybdenum signal. The E_M values for two molybdenum oxidation–reduction couples found were: E_M (Mo VI/Mo V) $= -415$ mV and E_M (Mo V/Mo IV) $= -530$ mV. This protein preparation had 12 Fe and 12 S* atoms per molecule, and Moura *et al.* proposed that six (2Fe–2S*) clusters existed in the protein. Double integration of the EPR intensities from centers IA and IB accounted for 3–4 of the two iron-type centers per molecules. Quantitative estimation of Fe-S type II could not be made because the main g values were obscured by other signals. Quantitation of the molybdenum signal accounted for approximately 0.85 spin per protein molecule. The EPR data suggested evidence for a heterogeneous population of molybdenum signals.

VI. Rubredoxin-Type Proteins

A. Rubredoxin

1. *Structure*

The presence of rubredoxin in several anaerobes, as well as in some aerobes, makes this small molecule ($M_r = 6000$) an ideal choice for phylogenetic studies in microorganisms. Odom *et al.* (1976) have shown that the NADH $+ H^+$ rubredoxin oxidoreductase from *D. gigas* presents a remarkable specificity for *D. gigas* rubredoxin (Table I). Although the tertiary structure of *D. gigas* rubredoxin is not known, Fig. 3 shows that interesting comparisons can be made between the amino acid sequences of the rubredoxins from *C. pastorianum*, *D. vulgaris*, and *D. gigas* (Bruschi, 1976a,b; Vogel *et al.*, 1977) as a large number of charged residues have changed from one protein to the other. Such changes are dramatic in a small protein and may well explain the differences of affinity for the reductase. That anchorage sites between two redox proteins are constituted by a complementary special arrangement of charges is suggested by the work of Salemme (1976) in the case of cytochrome c and cytochrome b_5. Figure 3 shows that the peptidic chain in close proximity of the active center remains essentially unchanged.

The structure of the rubredoxin from *D. vulgaris* has been established down to a 2 Å resolution (Adman *et al.*, 1977), and this allows interesting comparisons with the well known *C. pastorianum* rubredoxin structure (Herriot *et al.*,

TABLE I

K_m OF RUBREDOXINS OF DIFFERENT ORIGINS FOR NADH $+ H^+$ RUBREDOXIN OXIDOREDUCTASE[a]

Origin	K_m
Desulfovibrio gigas rubredoxin	6.2×10^{-6} *M*
D. vulgaris rubredoxin	5.3×10^{-5} *M*
Clostridium pastorianum rubredoxin	1.0×10^{-4} *M*

[a] J. M. Odom, and H. D. Peck, Jr., personal communication.

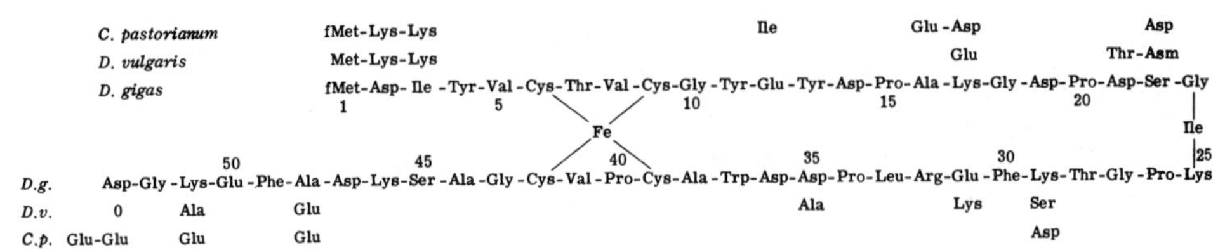

FIG. 3. Differences in the charged residues from the rubredoxins of *Clostridium pastorianum*, *D. vulgaris* and *D. gigas*. The "hairpin" structure is not the actual structure shown by X-ray analysis; it has been made to bring the four cysteine residues in close proximity.

1970; Watenpaugh *et al.*, 1973b). It shows, in particular, that the four Fe-S bounds range in length from 2.15 to 2.35 Å and do not differ significantly from the mean value of 2.29 Å.

2. *Physiological Activity*

Considering the efforts that have been made to resolve the structure of rubredoxins from strict anaerobes and the success that has followed them, it is paradoxical that their physiological function is still unknown. The relatively high redox potential of rubredoxin (−50 mV) makes it difficult to fit it in the physiological reactions which are commonly utilized to check these activities in sulfate-reducing bacteria (see Fig. 1). Probst *et al.* (1978) have isolated a rubredoxin from *Drm. acetoxidans.* Its amino acid composition shows that it is related to *Desulfovibrio* rubredoxins.

B. DESULFOREDOXIN

A new rubredoxin-like protein has been found in *D. gigas* by Moura *et al.* (1977a). It contains two iron atoms and eight cysteine residues per molecule. Although its optical spectrum is reminiscent of rubredoxin, it presents important differences. The polypeptide chain is constituted of 73 amino acid residues and lacks six amino acids, namely: histidine, arginine, proline, isoleucine, phenylalanine, and tryptophan. The N-terminal sequence, determined up to 35 residues, shows no homology with that of other nonheme iron protein. The magnetic properties of desulforedoxin, investigated by Moura *et al.* (1978a), indicate the presence of spin–spin interaction between the two iron centers. The protein undergoes reversible oxidoreduction process with $E_0' = -35$ mV, but it is sensitive to repeated redox cycles. No physiological function is known for desulforedoxin. The data indicate that the protein represents a new class of nonheme iron protein. It is of interest to note here the problem of redox protein nomenclature as far as trivial names are concerned. The utilization of the term desulforedoxin was first accepted (Moura *et al.*, 1977a), then refused (Moura *et al.*, 1978b) on the ground that it seems to mean "a protein without sulfur," a reasoning that could be applied to the name of sulfate-reducing bacteria themselves. Anyhow, the authors seem to have exhausted all Latin and Greek roots meaning red or purple.

VII. Hydrogenase

Extracts of the sulfate-reducing bacteria, genus *Desulfovibrio*, exhibit hydrogenase in high specific activity, and the enzyme has been studied in many laboratories over the past 20 years (LeGall and Postgate, 1973). Owing to the energy crisis, there has recently been a renewed interest in hydrogenase, and only the newer information concerning hydrogenase will be discussed.

Yagi *et al.* (1976) reported extensively on the properties of the solubilized particulate hydrogenase from *D. vulgaris*, strain Miyazaki. In contrast to other species of *Desulfovibrio*, this hydrogenase is largely particulate but can be solubilized by treatment with trypsin. The purified enzyme contained 7–9 iron atoms and 7 or 8 labile sulfides, had the spectrum of a nonheme iron protein and a molecular weight of 89,000, and was composed of two subunits having molecular weights of 28,000 and 59,000. They report a specific activity of 610 units/mg and a specificity for the reduction of cytochrome c_3 rather than ferrodoxin, as reported earlier (Yagi *et al.*, 1968). No EPR data were reported, but the enzyme did catalyze the deuterium–hydrogen exchange reaction, which was stimulated by cytochrome c_3, and the hydrogenase was reduced in the presence of H_2 only upon the addition of cytochrome c_3. LeGall *et al.* (1971a) purified the soluble hydrogenase of *D. vulgaris* to apparent homogeneity. It contained 3.5 iron atoms and 3.2 labile sulfides per molecular weight of 60,000 and was reported to contain two subunits of molecular weight 30,000. The enzyme had a specific activity of 53 but exhibited a typical nonheme iron spectrum and was reduced in the presence of H_2. EPR studies indicated the presence of a $g = 1.86$ signal upon reduction. These two reports strongly suggested the nonheme iron nature of the hydrogenase from *Desulfovibrio*, but the preparations were probably not pure.

Taking advantage of the report by Bell *et al.* (1974) indicating that the hydrogenase of *D. vulgaris* was largely located in the periplasmic space, Van der Westen *et al.* (1978) purified the hydrogenase of *D. vulgaris* to a specific activity of 3800 units/mg and demonstrated that it contained 12 iron atoms and 12 labile sulfides per molecular weight of 50,000. It thus appeared to contain three 4Fe–4S* clusters, but no information regarding the paramagnetic properties of the iron-sulfur centers was presented. The hydrogenase from *D. vulgaris* (Van der Westen *et al.*, 1978) is quite stable in the presence of oxygen, losing only 7% of its specific activity when stored in air plus bovine serum albumin for 2 weeks at 4°C. This is in strong contrast to the highly oxygen-labile hydrogenase from *C. pastorianum* (Chen and Mortenson, 1974). The hydrogenase from *D. vulgaris* is similar structurally to that from *C. pastorianum* which has a single polypeptide chain with a molecular weight of 60,500.

The hydrogenase from *D. gigas* has recently been purified by Hatchikian *et al.* (1978) and was reported to have a molecular weight of 89,500 and to be composed of two different subunits of molecular weight 62,000 and 26,000. The enzyme had a specific activity of 90 in the evolution assay, which is low compared to the specific activities of the hydrogenase from *D. vulgaris* and *C. pastorianum.* However, the enzyme contained 12 iron atoms and 12 labile sulfide, and exhibited a spectrum characteristic of nonheme-iron proteins. Quantitative extrusion of the Fe–S* cores indicated that they are of the 4Fe–4S* type and that there are three per molecule. Hydrogenase activity in the hydrogen evolution assay was not stimulated by the addition of cytochrome c_3 (M_r = 13,000). Although the kinetic and structural properties of hydrogenases from *D. vulgaris* and *D. gigas* appear to be significantly different, further characterization of the enzymes is essential to establish this point. The recent report of a unidirectional H_2-oxidizing hydrogenase suggest that these differences may have a physiological significance (Chen and Blanchard, 1978).

Hydrogenases from the earlier preparations from *D. vulgaris* and *D. gigas* (Bell *et al.*, 1974; Yagi *et al.*, 1968; LeGall *et al.*, 1971a), were reported to reduce the multiheme cytochrome c_3 (M_r = 13,000) and cytochrome c_3 (M_r = 26,000) but not ferredoxin, flavodoxin, or rubredoxin; however, the latter three electron carriers are reduced in the presence of cytochrome c_3 (Bell *et al.*, 1978). These are equilibrium-type studies, and no evidence is yet available from rapid-reaction kinetic studies involving optical or EPR stopped-flow methods. In contrast, hydrogenase from *C. pastorianum* is active only with ferredoxin. The specificity for natural electron carriers has not been confirmed for the most recent and highly purified hydrogenases from *D. vulgaris* and *D. gigas.*

VIII. Adenylyl Sulfate Reductase (APS Reductase)

APS reductase from *D. vulgaris* catalyzes the reduction of adenylyl sulfate (APS) to sulfite and AMP in the presence of reduced methylviologen and also the oxidation of sulfite in the presence of AMP and an electron acceptor, such as $Fe(CN)_6^{-3}$ or cytochrome *c*, to APS. The most thorough recent study on the purification and properties of the homogeneous enzyme is found in the report of Bramlett and Peck (1975). They reported that APS reductase has a molecular weight of 220,000 and contained 1 FAD and 12 iron and labile sulfide atoms per molecule. Gel electrophoresis in the presence of sodium dodecyl sulfate resulted in the detection of subunits of M_r 72,000 and 20,000. From the molecular weight and concentration of subunits, it was suggested that the native enzyme is composed of 3 subunits of M_r 72,000 and one of M_r 20,000. A dimeric form of APS reductase with a molecular weight of 440,000 was found in the presence of potassium phosphate buffer. The dimeric form could be dissociated to the monomer form of M_r 220,000 by changing the buffer system to Tris-maleate or on the addition of AMP, *d*AMP, AMP, or APS. A correlation of this change with a change in the kinetic properties of the enzyme could not be detected. A proposed scheme for the mechanism of action and flow of electrons through APS reductase is shown in Fig. 4.

Michaels *et al.* (1971) demonstrated that the reaction of sulfite with APS reductase resulted in the formation of an FAD-sulfite adduct with bleaching

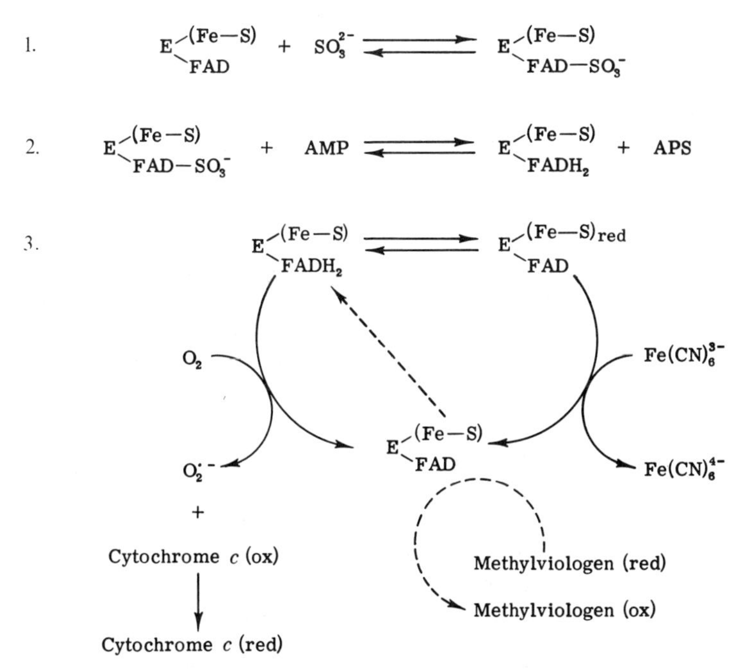

FIG. 4. Proposed scheme for the mechanism of action and overall flow of electrons through adenylyl sulfate (APS) reductase.

of the FAD and a formation of a characteristic maximum at 320 nm reflecting a reaction of sulfite at the N-5 position of the isoalloxazine ring system of FAD [Reaction (1)]. The subsequent addition of AMP to the enzyme–sulfite system results in a decrease in absorbance at 320 nm, the partial reduction of honheme iron, and the formation of APS [Reaction (2)]. The enzyme shows by low-temperature EPR spectroscopy a high-potential iron-sulfur cluster EPR signal at $g = 2.01$ in the oxidized state and a low-potential type iron-sulfur signal in the $g = 1.94$ region in the reduced state (R. Bramlett, H. D. Peck, Jr., and D. DerVartanian, unpublished observations). Addition of sulfite and AMP, but neither alone, leads to the formation of a $g = 1.94$ signal, thus confirming the reduction of nonheme iron observed spectrally. Sulfhydryl reagents inhibit the reduction of $Fe(CN)_6^{-3}$ and eliminate the nonheme iron component of the optical spectrum but do not affect cytochrome *c* reduction in the presence of sulfite and AMP or the reduction of APS with reduced methylviologen. Anaerobiosis was found to inhibit by 100% the reduction of mammalian cytochrome *c* while, in the presence of oxygen, increasing concentration of superoxide dismutase resulted in increasing inhibition reaching a maximum of 84% inhibition at a superoxide dismutase concentration of 1 m*M*. The results suggest that the enzyme-bound FAD can react directly with oxygen to form the superoxide radical and react with reduced methylviologen in the absence of nonheme iron, but that nonheme iron and presumably FAD are required for the reduction of $Fe(CN)_6^{-3}$ as shown in Fig. 4. The catalytic roles of the $g = 2.01$ and $g = 1.94$ EPR signals have not been determined. APS reductase has been found only in bacteria whose respiration involves the reduction of sulfate or the oxidation of reduced inorganic sulfur compounds. Dickerson and Timkovich (1975) have pointed out that the flavin-bound sulfite of APS reductase is energetically primed for subsequent reactions in a manner that a solitary sulfite ion is not. This observation may have a bearing on the mechanism of respiratory sulfate reduction as well as the oxidation of reduced sulfur compounds by the thiobacilli and the photosynthetic bacteria.

IX. Siroheme-Type Enzymes (Bisulfite and Sulfite Reductases)

The sulfate-reducing bacteria contain four types of sulfite-reducing enzymes, which can be differentiated on the basis of their spectral properties, substrates, and products. Three of these enzymes appear to reduce bisulfite (HSO_3^{-1}), as their pH optimum is around 6.0, and they have been termed bisulfite reductases. The bisulfite reductase found in most of the species of *Desulfovibrio* was first called desulfoviridin, that found in *D. desulfuricans*, strain Norway 4, was named desulforubidin, and the bisulfite reductase isolated from the *Desulfotomaculum* group was initially termed CO-reacting pigment or P_{582}. The high levels of these reductases (6–30% of the soluble protein) suggest that they are involved in respiratory-sulfate reduction; however, the multiplicity of products, sulfide, thiosulfate, and the trithionate, leaves the exact pathway of sulfite reduction unresolved. The fourth reductase is of the biosynthetic type, utilizing sulfite as substrate, and has been reported only in *D. vulgaris*.

A. DESULFOVIRIDIN

The green pigment desulfoviridin was first described by Postgate (1956) from *D. desulfuricans* and identified as a bisulfite reductase by Lee and Peck (1971). Lee and Peck purified the enzyme to homogeneity and found its molecular weight to be 226,000. The enzyme is a tetrameter composed of two different types of subunits with molecular weights of 42,000 and 50,000. The enzyme revealed absorption maxima at 628,580, 408,390, and 279 nm. The reductase did not form an alkaline hemochromogen, nor was it reducible by dithionite or borohydride. It also did not react with CO or other strong ligands. The major product formed from bisulfite reduction was found to be trithionate. Upon treatment of bisulfite reductase with 8 *M* urea, there appeared loss of the 628 and 580 nm absorption peaks, which were replaced by a new peak at 590 nm characteristic of the assimilatory sulfite reductases. These spectral events were interpreted by Lee *et al.* (1973a) to indicate that the chromophores of the assimilatory and dissimilatory sulfite reductases showed chemical similarities. Murphy and Siegel (1973) reported that acetone–HCl extraction of desulfoviridin resulted in the detection of sirohydrochlorin rather than a siroheme. Murphy *et al.* (1973) reported a highly distorted but quantitatively significant high-spin ferric heme EPR absorption for desulfoviridin indicating the presence of siroheme with the native enzyme. Murphy and Siegel (1973) proposed that these novel features of desulfoviridin could be rationalized if the iron of the porphyrin ring of the siroheme-containing bisulfite reductase was distinctly out-of-plane and was liberated as free iron on denaturation of the enzyme. The relative nonreactivity of substrate, reducing agents, and inhibitors with the enzyme from *D. vulgaris* raises important questions regarding the structure and mechanism of action of the enzyme.

B. DESULFORUBIDIN

Desulfoviridin is absent in the Norway strain of *D. desulfuricans* even though the organism is able to reduce sulfite to sulfide at normal rates.

Lee *et al.* (1973b) reported the presence of a new pigment in this strain and called it desulforubidin. They purified the protein and found that desulforubidin catalyzed the reduction of bisulfite to trithionate. Desulforubidin has absorption peaks at 580, 545, and 392 nm with a weak peak at 620 nm, and the molecular weight of 225,000 was estimated from the $S_{20,w}$ value of 9.8. The enzyme formed an alkaline pyridine hemochromagen but was not bleached by sodium dithionite. It appeared to have a hemelike spectrum analogous to sulfite reductase from *E. coli* (Siegel *et al.*, 1973) and, like sulfite reductase (Murphy *et al.*, 1973), showed characteristic high-spin rhombically distorted ferric-heme EPR resonances. This enzyme is therefore related to desulfoviridin and other sulfite reductases in having a sirohemelike chromophore and is analogous to desulfoviridin both in structure and function. The significance of the occurrence of desulforubidin in this single strain is not understood.

C. CO-Reacting Pigment, or P_{582}

In the spore-forming genus of the sulfate-reducing bacterium *Desulfotomaculum*, yet another type of bisulfite reductase has been found by the Trudinger and Akagi groups. This bisulfite reductase purified from *Dm. nigrificans* catalyzes the reduction of bisulfite to trithionate, thiosulfate or sulfide depending on the concentrations or substrate and reductant. It has also been called CO-reacting pigment of P_{582} (Trudinger, 1970; Akagi and Adams, 1973) because of its characteristic absorption maximum at 582 nm (as well as at 392 and 700 nm). It also reacts spectrally, unlike bisulfite reductases from *D. vulgaris* and *D. desulfuricans*, with strong ligands in the presence of sodium dithionite. P_{582} has been found by Murphy and Siegel (1973) to give, on extraction with acetone-HCl, a sirohemelike chromophore. P_{582} also contains iron-sulfide clusters in variable amounts and has a molecular weight of approximately 145,000.

It is clear that the three dissimilatory bisulfite reductases found in the sulfate-reducing systems described here share considerable similarities in the structure of the sirohemelike chromophore and in the fact that, depending on reductant and bisulfite concentrations, the products formed can be trithionate, thiosulfate, or sulfide. The spectral differences suggest that there are significant kinetic differences among these three reductases, but major differences have not yet been reported.

D. Sulfite Reductase

Lee *et al.* (1973a) reported the purification from *D. vulgaris* of a biosynthetic sulfite reductase that catalyzed the six-electron reduction of sulfite to sulfide. This enzyme had absorption maxima and physical properties different from the bisulfite reductases. The absorption maxima were at 590, 545, 405, and 275 nm. The enzyme had a molecular weight of 26,800 and consists of a single polypeptide chain based on disc electrophoresis studies that showed a single greenish-brown band corresponding to both activity and protein. The presence of subunits was not detected. Drake and Akagi (1976) have isolated a reductase identical to that isolated by Lee *et al.* (1973a) in spectral properties; however, it reduced bisulfite to thiosulfate and sulfide without the formation of trithionate. The relationship of these activities have not been resolved, and their physiological significance remains obscure.

E. Physiological Significance

As discussed in three recent reviews (Siegel, 1975; Dickerson and Timkovich, 1975; Thauer *et al.*, 1977) on sulfur metabolism in the sulfate-reducing bacteria, the reduction of bisulfite to sulfide (termed dissimilatory or respiratory sulfite reduction) occurs by one of two different mechanisms: (1) by direct six-electron reduction of bisulfite to sulfide catalyzed by bisulfite reductase or (2) by three successive two-electron reductions involving three different enzymes. These three steps are: (a) reduction of bisulfite to trithionate by bisulfite reductase; (b) reduction of trithionate to thiosulfate by trithionate reductase; and (c) reduction of thiosulfate to sulfide by thiosulfate reductase. The first mechanism is supported by observations of Chambers and Trudinger (1975) indicating that the putative intermediates, thiosulfate and trithionate, do not exchange with external pools of these substrates and the fact that the intracellular concentration of sulfite is probably quite low. The second mechanism is supported by the fact that it explains several bioenergetic questions, and by the existence of a thiosulfite reductase that has been purified and extensively studied by Hatchikian (1975). Although crude extracts reduce trithionate, a trithionate reductase has not been isolated.

The three bisulfite reductase have been recently purified from *D. gigas*, *Dm. ruminis* and *D. desulfuricans* (Norway 4) (Liu, DerVartanian and Peck, unpublished results). They were each shown to contain 14 nonheme irons and an identical number of acid labile sulfurs per mol. Desulforubidin was found to contain 2 sirohemes per mol; P_{582}, 1–2 sirohemes per mol and desulfoviridin, 2 sirohydrochlorins per mol. By EPR spectroscopy, all reductases exhibited a major complex rhombic high-spin ferric heme resonance in the $g = 6$ region which was lost in the presence of reduced methyl viologen, indicating reduction of the heme. In the presence of dithionite and methyl viologen, a $g = 1.94$ signal was observed with all reductases indicating the presence of nonheme iron of the low potential type. Comparable changes were also observed in the absorption spectra of the enzymes. The pattern of reduction with the two reducing systems observed with the three bisulfate reductases is remarkable considering the complexity of the enzymes involved. This strongly suggests that the functional aspects of the sirohemes and iron sulfur centers of each bisulfite reductase are extremely similar.

Two possibilities exist for the removal of trithionate. First, the split of trithionate by thiols is a well known reaction and Drake and Akagi (1977a) have reported a split of trithionate to sulfite and thiosulfate in the presence of sulfite, which they termed a "thiosulfate-forming" enzyme. Alternatively, Kobayashi *et al.* (1974) and Drake and Akagi (1977b) have suggested the presence of enzyme-bound intermediates, which at low concentrations of sulfite are reduced directly to sulfide but at high concentrations of sulfite can react irreversibly to produce both trithionate and thiosulfate. The bioenergetic problems discussed by Thauer *et al.* (1977) could be resolved in terms of different electron donors for the reduction of the enzyme-bound intermediates or in terms of the redox properties of cytochrome c_3, flavodoxin, and ferredoxin discussed here previously. As enzymological and biochemical approaches to the pathway of respiratory sulfite reduction have not produced definitive results, other approaches to this problem, such as the genetic, will probably be required.

X. Bioenergetics of Respiratory Sulfate Reduction

Sulfate-reducing bacteria of the genus *Desulfovibrio* were the first anaerobic nonphotosynthetic microorganisms in which an electron-transfer coupled phosphorylation was clearly established (Barton *et al.*, 1972). This conclusion is based upon the requirement of ATP for the reduction of sulfate with hydrogen in extracts but not in whole cells, the inhibition of sulfate reduction in whole cells by classical uncoupling agents, the demonstration of phosphorylation coupled to hydrogen oxidation in extracts, and the theoretical necessity for electron-transfer phosphorylation during growth in the presence of sulfate and an electron donor such as lactate, formate, or ethanol. Growth yields obtained with various combinations of electron donors and acceptors have substantiated these observations (MaGee *et al.*, 1978; Badziong and Thauer, 1978). Current areas of active interest concern the number of electron-transfer coupled phosphorylations in respiratory sulfate reduction, the specific electron donor/acceptor couples involved, the cellular topography of the electron donor/acceptor couples, and the composition of the electron-transfer chain.

The reduction of sulfate to sulfide with molecular hydrogen has a $\Delta G^{\circ\prime}$ of

-36.4 kcal/mol (Thauer *et al.*, 1977), which is theoretically sufficient for the formation of 4 mol of ATP from ADP and P_i. As whole cells can reduce sulfate to sulfide solely with hydrogen, it has been assumed that at least 2 mol of ATP must be synthesized by electron-transfer phosphorylation and utilized for the formation of APS, i.e., ATP + $SO_4^{2-} \rightarrow$ APS + PP_i; however, it has not been determined whether the pyrophosphate can be used for the formation of ATP rather than simply being hydrolyzed to orthophosphate. The common experience has been that the reduction of sulfate with hydrogen does not provide energy for growth, which is supported by the observation that supplementation of growing cultures with hydrogen does not increase cell yields (Khosrovi *et al.*, 1971). Recently, two new strains of *D. vulgaris*, the Marburg and Madison strains, have been isolated that are capable of net growth on H_2, SO_4^{2-} acetate, and CO_2 (Badziong *et al.*, 1978). Growth yields on hydrogen and thiosulfate indicate that 3 mol of ATP are formed per thiosulfate reduced, while growth yields on hydrogen and sulfate show that only 1 mol of ATP is available for growth per sulfate reduced. Thus, it was suggested (Badziong and Thauer, 1978) that ATP formation was only coupled to the reduction of sulfite (produced by thiosulfate reductase), with 2 mol of ATP being utilized for the formation of APS. One can cautiously generalize that (1) the pyrophosphate generated by ATP sulfurylase in the formation of APS is not utilized for ATP formation, and (2) the reduction of APS to sulfite is not coupled to electron-transfer phosphorylation; however, there is sufficient energy in the H_2/APS couple for the formation of at least 1 mol of ATP. MaGee *et al.* (1978) have reported the formation of 1 mol of ATP when cells were grown on formate and sulfate; formate can be regarded as the equivalent of hydrogen, and this result agrees with that of Badziong and Thauer (1978). Yields of cells grown on formate and sulfite suggested that 2 mol of ATP were formed per sulfite reduced, but sulfite may be toxic (Badziong and Thauer, 1978) and lower the growth yield.

Growth yield studies using lactate and sulfate as substrates suggest that there is net formation of 2 mol of ATP per sulfate reduced (Senez, 1962; Vosjan, 1970). Generalizing from the results obtained with cells grown on hydrogen and sulfate, one can suggest that 2 mole of ATP are formed by electron-transfer phosphorylation. One could further speculate that only electrons derived from pyruvate are involved in electron-transfer phosphorylation, those from lactate going by way of a nonphosphorylative pathway. An obligatory coupling between lactate oxidation and APS reduction might explain the failure of growth-yield studies to detect phosphorylation in the first step in the reduction of sulfate. Information regarding the actual pathway of sulfite reduction is required before it is possible to speculate on the linking of specific electron acceptors with the two electron-transfer chains. (See Section IX, on siroheme reductases.)

Hydrogenase is located in the periplasmic space of most sulfate-reducing bacteria with the exception of *D. vulgaris*, strain Miyazaki, where it is membrane bound. The lactate and formate dehydrogenases are also membrane-bound (L. L. Barton and H. D. Peck, Jr., unpublished observations). The enzymes of pyruvate metabolism are soluble as are all the enzymes of sulfate reduction. The status of the ethanol-oxidizing sequence of enzymes is unknown, but it might be anticipated that the ethanol dehydrogenase will be membrane-bound and that the aldehyde dehydrogenase involved in a substrate phosphorylation will be located in the soluble fraction. These observations suggest the possibility that proton translocation at the substrate level is important for the bioenergetics of the sulfate-reducing bacteria and merits further investigation.

Membrane vesicles capable of electron-transfer phosphorylation have been prepared only from *D. gigas*. The membrane vesicles are rich in cytochrome *b* and cytochromes *c*; however, because of the spectral similarity of the *c*-type cytochromes, it has not been possible to establish an electron-transfer sequence utilizing spectrophotometric techniques. However, the detailed studies of the physical properties of the *c*-type cytochromes should allow this to be accomplished utilizing optical measurements in conjunction with EPR spectrometry.

XI. Conclusion

If important progress has not been made in the last half-decade concerning the physiology of sulfate-reducing bacteria, much information has been obtained on the structure and, in some cases, the function of isolated redox proteins. It appears that it will be difficult to extrapolate this information from one sulfate reducer to another. For example, Bruschi *et al.* (1977) have shown that cytochrome c_3 (M_r = 16,000) of *D. desulfuricans*, Norway 4, couples the reduction of thiosulfate with molecular hydrogen, but the homologous cytochrome c_3 from *D. gigas* is inactive. The ferredoxin from the Norway 4 strain couples the same reaction in extracts of *D. gigas*, but not in active extracts from the Norway strain. Examples of such apparent anomalies are numerous. Thus, cytochrome c_{553} is reduced by NADH + H^+ in the presence of rubredoxin oxidoreductase, a set of reactions that mimic a small electron-transfer chain. But the reductase is found only in *D. gigas*, and cytochrome c_{553} only in *D. vulgaris* (LeGall and Postgate, 1973). Such results are probably due to the fact that the sulfate-reducing bacteria have undergone tremendous evolutionary changes, involving a complete change in the function of some redox proteins, during the time since the first sulfate reducers appeared (Ault and Kulp, 1959). This postulated antiquity of the process of sulfate reduction has prompted many studies of the phylogenetic relationships between these organisms and other "primitive" groups, such as fermentative and photosynthetic bacteria. The primary structure of such proteins as cytochrome *c*, ferredoxin, rubredoxin, and flavodoxin have been taken as references. The work of Vogel *et al.*, (1977) and the more recent publication of Schwartz and Dayhoff (1978) show that simple views in this type of work are not realistic and that an accurate picture of the evolution pattern will be reached step by step when sufficient information is available from different families of proteins and ribosome RNA sequences.

An example will show that the evolutionary pattern within a family of proteins is much more complex than is generally thought. The development of ferredoxins from obligate anaerobic bacteria to aerobic microorganisms and plants has been proposed (Buchanan and Arnon, 1973): the one 4Fe–4S* ferredoxins constituting a link between the two 4Fe–4S* cluster ferredoxins of clostridia and photosynthetic bacteria and the 2Fe–2S* ferredoxins of blue-green algae and plants. According to that scheme bacteria possessing one-4Fe–4S* cluster ferredoxins would represent a transition between anaerobic and aerobic environments. The discovery of a two 4Fe–4S* cluster ferredoxin in *Drm. acetoxidans* by Probst *et al.* (1977), together with the fact that a one 4Fe-4S* cluster ferredoxin has been found in *C. thermoaceticum* by Yang *et al.* (1977), shows that such evolutionary models are still very simplified.

A very puzzling question remains regarding the unique structure and occurrence of cytochrome c_3. Since it belongs to organisms whose particularity it is to reduce sulfates or sulfur, the obvious answer would seem to be that it is essential to this kind of respiration. *Desulfotomaculum* species possess the same respiratory system based on sulfate, and yet they are absolutely devoid of cytochrome c_3. So, the presence of cytochrome c_3 in desulfovibriones and the reasons for the existence of such a sophisticated molecule are still unresolved questions. In view of the specificity of cytochrome c_3 to hydrogenase (Yagi *et al.*, 1968; Bell *et al.*, 1978) a possible explanation could be that it is necessary for the rapid hydrogen exchange that allows interspecies syntropies such as the one that has been shown by Bryant *et al.* (1977) to occur between *D. vulgaris* and *Methanobacterium formicicum.* If this is the case, *Desulfotomaculum* species would not be efficient in interspecific hydrogen transfer owing to the lack of cytochrome c_3.

Another interesting question is the reason for the presence in sulfate-reducing bacteria of two proteins thought to be characteristic of the aerobic world, namely catalase and superoxide dismutase (Hatchikian *et al.*, 1977). The results of Hatchikian and Henry (1977) and Bruschi and Hatchikian (1977) have shown that the N-terminal sequence of this superoxide dismutase exhibits a high degree of homology with other superoxide dismutases.

Different explanations for the presence of these enzymes are possible: protection against accidental contact with molecular oxygen or protection against radicals that might be produced during the respiratory process of sulfate reduction, such as $SO_2^{\cdot}$ or $SH^{\cdot}$. Superoxide dismutase and catalase must have a physiological role, but it still remains obscure and difficult to approach experimentally.

ACKNOWLEDGMENTS

D. V. D. gratefully acknowledges research support from the National Institute for General Medical Sciences (Research Grant No. GM-18895). D. V. D. and H. D. P. acknowledge support from the National Science Foundation (Research Grant No. PCM-7622410), J.L. acknowledges a grant from the CNRS Solar Energy Program, PIRDES, and H.D.P. acknowledges a grant from the Department of Energy (Contract No. E-38-1-888).

REFERENCES

Adam, E. T., Sieker, L. C., Jensen, L. H., Bruschi, M., and LeGall, J. (1977). *J. Mol. Biol.* **112**, 113–120.

Akagi, J. M., and Adams, V. (1973). *J. Bacteriol.* **116**, 392–396.

Almassy, R. J., and Dickerson, R. E. (1978). *Proc. Natl. Acad. Sci. U.S.A.* **75**, 2674–2678.

Ault, M. W., and Kulp, J. L. (1959). *Geochim. Cosmochim. Acta* **16**, 201–210.

Badziong, W., and Thauer, R. K. (1978). *Arch. Microbiol.* **117**, 209–214.

Badziong, W., Thauer, R. K., and Zeikus, J. G. (1978). *Arch. Microbiol.* **116**, 41–49.

Barton, L. L., and Peck, H. D., Jr. (1970). *Bacteriol. Proc.* p. 134.

Barton, L. L., LeGall, J., and Peck, H. D., Jr. (1972). *In* "Horizons in Bioenergetics" (A. San Pietro and H. Gest, eds.), pp. 33–51. Academic Press, New York.

Bell, G. R., LeGall, J., and Peck, H. D., Jr. (1974). *J. Bacteriol.* **120**, 994–997.

Bell, G. R., Lee, J. P., Peck, H. D., Jr., and LeGall, J. (1978). *Biochemie* **60**, 315–320.

Bianco, P., and Haladjian, J. (1977). *Biochem. Biophys. Res. Commun.* **78**, 323–327.

Bianco, P., and Haladjian, J. (1979). *Biochim. Biophys. Acta* **545**, 86–93.

Biebl, M., and Pfennig, N. (1977). *Arch. Microbiol.* **112**, 115–117.

Bramlett, R. N., and Peck, H. D., Jr. (1975). *J. Biol. Chem.* **250**, 2979–2986.

Bruschi, M. (1976a). *Biochim. Biophys. Acta* **434**, 4–17.

Bruschi, M. (1976b). *Biochem. Biophys. Res. Commun.* **70**, 615–621.

Bruschi, M., and Hatchikian, E. C. (1977). *FEBS Lett.* **76**, 121–124.

Bruschi, M., and LeGall, J. (1972). *Biochim. Biophys. Acta* **271**, 48–60.

Bruschi, M., Hatchikian, E. C., LeGall, J., Moura, J. J. G., and Xavier, A. V. (1976). *Biochim. Biophys. Acta* **449**, 275–284.

Bruschi, M., Hatchikian, C. E., Golovleva, L. A., and LeGall, J. (1977). *J. Bacteriol.* **129**, 30–38.

Bryant, M. P., Campbell, L. L., Reddy, C. A., and Crabill, M. R. (1977). *Appl. Environ. Microbiol.* **33**, 1162–1169.

Buchanan, B. B., and Arnon, D. L. (1973). *Adv. Enzymol. Relat. Areas Mol. Biol.* **33**, 119–176.

Cammack, R., Rao, K. K., Hall, D. O., Moura, J. J. G., Xavier, A. V., Bruschi, M., LeGall, J., Deville, A., and Gayda, J. P. (1977). *Biochim. Biophys. Acta* **490**, 311–321.

Campbell, L. L., and Postgate, J. R. (1965). *Bacteriol. Rev.* **29**, 359–363.
Carter, C. W., Jr., Kraut, J., Freer, S. T., Alden, R. A., Sieker, L. C., Adam, E. T., and Jensen, L. H. (1972). *Proc. Natl. Acad. Sci. U.S.A.* **69**, 3526–3529.
Chambers, L. A., and Trudinger, P. A. (1975). *J. Bacteriol.* **123**, 36–40.
Chen, J. S., and Mortenson, L. E. (1974). *Biochim. Biophys. Acta* **371**, 283–298.
Chen, J. S., and Blanchard, D. K. (1978). *Biochem. Biophys. Res. Commun.* **84**, 1144–1150.
Cookson, D. J., Moore, G. R., Pitt, R. C., Williams, R. J. P., Campbell, I. D., Ambler, R. P., Bruschi, M., and LeGall, J. (1978). *Eur. J. Biochim.* **83**, 261–275.
DerVartanian, D. V., and LeGall, J. (1974). *Biochim. Biophys. Acta* **346**, 79–99.
DerVartanian, D. V., and LeGall, J. (1978). *Biochim. Biophys. Acta* **502**, 458–465.
DerVartanian, D. V., Xavier, A. V., and LeGall. J. (1978). *Biochimie* **60**, 321–325.
Dickerson, R. E., and Timkovich, R. (1975). *In* "The Enzymes" (P. D. Boyer, ed.), 3rd Ed., Vol. 11, pp. 397–547. Academic Press, New York.
Dobson, C. M., Hoyle, N. J., Geraldes, C. F., Bruschi, M., and LeGall, J. (1974). *Nature* (*London*) **349**, 425–429.
Drake, H. L., and Akagi, J. M. (1976). *Biochem. Biophys. Res. Commun.* **71**, 1214–1219.
Drake, H. L., and Akagi, J. M. (1977a). *J. Bacteriol.* **132**, 132–138.
Drake, H. L., and Akagi, J. M. (1977b). *J. Bacteriol.* **132**, 139–143.
Dubourdieu, M., LeGall, J., and Fox, J. L. (1973). *Biochim. Biophys. Res. Commun.* **52**, 1418–1425.
Dubourdieu, M., LeGall, J., and Favaudon, V. (1975). *Biochim. Biophys. Acta* **376**, 519–532.
Favaudon, V., LeGall, J., and Lhoste, J. M. (1979). *In* "Flavins and Flavoproteins" (K. Yagi and T. Yamano, eds.), in press.
Frey, M., Haser, R., Pierrot, M., Bruschi, M., and LeGall, J. (1976). *J. Mol. Biol.* **104**, 741–743.
Hall, D. O. (1977). *Adv. Chem. Ser.* No. 162, 227–250.
Hatchikian, E. C. (1975). *Arch. Microbiol.* **105**, 249–256.
Hatchikian, E. C., and Henry, Y. A. (1977). *Biochimie* **59**, 153–161.
Hatchikian, E. C., and LeGall, J. (1970). *Ann. Inst. Pasteur, Paris* **118**, 288–301.
Hatchikian, E. C., and LeGall, J. (1972). *Biochim. Biophys. Acta* **267**, 479–484.
Hatchikian, E. C., Bell, G. R., and LeGall, J. (1977) *In* "Superoxide and Superoxide Dismutases" (I. Fridovich, J. M. McCord, and A. M. Michelson, eds.), pp. 159–172. Academic Press, New York.
Hatchikian, E. C., Bruschi, M., and LeGall, J. (1978). *Biochem. Biophys. Res. Commun.* **82**, 451–461.
Herriott, J. R., Sieker, L. C., Jensen, L. H., and Lovenberg, W. (1970). *J. Mol. Biol.* **50**, 391–406.
Irie, K., Kobayashi, K., Kobayashi, M., and Ishimoto, M. (1973). *J. Biochem.* (*Tokyo*) **73**, 353–360.
Jones, H. E. (1971). *Arch. Mikrobiol.* **80**, 78.
Khosrovi, B., MacPherson, R., and Miller, J. D. A. (1971). *Arch. Microbiol.* **80**, 324–337.
Knight, E., Jr., and Hardy, R. W. F. (1966). *J. Biol. Chem.* **241**, 2752–2755.
Kobayashi, K. Y., Seki, Y., and Ishimoto, M. (1974). *J. Biochem.* (*Tokyo*) **75**, 519–529.
Laishley, E. J., Travis, J., and Peck, H. D., Jr. (1969). *J. Bacteriol.* **98**, 302–305.
Lee, J. P., and Peck, H. D., Jr. (1971). *Biochem. Biophys. Res. Commun.* **45**, 583–589.
Lee, J. P., LeGall, J., and Peck, H. D., Jr. (1973a). *J. Bacteriol.* **115**, 529–542.
Lee, J. P., Yi, C., LeGall, J., and Peck, H. D., Jr. (1973b). *J. Bacteriol.* **115**, 453–455.
LeGall, J., and Dragoni, N. (1966). *Biochem. Biophys. Res. Commun.* **23**, 145–149.
LeGall, J., and Hatchikian, E. C. (1967). *C. R. Acad. Sci.* **264**, 2580–2583.
LeGall, J., and Postgate, J. R. (1973). *Adv. Microb. Physiol.* **10**, 81–133.
LeGall, J., DerVartanian, D. V., Spilker, E., Lee, J. P., and Peck, H. D., Jr. (1971a). *Biochim. Biophys. Acta* **234**, 525–530.

LeGall, J., Bruschi-Heriaud, M., and DerVartanian, D. V. (1971b). *Biochim. Biophys. Acta* **234**, 499–512.
LeGall, J., Bruschi, M., and Hatchikian, E. C. (1975). *Proc. FEBS Meet. 10th* **40**, 277–285.
McDonald, C. C., Phillips, W. D., and LeGall, J. (1974). *Biochemistry* **13**, 1952–1959.
MaGee, E. L., Jr., Eusley, B. D., and Barton, L. L. (1978). *Arch. Microbiol.* **117**, 21–26.
Mayhew, S. G., and Ludwig, M. L. (1975). *In* "The Enzymes" (P.D. Boyer, ed.), 3rd Ed., Vol. 12, pp. 57–118. Academic Press, New York.
Michaels, G. B., Davidson, J. T., and Peck, H. D., Jr. (1971). *In* "Flavins and Flavoproteins" (H. Kamin, ed.), pp. 555–580. Univ. Press, Baltimore, Maryland.
Moura, J. J G., Xavier, A. V., Bruschi, M., LeGall, J., Hall, D. O., and Cammack, R. (1976). *Biochem. Biophys. Res. Commun.* **72**, 782–789.
Moura, I., Bruschi, M., LeGall, J., Moura, J. J. G., and Xavier, A. V. (1977a). *Biochem. Biophys. Res. Commun.* **75**, 1037–1044.
Moura, J. J. G., Xavier, A. V., Bruschi, M., and LeGall, J. (1977b). *Biochim. Biophys. Acta* **459**, 278–289.
Moura, J. J. G., Xavier, A. V., Cookson, D. J., Moore, G. R., Williams, R. J. P., Bruschi, M., and LeGall, J. (1977c). *FEBS Lett.* **81**, 275–280.
Moura, I., Xavier, A. V., Cammack, R., Bruschi, M., and LeGall, J. (1978a). *Biochim. Biophys. Acta* **533**, 159–162.
Moura, J. J. G., Xavier, A. V., Hatchikian, E. C., and LeGall, J. (1978b). *FEBS Lett.* **89**, 177–179.
Moura, J. J. G., Xavier, A. V., Cammack, R. Hall, D. O., Bruschi, M., and LeGall, J. (1978c). *Biochem. J.*, in press.
Murphy, M. J., and Siegel, L. M. (1973). *J. Biol. Chem.* **248**, 6911–6919.
Murphy, M. J., Siegel, L. M., Kamin, H., DerVartanian, D. V., Lee, J. P., LeGall, J., and Peck, H. D., Jr. (1973). *Biochem. Biophys. Res. Commun.* **54**, 83–88.
Niki, K., Hagi, T., Inokuchi, H., and Kimura, K. (1977). *J. Electrochem. Soc.* **124**, 1889–1891.
Odom, J. M., Bruschi, M., Peck, H. D., Jr., and LeGall, J. (1976). *Fed. Proc., Fed. Am. Soc. Exp. Biol.* **35**, 1360.
Pfennig, N., and Biebl, H. (1976). *Arch. Microbiol.* **110**, 3–12.
Postgate, J. R. (1956). *J. Gen. Microbiol.* **14**, 545–572.
Postgate, J. R., and Campbell, L. L. (1966). *Bacteriol. Rev.* **30**, 732–738.
Probst, I., Bruschi, M., Pfennig, N., and LeGall, J. (1977). *Biochim. Biophys. Acta* **460**, 58–64.
Probst, I., Moura, J. J. G., Moura, I., Bruschi, M., and LeGall, J. (1978). *Biochim. Biophys. Acta* **502**, 38–44.
Salemme, F. R. (1976). *J. Mol. Biol.* **102**, 563–568.
Schwartz, R. M., and Dayhoff, M. D. (1978). *Science* **199**, 395–403.
Senez, J. C. (1962). *Bacteriol. Rev.* **26**, 95–107.
Siegel, L. M. (1975). *In* "Metabolism of Sulfur Compounds" (D. M. Greenberg, eds.), 3rd Ed., Vol. 7, pp. 217–286. Academic Press, New York.
Siegel, L. M., Murphy, M. J., and Kamin, H. (1973). *J. Biol. Chem.* **248**, 251–264.
Thauer, R. K., Jungermann, K., and Decker, K. (1977). *Bacteriol. Rev.* **41**, 100–180.
Trudinger, P. A. (1970). *J. Bacteriol.* **104**, 158–170.
Van der Westen, H. M., Mayhew, J. G., and Vereger, C. (1978). *FEBS Lett.* **86**, 122–126.
Vogel, H., Bruschi, M., and LeGall, J. (1977). *J. Mol. Evol.* **9**, 111–119.
Vosjan, J. H. (1970). *Antonie van Leeuwenhoek; J. Microbiol. Serol.* **26**, 584–586.
Watenpaugh, K. D., Sieker, L. C., Jensen, L. H., LeGall, J., and Dubourdieu, M. (1972). *Proc. Natl. Acad. Sci. U.S.A.* **69**, 3185–3188.
Watenpaugh, K. D., Sieker, L. C., and Jensen, L. H. (1973a). *Proc. Natl. Acad. Sci. U.S.A.* **70**, 3857–3862.

Watenpaugh, K. D., Sieker, L. C., Herriot, J. R., and Jensen, L. H. (1973b). *Acta Crystallogr., Sect. B* **829**, 943–956.
Weitzman, P., Kennedy, J., and Caldwell, R. (1971). *FEBS Lett.* **17**, 241.
Widdel, F., and Pfennig, N. (1977). *Arch. Microbiol.* **112**, 119–122.
Xavier, A. V., Moura, J. J. G., LeGall, J., and DerVartanain, D. V. (1979). *Biochimie*, in press.
Xavier, A. V., and Moura, J. J. G. (1978). *Biochimie* **60**, 327–338.
Yagi, T., Honya, M., and Tamiya, N. (1968). *Biochim. Biophys. Acta* **153**, 629–698.
Yagi, T., Kimura, H., Daidoji, H., Sakai, F., Tamura, S., and Inokuchi, H. (1976). *J. Biochem. (Tokyo)* **79**, 661–671.
Yang, S. S., Ljungdahl, L. J., and LeGall, J. (1977). *J. Bacteriol.* **130**, 1084–1090.
Yoch, D. C., and Carithers, R. P. (1978). *J. Bacteriol.* **136**, 822–824.

CURRENT TOPICS IN BIOENERGETICS, VOLUME 9

Applications of the Photoaffinity Technique to the Study of Active Sites for Energy Transduction

RICHARD JOHN GUILLORY
John A. Burns School of Medicine
University of Hawaii
Honolulu, Hawaii

ISBN 0-12-152509-0

I. Introduction

A. Energy-Transducing Systems

This review will center principally upon recent applications of photo-affinity probes to the investigation of the active sites of bioenergetic systems.

The current concepts of energy in biological systems developed from considerations of the turnover of phosphate bonds in living organisms and the relationships of such turnover to the energy requirement of the cell. These concepts were given an easily understandable interpretation by Lipmann in his classic paper on the metabolic "Generation and Utilization of Phosphate bond Energy" (*1*). The concept of phosphorylated intermediates of metabolic pathways being either high, low, or intermediate with respect to phosphate transfer potential had a fundamental and potent influence on biochemical thinking. On the other hand, Lipmann clearly indicated that, although ATP could be considered as the direct currency of metabolism and had a pivotal role in cellular energetics, the utilization and synthesis of this nucleotide were coupled to energy transformers (i.e., transducers) directly responsible for the flux of biological energy. Thus the balance of the phosphate potential is regulated by those systems involved in the synthesis and utilization of the adenine nucleotides or their metabolic equivalents. These transducer systems function in converting biological oxidative energy from the metabolic process of catabolism into a variety of useful forms of potential or kinetic energy. They are the principal focus of this review.

Thirty years ago, except for some general concepts concerning the contractile system, little was understood about the actual molecular mechanism by which biological transducers function. Since that time our understanding has been broadened by an appreciation of the biological structure of such systems. Such understanding is nevertheless still fundamentally based upon static representations and is consequently limited. The mitochondrial electron transport and associated phosphorylation components, the proteins making up the contractile system, the membrane systems responsible for specific ligand translocations from one side of a membrane to another, the gating system of excitable tissues, and the machinery of protein synthesis, to name only a few transducing systems, are all arrangements of basic biochemical constituents. Each component is capable of undergoing time-dependent changes in positional relationships as dictated by physiological conditions. Thus, in spite of the static view of such systems as presented by most experimental approaches, the organization of energy-transducing systems has revealed certain facts that we may generalize as being dynamic characteristics of the group as a whole.

Important generalizations at our current level of understanding are as follows: first, biological transducers appear to be at least partly composed of self-assembling membrane units; second, consideration of the dynamics of energy transduction in a large number of organized bioenergetic transducers reveals that there is generally an interaction of a small molecular weight effector (or substrate) with a receptor component associated with a membrane or a membranelike structure. In some cases, as with mitochondrial F_1ATPase, the receptor (ATPase) may be experimentally dissociated from the particulate membrane (*2,3*); in others, such resolution has yet to be accomplished. The important consequence of this interaction is that it results in a specific and distinctive membrane change, often coupled directly, but sometimes indirectly, to energy dissipation (i.e., ATPase activity). It is this distinctive membrane change that is at the heart of energy transduction and may take many forms from the opening of specific K^+ gates in excitable tissue to the translocation of amino acids in *Escherichia coli* membranes. The molecular mechanisms for such ligand-induced membrane energy transductions are still much of a mystery.

B. Types of Interactions of Interest in the Application of Photoaffinity Labeling Techniques

1. *Ligand and Receptor Site Structure and Function*

One of the major metabolic controlling devices in bioenergetic systems, and perhaps the one most easily amenable to current investigative methodologies of photoaffinity labeling, is ligand interaction with receptor sites on membrane surfaces. In this case, similar problems present themselves as with enzyme catalysis. In the latter, the major unresolved problem is the relationship of ligand structure to enzymic reactivity, and with energy-transducing systems one wishes to know as well the molecular aspects of the specificity of interaction. Why do certain effectors interact with restricted sites, allowing for unique energy transduction? In theory the use of photoaffinity probes should aid in revealing topographical arrangements about specific binding sites, but, more important, it has the capability of revealing both spatial and time-dependent aspects of the structural topography. Do these restricted sites remain stationary with respect to other membrane components? If dynamic considerations apply, how are they controlled by physiological conditions?

A premise of modern biology is that structure and function are interrelated. Thus knowledge of the three-dimensional aspects of a ligand binding site and the study of the dynamic aspects of the interaction with specific ligands should result in understanding of the chemical reasoning behind specificity. In the case of enzymes, knowledge of the arrangement about a substrate binding site assisted in the interpretation of the mechanism of enzyme action (*4,5*). It is hoped that such information in the case of membrane-ligand reactivity will aid in our understanding of the mechanism of membrane transduction. Is there a common biological mechanism for energy transduction, i.e., proton translocation (*6*) or conformational transduction (*7*), or has nature actually devised a variety of mechanisms dependent upon the particular job to be done and the structure of the transducing machine evolved to carry it out?

2. *Allosteric Effector Sites*

In addition to a specific binding site that induces a membrane transduction, it is probable that membrane systems have as well secondary binding sites for effectors. Such sites, which might modulate the action of the binding site, can be considered to act much as allosteric effector sites on proteins, controlling and regulating the reactivity of membrane transduction. In excitable tissue the regulation and control of the variety of gating currents may be modulated by such regulatory sites. Experimentally, in the intact cell, as well as with isolated mitochondria, it is possible to measure a characteristic control in which the rate of oxygen utilization is experimentally a function of both the substrate being oxidized and the availability of a phosphate acceptor (*8*). The still unresolved molecular mechanism of such respiratory control may be partially explicable on the basis of the different nucleotide binding sites present on the F_1ATPase (*9*). Some of these sites may control the coupling of oxidation to phosphorylation; others may modulate this coupling. Some may as well modulate the interaction of the ATPase protein with its binding site on the mitochondrial membrane.

There may be yet undetected controls on the rate of electron flow within the electron-transport chain. The possible generation of local pH gradients at the site of interaction of two electron carrier proteins, such as cytochrome *c* and cytochrome oxidase, could result in a time-dependent modulation of the interaction of the two proteins, resulting in a physiological control of electron transport.

Such uncharted areas of investigation have within the past few years started to be investigated at the molecular level with the advantages of the photoaffinity labeling technique (*10,11*).

3. *Protein–Protein Interactions*

There is for all energy transducing systems the potential for modulation of concerted enzymic reactions by means of the interaction of one protein with another. It is clear from investigations on the resolution and reconstitution of energy-transducing systems that interacting membrane proteins need not have specific enzymic reactivities of their own, but may modulate the reactivity of a coreactant either with respect to direct ligand binding or perhaps by influencing a particular transition state of the interacting protein (*12*). One must consequently visualize the possibility of specific conformational states of protein that interact in a specific and reversible manner with another protein. The interaction between adjacent proteins of the respiratory chain, or the possible change from an opened to a closed conformation of a gating protein, might be considered potential representatives of such systems.

4. *Phospholipid–Protein Interactions*

A molecular interaction of import in bioenergetic systems, which will surely become an area for photoaffinity investigations, is that between phospholipids and specific membrane proteins. The experimental evidence indicating a boundary lipid for some membrane proteins, i.e., a specific organization of lipid surrounding integral membrane proteins (*13*) and the dependency of enzymic activity on both composition and quality of lipid associated with certain proteins (*14*), indicates clearly the importance of such lipid–protein interaction. The membrane as a lipid matrix is visualized as having a profound influence on the reactivity and physical state of the membrane's protein components. It is also obvious that the proteins that make up the membrane have a major influence on the structural and physical properties of the membrane.

With the chemical synthesis of photoaffinity probes of fatty acids and phospholipids (*15*), such specific protein–phospholipid interactions become capable of investigation at a molecular level. The investigation of the specificity of interaction of lipids and phospholipids with restricted regions of proteins will surely be an important aspect of such studies. However, other areas of investigation exist; there is the possibility of the creation of phospholipid vesicles containing specifically organized photoaffinity phospholipids. Incorporation of various proteins into such a matrix, followed by photoirradiation and reisolation of the protein component, can result in the identification of specific phospholipid binding sites on the proteins. This will as well yield information as to how various proteins are actually oriented within the fluid matrix of the membrane. In addition to such static studies, one can carry out dynamic studies involving photoinsertion reactions as a function of changes in temperature and physiological environment. While much information has been collected concerning the changes in phospholipid conformation within the membrane under varying experimental conditions (*16*), aspects of protein dynamics within membranes have to date received little direct attention.

In addition to the use of artificial membranes, one can introduce within natural membranes phospholipid or lipid photoprobes by subjecting appropriate mixtures to temperature variations about the phase transition temperature. Under carefully controlled conditions such incorporation could be specific and reproducible. Subsequent photolysis and isolation of protein components would reveal characteristics of phospholipid–protein interactions under more nearly "natural" conditions. Such "natural" but modified membranes could as well be utilized in investigations of ligand-induced dynamics of membrane proteins, with the endogenous photoaffinity phospholipid as a monitoring probe.

C. Fundamental Problems of Energy Transduction being Investigated by Photoaffinity Techniques

1. Transduction; Membrane Regions Responsible for Ligand Interaction

A review of a number of energy-transduction systems reveals that they presently present two common unanswered questions. The first question is one of specificity of interaction, which by itself is similar to asking about the mechanism of substrate specificity for enzymic reactions. The second question involves a less directly approachable concept: How do the specific ligand interaction and binding induce the transduction of energy? How is such ligand binding coupled to energy dissipation reactions? The current vogue of introducing protein conformational changes may offer a possible unifying explanation; however, within the lipid matrix in which the proteins of the transducing system find themselves may rest other yet unrealized means of explaining transduction. Reorientation of lipid matrix or of boundary lipid due to charge redistribution may be a potential mechanism. Is there by way of example a possible control of gating in excitable tissue via electrostatic charge differentiation within the lipid bilayers?

2. Oxidative Phosphorylation

The mechanism linking the oxidative or electron transport sequence to the phosphorylation reaction in oxidative phosphorylation is still an area of conflict and intensive investigation (*17*).

A photoaffinity analog of the classical uncoupling agent 2,4-dinitrophenol has been utilized in an attempt to localize the specific site for the uncoupling reaction(s) (*18,19*). The initial apparent success of this probe appears to offer a means of investigating the molecular mechanism for coupling. While this would be expected to be a revealing experimental direction, one of the major difficulties of the photoaffinity technique, that of nonspecific binding, complicates work with this probe. A cautious attitude is in order, especially in view of the hydrophobic nature of many of the reagents being used as photoaffinity probes and the hydrophobic nature of membrane systems.

A start has been made in the study of the interaction of mitochondrial F_1ATPase subunits with photoaffinity probes of the adenine nucleotides and their derivatives (*20–23*). Determination of the positioning of the different ligand binding sites on this ATPase may reveal the reason for the numerous nucleotide binding sites on the same protein. A comparison of the nucleotide labeling pattern of the enzyme when free (i.e., in solution) compared to that observed in its more native state (i.e., when present on the membrane and under different physiological conditions) may reveal essential characteristics of its role as the central component of mitochondrial energy translocation.

In this regard, the investigation of the allosteric sites of glutamic dehydrogenase with the aid of photoaffinity probes of GTP and ADP have revealed that, once bound, these analogs freeze the conformation of the protein into a particular state (*24*). A similar situation may occur with membrane-bound components in which the membrane, acting as an allosteric effector, influences reactivity of the protein by virtue of inducing a specific conformation. Such an influence would be accessible to investigation with the photoaffinity technique. The special aspect of "allosteric" membrane control at the level of protein membrane interaction has been given the term *allotopy* (*25*).

The availability of photoaffinity probes of the different pyridine nucleotides represent specific probes of components of the electron transport system. With Complex I preparations in combination with photoaffinity analogs of NAD^+, $NADP^+$, NMN, and 3-acetyl NAD^+, work is directed toward the total resolution and independent characterization of the NADH and NADPH dehydrogenase and the NADH and NADPH specific transhydrogenase activities (*26,27*).

3. *Organization of the Components of the Respiratory Chain*

One may envision the potential synthesis of photoaffinity analogs of the classical inhibitors of the respiratory chain, such as amytal and rotenone, as an aid in the understanding of the characteristics of respiratory chain electron transport. A photoprobe for the antimycin binding site has already been prepared (*28*).

One approach to the utilization of photoaffinity probes involves the synthesis of a specific photoaffinity protein (i.e., 5-azido-4,6-dinitrophenyl cytochrome *c*) which can be photoirradiated in the presence of its partner (cytochrome oxidase) within a multienzyme system (*11*). In this case one utilizes a protein component of an interacting system as a reagent directed toward a specific site on its interacting partner. Since the photoinsertion reaction is under the control of the investigator, covalent bonding of one protein to another can be induced as a function of the time of initiation of electron transport as well as a function of varying external physiological conditions (i.e., pH, temperature, ionic strength).

4. *Ion Flux through Excitable Membranes: Gating Processes*

Ion channels of excitable membranes consist of two functional components, which may or may not correspond to separate entities. There is an ion translocator that determines the type of ion having access to the channel, and a gate, which regulates the ion flow by opening or closing. In excitable tissues this gate is under the control of the membrane potential. One of the major experimental difficulties in the examination of the structure of these channels has been the lack of functional assays for their components following extraction from a functional environment.

The development of specific photoaffinity probes for the Na^+ and K^+ channels of excitable tissues provides for the possibility of resolution and reconstitution of these specific and highly reactive regions (*29,30*). In such studies, one photolabels the binding site with a radioactive marker and then subjects the photolabeled system to dissociation in order to resolve the binding components from accessory structures. Because the total number of potential binding sites is small, a major present problem in such studies is the amount of label one is able to incorporate into the analog on the basis of current synthetic methods. With tritiated tetrodotoxin, the number of binding sites on an axon membrane has been estimated as 30 per square micrometer (*31*).

The possibility of nonspecific binding in the case of the photoprobes used with these tissues has not been evaluated adequately and if present would pose a major problem, considering the low level of binding sites. Three types of drugs affecting different states of the sodium channel are available. Tetrodotoxin and saxitoxin block the open channel (*32*), and sea anemone toxin and scorpion toxin prevent inactivation of the channel (*33,34*). Substances such as grayanotoxin, batrachotoxin, and vertridin stimulate the opening of the channel. All three classes bind with high affinity to the axonal membrane but do not apparently compete for a common binding site.

Modification of the sea anemone toxin with photoaffinity adjuncts would appear feasible. On the basis of the accomplished chemical modification and required structural composition of the toxin for the maintenance of biological activity, such probes would probably retain their physiological activity.

As indicated above, one of the advantages of the photoaffinity technique is the ability to bring about modification of specific component parts of a complex multienzyme system as a function of time and under various metabolic states. Once the individual components of a multienzyme system are recognized, one can apply cross-linking reagents in order to evaluate the juxtapositioning of the component parts. The utilization of similar procedures under differing metabolic conditions would be a potential mean of understanding those factors influencing interactions within the multienzyme system.

The above considerations apply as well to the investigations of membrane transport. The coupling between biological energy utilization and transport is still much of a mystery; we do not know, by way of example, whether ligand binding sites for transport on membranes rest within static (immobile) regions or are mobile and positioned as a function of specific external conditions.

D. The Background to Photoaffinity Studies

1. *Affinity Techniques: Development of the Concept of Active Site-Directed Reagents*

Chemical experiments have been aimed at the elucidation of the mechanism of specific enzyme reactions by utilizing amino acid reagents directed toward particular side-chain residues. Inactivation by such "unique" group labeling reagents if shown to be associated with the modification of a single residue would then be considered as reacting at an active site. Inactivation was, however, often accompanied by the destruction of a number of residues, and the relationship of inhibition to interaction at the catalytic site was often uncertain. Competitive inhibitors, in contrast, while demonstrating clear interaction at specific binding sites, have dissociation constants often higher than that of the natural substrate and are consequently of little use for labeling studies (*35*).

The idea that potential amino acid modifying reagents and substrate analogs could be integrated into the same molecule, with active site binding potential directing a specific chemical modification (affinity labeling), was advanced initially by a number of laboratories (*36*). The general concept was to prepare compounds that were similar structurally to a substrate or product of a particular enzymic reaction and carrying a reactive grouping, most often electrophilic in nature. This compound would then be capable of utilizing the binding characteristics of the active site for its natural ligand, which would carry the reactive group to the vicinity of the binding site. If such an analog were capable of being bound at this site, then nucleophilic amino acid residues in the vicinity of the binding site would carry out displacement reactions resulting in covalent binding of the reagent. Subsequent degradation of the inactive protein would result in the identification of the modified residue and thus provide information as to the mechanism of the specific enzymic reaction.

It is now an experimental fact that the affinity-labeling technique can be applied to all molecules possessing a site that interacts with another with some affinity based upon specificity. Photoaffinity labeling represents a special category of the general affinity-labeling technique in which the reactive chemical grouping is replaced by an inert but photoactive reagent. The application of such photoaffinity probes to active-site labeling has recently received a great deal of attention, owing primarily to the pioneering work of Westheimer and his group, who in 1962 (*37*) initiated experiments utilizing diazoacetyl precursors of active carbene, and of Fleet *et al.* (*38*) in the application of nitrene precursors to the investigation of the interacting site for antibody recognition. These two initiating studies were instrumental in pointing out some of the advantages and weaknesses of the photoaffinity approach.

In the experiments of Singh *et al.* (*37*) *p*-nitrophenyl diazoacetate was allowed to react with α-chymotrypsin to form the monodiazoacetyl derivative of the enzyme. The *p*-nitrophenyl diazoacetate is both a carbene intermediate and a bifunctional quasisubstrate for chymotrypsin. The interaction resulted in acylation of a unique seryl residue in the active site of the enzyme and the liberation of the *p*-nitrophenol [Eq. (1)].

$$\text{Chymotrypsin-CH}_2\text{OH} + \text{O}_2\text{N}-\text{C}_6\text{H}_4-\text{O}-\overset{\text{O}}{\overset{\|}{\text{C}}}-\text{CHN}_2 \Downarrow \text{Chymotrypsin-CH}_2-\text{O}-\underset{\text{O}}{\underset{\|}{\text{C}}}-\text{CHN}_2 + \text{O}_2\text{N}-\text{C}_6\text{H}_4-\text{OH} \qquad (1)$$

The diazo group introduced into the enzyme as an adjunct of the acyl residue, when photolyzed, formed a reactive carbene free radical and liberated molecular nitrogen [Eq. (2)].

$$\text{Chymotrypsin-CH}_2\text{—O—}\overset{\overset{\text{O}}{\|}}{\text{C}}\text{—CHN}_2 \xrightarrow{h\nu} \text{Chymotrypsin-CH}_2\text{—O—}\overset{\overset{\text{O}}{\|}}{\text{C}}\text{—}\dot{\underset{\cdot}{\text{C}}}\text{H} + \text{N}_2 \qquad (2)$$

The formed radical has the potential to react instantaneously with a great variety of residues in its immediate vicinity. In the reported experiments, the initial acylation of the enzyme resulted in complete inactivation. However, upon photolysis 75–80% of the activity was recovered together with the release of glycolic acid, presumably from the interaction of the carbene with water [Eq. (3)]

$$\text{Chymotrypsin-CH}_2\text{—O—}\overset{\overset{\text{O}}{\|}}{\text{C}}\text{—}\dot{\underset{\cdot}{\text{C}}}\text{H} \xrightarrow{H_2O} \text{Chymotrypsin-CH}_2\text{—O—}\overset{\overset{\text{O}}{\|}}{\text{C}}\text{—CH}_2\text{OH} \qquad (3)$$

and subsequent hydrolysis of the ester bond (Eq. 4).

$$\text{Chymotrypsin-CH}_2\text{—O—}\overset{\overset{\text{O}}{\|}}{\text{C}}\text{—CH}_2\text{OH} \xrightarrow{H_2O} \text{Chymotrypsin-CH}_2\text{OH} + \text{HO—}\overset{\overset{\text{O}}{\|}}{\text{C}}\text{—CH}_2\text{OH} \qquad (4)$$

The remaining 20–25% of the enzyme was irreversibly inactivated. Hydrolysis of the modified enzyme revealed the presence of *O*-carboxymethylserine, which was postulated as being due to an internal rearrangement of the carbene and subsequent reaction with water [Eq. (5)], together with some 1-carboxymethylhistidine and *O*-carboxymethyltyrosine(*37,39,40*).

$$\text{Chymotrypsin-CH}_2\text{—O=CH—C=O} \Longrightarrow \text{Chymotrypsin-CH}_2\text{—O—CH}_2\overset{\overset{\text{O}}{\|}}{\text{C}}\text{—OH} \qquad (5)$$

Thus it was obvious that the major reactivity of the carbene species formed upon photolysis was, with water, presumably present within the vicinity of the active site.

Subsequent investigations by Vaughan and Westheimer (*41*) utilizing a ^{14}C-labeled diazomalonyl derivative of trypsin showed that irradiation resulted in the formation of [^{14}C]glutamic acid. This labeled amino acid was reasoned to have originated from carbene insertion into the methyl side chain of an alanine residue at the active site. This finding represented the first successful labeling of a hydrocarbon amino acid in a protein. Although the yield of the insertion product was of the order of 1–3%, it established the possibility of insertion into amino acid side chains previously considered to be unreactive.

Since these initiating experiments, a number of excellent reviews on the subject of photoaffinity labeling have been published and should be consulted for information on the interaction of such probes with enzyme systems. Knowles (*42*) has detailed the criteria for effective photochemical probes covering the literature up to 1972, and Creed (*43*) has done so up to the end of 1973. Cooperman (*44*) has outlined the photoaffinity labeling procedures and reagents developed up to 1977. The publication of a volume containing a variety of methods utilized up to 1977 for photoaffinity labeling experiments contains a wealth of material concerning specific applications and methodologies (*45*).

2. *Affinity Labeling: General Considerations*

a. Basic Theory: Ligand Binding. The basic theory for the method of affinity labeling was outlined by Wolsy *et al.* (*46*). In their analysis it was assumed that there was at the active site a particular residue capable of being chemically modified by appropriate reagents (R = active site receptor), and that there were in addition a number of other residues (r = nonactive site residues) on the molecule also capable of being modified by the same reagent. Reaction of the protein with a specific active site-directed ligand possessing an electrophilic reactive group X results in the initial formation of a noncovalent reversible complex, similar to the Michaelis complex of the natural ligand [Eq. (6)].

$$\mathrm{R} + \mathrm{L\text{-}X} \underset{k_2}{\overset{k_1}{\rightleftharpoons}} \mathrm{R \cdot L\text{-}X} \tag{6}$$

A secondary reaction by group X results in a covalent attachment depending upon the reactivity of X and that of the residue at the bonding site near X. With photoaffinity probes this secondary reaction would be light-induced [Eq. (7)] and represents the affinity labeling.

$$\mathrm{R \cdot L\text{-}X} \xrightarrow[h\nu]{k_3} \mathrm{R\text{-}X'\text{-}L} \quad \text{(affinity labeling)} \tag{7}$$

The covalent insertion reaction (k_3) at low ligand concentrations would be a first-order process, assuring specific attachment of the ligand at the reactive site R. Similar reactivity at secondary binding sites (r) would be independent of the ligand binding potential and would consequently be a second-order reaction [Eq. (8)].

$$\mathrm{r} + \mathrm{X\text{-}L} \xrightarrow{k_4} \mathrm{r\text{-}X\text{-}L} \tag{8}$$

One achieves an increased concentration of the reactive species X at the locus R utilizing the natural binding forces for L inherent in R. With a photoaffinity probe in addition to the active site-directed ability, one has the advantage of being able to decide at what point in time one wishes to bring about covalent insertion. This approach simulates the natural situation with a reagent that structurally mimics the natural ligand and thus preserves the specificity of interaction.

b. Analysis of Labeling Data. Previous reviews dealing with the labeling of the active site of proteins (*47,48*) were based upon electrophilic labeling and considered three primary qualities as being required for affinity labeling: (i) the reagent must form a noncovalent complex prior to covalent bond formation; (ii) the complex formed should mimic that formed with the natural ligand; (iii) covalent bond formation should be equal to the loss of the reversible binding sites of the natural ligand.

The second of these "requirements" is usually tested for kinetically by examining whether the natural ligand blocks either noncovalent binding or covalent incorporation of the affinity label. The third point is often difficult to demonstrate, since it requires that there be a reliable quantitative assay for natural ligand binding.

The ultimate object of such labeling experiments is the subsequent isolation of a peptide (binding region) distinguished by a large amount of label in comparison to all other peptides. Singer and his group have defined a useful term to which they have given the name enhancement (*49*). This represents the ratio of ligand bound to the R site as compared to ligand bound to r sites. In their analyses they arrive at four conclusions with respect to the specificity of the affinity labeling process: (i) The more dilute the solution in which the affinity labeling is carried out, the more nearly the equilibrium concentration of free sites is equal to the total concentration of active sites and the higher is the specificity of the labeling reaction. (ii) Specificity increases with increase in the equilibrium constant for the formation of the reversible complex between receptor and ligand. (iii) Specificity does not depend upon the rate constant for the formation of the covalent bound ligand (k_3), but depends on the ratio of k_3 to k_4, where k_4 is the rate constant for the formation of nonspecific product outside the active site. (iv) Under favorable circumstances the labeling of an active-site group of only average reactivity can occur with an enhancement of the order of magnitude of k_3.

Additional support for the postulate that the reactivity of the active group at the binding site that becomes labeled need not be an unusually reactive residue has been derived from the antibody studies of Singer and his group (*50*). Azo linkages to the tyrosol residues of antibodies are formed with the labeling reagent *p*-arsonic acid-benzene diazonium fluoroborate (Compound 1), and protection against this label is afforded by *p*-nitrobenzene arsonate. The studies show stoichiometric inactivation of antibodies and formation of azo linkages that follow first-order kinetics in the reversible complex.

N_2^+ — C_6H_4 — $As(=O)(OH)_2$ BF_4^-

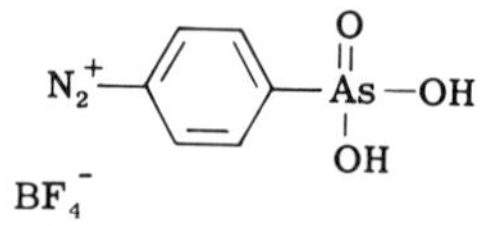

Compound 1:
p-Arsonic acid benzene
diazonium fluoroborate

c. In the Active Site or Near the Active Site? Considerations of Design of Photoaffinity Labels. Theoretical consideration for photoaffinity labeling, as with general affinity labeling, indicates that at least four criteria must be satisfied in order to assure that the design of an active site-directed irreversible inhibitor will function in the expected manner (*51*). (i) It is necessary to maintain specificity of binding interaction with the particular binding site; i.e., the reagent must contain structural elements required for binding at the active site. A high local concentration of reagent at the binding site ensures preferential reaction. (ii) The reagent should not react too rapidly with water or with protein amino acids in order not to nullify the effect of requirement i. (iii) The reagent should be designed with a knowledge of the enzyme's probable mechanism. (iv) The intracomplex reaction that labels the receptor covalently following formation of the reversible complex should be indiscriminatory.

Such considerations were used by Jeng and Guillory (*52*) in their search for a general reagent that might be utilized in a comparative study of the nucleotide binding region of a variety of energy-transducing systems. It was realized that a valuable approach to the study of the mechanism of the myosin ATPase was the prior utilization of a large number of structural analogs of ATP. The study of the interaction of such analogs with the myosin protein resulted in the accumulation of rather specific information concerning those portions of the adenine nucleotide molecule that are important for proper orientation with the contractile protein. Such studies allowed for an indirect assessment of the components of the nucleotide necessary for nucleotide binding (*52*).

The ATP molecule contains three distinct structural groupings: the adenine base, the ribose sugar, and the triphosphate component. On the basis of the interaction of actomyosin with ATP analogs modified in these positions, a model of the enzyme–substrate complex has been postulated

(*53*). This model is based upon evidence suggesting that the amino grouping at the 6 position of the adenine base and the 3′ hydroxyl of the ribose ring are hydrogen bonded to the protein. The nucleotide binding is thought to involve a specific NH group as well as conjugation of triphosphates with asparagine via metal interaction (*54*) and binding of the terminal pyrophosphate to a sulfhydryl and guanidinium group at the active site (*55*). An important aspect of the actomyosin-nucleotide analog studies was the finding that the ribose ring can be drastically modified without affecting markedly the rate of myosin-catalyzed hydrolysis. Such considerations have been reexamined and confirmed (*56*).

In view of these facts it was decided to develop methods by which an active photosensitive adjunct could be attached to the ribose portion of the nucleoside triphosphate. It was anticipated that coupled to substrate specificity would be the possibility of a controlled labeling of the active site by the photogeneration of an extremely active chemical species.

The anticipated properties of the synthesized reagent 3′-*O*-{3-[*N*-(4-azido-2-nitrophenyl)amino]propionyl} adenosine-5′-triphosphate (Compound 2)

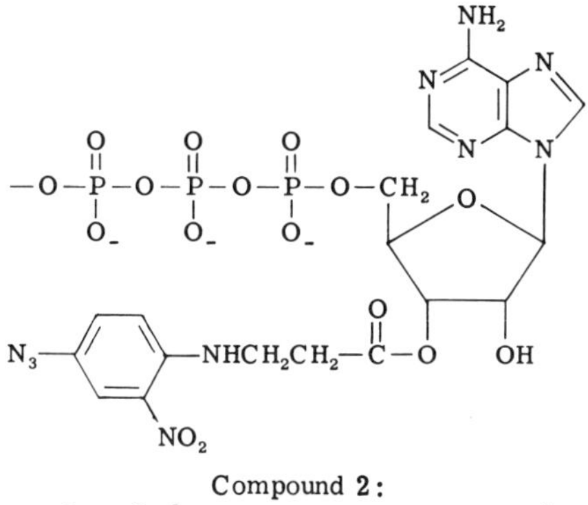

NH_2 N N N N O O O $-O-P-O-P-O-P-O-CH_2$ O O^- O^- O^- O N_3 $-NHCH_2CH_2-C-O$ OH NO_2

Compound **2**:
3′-*O*-{3-[*N*-(4-azido-2-nitrophenyl)-]propionyl}adenosine-5′-triphosphate; arylazido-β-alanyl ATP

were substantiated with the demonstration of the photodependent inhibition of subfragment (S_1) ATPase, an almost stoichiometric covalent binding of the analog to the ATPase protein, and a dark reaction in which the analog underwent a myosin-catalyzed dephosphorylation.

A question arises whether the group on the enzyme that reacts with a photoactivated reagent is within or simply "close to" the active site. If the insertion group is small, one might expect that it could be so positioned as to be encompassed within an active-site volume. This involves considerations of a distinction between a binding-site and an active-site region, two positions that need not necessarily be identical. The enzymic reactivity of the arylazido-β-alanyl ATP analog and the distance between the cleaved pyrophosphate unit and the aryl group as revealed by space-filling models make it clear that the insertion reaction with myosin is probably not at the catalytic site. Insertion directly at a ligand binding site in the case of such arylazido analogs is most probably an impossibility.

Such specificity, however, need not be required, nor is it necessarily desirable. A great deal of information concerning the three-dimensional topographical structure of the binding site can be theoretically obtained by having the insertion reaction take place at a number of positions. This would be dictated by the proximity of the particular residue to the arylazido reagent at the moment of activation. Such insertions would presumably be equidistant (as a first approximation) to the region at which the ligand is initially noncovalently bound. With sufficient variations a three-dimensional structure of the binding site could be reconstructed on the basis of the presumed sweep of the azido adjunct within the binding-site region (*52*).

Such an analysis as a function of the temperature at which insertion took place might be revealing with respect to residue mobility within the binding site. As long as the specificity of the insertion reaction under given experimental conditions is assured, such studies on the mapping of a ligand binding site need not require complete saturation of the site.

Uniform variations in the distance between the binding site and the point of insertion can be theoretically accomplished by the proper design of a family of affinity probes. In the case of the arylazido ATP analogs synthesized by Jeng and Guillory (*52*), variation in the distance of the azido group from the 3′ hydroxyl of the nucleotide ribose has been obtained by the synthesis of arylazido-4-aminobutyric ATP (Compound 3) and arylazido-6-aminocaproic ATP (Compound 4) in addition to arylazido-β-alanyl ATP. All the analogs were found to inhibit myosin ATPase activity, the extent of photoinactivation being dependent upon the concentration of the analog and

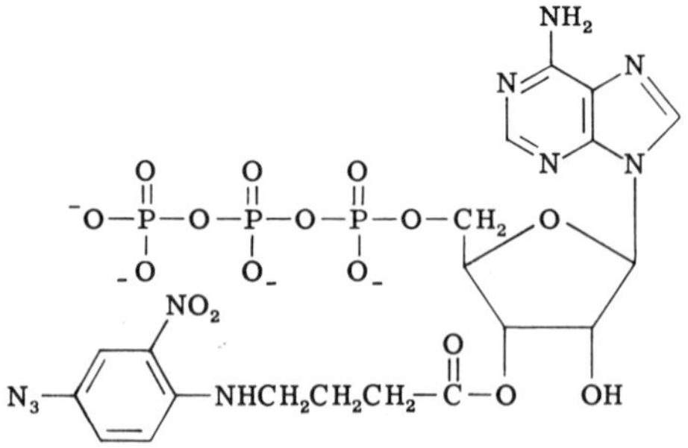

Compound 3:
Arylazido-4-aminobutyric ATP

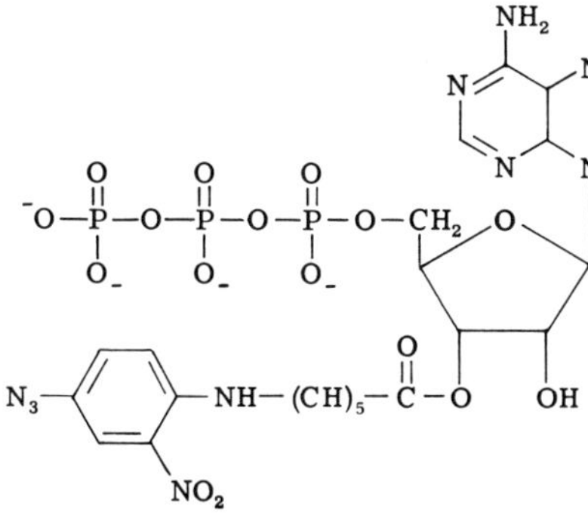

Compound **4**:
Arylazido-6-aminocaproic ATP

showing saturation kinetics (*52*). The availability of reactive analogs with such variations in the distance of the azido group from the 3′ hydroxy position is anticipated to be a powerful tool in the mapping of nucleotide binding regions of ATP dependent proteins.

With respect to interaction with a ligand, two major classes of amino acid residues on protein have been defined (*51*): contact and noncontact residues. A demarcation set at 2 Å distance between a residue and the reversibly bound ligand defines a contact residue. With respect to the three-dimensional attitude of proteins there are a number of potential contact positions, which might not necessarily be on the same polypeptide chain. Thus we can have contact residues on the continuous sequence and on proximal sequences of the same polypeptide chain or on remote sequences of two different chains. Such considerations are important with respect to photoaffinity labeling in which the insertion reaction can in fact take place at a position remote, in terms of amino acid sequence, from the actual noncovalent binding region. Covalent insertion can thus take place with a contact residue playing a role in either the binding or catalytic event. This may be remote from a linear peptide sequence representing the principal binding region.

The noncontact residues are defined as being at a distance in excess of 2 Å from the ligand binding site. With such proteins as myosin ATPase, one of the principal binding areas is functioning at the nucleotide base while enzymic reactivity takes place at the pyrophosphate locus removed from the principal recognition site for binding. Consequently, a contact residue need not be situated at the region of the active site. In addition, residues removed from the binding site can still be of importance to the functioning of the active site by interacting directly with contact residues, by intramolecular transfer, by providing a particular microenvironment within the catalytic or binding site, or by maintaining the proper conformation for contact residues.

The active site and the region defined as the binding site may undergo, in addition to normal variations in the relative positions of residues created by thermal energy, larger shifts in the position of residues due to varying conformational states. As pointed out by Singer (*51*), the static and rigid representation of the active site (and thus also the binding site) is a limiting case, and there are dynamic aspects for this region—aspects that would influence the interpretation of photoaffinity labeling. The conformation of the active site can differ somewhat with the binding of different ligands. A modified ligand such as an affinity label, while being bound and even while exhibiting "proper" biological functions of activation and/or inhibition of enzyme activity, may in fact label a particular residue brought into the vicinity of the binding site. Such a residue could conceivably be distant from the binding site in the presence of the natural ligand. One must consequently utilize, as suggested, the concerted application of a variety of labeling methods or reagents in order to give some degree of confidence in the "mapping" of the three-dimensional structure of the binding site.

In the testing of active-site labels, one normally thinks in terms of inhibition of enzymic activity. As a measure of effective covalent insertion, inhibitory action is dependent upon the distance separating the catalytic region from the point of insertion. With analogs having a sufficiently large distance between these two, one might expect to see a loss in the inhibitory potential for the analog. Such reagents would nevertheless be important in a comparative study to ascertain the three-dimensional aspects of the binding region.

In addition to designing a labeling reagent so as to allow for steric complementarity with the active site, one must consider as well the design in view of what may be the three-dimensional aspects of the binding site.

3. *Kinetic Analysis of Active Site-Directed Irreversible Inhibitors: Application to Photoaffinity Probes*

There are kinetic methods to test whether a particular reagent is truly a photoaffinity, active site-directed reagent. Such kinetic "proofs" are similar to those utilized for general affinity labels. On the basis of assumed steady-state conditions, Meloche (*57*) has derived kinetic expressions for enzyme active site-directed inactivator complex formation and for substrate inactivator competition. In the specific case cited, the reaction of monobromopyruvic acid with 2-keto-3-alkoxy-6-phosphogluconic aldolase was tested. The kinetic analysis was used to show competition between substrate (pyruvate) and inactivator (monobromopyruvic acid), one of the criteria for active site-directed inhibitors.

Meloche (*57*) assumed that the enzyme and bromopyruvate formed a complex prior to inactivation, inactivation being a result of alkylation by monobromopyruvic acid [Eq. (9)].

$$E + I \underset{k_{-1}}{\overset{k_1}{\rightleftharpoons}} C_{inact} \xrightarrow{k_2} E_{inact} \tag{9}$$

In the case of photoaffinity labeling, the reactions leading to the formation of the inactivated enzyme (E_{inact}) would be light-dependent. It was assumed that the concentration of the inhibitor enzyme complex (C_{inact}) was in a steady state so that $dC_{inact}/dt = 0$. In addition, it was assumed that the total enzyme concentration could be represented by a sum of three concentrations: that for unreactive enzyme [E], that for enzyme-inactivator complex [C_{inact}] and that for inactivated enzyme [E_{inact}]. It was also assumed that the inactivation velocity, v_{inact}, at finite inhibitor concentration is equal to the product $k_2[C_{inact}]$. V_{inact} is the inactivation velocity when [I] is infinite. From these assumptions the derived rate equation [Eq. (10)] was

$$v_{inact} = \frac{V_{inact}}{(K_{inact}/[I]) + 1} \tag{10}$$

where K_{inact} is $(k_{-1} + K_2)k_1$ representing the concentration of affinity inhibitor giving half-maximal inactivation, i.e., half-saturation of the enzyme. The time period in which inactivation is pseudo first order can be taken as representing initial rates. With aldolase the time required for the inhibitor to cause a 50% loss of enzyme activity, i.e., the inactivation half-time (τ), was proportional to $1/v_{inact}$ and T was proportional to $1/V_{inact}$.

The inactivation half-time (τ) is difficult to define with photoaffinity inhibitors, at any rate with the photolytic methodology being utilized in most laboratories. During a prolonged period of irradiation a certain proportion of the photoaffinity inhibitor is degraded to photolytic products. The reagent that is free in solution is thus converted to reagents that, while no longer capable of binding covalently to the specific binding site, can act as competitive inhibitors. Thus one has a complex kinetic system almost from the initial point of irradiation. Cosson and Guillory (*58*) have investigated the use of arylazido-β-alanyl ATP as a photoaffinity label for the isolated and membrane-bound mitochondrial ATPase complex. In their experiments photoirradiation of the samples were carried out using the flash generated by a xenon lamp with flashes of less than 1 msec duration. In addition to preventing heating of the solution, the short flash duration allows for control of the experimental procedures. When arylazido-β-alanyl ATP was added in several aliquots, each addition being followed by photoirradiation, the

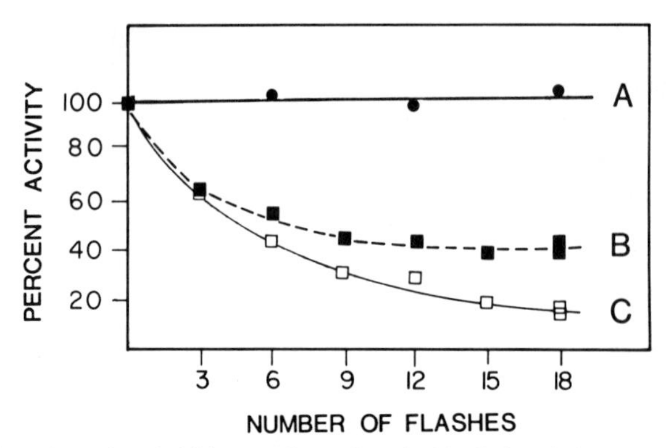

FIG. 1. Photodependent inhibitory effect of arylazido-β-alanyl ATP accomplished by a successive addition procedure. During photoirradiation submitochondrial particles were at 1.55 mg of protein per milliliter in 25 mM Tris-SO_4, pH 7.5. After irradiation of a specific interval as indicated by the number of flashes on the abscissa, an aliquot of the irradiated mixture was tested for ATPase activity and a comparison made with the activity of a dark control taken to represent 100% activity. Under condition A, no arylazido-β-alanyl ATP was present during irradiation; for condition B, the nucleotide analog was added to 15 μM and the mixture was subjected to a 3-flash irradiation. An aliquot was then taken and kept in the dark for subsequent ATPase determinations while the remaining mixture was irradiated again with sets of 3 flashes up to a total of 18 flashes. Under condition C, arylazido-β-alanyl ATP was added at 2.5 μM and the mixture was irradiated with 3 flashes. After an aliquot was taken for analysis, an additional 2.5 μM nucleotide analog was added, the mixture was subjected to an additional 3 flashes and so on up to 18 flashes, giving a final arylazido-β-alanyl ATP concentration of 15 μM. Control: ATPase, 2.55 μmol P_i per minute per milligram of protein. From Cosson and Guillory (*183*).

efficiency of the photodependent inhibitor was increased over that system in which the adenine nucleotide analog was added in a single addition and the mixture then subjected to the same number of irradiation flashes (Fig. 1). In the latter case, from the initial flash the free photolyzed product of arylazido-β-alanyl ATP competes with the nonirradiated arylazido-β-alanyl ATP. A lower concentration of this product permits an increased efficiency of interaction of the additionally added nucleotide. With the successive addition procedure one can evaluate a linear region along the percentage of inactivation curve, which could be used for kinetic evaluations. In this representation the time coordinate would be replaced by the number of flashes.

In the Meloche representation, substitution of the values for the half-inactivation of the rate expression yields Eq. (11),

$$\tau = \frac{1}{[\mathrm{I}]} \cdot [T \cdot K_{\mathrm{inact}}] + T \qquad (11)$$

from which a plot of τ against $1/[\mathrm{I}]$ yields a straight line with intercept at T, the minimum inactivation half-time at infinite inactivation concentration.

The data observed indicated that inactivation obeyed saturation kinetics and that the inhibitor and enzyme form a dissociable complex prior to inactivation.

Steady-state kinetics were also used to derive an equation for competition between substrate and inactivator under conditions of no product formation, i.e., a single substrate for a two-substrate enzyme (one could perhaps consider in some cases using the same relationships with nucleotide-independent reactions in the absence of Mg^{2+} or other activators) [Eq. (12)],

$$E + S \underset{k_{-3}}{\overset{k_3}{\rightleftharpoons}} C \tag{12}$$

where S is a single substrate and C represents the enzyme substrate complex. The sum of all the enzyme forms is given in Eq. (13).

$$[E_t] = [E] + [C_{inact}] + [C] + [E_{inact}] \tag{13}$$

and the rate equation derived is Eq. (14),

$$v_{inact} = \frac{V_{inact}}{K_{inact}/[I] + K_{inact}/[I] \cdot ([S]/K_S) + 1} \tag{14}$$

where $K_S = k_{-3}/k_s$ represents the apparent dissociation constant of the enzyme substrate complex. The linear form of this expression expressed as half-time is Eq. (15),

$$\tau = 1/[I] \cdot T(K_{inact} + K_{inact}[S]/K_s) + T \tag{15}$$

which states that in a plot of τ vs $1/[I]$ at constant substrate levels, straight lines should result whose slopes are determined by the substrate concentration and whose lines should intercept the ordinate at the minimum inactivation time. From the slopes of the lines one can calculate the K_s value for the substrate. Application of this relationship to the kinetics of the aldolase reaction provided kinetic evidence that both pyruvate and bromopyruvate compete for the same site on the enzyme (*57*).

Further evidence for competition between the two reagents was obtained by measuring inactivation half-times at constant [I] and varying substrate concentration. Rearrangement of Eq. (15) results in the following expression [Eq. (16)].

$$\tau = [S] \cdot T \cdot 1/[I] \cdot (K_{inact}/K_s) + (T + T \cdot K_{inact}/[I]) \tag{16}$$

From a plot of τ vs [S] a straight line results whose slope is proportional to $1/[I]$ and whose intercepts are influenced by the inhibitor concentration: $T + T \cdot (K_{inact}/[I])$. The data obtained provided kinetic evidence that, in addition to pyruvate and bromopyruvate competing for the same site on the enzyme, the substrate acts as a competitive inhibitor of inactivation and inactivation obeys saturation kinetics. Both conditions are necessary criteria for active-site labeling.

E. The Utility and Limitations of the Photoaffinity Technique

1. Criteria of Active-Site Specificity

As indicated in the above kinetic analysis, two necessary criteria for active-site labeling are that the substrate act as a competitor against inactivation and that the inactivation obey saturation kinetics.

The requirement that inactivation obey saturation kinetics and show pseudo first-order reactivity is not often experimentally achieved. In many systems inactivation kinetics have been described as the summation of two different rates representing multiple-site interaction. With photoaffinity probes such multiple reactivities may in some cases be explained on the basis of a single-site reactivity with the nonbound photoaffinity product competing for the binding site with the still active photoaffinity probe.

In addition to the above, two added criteria are important in the assessment of active site specificity: complete inactivation and the demonstration that covalent incorporation into the binding site is stoichiometrically related to the number of binding sites as deduced from other criteria.

Complete inactivation is often not achieved for a variety of reasons, especially with the use of photoaffinity labeling. Depending upon the type of photoaffinity probe and the purpose of the labeling experiment, this may or may not be an important consideration. If the possibility that the presence of a modified reagent is responsible for incomplete inactivation has been ruled out, there are a number of other interpretations possible: (a) there may in fact be two or more interacting sites, one of which influences the catalytic activity of the others; (b) during inactivation a small percentage of the enzyme may be converted to a form that reacts with the inhibitor slowly; and (c) the enzyme may catalyze a secondary reaction with the inhibitor itself.

While complete inactivation should occur when the number of moles of label attached per mole of protein equal the number of active sites per protein molecule, one has to be careful about the interpretation of loss of activity. A reduction in the rate of an enzymic reaction may result from an increased k_m of a modified enzyme for the particular substrate being used (*59*). Such a reduction in rate may be due to partial impairment of all the active sites or to a relatively complete impairment of a fraction of them. One may have as well steric interference without modification of contact residues. In addition, attachment of a label to a contact residue if small may not interfere appreciably with the binding of a specific ligand.

Covalent incorporation into protein in a stoichiometric manner, related to the number of binding sites deduced by other criteria, is often used as a vindication of active site-directed reactivity. Also, in the absence of specific evidence concerning active-site numbers, photoaffinity insertion can be used to distinguish between a number of alternatives. In order to be completely assured of the authenticity of active-site labeling, it is important to show that inactivation is proportional to the amount of label incorporated.

A method that is useful in assuring specificity of labeling is that of differential labeling, a procedure based upon the protective effect of natural substrates or reagents with a substantially lower binding constant than that of the labeling reagent being used. In this procedure the enzyme is pretreated with a nonlabeled analog in the presence of a large excess of substrate or competitive inhibitor. Complete modification by the reagent of all reacting residues on the outside of the active site is achieved in this first stage of the reaction. The protective agent is then removed, and a radioactively tagged form of the photoreagent is added. It is assumed that this now reacts preferentially with residues at the active site previously protected.

The differential labeling technique can be utilized to label specific reactive sites in complex mixtures. Cohen and Warringa (*60*) utilized it for the labeling of cholinesterase in a crude enzyme preparation. Protection of cholinesterase was afforded by butyrylcholine during labeling of nonspecific centers by DFP. After the reaction the mixture was dialyzed and ^{32}P-labeled DFP added to specifically label the now unprotected seryl residues at the active site of the true cholinesterase.

It is, of course, not possible to obtain stoichiometric inactivation in the second-stage reaction if exhaustive modification of nonspecific groups has not been achieved in the first stage. In such experiments where dialysis or other separative methods are used to effect release of the protecting agent, caution should be exercised in assuming that there is complete clearing of proteins of bound ligands under what might be considered reasonable conditions. Silhavy *et al.* (*61*) have shown that during dialysis proteins are able to retain bound ligand, especially when the protein concentration is in large excess over the total ligand concentration. This retention effect is considered to occur whenever high local concentrations of binding sites exist in a volume where boundaries allow for diffusion of the ligand, as during chromatography, electrophoresis, or ultacentrifugation of proteins. The theoretical considerations were verified by showing the retention of ligand by the periplasmic maltose-binding protein of *Escherichia coli*.

A modification of the differential labeling method has been developed by Pressman and Roholt (*62*). In their approach, protected and unprotected samples of protein are allowed to react with either one of two tagged forms of the same labeling reagent; the two samples are then mixed, and the protein is degraded to peptide fragments. Upon resolution of the peptide fragments, the ratio of the two tagged labels in each peptide fragment is determined. Peptides from the active-site region should show an elevated ratio of labeling in nonprotected peptides: protected peptides. There was, however, a depressed ratio with some peptides, for which there is no clear explanation.

The protective effect utilized in differential labeling is not always absolute; the closeness of the fit of the protector to the active site, the size of the modifying reagent, possible extensive modification outside the active site region—all may influence the conformation at the active site. The protecting agent itself may influence the active-site conformation. With all considerations it is important to demonstrate stoichiometric inactivation with differential labeling.

An indication that two linearly remote portions of a single polypeptide chain are in close array has been found in the experiments of Crestfield *et al.* (*63,64*) and Henrikson *et al.* (*65*). In these experiments the active site of bovine pancreatic ribonuclease was labeled with iodoacetic acid. The iodoacetate at pH 5.5 labeled histidyl 12 in the imidazole ring position 3 and histidyl 119 in position 1. Both were specifically modified, but on different ribonuclease molecules. The ratio of the two residues modified was found to be 1 : 8. Both active sites were specifically affected by the reagent, thus fulfilling the criteria of stoichiometric inactivation. The two histidyl residues reacted at different rates with different unbranched α and β haloacids. Binding is predominantly at one or the other histidyl residue, depending upon the chain length, the position of halogen relative to carboxylate, and the optical configuration at the carbon of labeling reagent.

From the above considerations it is clear that in the analysis of labeled peptides one obtains an understanding of the linear amino acid sequence in the vicinity of a labeled residue. Such one-dimensional information must then be translated into a three-dimensional structure. Labeling with photoaffinity reagents permits the controlled simultaneous labeling of two or more residues within an active site.

2. *Advantages of the Photoaffinity Technique*

Most chemical reagents used to modify proteins are designed to be weakly electrophilic in order to avoid being rapidly consumed by water, and consequently they react preferentially with nucleophiles at active sites (*47,48*). Knowles (*42*) and Bayley and Knowles (*66*) have discussed the advantages of photoaffinity labels in comparison to other active site-directed labeling methods and have pointed out that there are no *a priori* reasons to expect reactive nucleophilic centers at ligand binding sites. It is clear that for labeling of active sites one requires a more general reagent as typified by the photoaffinity probes. The great utility of affinity labeling is enhanced by photoaffinity techniques because of the generality of the insertion reaction, which decreases specificity requirements for covalent binding. On the other hand, one may consider that, owing to their general and nonspecific interaction, the photogenerated species has a great potential to act with the solvent system. If the photoaffinity probe exhibits a strong binding potential to the active site or binding site, the degree of solvent reactivity may be unimportant. Such considerations are dependent upon the topographical relationship of the binding site and the proximity of the photogenic species amino acid residues within the binding site. As a first approximation one must assume that the activated portion of a photoaffinity probe must be within 2 Å distance of a residue with which it will react (*51*). If a water or solvent molecule is centered between it and a potential reactive amino acid residue, one would not expect insertion to take place.

Nevertheless, the finding that in a number of cases apparent insertion takes place in a rather indiscriminate manner clearly indicates the advantage of photoaffinity probes over electrophilic affinity labeling. It is this apparent success that is responsible for the current widespread use of photoaffinity labeling.

3. *Criteria That an Affinity-Labeling Reagent Must Meet in Order to Adequately Do Its Job*

The general criteria that give to photoaffinity probes their special advantage have been summarized by Knowles (*42*).

a. The Reagents upon Irradiation Give Rise to Reactive Species When Situated at the Binding Site. The active species should be generated while noncovalently bound to the receptor. In the design of photoaffinity probes, the total binding-force energy is generally considered to be sufficient to allow for some flexibility in the modification of photoaffinity probes from substrates or from competitive inhibitors (*42*). Since an appreciation of the composition of the binding site for a particular ligand is not always at hand, the closest approximation is the structure of the substrate. The maintenance of enzymic reactivity of a substrate following modification with a photoaffinity adjunct is probably the best portent for success with the photoaffinity probe. However, even in cases where modification of the structure of a natural ligand does not sacrifice either selectivity or binding interaction, one must still be concerned with the position of the photolytic portion relative to the topology of the binding site. If the photoaffinity probe is directed away from the binding site and into the solvent phase, one would expect no covalent insertion to take place.

b. The Active Photogenic Species Must Have a Short Lifetime. The short lifetime is necessary in order to allow for the covalent reaction at the site(s) of generation to take place at a rate comparable to equilibrium among the sites. Classical reactive groups have lifetimes that probably are generally long compared with the time required for enzyme molecules to interconvert between different conformational states. With such reagents there is always a question whether a major or minor conformational form of the enzyme is labeled, i.e., is a typical or nontypical state reactive with the reagent. With photoaffinity probes this is perhaps less of a problem. In order to assure this upon photolysis, the intermediate product should not rearrange to something less reactive. This latter point is an important consideration with carbenes and alkyl nitrenes. With respect to the chemical reactivity of the labeling reagent, a rather broad range of reactivity can be tolerated. This is because the specificity with which labeling of the active site occurs is not a matter of the absolute value of the rate constant for insertion, but of the ratio of the rate constant for insertion at the active site to that for nonspecific reactivity (*51*).

The practical upper limit on the reactivity of a reagent is determined by the relative rate of its reaction with a residue within an active site as compared to its rate of reaction with solvent. The lower limit is usually determined by the requirement that its reaction with a residue in the active site occur under conditions that preserve the natural structure of the protein. If this latter condition does not hold, the operational criteria of stoichiometric inactivation and protection cannot be employed.

c. Photoaffinity Reagents Are Much Less Selective toward Covalent Bond Formation. Carbenes and nitrenes are the principal reagents utilized up to the present for photoaffinity studies in biological systems. These can be produced photolytically, are much less selective toward covalent bond formation than electrophiles (*67*), and are very reactive and capable of carbon–hydrogen bond insertions. The property of thermolytic production of these free radicals has not been extensively utilized as yet in biological systems. By appropriate reagent design, the photolytic activation is potentially capable of being carried out away from damaging irradiation wavelengths.

Nitrenes are the nitrogen analogs of carbenes (*68*). They are neutral compounds of univalent nitrogen. The arylnitrene-generating reagents have been suggested as being the most ideal photoaffinity analogs for use in the study of the active sites of enzymes (*38,70,71*).

In addition to direct carbon-hydrogen bond insertion, carbenes and nitrenes are able to participate in abstraction reactions (normally of hydrogen from carbon) cycloaddition, attack by nucleophiles, and rearrangement reactions (*68*).

4. *Limitations of the Photoaffinity Labeling Methods*

The principal problem with the photoaffinity technique revolves about the question whether the labeling patterns obtained are a reflection of the actual native structure of the protein or a representation of an altered structure.

a. Low Yields of Photodependent Insertion. In comparison with electrophilic affinity labeling, photoaffinity labels have in the past been associated with low yields. This reflects the problem arising from the reaction of the photogenerated species with solvent. Methods have been used in attempts to increase such yields to stoichiometric levels (*72,73*).

In the case of arylazido-β-alanyl NAD^+, Chen and Guillory (*26,27*) have shown that photolysis of yeast alcohol dehydrogenase in the presence of the nucleotide analog results in an irreversible binding of the analog to the enzyme. The stoichiometry indicated 1.9 mol of the azido analog to be covalently bound to 1 mol of the enzyme associated with a 91 % inhibition of enzymic activity. In another study the photoinactivation of myosin subfragment 1 and myosin with arylazido-1-[^{14}C]-β-alanyl ATP resulted in the labeling of the protein with approximately stoichiometric amounts of radioactivity (*52,74*). Recent experiments with heavy meromyosin substantiated the initial data. In a third study the arylazido-β-alanyl ATP was found to bring about a photodependent inhibition of mitochondrial F_1ATPase and an associated specific covalent labeling of the soluble enzymes. About 0.8 μmol of arylazido ATP analog was bound per micromole of F_1(ATPase) protein associated with 80–90 % inhibition of enzymic activity (*20*).

In each of the above referred to studies, the photoaffinity analogs are able to act as substrates for the respective enzymic activities. The stoichiometric labeling pattern may consequently represent an indication of a very specific initial interaction of a noncovalent nature, which so orients the arylazido that direct insertion is the only course open for the bound analog. One may consider that a certain similarity of the nucleotide binding sites exists in these three enzymes.

b. Photosensitized Destruction by the Photolabile Ligand. There is the possibility that destruction of a particular receptor site can take place even when irradiated at wavelengths where the receptor itself does not absorb. This could be ascribed to the possibility of photosensitized destruction by the photolabile ligand. Under such conditions the ligand photoaffinity binding site would be modified and not representative of the natural state of the protein. It has been suggested that this possibility could be checked by investigation of the labeling patterns as a function of the activating light intensities. Presumably a light-induced conformational change in the receptor would be reflected by changes in the labeling pattern. To date no complete studies of this type have been reported.

c. Elucidation of the Photochemical Mechanism Giving Rise to Incorporation. In order to examine the photochemical mechanism that gives rise to insertion, a knowledge of the properties of the labeling species is absolutely necessary. Unfortunately, most receptors are complex, and so are the photolysed reaction products. This is especially true with the arylazido analogs, which appear to undergo multiple forms of degradation dependent upon the type of light used for irradiation (unpublished observations, R. Guillory). Thus, diversified means of activation of photoaffinity probes could result in significant differences in labeling due to potential different intermediates in the activation process.

In spite of statements to the contrary, it is still possible that with azido interaction there may be potentially enhanced reactivity of specific groups toward photogenic insertion. There is presently not enough complete work accomplished to test this possibility. An interesting report concerning the reduction of arylazides by a nonphotolytic reaction involving thiols has appeared (*75*). In this study, the pH dependency of the reduction reaction (maximum at pH 10.0 with a calculated reduction of less than 1% at pH 7) indicates little possibility of such reactivity under normal conditions of photolytic experimentation. However, our uncertain knowledge of the environment of the photoaffinity group when bound noncovalently (i.e., its hydrophobic nature or the presence of sulfhydryl groups) makes such nonphotolytic reactivity potentially possible. Complete work with model systems is obviously of prime importance in this field. One consideration deserving special attention is the possibility that residues at the active site may have a depressing rather than an enhancing reactivity toward photodependent insertion.

d. Specific vs. Nonspecific Reactivity. A difficulty with photoaffinity labeling is the inability to distinguish between nonspecific labeling and that due to binding at specific loci. Such problems can be treated in a number of ways, of which the differential labeling technique discussed above is one method. The best security at present against nonspecific interaction is a rapid insertion, as is approached with analogs having strong binding characteristics to the ligand site. The possibility is always present that the label covalently binds to a site different from the "true" receptor site and blocks natural ligand binding by steric interference or by allosteric effects.

With general affinity labeling, stereochemical restrictions favor the use of highly reactive labeling reagents. The more reactive, generally, the less selective is the reagent and the wider the spectrum of amino acid residues with which it can react under physiological conditions. With photoaffinity labeling the instantaneous reactions of the free radical in an active site may have what Singer (*51*) has described as a "shotgun" effect; that is, a number of different residues in the vicinity may be simultaneously modified. The diversity of reaction products makes it difficult to evaluate the results of such labeling experiments through peptide fragment studies. On the other hand, with respect to attempts at working out the binding site topography such multiple sites of insertion may be of aid. In such cases structural knowledge from X-ray studies could be of significant help. Again, however, caution concerning any interpretation is necessary, since the differential labeling may result in an altered conformation of a particular site.

F. Types of Photoaffinity Reagents in Current Use

In general the probe to be used would be dependent upon the particular system to be examined. The underlying fact is that only two photochemically active species have the physical, chemical, and visual characteristics enabling them to be utilized effectively in such systems. Consequently, the use of photoaffinity probes to explore the active site for energy transduction has up to the present time been concentrated on these two types of analogs, those which upon photolytic reaction result in the formation of either a carbene (α-diazocarbonyles) or nitrene (arylazides) free radical. Both photoderived species display broad spectra of reactivity including insertion to CH bonds; both can as well be generated while noncovalently bound to receptors if equilibrium is allowed to proceed prior to photolysis.

None of the photolabile reagents are without some inherent problem; however, they are in most cases superior to electrophilic reagents, and good prospects exist for improvement through studies correlating selectivity and reactivity with structures.

1. Carbenes

Carbenes are compounds containing a carbon atom with two nonbinding (i.e., free) electrons. Because of the free electron pair, carbenes are in general very active; for example, methylene ($:CH_2$), the simplest carbene, reacts with the various carbon hydrogen bonds in *n*-pentane with little selectivity and at $-75°C$ (*76*).

Carbenes can be generated by the photolysis of diazoalkanes, ketenes, diazirines, and α-ketodiazo compounds with concurrent loss of nitrogen. Carbene itself is produced from diazomethane (CH_2N_2). Because their major absorption bands are in the ultraviolet (UV) region, the carbene precursors must be irradiated in this region for effective generation of the carbene. Carbonyl diazo compounds have two absorption maxima, an intense 250–270 nm band ($\varepsilon \sim 10^4$) and a broad–weak band centered at about 350 nm (ε from 10 to 40). The carbenes generated at the shorter wavelength are more reactive and thus more desirable as labeling reagents (*68*). Unfortunately, the destruction of proteins during ultraviolet irradiation limits the usefulness of these derivatives. Carbenes as well have the property of readily undergoing intramolecular rearrangements resulting in loss of nonselectivity. For example, α-ketocarbene, generated from α-diazoketones and esters, undergoes Wolff rearrangement to ketene (*69*) much less reactive than the parent carbene (Eq. 17).

$$\underset{\text{an }\alpha\text{-diazo-carbonyl}}{R-\overset{O}{\overset{\|}{C}}-CH=N_2} \rightleftharpoons R\cdot\overset{O}{\overset{\|}{C}}-\overset{-}{C}H-\overset{+}{N_2} \xrightarrow{h\nu} \underset{\text{an }\alpha\text{-carbonyl-carbene}}{:CH-\overset{O}{\overset{\|}{C}}-R} + N_2\uparrow \quad \xrightarrow{\text{a ketene}} O=C=CH-R \tag{17}$$

Thus, the potential nonselectivity of α-carbonylcarbenes is lost, since the ketene formed has the selective characteristics of any electrophilic reagent (*69*). The diazocarbonyl compounds are not particularly stable, especially at low pH. However, electron withdrawing groups strategically arranged have been reported to offer some degree of percursor stability in the case of diazomalonyl derivatives (Compound 5).

$$R''-\overset{O}{\overset{\|}{C}}-\underset{COOR'}{\underset{|}{C}}=\overset{+}{N}=\overset{-}{N}$$

Compound 5

While the derived carbene is less susceptible to the Wolff rearrangement, the electrophilic nature of α-ketocarbenes causes a preference for O—H bonds, i.e., H_2O, even in the presence of neighboring CH bonds. Substitution of trifluoromethyl on the carbene carbon stabilizes the diazo compound and decreases the tendency toward Wolff rearrangement (Compound 6).

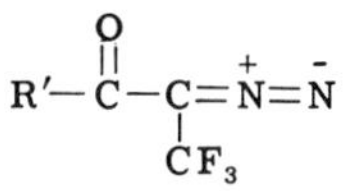

$$R'-\overset{\overset{\displaystyle O}{\|}}{C}-\underset{\underset{\displaystyle CF_3}{|}}{C}=\overset{+}{N}=\overset{-}{N}$$

Compound 6

In contrast to the alkyl derivatives, the aryldiazomethanes are chemically very reactive. Possibly proper substitution on the aryl group could result in modification of this reactivity. It has been proposed that aryldiazirines, since they are more stable, may be useful as a source of arylcarbenes (*66*).

Other reactions open to carbenes in addition to carbon–hydrogen insertion reactions are abstraction yielding two free radicals, which may then couple; addition to multiple bonds such as insertion into aromatic systems; and coordination to nucleophilic centers giving carbonions.

2. *Nitrenes*

Nitrenes, the nitrogen analogs of carbenes, can be generated photolytically by reactions analogous to carbene formation from alkyl and arylazido, isocyanates, and carbonylazides (*68*). Compared to alkylnitrenes, which are activated at wavelengths in the UV region, arylnitrenes can be generated from their precursors by photolysis at wavelengths above 350 nm, sufficiently removed from protein absorption bands. Suitably substituted arylazides have in fact strong absorption bands above 400 nm and can be thus photolyzed to nitrenes at wavelengths where photoinduced changes should be minimal (*77*). The alkyl and arylazides, isocyanates, and acylazide nitrenes, however, are more selective than carbenes, a difference in reactivity reflected in the relatively long lifetime of 10^{-4} second found for simple nitrenes situated in a "soft" polystyrene matrix. Staros and Richards (*78*) and Staros *et al.* (*79*) made use of the relative stability of arylnitrenes in their study of the topology of red blood cell membranes.

An arylnitrene was first used as a labeling reagent for a specific antibody (*38*) with indications that nitrene interaction was responsible for the modification of an alanine and a cysteine of the heavy-chain peptide in the antibody combining site.

Arylnitrene half-lives vary with substituents and increase by nearly two orders of magnitude when going from *p*-acetyl (short-lived) to a *p*-*N*-morpholino (long-lived) derivative (*80*). A range of reagents of different reactivities can therefore in principle be synthesized. While electron-withdrawing groups attached to arylnitrenes increase nitrene reactivity, they presumably also decrease selectivity; however, this point has not yet been systematically investigated. Presumably nitro group attachment would decrease the half-life, i.e., increase the reactivity, of the nitrene. In comparison to nitrene probes, which decay in the millisecond range, the upper limit of carbene lifetimes is of the order of 1 μsec. Singlet nitrenes react preferentially by insertion into oxygen–hydrogen or nitrogen–hydrogen bonds and show electrophilic character. Electron-withdrawing groups, particularly para to the nitrene, increase the electrophilic character, and hence reactivity, of the singlet nitrene. Electron-releasing substituents decrease the electrophilicity and enhance the yield of triplet-state reactants (*81*). In the less reactive state, insertion into carbon–hydrogen bonds becomes more probable.

As with the carbenes, direct insertion into C—H bonds is only one of a number of pathways available for nitrene interaction. Comparable reactions of abstraction, cycloaddition, attack by nucleophiles, and rearrangement are possible. Of these reactive pathways, direct insertion, abstraction coupling, or addition reactions would be expected to result in covalent attachment of label to binding sites. Although direct intramolecular insertion reactions are considered rare, if a nitrene is generated *in situ* at a binding region, intramolecular insertion could represent the major direction of reactivity.

The alkylazides have absorption maxima at about 290 nm, so that in addition to rearrangement reactions the wavelength necessary for activation reduces the effectiveness of these reagents. Arylnitrenes by contrast are much less susceptible to photochemical rearrangement reactions (*82,83*) and are chemically stable at 37°C. Thus the three criteria of *chemical inertness of the precursor*, *lack of rearrangement of reactive species*, and *suitable absorptive characteristics* suggest that arylnitrene-generating reagents are the most ideal photoaffinity reagents at the present time (*38,84*).

The work of Breslow *et al.* (*85*) indicates that phosphoryl nitrenes may have potential applications to bioenergetic systems. Intramolecular insertion reactions of phosphorylnitrenes react exclusively via direct insertion processes without rearrangement and with low selectivity toward tertiary versus primary C—H insertion processes.

A word of caution concerning the reaction specificity of arylnitrenes. These reagents are quite selective with respect to C—H, inserting into tertiary C—H one hundred times faster than into primary C—H bonds. Consequently, these reagents do not necessarily meet the requirement that insertion should be completely nonselective in order to assure proper affinity reactivity. The half-life of the arylnitrenes is of the order of 10^{-3} second; thus, except for very tight binding sites, these reagents might not necessarily meet another requirement for photoaffinity labeling: that is, that their lifetime should be short enough to enable covalent interaction to take place at rates fast in comparison to possible equilibrium among sites.

On the other hand, labeling by a long-lived reagent in which the lifetimes are sufficient to equilibrate between receptor and solution can be of an advantage in demonstrating experimentally competitive reactivity with normal substrates. Rucho *et al.* (*86*) have shown that labeling under these conditions can be suppressed through appropriate use of scavengers.

II. Photoaffinity Analogs of Ligands Used in Energy-Transduction Systems

A. Membrane Systems

1. *Antibody Binding Sites*

Fleet and co-workers first introduced arylazides as photochemical labeling reagents in 1969 (*38*). Their first report was followed by an elegant study of photoaffinity labeling of the 4-azido-2-nitrophenyl binding site antibody (*84*). Photodependent binding of the arylnitrene, *N*-ε-(4-azido-2-nitrophenyl)-lysine (Compound 7), was found to be associated with the labeling of heavy-chain cysteine residue number 92 and alanine residue number 93.

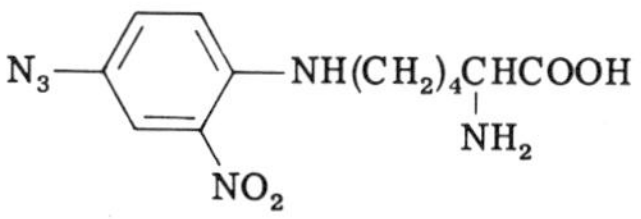

N_3—(ring)—$NH(CH_2)_4CH(NH_2)COOH$, with NO_2 on the ring

Compound 7:
N-ε-(4-Azido-2-nitrophenyl)lysine

Richards and his co-workers (*87–90*) have described the use of three photoaffinity labels for a homogeneous mouse immunoglobin. Both 2,4-dinitrophenylalanine diazoketone (Compound 8) and 2,4-dinitrophenyl-1-azide (Compound 9), carbene and nitrene precursors, respectively, competitively inhibit the binding of the haptene, ε-2,4-dinitrophenyllysine, in

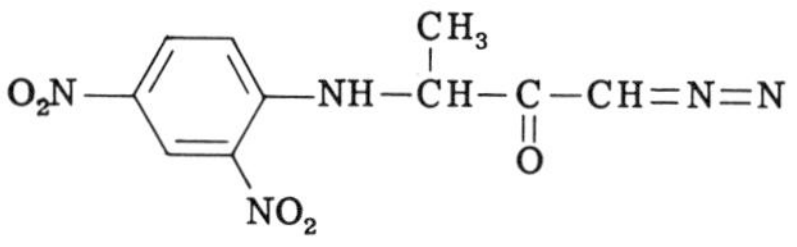

O_2N—(ring)—$NH-CH(CH_3)-C(=O)-CH=N=N$, with NO_2 on the ring

Compound 8:
2,4-Dinitrophenylalanine diazoketone

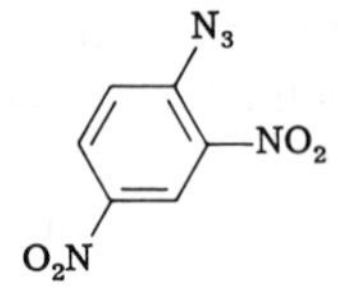

N_3 —NO$_2$ O$_2$N

Compound 9:
2, 4-Dinitrophenyl-1-azide

the dark and label the immunoglobulin upon irradiation. The label is found to be inserted into the heavy-chain tyrosine residue 33 and tyrosine residue 88. Purified anti-dinitrophenyl antibodies labeled with the same reagents were found to be modified at a phenylalanine residue in one case and at an alanine residue in the other (*91*).

A menadione analog, 2-methyl-3-thioglycoldiazoketone-1,4-naphthoquinone (Compound 10), labels over 60% of the possible hapten sites upon irradiation (*92*).

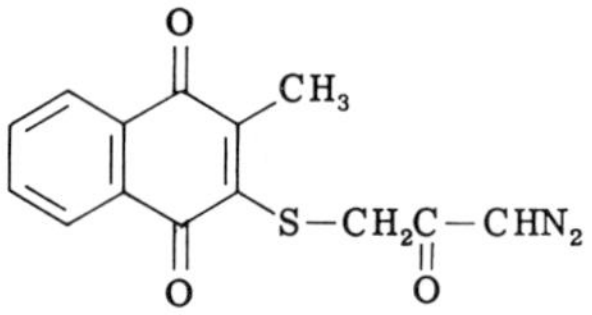

O CH$_3$ S—CH$_2$C—CHN$_2$ O O

Compound 10:
2-Methyl-3-thioglycoldiazo-ketone-1, 4-naphthoquinone

2. *Transport Systems*

The lactose transport system of *E. coli* is being investigated with the aid of two monosaccharide azido analogs: *N*-4-Azido-2-nitrophenylaminoethyl-1-thio-β-galactopyranoside (Compound 11) and 4-azido-2-nitrophenyl-1-

CH$_2$OH N$_3$— —NH—CH$_2$—CH$_2$—S O OH H H HO NO$_2$ H H HO H

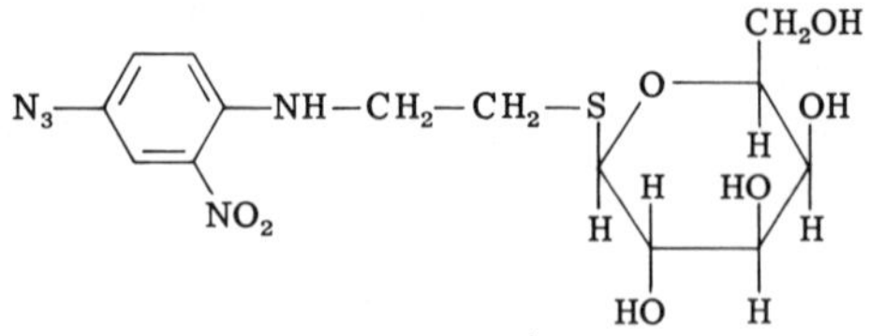

Compound 11:
N-4-Azido-2-nitrophenylamino-ethyl-1-thio-β-galactopyranoside

thio-β-D-galactopyranoside (Compound 12). The former analog binds to the *E. coli* membrane; however, it is not transported. In this case, in the presence of the analog, there is a photoinduced inactivation of transport, which is protected against by lactose. The second analog is transported with kinetics showing competition with lactose. Irradiation inactivates transport in the presence of the azido reagent and D-lactate (*93,94*). The synthesis of a number of other arylazido carbohydrate derivatives has been reported (*95,96*).

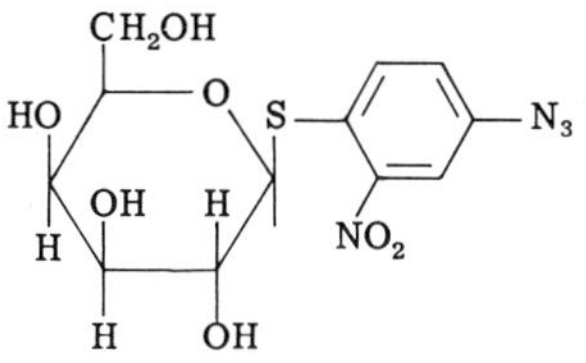

Compound 12:
4-Azido-2-nitrophenyl-1-thio-β-D-galactopyranoside

3. *Membrane Channels*

The analog *N*-(4-azido-2-nitrophenyl)taurine (Compound 13) has been used in an interesting series of experiments designed to label differentially the two sides of the erythrocyte membrane (*78,79,97*). Labeling of the outer membrane surface results in the insertion of the analog into eight distinct proteins. After transport of the reagent to the internal compartment, labeling of the cytoplasmic membrane proteins was observed. The labeling pattern obtained with resealed erythrocyte ghosts was decidedly different from that observed with the intact membrane.

The compound has been found to undergo photodependent binding to an ion channel of the erythrocyte membrane (*98,99*).

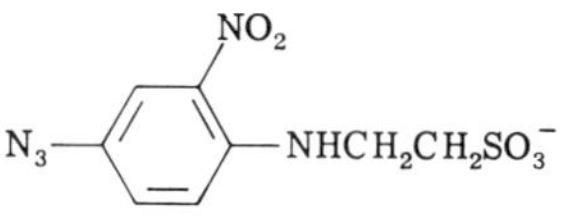

Compound 13:
N-(4-Azido-2-nitrophenyl) taurine

Kiefer *et al.* (*100*) have synthesized a group of azido reagents (Compounds 14–16) as specific photoaffinity labeling probes for the cholinergic ligand binding sites in frog sartorius muscle.

While the application of Compound 15 results in inactivation of cholinesterase, there is much nonspecific labeling observed. Attempts were made

Compound 14:
4-Azido-2-nitrobenzyl-trimethylammonium salt

Compound 15:
4-Azido-2-nitrobenzyl-triethylammonium salt

Compound 16:
Diazido propidium salt

to minimize the nonspecific labeling of the erythrocyte acetylcholinesterase by irradiation in the presence of 4-aminobenzoate, a potential scavenger for the photolysis products not formed at the receptor sites. Both 4-aminobenzoate and soluble proteins were found to prevent photoinactivation of the acetylcholinesterase by the azides.

Interestingly, preirradiation of Compound 14 did not prevent its effective inhibition of acetylcholinesterase unless irradiation was carried out in the presence of 4-aminobenzoate prior to the addition of the erythrocyte membrane. Kiefer *et al.* propose that the irradiation of the azide results in the formation of a relatively long-lived intermediate, which then reacts with the enzyme. They term this pseudophotoaffinity labeling (see below).

Diazido ethidium (Compound 17) and diazido propidium (Compound 16) are fluorescent photoaffinity labels, and the latter reagent has been shown to irreversibly block the cholinergic synapses from *Torpedo californica* electric tissue (*101*). Hucho *et al.* (*102*) have reported that the label reversibly

Compound 17:
Diazido ethidium bromide;
[2 3(8), 8(3)]-diamino-5-ethyl-6-phenylphenanthridinium bromide

inhibited the potassium current of the nodal membrane of frog sciatic nerve fibres without an appreciable influence on the sodium current. After irradiation with 4 m*M* reagent, inhibition of sodium current was at the level of 10% and there as an 80% decrease in potassium current. The potassium current inhibition was shown to be irreversible and to arise from a photoreaction of the molecules bound to a receptor in the potassium channels rather than from the action of photolytic products of the reagent. Hucho and Schiebler (*29*) have recently summarized their observations and thoughts on investigations of ionic channels in excitable membranes.

The voltage-dependent sodium channel is known to be specifically blocked by tetrodotoxin (*32*). Guillory *et al.* (*30*) have prepared an arylazidotetrodotoxin derivative by the application of the carbodiimidazole-catalyzed esterification reaction utilized by Jeng and Guillory (*52,103*) for the esterification of the ribose hydroxyls of nucleotides. The tetrodotoxin is covalently bound to receptor sites associated with the sodium pores of excitable membranes. The biological activity of the toxin analog, which does not differ appreciably from that of the underivatized toxin, is retained following covalent binding to the membrane system. The region of the tetrodotoxin molecule modified by the arylazido-β-alanyl conjugate has yet to be delineated.

4. *Steroid Analogs*

Katzenellenbogen *et al.* (*104–106*) have described the synthesis of a number of diazoacetate derivatives of estrogens as photoaffinity labels for the estrogen receptor protein of rat uterus. The compounds azidoestrone (Compound 18), azidoestradiol (Compound 19), and azidohexestrol (Compound 20) have been synthesized and shown to be photoattached to the estrogen receptor (*104,105*). Use of the photoactive hormone analog has been instrumental in the isolation of the estrogen receptor (*106*).

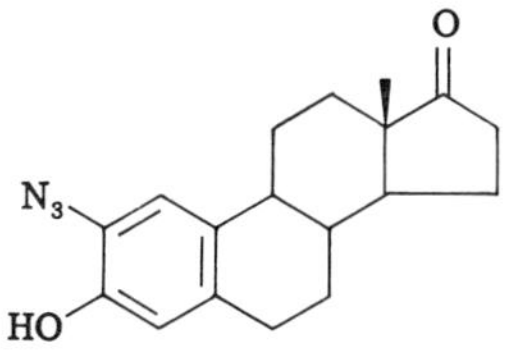

Compound 18: 2-Azidoestrone

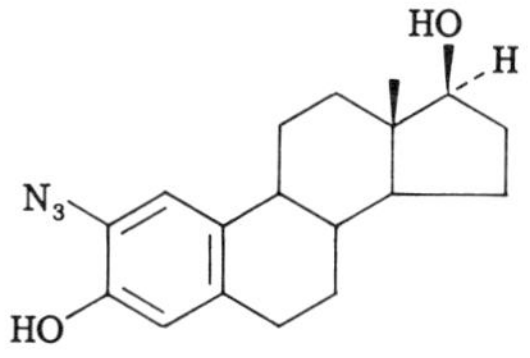

Compound 19: 2-Azidoestradiol

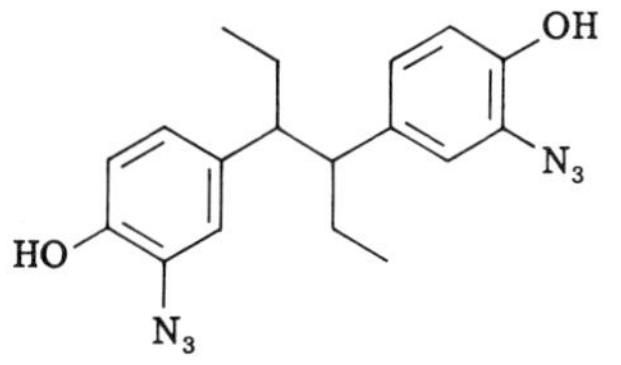

Compound 20: 3, 3-Diazidohexestrol

5. *Opiate Receptor*

An arylazide derivative of the alkaloid norlevorphanol, *N*-4-azidophenylethylnorlevorphanol (Compound 21), has been synthesized with potent opiate pharmacological properties (*107*). The analog was found to be irreversibly incorporated into brain tissue upon photolysis *in vitro*; however, the bonding was not restricted to specific sites.

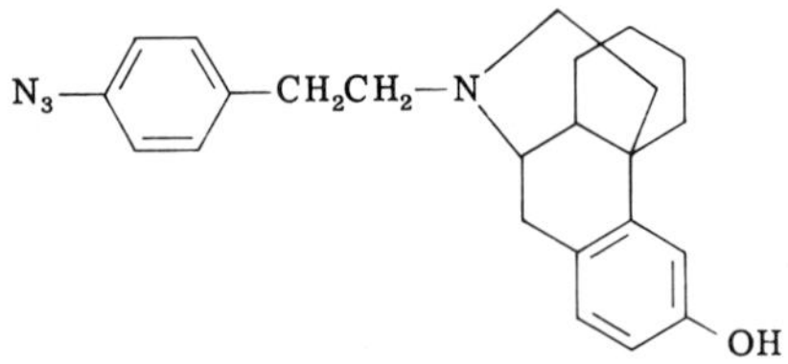

N_3—C₆H₄—CH_2CH_2—N ... OH

Compound **21**:
N-2-*p*-Azidophenylethylnorlevorphanol

6. *Insulin Receptor*

Several biologically active derivatives of insulin containing the 3-nitrophenylazido chromophore [*N*-(4-azido-2-nitrophenyl insulins)] have been prepared as photoaffinity probes for the study of the location and function of the parent hormone (*108*).

7. *Erythrocyte Membrane*

Strophanthidin modified with a 3-azidoacetyl adjunct (*109*) (Compound 22) has been used to study the ($Na^+ + K^+$) ATPase of the erythrocyte membrane.

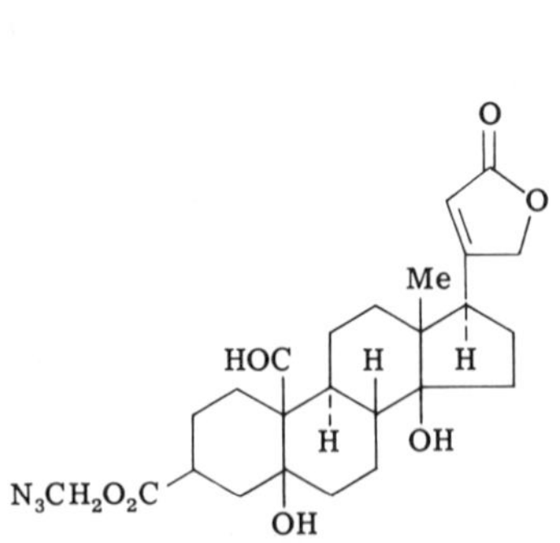

O, O, Me, HOC, H, H, H, OH, $N_3CH_2O_2C$, OH

Compound **22**:
3-Azidoacetylstrophanthidin

8. Adipocyte Plasma Membranes

The adipocyte plasma membranes have been labeled with the glucose analog *N*-(4-azido-2-nitrophenyl)-2-amino-2-deoxy-D-glucose (Compound 23) (*108*).

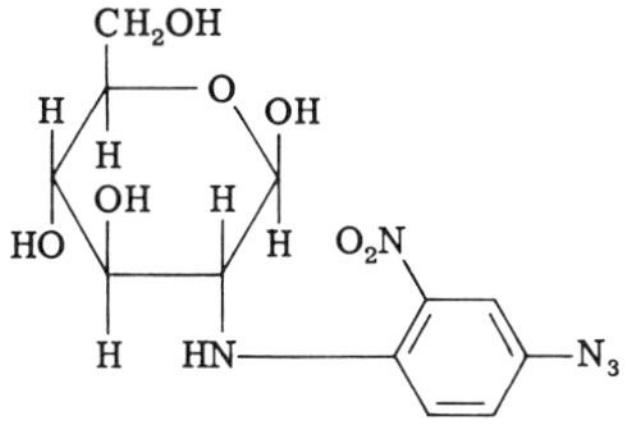

Compound **23**:
N-(4-Azido-2-nitrophenyl)-
2-amino-2-deoxy-D-glucose

B. Lipids and Liposomes

The reported synthesis of a variety of alkyl and arylazido fatty acids and their biosynthetic incorporation into phospholipids has exciting potential for future work (*110,111*). There are possibilities for the formation of a unique class of phospholipid vesicles with strategically oriented photoaffinity probes constituting a portion of the membrane. The possibility for control labeling of proteins as they are organized within the membrane (either as an original component or as an exogenously introduced component) would be extremely useful in investigations of the organization and orientation of proteins within membranes.

The reagents 1-azidonaphthalene (Compound 24) and 4-iodophenylazide (Compound 25) are able to label liposomes and sarcoplasmic reticulum membranes showing a distribution of about 50 % toward membrane proteins and 50 % toward fatty acid chains (*112*).

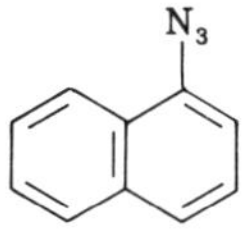

Compound **24**:
1-Azidonaphthalene

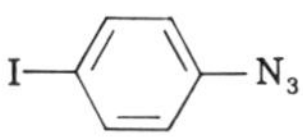

Compound **25**:
4-Iodophenylazide

C. Mitochondrial Systems

1. *Antimycin Binding-Site Label*

Azidodeformamido antimycin A (Compound 26) has been used in an attempt to characterize the antimycin A binding site in Complex III of the electron-transport chain (*113*). The binding capacity of the analog would appear to suggest the labeling of a single protein.

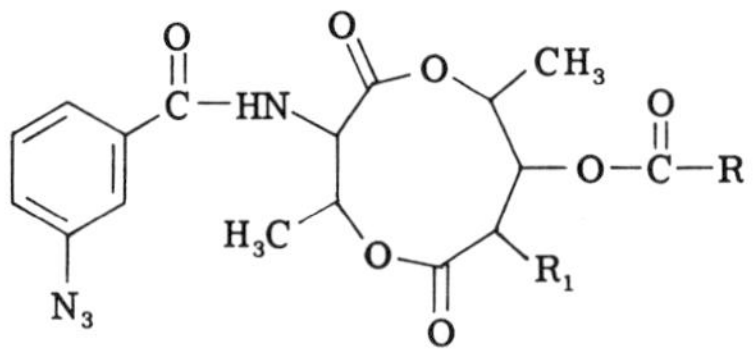

Compound 26:
Azidodeformamido antimycin A
$R = CH_2CH(CH_3)_2$
$R' = n$-hexyl

2. *Cytochrome c*

One of the most potentially useful approaches to the study of the electron-transport system with photoaffinity probes has been the synthesis of a specific active protein containing a photoaffinity adjunct. The reagents 2,4-dinitro-5-fluorophenylazide and *p*-azidophenylacylbromide have been used to effect the synthesis of 3-azido-4,6-dinitrophenyl-cytochrome *c*. This protein reagent has been found to be able to bind to cytochrome oxidase in a photo-dependent reaction (*114–116*). The advantages of such a probe in the investigation of the mechanism of electron transfer and in the study of protein–protein interactions are obvious.

3. *Arylazido Nucleotide Probes*

In 1973 Jeng and Guillory (*52*) reported on the utilization of carbodiimidazole to facilitate the formation of an activated carboxylic acid containing an arylazido adjunct. Synthetic schemes utilizing this material resulted in esterification of adenosine and diphosphopyridine nucleotides (*74*). Such esterification results in the formation of nucleotide analogs containing an azido function on the ribose molecule (*103*). Since that time the interaction of the ATP analog (Compound 2) has been reported for myosin and its proteolytic formed components (*52*) and for F_1 beef heart mitochondrial ATPase (*20*).

A full report on the preparation and properties of arylazido-β-alanyl NAD^+ has appeared (*26*). Evidence is presented for the structural assignment of A3′-*O*-{3-[*N*-(4-azido-2-nitrophenyl)amino]propionyl} NAD^+ (Compound 27). The pyridine nucleotide analog acts as a substrate for yeast alcohol dehydrogenase when incubated with the enzyme in the dark (k_m, 0.052 μM; V_{max}, 4.4 μmol mg^{-1} min^{-1} at pH 7) and inhibits reduction of the natural substrate. Upon photolysis in the presence of the enzyme there is a potent photodependent inhibition, which is prevented by the presence of NAD^+ or NADH. Complete photodependent inhibition ($>95\%$) is associated with the incorporation of 2 mol of the analog per mole of enzyme.

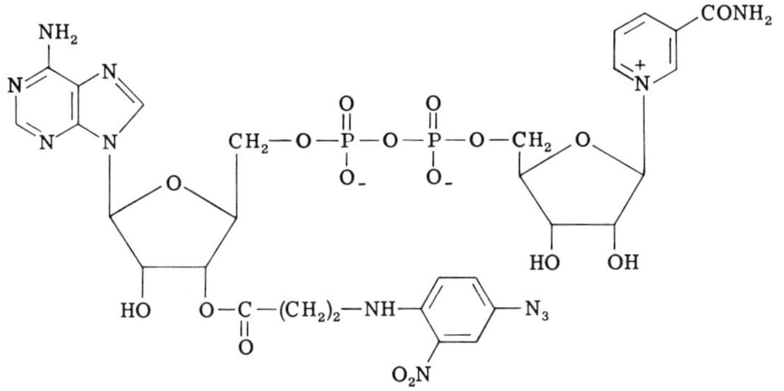

Compound **27**:
Arylazido-β-alanyl NAD^+;
A 3′-*O*-{3-[*N*-(4-azido-2-nitrophenyl) amino] propionyl}NAD^+

An interesting aspect of the interaction of arylazido-β-alanyl nucleotide derivatives with the enzymes examined is their property of being reasonably good substrates for specific reactions. They exhibit competitive influences in the dark and become irreversible inhibitions upon photoirradiation. Irreversible inhibition upon photolysis appears to be correlated with a stoichiometric labeling pattern.

Utilizing the synthetic scheme as described by Jeng and Guillory (*52*), the synthesis of arylazido atractylosides (Compound 28) and arylazido nitrophenylaminobutyryl ADP (Compound 29) has been reported (*117,118*). The atractyloside analog has been used as a photoaffinity reactant for the mitochondrial adenine nucleotide carrier (*117*) and Compound 29 as a reactant for the mitochondrial ATPase (*118*) and the nucleotide carrier (*119*).

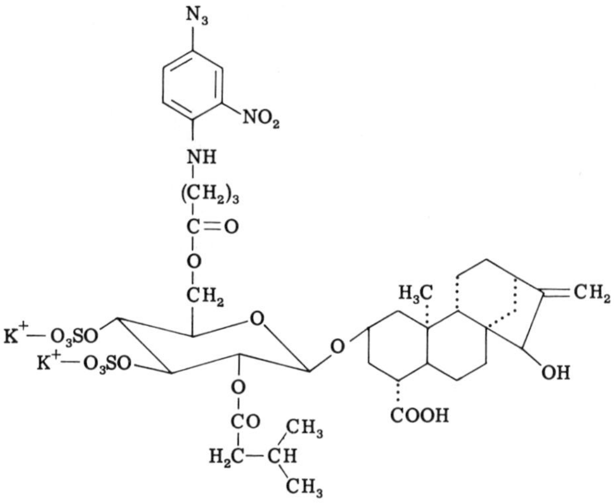

N_3
NO_2
NH
$(CH_2)_3$
C=O
O
CH_2
$K^+—O_3SO$
$K^+—O_3SO$
O
O
H_3C
$=CH_2$
OH
O
CO
CH_3
$H_2C—CH$
CH_3
COOH

Compound 28:
4-Azido-2-nitrophenyl-
aminobutyryl atractyloside

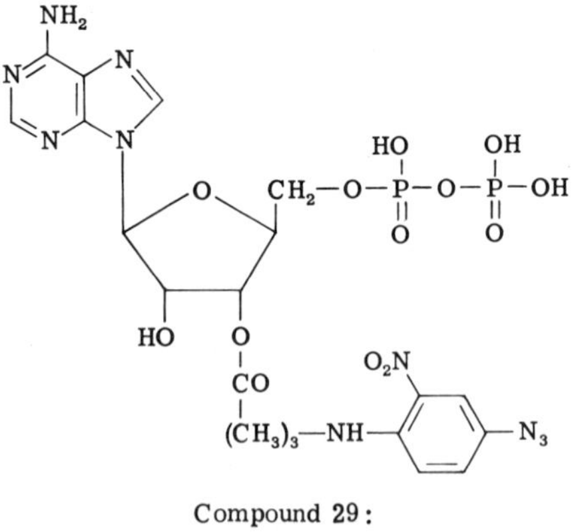

NH_2
N
N
N
N
HO OH
O
$CH_2—O—P—O—P—OH$
O O
HO O
O_2N
CO
$(CH_3)_3—NH—$
N_3

Compound 29:
N-4-Azido-2-nitrophenyl-
aminobutyryl ADP

4. Pyridine Nucleotide Photoaffinity Probes

The first synthesis of a photoaffinity analog of pyridine nucleotide was reported in 1971 by Westheimer and his colleagues (*120*). This compound, the 3-diazoacetoxymethyl analog of NAD^+ (Compound 30), is a carbene-generating analog. The $[^{14}C]$diazoacetate ester of 3-hydroxymethylpyridine was synthesized

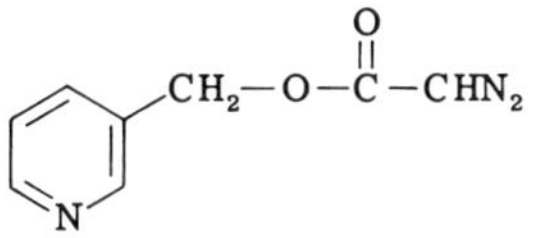

Compound 30a: Diazoacetate ester of 3-hydroxymethylpyridine

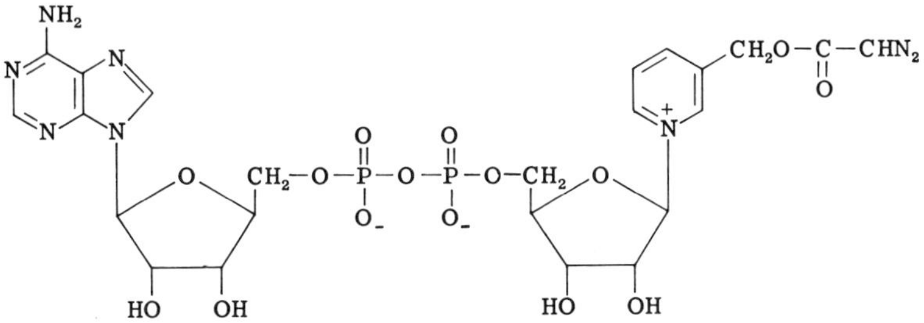

Compound 30b: 3-Diazoacetoxymethyl analog of NAD^+

and exchanged enzymically with NAD^+ to form the 3-diazoacetoxymethyl analog of NAD^+. This material was shown to inhibit yeast alcohol and pig heart malic dehydrogenase competitively ($K_i = 3 \times 10^{-4}$) with association constants comparable to that of the natural coenzyme, NAD^+ (i.e., 5×10^{-4} M). Upon photolysis it is converted to a carbene derivative allowing for covalent insertion into the protein.

When photolysis of the enzyme–analog complex was carried out (150 minutes at 4°C within the range of 310 to 420 nm), there resulted a covalent attachment of the radioactive label to the protein. The degree of binding of the analog to various enzymes, however, was only modest. It is theorized that attachment of the analog to the active site occurs principally via the adenine residue, accounting for the lack of specific reaction products during photolytic decomposition. Of a number of labeled residues found, one product was identified as *S*-carboxymethylcysteine (Compound 31). Its formation is possibly due to interaction of the analog with an —SH group at the active site of yeast alcohol dehydrogenase.

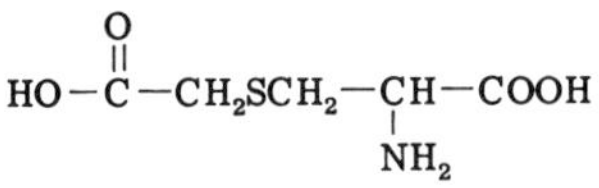

$$HO-\overset{\displaystyle O}{\overset{\|}{C}}-CH_2SCH_2-\underset{\displaystyle NH_2}{\underset{|}{C}H}-COOH$$

Compound 31: *S*-Carboxymethylcysteine

The synthesis of a nitrene-generating analog of NAD^+ (Compound 32) has been described by Hixson and Hixson (*121*). The analog 3-azidopyridine adenine dinucleotide, which is relatively stable to heat in contrast to the diazoacetoxy analog of NAD analog (Compound 30), was shown to be a competitive inhibitor of NAD^+ reduction by yeast alcohol dehydrogenase (K_1 value of 2.4×10^{-4} M). Upon photolysis of the ^{3}H-labeled analog with the enzyme, a limited amount (7%) of the nucleotide analog was covalently bound to the enzyme; in the presence of NAD^+ only 4% was found incorporating, making it difficult to distinguish what might be the specifically labeled (i.e., active site) portion of the enzyme from the randomly labeled ones. The authors make the important point that photochemical labeling reagents would be most profitably used when they can be first bound strongly via noncovalent interactions with K_i of the order of 10^{-6} M.

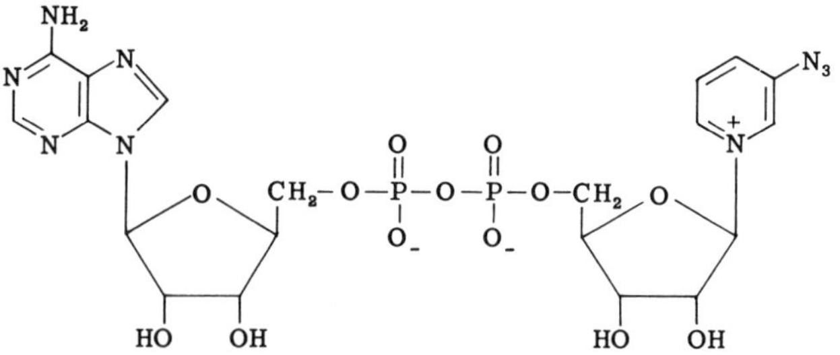

NH2 N N N N O CH2–O–P–O–P–O–CH2 O O O O- O- N3 N+ HO OH HO OH

Compound 32:

3-Azidopyridine adenine dinucleotide

Koberstein (*122*) has described the synthesis of the 8-azido adenine analogs of NAD^+ (azido NAD^+), Compound 33, and FAD (azido FAD) from 8-azidoadenosine 5′-phosphate and NMN or FMN. Evidence is presented for a folded structure for both azido derivatives. The K_M (azido

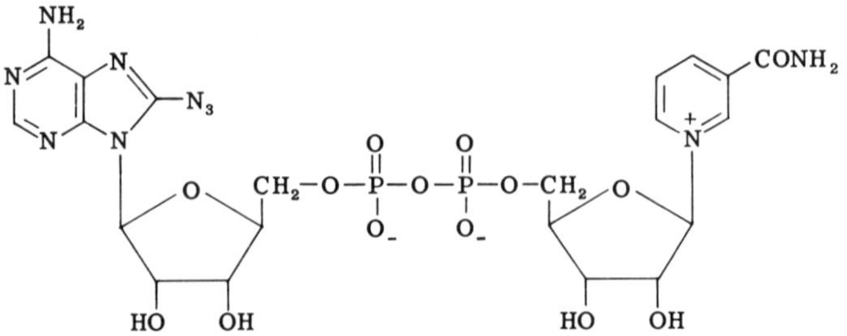

NH2 N N N3 N N O CH2–O–P–O–P–O–CH2 O O O O- O- CONH2 N+ HO OH HO OH

Compound 33:

8-Azido NAD^+

NAD^+) demonstrated for lactate, glutamate, and alcohol dehydrogenase is 1.7-, 3.7-, and 3-fold higher than for NAD^+. Azido FAD was shown to be a coenzyme for apoglucose oxidase but was completely inactive with apo-D-amino acid oxidase. The cofactor is however specifically bound to the latter enzyme as indicated by difference spectra. The possible photodependent effect of the analogs toward enzymes was not reported.

5. *Azido Probes for Uncoupler Binding Site(s)*

The azido analog of 2,4-dinitrophenol, i.e., 2-azido-4-nitrophenol (Compound 34), has been used by Haustein and Hatefi (*123–126*) in their study of the mitochondrial uncoupling mechanism. This work and that of Ramakrishna Kurup and Sanadi (*127*) and Cybron and Dryer (*128*) indicates that the azido analog is a more effective uncoupler than is 2,4-dinitrophenol and that after photolysis it becomes covalently bound to the inner mitochondrial membrane. In addition to a protein of molecular weight (MW) 31,000, the analog is covalently bound to the α subunit of the F_1ATPase (MW 56,000). Senior and his co-workers (*129*) have approached the same problem using the analog 4-fluoro-3-nitrophenylazide (Compound 35).

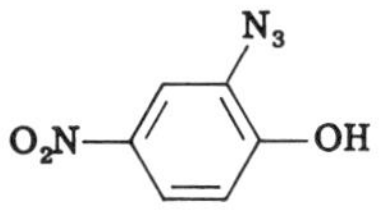

N_3 — O_2N — OH

Compound 34:
2-Azido-4-nitrophenol

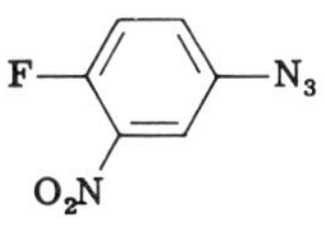

F — N_3 — O_2N

Compound 35:
4-Fluoro-3-nitrophenylazide

6. *8-Azido Adenine-nucleotide Probes*

In 1974, Haley and Hoffman (*130*) first reported on the synthesis of 8-azido ATP (Compound 36), a photoaffinity analog of ATP. The ATP analog was prepared by heating 8-bromo AMP at 75°C in anhydrous dimethylformamide containing tri-*n*-octylamine azide followed by coupling pyrophosphate to the 8-azido AMP. This analog was initially used in the study of the erythrocyte membrane, in which it was shown that both the Mg^{2+} and the Na^+–K^+ ATPase activities are inhibited in a photodependent manner. The natural substrate ATP was effective in preventing this inhibition.

Similar methods have been utilized for the synthesis of 8-azido ADP (*131*) and 8-azido 3′,5′-cAMP (*131,133*). The 8-azido ADP was initially used in an investigation of the allosteric activation site of glutamic dehydrogenase,

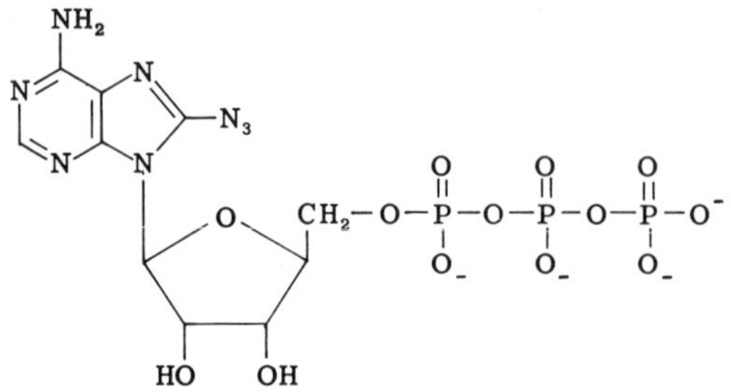

Compound 36:
8-Azido ATP

and the 8-azido 3′,5′-cAMP was used in an investigation of the nucleotide binding site of a protein kinase.

The synthesis and utilization of 8-azido adenine analog of acyl-CoA (Compound 37) with acyl-CoA : glycine-*N*-acyltransferase have been reported (*134,135*).

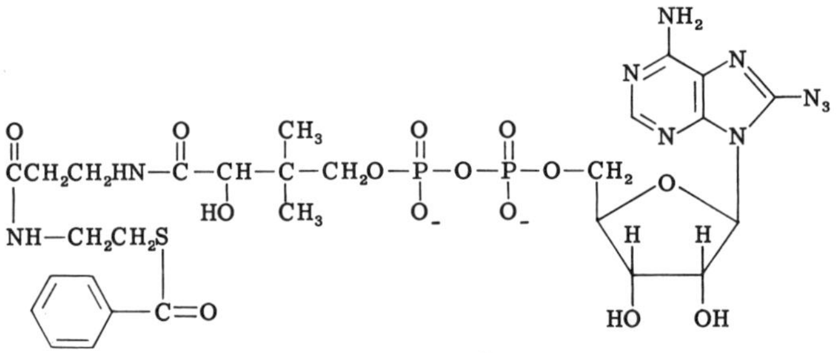

Compound 37:
Benzoyl-(3′-dephospho-8-azido)-CoA

The carbodiimidazole-catalyzed esterification reaction applicable to the synthesis of adenine nucleotide arylazido-β-alanyl photoaffinity analogs (*103*) has been accomplished with acetyl-CoA (unpublished observations of R. Guillory). Although one of the reaction products exhibits photoaffinity labeling characteristics with acyl-CoA : glycine-*N*-acyltransferase, the structural characteristics of the molecule have not yet been worked out.

D. Reagents for the Study of the Structure of RNA and the Mechanism of Protein Synthesis

The influence of monoazido ethidium bromide (*136,137*) (Compound 38) and diazido ethidium bromide (*138*) (Compound 17) has been examined with respect to yeast mitochondrial development. Compound 38 causes an

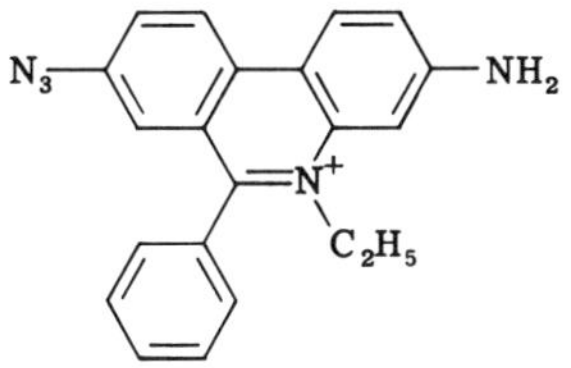

Compound 38:
Monoazido ethidium bromide

increase in the level of petite mutants following irradiation, and diazido ethidium bromide appears to label subunit 9 of the yeast membrane ATPase.

The reported synthesis of the azido derivatives of chloramphenicol (Compound 39 and Compound 40) should have direct use in studies relating to ribosomal protein synthesis (*139,140*).

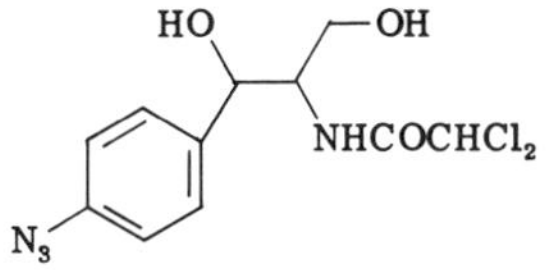

Compound 39:
2-Dichloroacetamido-1-(4-azidophenyl)-1, 3-propandiol

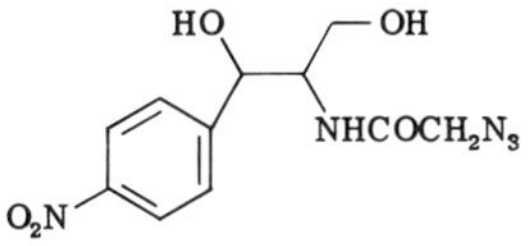

Compound 40:
2-Azidoacetamido-1-(4-nitrophenyl)-1, 3-propanediol

The synthesis of a peptidyl-tRNA analog, ethyl-2-diazomalonylphenylalanyl tRNA (Compound 41), has been reported by Bispink and Matthali (*141*). The label binds covalently to the 23S ribosomal RNA of *E. coli* upon irradiation.

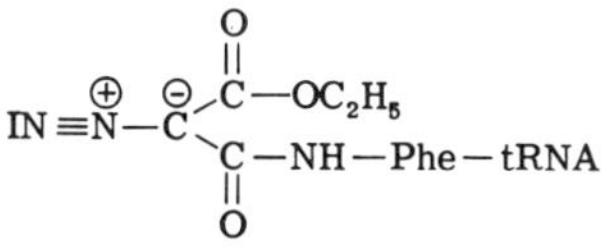

Compound 41:
Ethyl-2-diazo-malonyl-Phe-tRNA

III. Current Investigations on Specific Energy-Transducing Systems

A. Mitochondrial Electron Transport

1. General Photochemical Reduction

Aggarwal *et al.* (*142*) have investigated the influence of visible light irradiation on the electron-transport and energy-coupling enzymes of rat liver mitochondria. The mitochondria were exposed to a light intensity of 300 $\mu W/cm^2$ via a battery of 250 W quartz iodide lamps maintaining irradiation at wavelengths above 400 nm. Over a 12-hour irradiation period, there was observed a time-dependent series of changes in which an initial stimulation of respiration coupled to ATP synthesis was followed by a decline in ATP synthesis, an inactivation of respiration, and a final loss of membrane potential. The loss of respiratory activity was due primarily to the inactivation of dehydrogenases, succinic dehydrogenase, choline, and NADH dehydrogenase being inactivated in that order. Flavins and quinones were most susceptible to the destructive effects of illumination, while the iron sulfur centers were resistant to such damage. The redox reactions of the cytochromes and cytochrome oxidase activity were unaffected under the conditions studied.

Inactivation of the electron transport system was O_2 dependent, prevented by anaerobiosis, and most effectively accomplished at wavelengths of 400–500 nm. On the basis of their findings and the fact that the presence of free flavins greatly enhanced inactivation of all the mitochondrial activities investigated, the authors postulate that visible light mediates a flavin-photosensitized reaction that initiates damage and involves participation of an activated species of oxygen in the damage propagation.

The chemical properties of flavins in relation to flavoprotein catalysis has been previously reviewed (*143*), and Massey and Hemmerich (*144*) have described a photochemical procedure for reduction of oxidation–reduction proteins, such as flavoproteins, heme proteins, and iron sulfur proteins. Free flavins and 5-deazaflavins added in trace amounts behave as very efficient catalysts in the photoreduction. In the latter case the mechanism involves the photochemical generation of the highly reactive radical of 5-deazariboflavin.

2. *Antimycin-Binding Protein*

Gupta and Rieske (*113*) have utilized deformamidoazido antimycin A in the identification of a 11,500 MW peptide as the antimycin-binding site of Complex III of the electron-transport system. The azido antimycin analog was prepared from deformamido antimycin A by treatment with $NaNO_2$ in HBF_4 followed by the addition of sodium azide. The azido analog in the dark has the same inhibitory effects as antimycin A except that it is freely reversible upon dilution. Irradiation with UV light converts the analog to an irreversible inhibitor. The irreversible effect is associated with binding to a specific peptide as sown by sodium dodecyl sulfate (SDS) gel electrophoresis of Complex III after irradiation in the presence of the tritium-labeled analog.

3. *Pyridine Nucleotides*

Arylazido-β-alanyl NAD$^+$ *and Arylazido-β-alanyl* NADH

The above-mentioned pyridine nucleotide analogs (Section II,C,4) have yet to be systematically examined with respect to possible interactions with the mitochondrial electron-transport system. In contrast to the analogs having the photoreactive azido on the adenine or nicotinamide ring are those prepared by Chen and Guillory (*26,27*). These investigators utilized the methodology developed by Jeng and Guillary (*52*) for the carbodiimidazole-catalyzed esterification of arylazido-β-alanine to the ribose of ATP. Controlled incubation of *N*-4-azido-2-nitrophenyl-β-alanine and carbodiimidazole in the presence of nicotinamide adenine dinucleotide resulted in the formation of two reaction products containing the arylazido adjunct. One of the reaction products (arylazido-β-alanyl NAD$^+$) was found to be capable of undergoing reduction in the presence of ethanol and yeast alcohol dehydrogenase.

The initial report on experiments with arylazido-β-alanyl NAD (*26*) presented evidence for the structural assignment of A3′-*O*-{3-[*N*-(4-azido-2-nitrophenyl)amino]propionyl} NAD$^+$ (Compound 27). The structural assignment was indicated by the presence of a single arylazido-β-alanyl group per pyridine nucleotide unit as evidenced by base hydrolysis of the arylazido-β-alanyl from the nucleotide analog, followed by isolation and quantitative assessment of the hydrolyzed components. Enzymic hydrolysis of the analog with nucleotide pyrophosphatase resulted in the formation of two products, one of which has the characteristic arylazido absorption at 475 nm. Base hydrolysis of this latter material resulted in the formation of arylazido-β-alanine and adenosine monophosphate, establishing the adenine nucleotide ribose as the point of arylazido-β-alanyl esterification. Initially, considerations with respect to the stability of 2′ or 3′ esterification lead to the conclusion that the 3′ hydroxyl of the adenine ribose is the point of esterification under the conditions of the synthetic procedure. This assessment has been substantiated by nuclear magnetic resonance (NMR) studies on a variety of such nucleotide analogs (*145*).

The pyridine nucleotide analog acts as a substrate for yeast alcohol dehydrogenase when incubated with the enzyme in the dark and inhibits reduction of the natural substrate. Upon photolysis in the presence of the enzyme, the analog is converted to a potent inhibitor of yeast alcohol dehydrogenase. The photodependent inhibition is prevented by the presence of NAD$^+$ or NADH during photolysis, but high concentrations of pyridine nucleotide have no effect on reversing the photodependent inhibition, if added subsequent to photolysis.

The reduced arylazido-β-alanine NAD$^+$ formed by the reaction of arylazido-β-alanyl NAD$^+$ with yeast alcohol dehydrogenase in the dark can

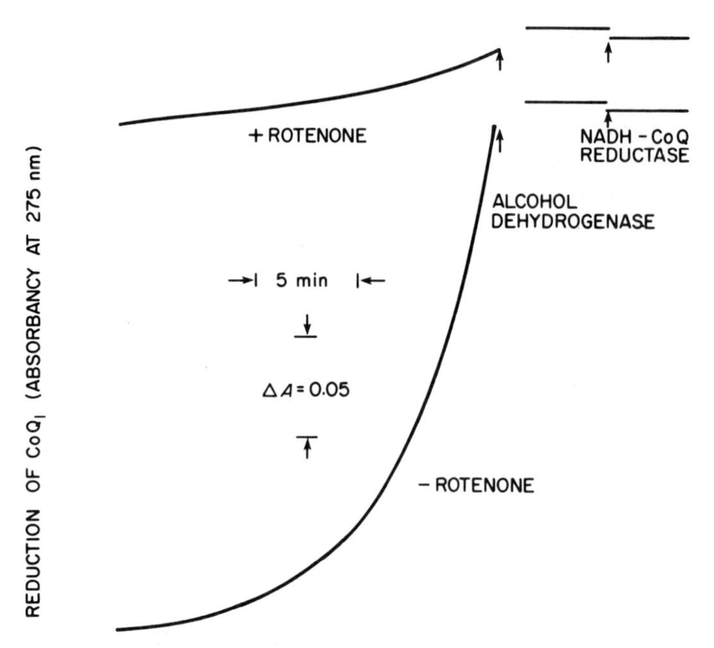

FIG. 2. Arylazido-β-alanine NADH as a substrate for NADH–CoQ reductase. The CoQ reductase activity was measured as a decrease in absorbance at 275 nm, in the presence of ethanol and alcohol dehydrogenase as a regenerating system for arylizado-β-alanine NADH. Control activity was estimated as 0.17 μmol of CoQ_1 reduced per minute per milligram of protein. The assay mixture contained, in a final volume of 1 ml, 0.05 *M* phosphate buffer (pH 7.0), 0.047 m*M* arylazido-β-alanine NAD^+, 0.02 m*M* CoQ_1, and 17.4 m*M* ethanol. NADH–CoQ reductase (20 μg) and yeast alcohol dehydrogenase (20 μg) were added sequentially as indicated in the figure. When present, rotenone was at 25 μ*M*. From Chen and Guillory (*26*).

be utilized as a substrate by the NADH–CoQ reductase of the electron-transport system of ox heart mitochondria. As can be seen from Fig. 2, the arylazido-β-alanine NADH–CoQ reductase activity is rotenone-sensitive. Both the oxidized and reduced pyridine nucleotide analogs inhibit the NADH–CoQ reductase in a photodependent manner, an inhibition that is completely prevented by the presence of pyridine nucleotides during photolysis. In contrast to the photodependent inhibition, the inhibition of NADH–CoQ reductase by arylazido-β-alanine NAD^+ observed in the dark can be reversed competitively by addition of NADH.

Arylazido-β-alanine of itself at the same concentrations does not inhibit yeast alcohol dehydrogenase or the NADH–CoQ reductase, indicating that the NAD^+ moiety of arylazido-β-alanyl NAD^+ is a required portion for directing the reagent to the active site of the enzymes. At 100-fold higher concentrations of arylazido-β-alanine the NADH–CoQ reductase is inhibited to a minor extent. At an arylazido-β-alanine to enzyme ratio of 180 or greater, some 40% maximal inhibition of NADH–CoQ_1 reductase activity could be obtained. Under the same conditions the NADH–$K_3Fe(CN)_6$ reductase is not influenced by arylazido-β-alanine. Such results suggest that the arylazido-β-alanine has a certain definitive specificity toward the different nucleotide-dependent activities. Photodependent inhibition in the presence of arylazido-β-alanine could not be prevented by the presence of either NADH or CoQ in the irradiation mixture, indicating that inhibition was probably not due to interaction at the reactive site of either of these cofactors.

Utilizing arylazido-β-alanyl NADH as a substrate in the NADH–CoQ reductase assay, a k_m value of 33 μM and a V_{max} of 2.6 μmol min^{-1} mg^{-1} was obtained. The arylazido-β-alanyl NADH is thus as active as the natural cofactor in the catalysis of NADH oxidation by NADH–CoQ reductase of Complex I.

In addition to eventually providing active-site structural information for a complex enzyme system, photoaffinity analogs can in theory provide information concerning functional properties. In the case of Complex I, both NADH and NADPH are utilized as substrates for oxidoreductase activities; there are as well pyridine nucleotide transhydrogenase activities associated with this enzyme complex. Because of the obvious complexity of this multienzyme complex and because of the uncertain relationship between pyridine nucleotide oxidoreduction and transhydrogenation, an interpretation of enzymic reaction pathways is difficult. Chen and Guillory (*27*) consequently explored with the aid of arylazido-β-alanyl NAD^+ the interrelationship between NADH oxidoreduction, NADPH oxidoreduction, and NAD(H)–NADP(H) transhydrogenation in Complex I. Two basic questions were posed: Are the NADH and NADPH oxidoreduction systems identical? What is the relationship between pyridine nucleotide reduction and the transhydrogenase systems?

The nucleotide analog (arylazido-β-alanyl NAD^+) was found to act, in the dark, as a competitive inhibitor with respect to both NADH and CoQ_1 for NADH–CoQ_1 reductase ($K_{i,app}$ of 68 μM and 171 μM, respectively). For NADH–$AcPyAD^+$ (3-acetyl-NAD^+) transhydrogenase activity, the analog was a noncompetitive inhibitor with respect to NADH and a competitive inhibitor with respect to $AcPyAD^+$ ($K_{i,pp}$, 9.5 μM) (Figs. 3A, 3B). Arylazido-β-alanyl NAD^+ had only a minor effect on the NADPH–$AcPyAD^+$ transhydrogenase activity and demonstrated a complex kinetic influence on NADPH–CoQ_1 reductase activity in the dark. With respect to this latter activity, the $K_{m,app}$ for NADPH (385 μM) was decreased to 81 μM in the presence of 14.3 μM arylazido-β-alanyl NAD^+ and the $V_{max,app}$ was reduced by 57%. This complex kinetic interaction can be explained as a

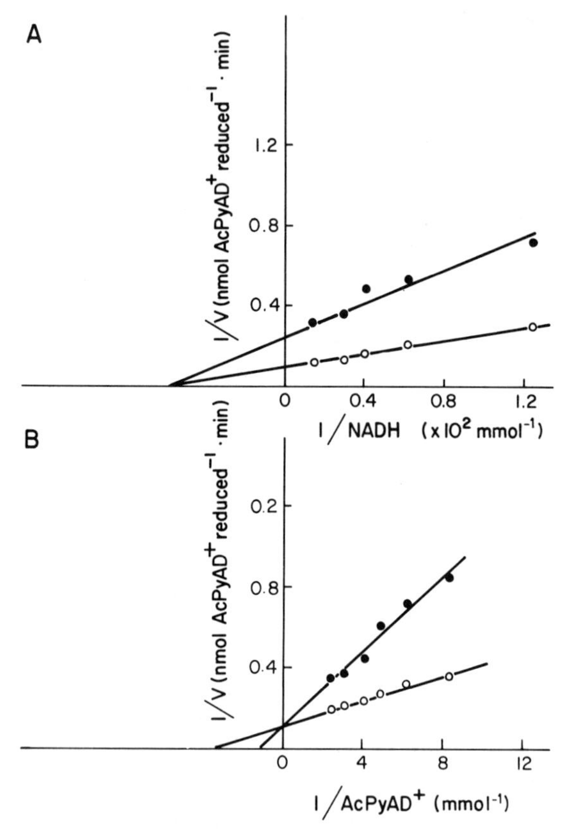

FIG. 3. Kinetic analysis of the arylazido-β-alanyl NAD^+ dependent inhibition on NADH–$AcPyAD^+$ transhydrogenase activity present in the NADH–CoQ reductase (Complex I) preparation. (A) No addition, ○; 19 μM arylazido-β-alanyl NAD^+, ●; the concentration of $AcPyAD^+$ was 200μM. (B) No addition, ○; 21 μM arylazido-β-alanyl NAD^+, ●; the concentration of NADH was 120 μM. Activity was assayed as indicated in the Methods section. The control activity was assayed at 2.86 μmol of $AcPyAD^+$ reduced per minute per milligram of protein. Chen and Guillory (*27*).

simultaneous expression of a transhydrogenase and CoQ reductase action in the presence of the nucleotide analog.

Oxidoreduction

$$\text{NADPH} + H^+ + \text{CoQ}_1 \xrightarrow[\text{Arylazido-}\beta\text{-alanyl NAD}^+]{\uparrow} \text{NADP}^+ + \text{CoQ}_1\text{H}_2 \quad (18)$$

Transhydrogenation

$$\text{Arylazido-}\beta\text{-alanyl NAD}^+ + \text{NADPH} \rightleftharpoons \text{arylazido-}\beta\text{-alanyl NADH} + \text{NADP}^+ \quad (19)$$

Oxidoreduction

$$\text{Arylazido-}\beta\text{-alanyl NADH} + \text{H}^+ + \text{CoQ}_1 \longrightarrow \text{arylazido-}\beta\text{-alanyl NAD}^+ + \text{CoQ}_1\text{H}_2 \quad (20)$$

Arylazido-β-alanyl NAD^+ can influence NADPH–CoQ_1 reductase in two ways. First, it may be a competitive inhibitor with respect to NADPH during the oxidoreduction reaction [Eq. (18)]. Alternatively, it can be an electron acceptor for the transhydrogenase, providing a pool for NADPH–arylazido-β-alanyl NAD^+ transhydrogenation activity [Eq. (19)]. Upon reduction the analog becomes an electron donor and reduces CoQ_1 by the arylazido-β-alanyl NAD^+-sensitive oxidoreduction reaction [Eq. (20)]. Thus the activity of the assay sample containing arylazido-β-alanyl NAD^+ is actually a combination of inhibited NADPH–CoQ_1 reductase [Eq. (18)] and arylazido-β-alanyl NADH–CoQ_1 reductase [Eq. (20)] activities. Since the activity of the NADPH–CoQ_1 reductase [Eq. (18)] is low (the specific activities for NADPH–CoQ_1 and NADH–CoQ_1 reductase activities being around 0.5–1 and 10–15, respectively), the minor NADPH–arylazido-β-alanyl NAD^+ transhydrogenation [Eq. (19)] and arylazido-β-alanyl NADH–CoQ_1 reduction [Eq. (20)] activities can influence the kinetic results significantly, especially at low NADPH concentration. While reliable quantitative values for the analog-dependent activities were not obtained, the involvement of transhydrogenase activity as postuated above was evaluated by the following experiment. When 6.3 μM palmitoyl-CoA was added to the NADPH–CoQ reductase assay mixture, the activity of the control, which did not contain arylazido-β-alanyl NAD^+, was little influenced (see Table I). On the other hand, the activity of the sample containing 14.3 μM arylazido-β-alanyl NAD^+ was inhibited 44%. Palmitoyl-CoA is known to be a potent competitive inhibitor of NADPH–NAD^+ transhydrogenase (*146*). This result suggests that some 50% of the activity of the sample containing arylazido-β-alanyl NAD^+ may be due to the coupled NADPH-arylazido-β-alanyl NAD^+ transhydrogenase and arylazido-β-alanyl NADH–CoQ_1 reductase activities.

Upon photolysis the NAD^+ analog was a potent irreversible inhibitor for both NADH–CoQ reductase and NADH–$AcPyAD^+$ transhydrogenase activities. A maximum inhibition of 70–75% of all the above activities excepting NADPH–$AcPyAD^+$ transhydrogenase could be obtained by irradiation of the enzyme complex with a 25-fold molar excess of the analog

TABLE I

EFFECT OF PALMITOYL-CoA AND ARYLAZIDO-β-ALANYL NAD^+ ON THE NADPH–CoQ_1 REDUCTASE ACTIVITY OF COMPLEX I[a,b]

Additions	Activity (nmol NADPH oxidized/min/mg)	Percent of control
Experiment 1		
a. NADPH (control)	40	100
b. NADPH		
Palmitoyl-CoA	40	100
c. NADPH		
Arylazido-β-alanyl NAD^+	41	103
d. NADPH		
Arylazido-β-alanyl NAD^+		
Palmitoyl-CoA	22.6	56

[a] Chen and Guillory (*27*).

[b] The specific activity of the NADPH–CoQ_1 reductase was 0.124 μmol of NADPH oxidized per minute per milligram of protein. NADPH was 160 μM. Palmitoyl-CoA, when added, was 6.3 μM and arylazido-β-alanyl NAD^+ was 14.3 μM.

(Fig. 4). The photodependent inhibitions could be prevented by the presence of the natural substrate, NADH, during irradiation. Under these concentration conditions there is no photodependent inhibition of the enzyme activities in the presence of arylazido-β-alanine or NAD^+. Table II summarizes the kinetic effects of arylazido-β-alanyl NAD^+ on the pyridine-dependent activities of Complex I from beef heart mitochondria.

The experimental results demonstrate that arylazido-β-alanyl NAD^+ is a specific active site directed reagent for both the NADH–CoQ_1 reductase and NADH–$AcPyAD^+$ transhydrogenase activities present in Complex I by the following criteria: (a) without irradiation arylazido-β-alanyl NADH is a substrate for NADH–CoQ_1 reductase; (b) in the dark the NAD^+ analog is a competitive inhibitor with respect to NADH and CoQ_1 for NADH–CoQ_1 reductase activity and with respect to $AcPyAD^+$ for NADH–$AcPyAD^+$ transhydrogenase; (c) the photodependent inhibition by arylazido-β-alanyl NAD^+ could be prevented by the presence, during the irradiation, of the natural substrate NADH; (d) kinetic analysis indicates that the photodependent inhibition brought about by the analog is noncompetitive. Under identical conditions arylazido-β-alanine or NAD^+ alone or in combination did not significantly inhibit the above activities. During photoirradiation an arylazido-β-alanyl NAD^+ concentration of 8.6 μM provided 50% inhibition of NADH–CoQ_1 reductase activity. This concentration is only 12.6% of the $K_{i,app}$ value for arylazido-β-alanyl NAD^+ as a competitive inhibitor of

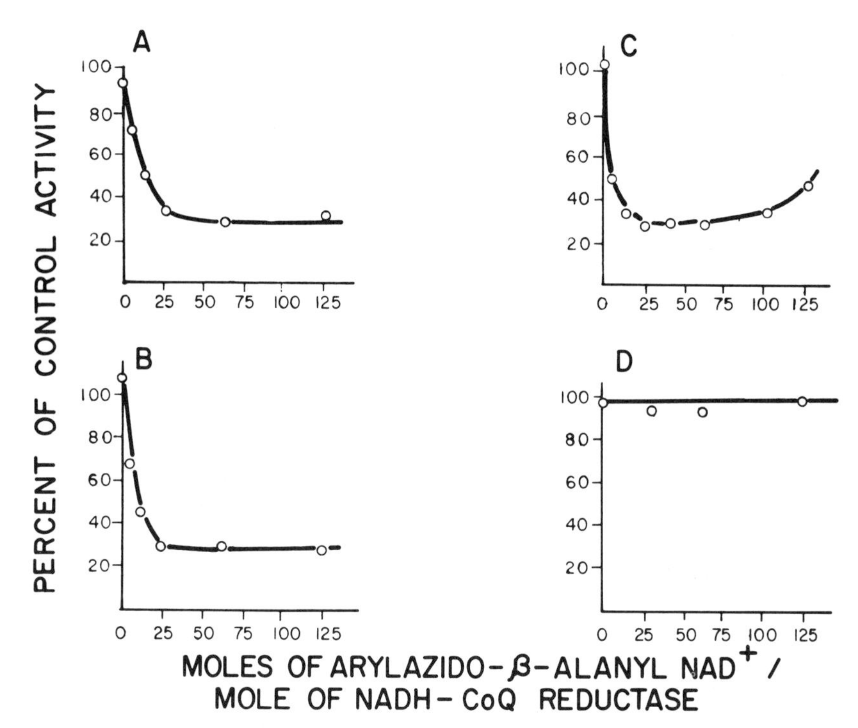

FIG. 4. Photodependent inhibition of arylazido-β-alanyl NAD^+ on the various activities present in NADH-CoQ reductase (Complex I) preparation. The samples were assayed prior to and following a 1-minute period of light irradiation, with the activity prior to photolysis taken as representing 100% activity. Assays were performed as described in the Methods section. For panels A (NADH–CoQ_1 reductase) and B (NADH–$AcPyAD^+$ transhydrogenase), the irradiation mixture contained 0.147 nmol of Complex I and arylazido-β-alanyl NAD^+ at a molar excess above the enzyme concentration as indicated in the respective figures. The volume of the irradiation mixture was 0.24 ml. For panel C (NADPH–CoQ_1 reductase), the irradiation mixture contained 0.441 nmol of Complex I and a concentration of NAD^+ analog at a molar excess over the enzyme concentration indicated in the figure. Irradiation volume was 0.18 ml. For panel D (NADPH–$AcPyAD^+$ transhydrogenase), the irradiation mixture contained 0.294 nmol of Complex I and an analog concentration at the molar excess over the enzyme indicated in the figure. The irradiation volume was 0.27 ml. From Chen and Guillory (*27*).

NADH (i.e., 68 μM). In addition, as a result of the dilution of the photolyzed preparation during enzymic assay, the expected free titer of the nucleotide analog would be only 0.34 μM. The low inhibition titer following photolysis combined with the specificity of interaction are strong indications of the covalent nature of the photolytic dependent interaction.

The finding by Hatefi and his group (*147*) that NAD^+ was not required for either NADPH–CoQ or NADPH–ferricyanide reduction eliminated the

TABLE II

EFFECTS OF ARYLAZIDO-β-ALANYL NAD^+ ON PYRIDINE NUCLEOTIDE-DEPENDENT ACTIVITIES OF COMPLEX I FROM OX HEART MITOCHONDRIA[a]

Reaction	Concentration of fixed substrate (μM)	Concentration of arylazido-β-alanyl NAD^+ (μM)	Inhibitory kinetic effect	$K_{m,app}$ (μM)	$K_{i,app}$ (μM)
NADH–CoQ_1 reductase	30 (CoQ_1)	147	Competitive	7.5 (NADH)	68
	80 (NADH)	140	Competitive	11.1 (CoQ_1)	171
NADH–$AcPyAD^+$ transhydrogenase	200 ($AcPyAD^+$)	19	Noncompetitive	16.7 (NADH)	—
	120 (NADH)	21	Competitive	278 ($AcPyAD^+$)	9.5
NADPH–CoQ_1 reductase	30 (CoQ_1)	14	Mixed	385 (NADPH)	—
NADPH–$AcPyAD^+$ transhydrogenase	300 ($AcPyAD^+$)	157	Competitive	71 (NADPH)	384

[a] From Chen and Guillory (*27*, *150*).

possibility that such reductions were catalyzed via the transhydrogenation system. Spectral studies with the soluble NADH dehydrogenase isolated following urea treatment of Complex I indicated that both NADH and NADPH were capable of independently reducing the flavin chromophore completely.

Hatefi and Galante (*148*) have shown that the soluble NADH dehydrogenase isolated by chaotropic resolution of Complex I as described by Hatefi and Stempel (*149*) has NADPH dehydrogenase in addition to NADPH–NAD^+ transhydrogenase activities. As a working model, they considered that the dehydrogenase (flavoprotein) contains two closely related active sites. The first site is the pyridine nucleotide site responsible for dehydrogenation and for NAD(H) binding involved in transhydrogenation. The second site is that responsible for NADPH(H) binding resulting in transhydrogenation.

Chen and Guillory interpret their study of the interaction of the arylazido analog of NAD^+ with Complex I as indicating the presence of three distinct pyridine nucleotide reactive sites within the preparation. One of these sites is that responsible for the oxidoreductase activities (i.e., NADH–CoQ reductase and NADPH–CoQ reductase activities). The data indicate that these two activities utilize an identical site, since the photodependent inhibitory profiles of arylazido-β-alanyl NAD^+ for both oxidoreductase activities are identical at low arylazido-β-alanyl NAD^+ concentrations. At higher analog concentrations it is suggested that the NADPH–NAD^+ transhydrogenase activity interferes with a correct evaluation of the NADPH–CoQ_1 reductase activity. Both oxidoreductase activities were inhibited more than 70% with a 25-fold molar excess of analog concentration over that of the enzyme. Second, the inhibitory picture clearly suggests a common binding site for both pyridine nucleotides in the oxidoreductase activities. The fact that NADH is much more effective than NADPH in protecting against inhibition of both activities indicates that the two activities arise from an identical site for nucleotide binding. If this were not the case, one would expect NADPH to be much more effective in protecting the NADPH–CoQ_1 reductase activity against photodependent arylazido-β-alanyl NAD^+ inhibition.

A second pyridine nucleotide reacting site indicated is that associated with the NADH–$NADP^+$ transhydrogenase. This activity responds differently toward arylazido-β-alanyl NAD^+ when compared to the response of the other activities associated with Complex I. In this case arylazido-β-alanyl NAD^+ is a noncompetitive inhibitor toward NADH and a competitive inhibitor with respect to $AcPyAD^+$. This result is consistent with an ordered mechanism for the reaction, with NADP(H) the first nucleotide bound to the reactive site.

The possibility of an identical catalytic site being responsible for NADH–AcPyAD$^+$ transhydrogenase and NADH dehydrogenase is ruled out by the fact that arylazido-β-alanyl NAD$^+$ is a competitive inhibitor with respect to NADH for the oxidoreductase activities (NADH–CoQ_1 reductase and NADH–$K_3Fe(CN)_6$ reductase activities) but not for the NADH–AcPyAD$^+$ transhydrogenation.

The third postulated independent nucleotide binding site is that associated with the NADPH–NAD$^+$ transhydrogenase. The experimental evidence pointing to this distinct site is indirect in that arylazido-β-alanly NAD$^+$ is not an effective inhibitor of this reaction in the dark or under photoirradiation.

Arylazido-β-alanyl NADP

In addition to arylazido-β-alanyl NAD$^+$ an NADP$^+$ analog, arylazido-β-alanyl NADP$^+$ a potential photoaffinity analog for some NADP$^+$-dependent enzyme systems, has been prepared by Chen and Guillory utilizing the same methods developed for adenine nucleotides and NAD$^+$ (*150*). The major azido-containing product formed by interaction of NADP with arylazido-β-alanine in the presence of carbodiimidazole was subjected to structural analysis using the procedure initially applied for the elucidation of the structure of the NAD$^+$ analog. From this analysis it is clear that the product has a single arylazido-β-alanyl group associated with the NMN portion of NADP$^+$. The structural assignment is *N*-3′-*O*-{3-[*N*-(4-azido-2-nitrophenyl)amino]propionyl} NADP$^+$ (Compound 42).

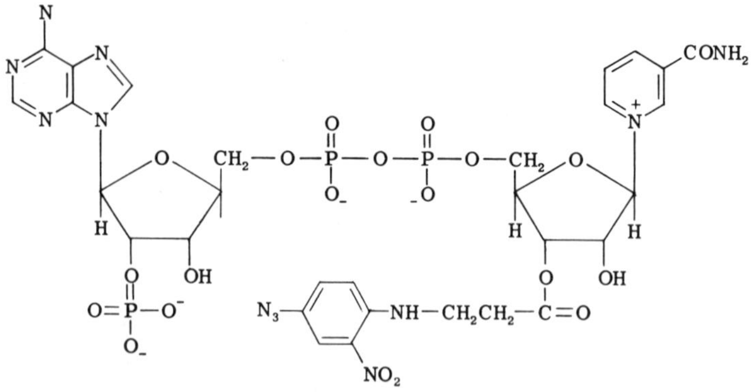

Compound 42:
Arylazido-β-alanyl NADP$^+$;
N-3′-*O*-{3-[*N*-(4-azido-2-nitrophenyl)amino]propionyl} NADP$^+$

TABLE III

PAPER CHROMATOGRAPHY OF NUCLEOTIDES AND THEIR ARYLAZIDO-β-ALANYL DERIVATIVES[a]

	R_f Values		
Compound	*n*-Butanol/water/ acetic acid (5/3/2) v/v	1 *M* Ammonia acetate/ethanol (3/7) v/v	Isobutyric acid/ ammonia/water (70/1/29) v/v
$NADP^+$	0.07	0.02	0.24
Arylazido-β-alanyl $NADP^+$	0.40	0.03	0.52
NMN	0.31	0.08	0.38
Arylazido-β-alanyl NMN	0.63	0.26	0.64
AMP	0.29	0.07	0.45
Arylazido-β-alanyl AMP	0.61	0.21	0.78
Arylazido-β-alanine	0.91	0.70	0.88
NAD^+	0.12	0.09	0.41
Arylazido-β-alanyl NAD^+	0.49	0.19	0.57

[a] From Chen and Guillory, R.J. (*27,150*).

In addition to the pyridine nucleotides Chen and Guillory have described the chemical synthesis of the arylazido-β-alanyl NMN and arylazido-β-alanyl AMP (*150*). Both of the arylazido analogs of NMN and AMP were determined to have a single *N*-(4-azido-2-nitrophenyl)-β-alanyl group attached to the hydroxyl group (3′) on the ribose portion of the molecules; they are named 3′-*O*-{3-[*N*-(4-azido-2-nitrophenyl)amino]propionyl} NMN and 3′-*O*-{3-[*N*-(4-azido-2-nitrophenyl)amino]propionyl} AMP, respectively. These compounds were utilized for the characterization of the structure of the intact pyridine nucleotide analogs. Table III presents some characteristic R_f values for these analogs in a number of solvent systems.

The $NADP^+$ analog has been utilized in a study of the NAD(P)H–CoQ_1 reductase, NADH–$AcPyAD^+$ transhydrogenase, and NADPH–$AcPyAD^+$ transhydrogenase activities present in the mitochondrial NADH–CoQ reductase (Complex I). Among these activities only NADPH–$AcPyAD^+$ transhydrogenase was competitively inhibited in the dark by arylazido-β-alanyl $NADP^+$ with respect to NADPH. The evaluated $K_{i,app}$ value for the $NADP^+$ analog in this later activity was 1.94 μM, which can be compared to the $K_{m,app}$ for NADPH of 90.9 μM (Fig. 5).

Arylazido-β-alanyl $NADP^+$ is as well a noncompetitive inhibitor with respect to $AcPyAD^+$. Table IV summarizes the kinetic results of the effect of arylazido-β-alanyl $NADP^+$ on the measured activities present in the NADH–CoQ reductase preparation. The results show that of all the Complex I activities assayed, only the NADPH–NAD^+ transhydrogenase

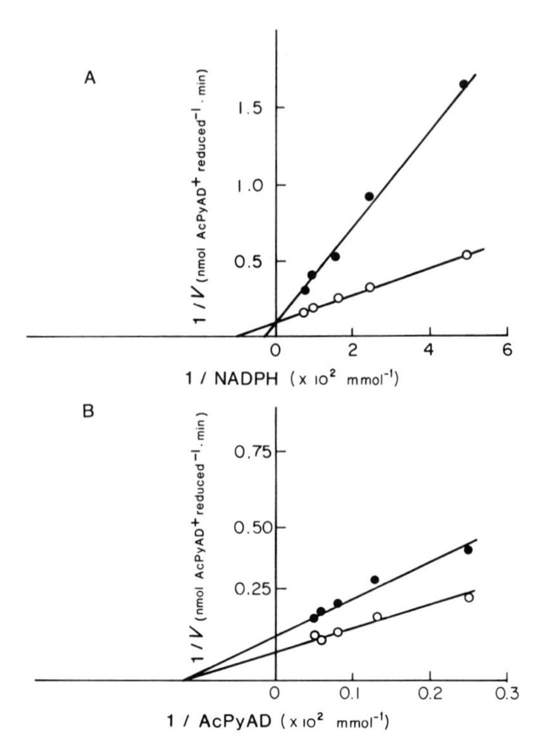

FIG. 5. Kinetic analysis of the inhibition by arylazido-β-alanyl $NADP^+$ of the NADPH–$AcPyAD^+$ transhydrogenase activity present in NADH–CoQ reductase (Complex I) preparation. (A) Without further additions, ○; in the presence of 5.17 μM arylazido-β-alanyl $NADP^+$, ●; the concentration of $AcPyAD^+$ was 300 μM. (B) Without further additions, ○; in the presence of 5.17 μM arylazido-β-alanyl $NADP^+$, ●; the concentration of NADPH was 200 μM. The control activity was assayed at 1.05 μmol of $AcPyAD^+$ reduced per minute per milligram of protein. From Chen and Guillory (*150*).

activity is influenced by arylazido-β-alanyl $NADP^+$. This analog thus has an effect opposite to that of arylazido-β-alanyl NAD^+, which is a potent inhibitor of all the activities measured in the Complex I preparation, except for the NADPH–NAD^+ transhydrogenase.

After a 1-minute photolysis of enzyme complex with 44 nmol of arylazido-β-alanyl $NADP^+$, only the NADPH–$AcPyAD^+$ transhydrogenase was found to be inhibited. Photolysis in the presence of $NADP^+$ or arylazido-β-alanine did not result in significant inhibition. However, the dark control, i.e., the enzyme mixed with an identical concentration of arylazido-β-alanyl $NADP^+$ but not photolyzed, revealed that NADPH–$AcPyAD^+$ transhydrogenase activity was inhibited to 41% without photolysis as compared to 54% after photolysis. The inability to demonstrate a clear indication of a

TABLE IV

EFFECT OF ARYLAZIDO-β-ALANYL NADP$^+$ ON THE PYRIDINE NUCLEOTIDE-DEPENDENT ACTIVITIES PRESENT IN THE NADH–CoQ REDUCTASE (COMPLEX I) PREPARATION FROM OX HEART MITOCHONDRIA[a]

Reaction	Concentration of fixed substrate (μM)	Concentration of arylazido-β-alanyl NADP$^+$ (μM)	Inhibitory kinetic effect	$K_{m,app}$ (second substrate) (μM)	$K_{i,app}$ arylazido-β-alanyl NADP$^+$ (μM)
NADH–CoQ$_1$ reductase	30 (CoQ$_1$)	183	None	12.3 (NADH)	—
NADH–AcPyAD$^+$ transhydrogenase	300 (AcPyAD$^+$)	155	None	13.3 (NADH)	—
NADPH–CoQ$_1$ reductase	30 (CoQ$_1$)	183	None	357 (NADPH)	—
NADPH–AcPyAD$^+$ transhydrogenase	300 (AcPyAD$^+$)	5.2	Competitive	90.9 (NADPH)	1.94
	200 (NADPH)	5.2	Noncompetitive	83.3 (AcPyAD$^+$)	—

[a] Chen and Guillory (*150*).

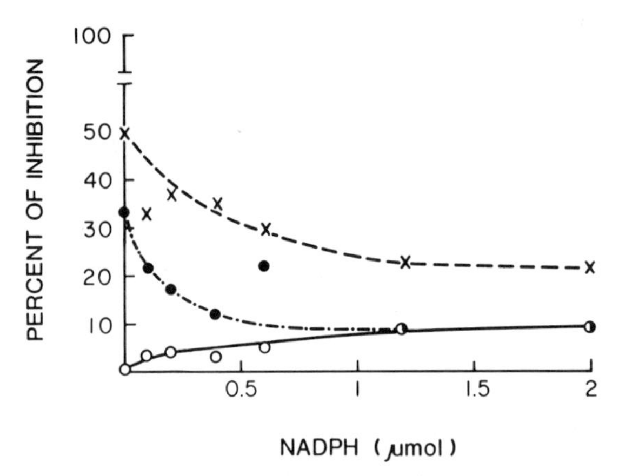

FIG. 6. NADPH reversal of arylazido-β-alanyl $NADP^+$ inhibition of NADPH–$AcPyAD^+$ transhydrogenase activity present in the NADH–CoQ reductase (Complex I) preparation. The mixture, 0.175 ml containing 60 nmol of arylazido-β-alanyl $NADP^+$, 0.3 mg of Complex I, and NADPH concentration as indicated in the figure, were assayed for NADPH–$AcPyAD^+$ transhydrogenase activity before (●) and after (×) a 2-minute photoirradiation. Control activity prior to photolysis was taken as 100%. The control containing Complex I and NADPH had identical activities when measured before or after (○) photolysis. From Chen and Guillory (*150*).

major photodependent effect was surprising. In all cases it was clear that the photolyzed sample was always inhibited 10–15% more than that without photolysis, even at a 400-fold molar excess of analog to enzyme, at which level there was an 80% inhibition. In view of the apparent strong affinity of arylazido-β-alanyl $NADP^+$ to the active site of the NADPH-dependent transhydrogenase present in Complex I, a higher concentration of substrate, NADPH, would be expected to be required in order to afford protection against analog inhibition. A substrate protection experiment was therefore performed with NADPH concentration up to 2 μmol present during irradiation (Fig. 6). Several important aspects of arylazido-β-alanyl $NADP^+$ inhibition on NADPH–$AcPyAD^+$ transhydrogenase activity were revealed by this experiment. The control sample containing increasing NADPH but no nucleotide analog was inhibited upon photolysis maximally only 9% at 2 μmol of NADPH. When 60 nmol of arylazido-β-alanyl $NADP^+$ were added to the enzyme solution, the NADPH-dependent transhydrogenase activity was inhibited to 33% without photolysis. This analog-dependent inhibition could be completely prevented by 1.2 μmol of NADPH. Upon photolysis of the enzyme analog mixture a 50% inhibition was obtained, which could be reduced to 24% inhibition in the presence of 2 μmol of NADPH.

In another experiment, when 0.18 μmol of arylazido-β-alanyl $NADP^+$

was added to 0.3 mg of the NADH–CoQ reductase preparation, there was a 65% inhibition in the case of the photolyzed sample and a 50% inhibition for the sample in which the analog was photolyzed prior to mixing with the enzyme. Both samples were then subjected to another 2-minute irradiation and similar percentages of inhibition were obtained, 68 and 64%, respectively. Clearly, maximal irreversible inhibition is obtained only upon photolysis of the enzyme complex in the presence of the nucleotide analog. The increase in inhibition following the second photolysis for that sample, in which the analog was photolyzed prior to mixing with the enzyme, could be prevented by the presence of 0.3 μmol of NADPH during the second photolysis, suggesting a specific active-site reactivity for the analog.

The photodependent covalent labeling by arylazido-β-alanyl $NADP^+$ has been further indicated recently by experiments utilizing radioactive analog arylazido-[3-^3H]β-alanyl $NADP^+$ (S. Chen and R. J. Guillory, unpublished observations).

The question arises why the difference between photodependent and non-photodependent inhibitory effects are so small. Since the $NADP^+$ analog is already a very potent inhibitor for NADPH–$AcPyAD^+$ transhydrogenase activity, such a small difference between two conditions might be expected. However, a concentration of arylazido-β-alanyl $NADP^+$ of 400-fold molar excess over that enzyme is required to provide 80% inhibition (cf. only a 25-fold molar excess of arylazido-β-alanyl NAD^+ over that enzyme concentration is able to inhibit the NADH–CoQ_1 reductase or NADH–$AcPyAD^+$ transhydrogenase activities, a comparable amount). In an attempt to answer this problem, the nature of the $NADP^+$ analog inhibition of the NADPH–$AcPyAD^+$ transhydrogenase following photolysis was investigated by a kinetic analysis. Surprisingly, a competitive inhibition was found for the enzyme with respect to NADPH following irradiation with arylazid-β-alanyl $NADP^+$. The $K_{i,app}$ values for arylazido-β-alanyl $NADP^+$ with and without light irradiation were 2.71 μM and 3.92 μM, respectively. Although the data indicate the analog to have a lower affinity in the latter case, the difference observed is within experimental error.

A possible explanation for these results is that the analog inserts itself within the area of the active site, but that there is sufficient flexibility provided either by the nucleotide derivative or the binding site itself to allow for reversible interaction of the natural nucleotide base with the catalytic site. Thus, while the analog competes with the natural substrate for the catalytic site, the positioning of the photoaffinity group is not of a proper orientation to allow for insertion although it effectively blocks the catalytic site. The bound analog consequently provides a steric hindrance against the reactivity of NADPH at the active site of enzyme accounting for the competitive-type inhibitory effect. Alternatively, there is the possibility that this enzyme has a secondary (allosteric) binding site with which the nucleotide analog binds preferentially, influencing reactivity at the catalytic site.

Recent investigations on Complex I utilizing radioactive photoaffinity analogs, i.e., arylazido-[3-^{3}H]β-alanyl NAD$^+$ and arylazido-[3-^{3}H]β-alanyl NADP$^+$, have been combined with fractionation with chaotropic agents and SDS gel electrophoresis. These studies have revealed a high order of specific interaction following photolytic labeling (S. Chen and R. J. Guillory unpublished observations). Peptides of the purified complex with molecular weights of 90,000, 76,000, 37,000, and 25,500 are labeled under conditions indicating that they are involved with the reactive site for the NADPH-dependent transhydrogenase, those of molecular weight 76,000 and 50,000 with the NADH-dependent transhydrogenase, and those of 50,000, 30,500, and 14,500 with the reactive site for NADH dehydrogenase.

Whether the overlapping labeling pattern (i.e., that the 76,000 peptide is involved with both NADH- and NADPH-dependent transhydrogenation and the 50,000 with the NADH dehydrogenase and NADH-dependent transhydrogenase) represents true structural relationships in the enzymic system is not yet clear. Cross-linking studies of the components of Complex I together with further evaluation of the binding pattern of arylazido-β-alanyl NAD$^+$ and the NADP$^+$ analog may provide us with a clearer interpretation for the specific enzymic role of each of the peptides of Complex I.

4. *Cytochrome c and Cytochrome Oxidase*

An important theoretical consideration for electron transfer is whether such transfer occurs between bound complexes that are rigidly attached to their positions within the mitochondrial membrane during electron transfer, or whether rotational or lateral diffusion of some or all of the carriers is required for the electron flow to occur. As a means of testing this consideration, recent work from a number of laboratories has dealt with the chemical modification of mitochondrial cytochrome *c* with adjuncts containing the azido function. It was reasoned that, because of the relative ease of reversible dissociation and reassociation of the cytochrome *c* molecule with the particulate cytochrome *c* resolved mitochondrial membrane, one would be able to reversibly bind modified cytochrome *c* molecules. After binding, irradiation of the protein membrane complex would cement the molecule into a rigid and presumably "natural" position relative to other electron carriers. Examination of the enzymic reactivity of such a preparation would then reveal aspects of the structural requirements for the carriers participating in electron transport.

Wilson *et al.* (*114*) prepared the azido derivative 2,3-dinitro-5-fluorophenylazide (DNFA) and characterized it with respect to the rate of nucleophilic displacement of the active fluorine by methylamine, histidine, and glycylglycine at 30°C and pH 10. The relative reactivities of 3-nitro-4-fluorophenylazide, 2,4-dinitro-5-fluoroaniline, 2,4-dinitrofluorobenzene, and 2,4-dinitro-5-fluorophenylazide were 1 : 140 : 1000 : 2000. The high reactivity of DNFA permits the photoaffinity label to be attached to free amino groups in a few minutes at temperatures below 30°C and pH values below 10.

Utilizing this azido analog, Erecinska *et al.* (*115*) were successful in preparing a photoaffinity-labeled derivative of horse heart cytochrome *c*. Two dinitrophenyl groups were bound per mole of cytochrome *c*, the modified cytochrome having a half-reduction potential identical to that of native cytochrome *c* (± 10 mV). The half-reduction potential of free unmodified cytochrome *c* is about 60 mV more positive than that of the bound form. In intact mitochondria this amounts to 0.235 V.

When added to cytochrome *c*-depleted mitochondrial membranes, the cytochrome *c*, after photolysis, was covalently bound. Although covalently bound, the cytochrome *c* was active in mediating electron transfer between *N,N,N′,N′*-tetramethyl-*p*-phenylenediamine (TMPD) (ascorbate) and cytochrome oxidase. Fractionation of the irradiated mitochondria by means of detergents showed that the labeled cytochrome *c* was bound to the cytochrome oxidase in a 1 : 1 molar complex. There was no indication of covalent linkage to cytochrome *c* reductase. That the protein itself is the actual site of interaction, rather than a lipid portion of the oxidase complex, is indicated by the fact that lipid-depleted cytochrome oxidase (*151,152*) binds one to two molecules of cytochrome *c* analog.

The specificity of covalent interaction with cytochrome oxidase suggests that one of the cytochrome *c* binding sites within the mitochondrial membrane is on cytochrome oxidase, and that the side that binds to the oxidase has different structural characteristics from the side that *must* be pointing toward the cytochrome *c* reductase. Comparison with the characteristics of a trinitrophenylated derivative of cytochrome *c* makes lysine 13 a contender for the cytochrome *c* interaction with DNFA. More extensively modified cytochrome *c* (with DNFA) is able to insert photolytically into what is believed to be the cytochrome *c* binding site on the bc_1 complex.

Since the cytochrome *c* covalently bound to cytochrome oxidase retains full activity with respect to electron transport between TMPD (ascorbate) and the oxidase, it is reasoned that dissociation of the cytochrome *c* from the membrane and equilibration with a pool of cytochrome *c* molecules is not required for such electron transfer.

The retention of succinate oxidase activity in the same preparation is taken as providing evidence that lateral diffusion of cytochrome *c* in place, within the membrane, is not required for the transfer of reducing equivalents from cytochrome C_1 to cytochrome oxidase. In this study, although the labeled cytochrome *c* bound covalently to cytochrome *c* oxidase in a 1 : 1 molar complex, the overall yield of this complex was less than 5%.

In a continuation of the work on photoaffinity derivatives of mitochondrial cytochrome *c*, Erecinska allowed horse heart cytochrome *c* to react with methyl-4-mercaptobutyrimidate hydrochloride.

$$HS{-}(CH_2)_3{-}\overset{\overset{NH_2^+}{\|}}{C}{-}O{-}CH_3$$

The free sulfhydryl group of the adjunct on the protein was then covalently linked to *p*-azidophenacylbromide, yielding a photoaffinity cytochrome *c* analog (*116*). The presence of a $=NH_2^+$ group maintains the same charge balance on the cytochrome *c* molecule, since the imidate reacts with ε-amino groups of lysine residues.

Under the conditions described, three lysine residues of horse heart cytochrome *c* were modified. The photoaffinity cytochrome *c* analog was bound by irradiation into a covalent complex with the cytochrome *c* oxidase of cytochrome *c*-depleted pigeon breast mitochondria. The K_d for photoaffinity analog of cytochrome *c* binding to the cytochrome-depleted membrane was approximately 1×10^{-8} *M* with two high-affinity binding sites apparent for the labeled cytochrome *c* per cytochrome *a* in the mitochondrial membrane. The modified cytochrome restored 70–75% of the oxygen uptake normally observed for native cytochrome *c*.

Disruption of the cytochrome *c* analog mitochondrial membrane complex with Triton X-100 followed by SDS gel electrophoresis indicates the presence of a cytochrome *c*-cytochrome oxidase 1 : 1 molar complex as well as a cytochrome *c*–cytochrome oxidase 2 : 1 complex; the latter was considered to be a cytochrome *c*–cytochrome *c*–cytochrome oxidase arrangement. Preliminary results indicate that the two smallest polypeptide chains (MW 9000 and 11,000) of the oxidase may be the site of cytochrome *c* binding.

The preparation of two other azido-modified derivatives of cytochrome *c* has been reported by Bisson *et al.* (*153*). Cytochrome *c* (horse heart) was incubated with 4-fluoro-3-nitrophenylazide, and the reaction products were separated by chromatographic procedures. Two mono-labeled derivatives P_2 and P_1 were subjected to analysis, and their interaction with purified cytochrome oxidase was evaluated. Evidence is presented indicating that P_2 has the lysyl 13 modified and P_1 is assumed from structural considerations to be labeled at residue 22.

The catalytic activity of P_1 is essentially maintained with respect to V_{max} and K_m when assayed in the presence of cytochrome oxidase, while P_2, in which lysyl 13 is modified, had a K_m some 4-fold higher than that of the native cytochrome *c*. When each was photolyzed in the presence of purified cytochrome *c* oxidase, P_1 formed a covalent complex containing 0.1 cytochrome *c* per mole of oxidase, while P_2 formed a 0.8 molar ratio of cytochrome *c* to cytochrome oxidase. In the absence of added native cytochrome *c*, the covalently bound cytochrome *c* was not able to function as an electron carrier. In the presence of saturating levels of added native cytochrome *c*, the cytochrome *c* oxidase complex of P_1 cytochrome *c* was fully active, i.e., catalyzed the same maximum reaction rate as with native cytochrome *c*; on the other hand, that complex formed with cytochrome *c* in which lysyl 13 was modified, i.e., P_2, was inhibited more than 50%.

The polypeptide profile of the two complexes as revealed by polyacrylamide gel electrophoresis in SDS buffer showed the complex with the cytochrome *c* modified at residue 22 to have the same profile as that of the control oxidase preparation. On the other hand, covalent labeling with the cytochrome *c* modified at lysyl 13 (i.e., P_2) showed a large diminution in cytochrome oxidase band II. This was associated with an increase in cytochrome oxidase band I. The results indicate that the labeling of polypeptide band II of the cytochrome oxidase by the cytochrome *c* photoaffinity analog is specific. It is concluded that the polypeptide band II (MW 23,700) may represent the natural binding site of cytochrome *c* on cytochrome *c*-oxidase.

The requirement for exogenous native cytochrome *c* is difficult to reconcile with the results of Erecinska *et al.* (*115*), who have accomplished the active covalent coupling between an arylazido cytochrome *c* derivative and cytochrome oxidase of cytochrome *c*-resolved mitochondrial membranes. On the basis of the above results, one might conclude that the covalent labeling examined by Bisson *et al.* (*153*) is perhaps less oriented than that of Erecinska *et al.* (*115*); i.e., that proper and complete orientation requires membrane constraints in addition to direct interaction with the oxidase. Alternatively, the cytochrome *c*-resolved mitochondrial membrane utilized by Erecinska *et al.* (*115*) may yet contain some endogenous cytochrome *c*. It would be interesting to know more about the mechanism by which exogenous cytochrome *c* stimulates the oxidative reactions within the cytochrome *c*–cytochrome oxidase complex used by Bisson *et al.* (*153*). It is difficult to compare the two reconstituted respiratory systems, since one utilizes a purified oxidase preparation [Bisson *et al.* (*153*)], and the other utilizes cytochrome *c*-resolved mitochondrial membranes.

Another point of uncertainty arises with respect to the particular subunit of the cytochrome oxidase to which cytochrome *c* binds under normal enzymic reactivity. A gel electrophoretic analysis of the photodependent reaction product of horse heart cytochrome *c*, modified at lysyl 13 by the addition of a 4-nitrophenylazide group, and purified cytochrome oxidase, indicated that polypeptide band II of the oxidase is labeled by the azido cytochrome *c* (*115*). No labeling is observed for subunits V, VI, and III, which are presumably the more exposed subunits. On the other hand, when Birchmeier *et al.* (*154*) allowed yeast cytochrome *c* oxidase to react with yeast iso-1-cytochrome *c* [whose single free sulfhydryl group at position 107 was activated with 5,5′-dithiobis(2-nitrobenzoate)], the derivatized cytochrome *c* specifically formed a mixed disulfide with subunit III of the oxidase (MW 24,000). Since subunit II contains a major portion of the active free sulfhydrl groups in cytochrome oxidase, the formation of a specific mixed disulfide between the modified cytochrome *c* and subunit III of cytochrome oxidase cannot be a random event. These investigators concluded that a specific noncovalent binding of thionitrobenzoate cytochrome *c* to the oxidase is a prerequisite for disulfide formation. Consequently, the hydrophobic mitochondrial subunit III of yeast cytochrome *c* oxidase is considered to be in close proximity to the cytochrome *c* binding site on the enzyme, at least on the same side of the mitochondrial inner membrane as cytochrome *c* itself.

The results of Birchmeier *et al.* (*154*) appear to be at variance with those reported by Vanderkooi *et al.* (*155*). The latter found that the binding of yeast iso-1-cytochrome *c* (carrying an electron spin label on cysteinyl residue 103) to cytochrome *c*-resolved membranes did not result in any measurable immobilization of the spin label. When horse heart cytochrome *c* spin-labeled at methionine 65 or cysteine 103 was used, the former label was immobilized upon binding to the membrane, but the latter was not. The immobilization of the methionine 65 spin label suggests that cytochrome *c* binds to the membrane at the side at which methionine 65 is located.

It is generally assumed that cytochrome *c* binding to the depleted mitochondrial membrane is similar for all cytochrome *c* molecules tested. This assumption is based upon similarity of biological properties, i.e., the concentration dependency for maximal reconstitution of electron transport rate. However, when compared to horse heart cytochrome *c*, the beef heart cytochrome *c* has only three amino acids that are different, but baker's yeast cytochrome *c* has nearly a third different (*156*). The methionine 65 is on the side of cytochrome *c* that contains a great deal of invariant amino acids. The differences in the immobilization of spin labels associated with cytochrome *c* from yeast and horse heart cytochrome *c* may yet be found to reside in their individual structures. Perhaps the specific folded configuration upon binding to the mitochondrial membrane is different owing to a specific requirement for integration into the multienzyme membrane. A possible indication of this is found in the experiments of Folin *et al.* (*157*), who were able to bring about a selective heme-sensitized photooxidation of either histidine 18 or methionine 80 of horse heart ferricytochrome *c*. In cytochrome *c*-depleted mitochondria, the histidine 18-modified cytochrome exhibited no enzymic activity in spite of very limited conformational changes due to photooxidation. On the other hand, methionine 80-modified protein, which undergoes large conformational rearrangement, has its heme orientation maintained as well as full enzymic properties. The authors discuss the possibility of a specific conformation of the protein being required for activity induced by the cytochrome–membrane interaction.

An approach to the study of the topography of membrane-bound protein is suggested by the interesting work of Jori *et al.* (*158*) on photosensitizer modification of proteins. In this technique proteins containing chromophores, which act as sensitizers for amino acid photooxidation, are modified in the amino acid side chains surrounding the photosensitizer. Irradiation of a ferrocytochrome *c*–cardiolipin complex at pH 8.8 with visible light leads to the photooxidative modification of histidine 18, tyrosine 48, and methionine 80. Comparison with the results of photooxidation of unbound ferrocytochrome *c* indicates that the interaction between cardiolipin and ferrocytochrome *c* provokes a perturbation of the protein conformation with significant alteration upon complexation in the geometry of the heme environment. The photooxidized amino acid residues of ferricytochrome *c* were histidine 18 and methionine 80, whereas for ferrocytochrome, tryptophan 59 and tyrosine 48 were modified as well. When bound to cardiolipin, the histidine 18, methionine 80, and tyrosine 48 residues of ferrocytochrome *c* were modified, whereas for bound ferricytochrome *c* only histidine 18 and tyrosine 48 underwent modification. What is clear is that the structural characteristics of cytochrome *c* free or bound to phospholipid microdispersions are different. The authors point out that studies of protein conformation in the crystal state or in aqueous solution may not give complete information with respect to *in vivo* condition, especially for membrane-bound proteins.

Folin *et al.* (*159*) studied the photooxidation of horse and sperm whale myoglobin sensitized via the heme group. They presented evidence that suggests that the main factor determining the photolability of the myoglobins is their spin state. Only low-spin myoglobins undergo chemical modification when irradiated with wavelengths directly absorbed by the prosthetic group. High-spin myoglobins are insensitive to visible light.

Cytochrome *c* has been chemically modified by methylene blue-mediated photooxidation (*160*), the methionine residues of the protein being specifically converted to methionine sulfoxide residues. No oxidation of any other amino acid residues or of the cysteine thioether bridges of the molecule were detected. Circular dichroism spectra indicate that the heme crevice of methionine sulfoxide ferri- and ferrocytochrome *c* is weakened relative to native cytochrome *c*. The redox potential of methionine sulfoxide cytochrome *c* is 184 mV compared to 260 mV for the native cytochrome *c*. The modified protein is a substrate for cytochrome oxidase, but is not very active with succinate–cytochrome *c* reductase. The authors conclude that the major role of the methionine 80 in cytochrome *c* is to preserve a closed hydrophobic heme crevice, which is essential for the maintenance of the redox potential of the protein.

Vanderkooi (*161*) has described an interesting reaction in which ferricytochrome *c* can be reduced by excited-state phenothiazine, although it is not a direct photoaffinity reaction. Cytochrome *c*, a positively charged molecule, adheres to the surface of negatively charged phospholipid dispersions, and the phenothiazine, which is solubilized by the phospholipid artificial membrane, reduces the cytochrome *c*. Under conditions in which cytochrome *c* is not bound to phospholipid, the rate of reduction by phenothiazine is greatly reduced. The reaction is discussed in terms of the interaction of the excited triplet state of phenothiazine with cytochrome *c* and a tunneling mechanism for electron transfer.

Examination of the photoreduction of cytochrome c_1 revealed that ferricytochrome c_1 was reduced completely by illumination under anaerobic conditions (*162*). Photoreduction was not affected by the ionic strength of the medium, was restricted to pH region 6 to 10, and did not take place in the presence of *p*-hydroxymercuric benzoate. When the photoreduced C_1 was reoxidized with ferricyanide, it could no longer be reduced upon illumination. The reductant is considered to be a specific sulfhydrl group of the subunit containing the heme of the cytochrome. This subunit was found to contain one less titrataible —SH group in the photoreduced as compared to the untreated preparation.

B. Oxidative Phosphorylation and Uncoupling Site(s) in the Inner Mitochondrial Membrane

The observation that 2,4-dinitrophenol and related phenols can act to bring about an uncoupling or dissociation of the oxidative from phosphorylative reaction in mitochondrial systems was initially made over 35 years ago (*163*). Since then a large and varied assortment of chemical agents have been found to possess the physiological properties of uncoupling agents. The mechanism considered for their action has varied from a postulated interaction with a specific respiratory component, or component of the energy transfer system, to that of acting as a reagent that functions by disrupting the normal permeability barriers of membrane systems. With the isolation and purification of coupling factor 1 from beef heart mitochondria, the nitrophenols were shown to have the property of stimulating the factors ATPase activity (*3*), a property now known to be characteristic of other enzymes (*164*).

Hanstein and Hatefi (*165*) first reported on the synthesis of 2-azido-4-

nitrophenol (Compound 34), a structural analog of the classical uncoupler 2,4-dinitrophenol. The analog is a potent water-soluble uncoupler of oxidative phosphorylation some 3-fold more potent than 2,4-dinitrophenol. Binding of the azido derivative is reversible and occurs in two distinct phases, initially a phase in which there is a relatively large uptake of the reagent at low uncoupler concentrations followed by a second phase characterized by weaker binding, which is a linear function of 2-azido-4-nitrophenol concentration. The initial phase is characterized by a maximal specific binding estimated at 0.56 nmol of analog per milligram of protein (3°C, pH 7.0). This is of the same magnitude as that estimated for the concentration of F_1ATPase (0.2 μmol) or electron carriers, excepting CoQ, (0.15–0.7 nmol mg^{-1}). The initial binding occurs without appreciable cooperativity (K_d 6 $\pm$ 3 μM at 3°C). The weaker binding takes place as a linear function of analog concentration without saturation and with binding estimated in excess of 10 nmol mg^{-1}. In contrast to the biphasic binding with beef heart mitochondria, the binding of 2-azido-4-nitrophenol to erythrocyte ghosts is monophasic up to 120 μM reagent.

Upon irradiation of the tritium-labeled analog in the presence of mitochondria, some 40% of the radioactivity becomes covalently associated with protein bands in the molecular weight region of 20,000 to 30,000. Binding of the analog is competitively inhibited by other uncouplers, such as sodium azide, 2,4-dinitrophenol, pentachlorophenol carbonyl cyanide *m*-chlorophenyl hydrazone, and 5-chloro-3-*t*-butyl-2′-chloro-4′-nitrosalicylanilide. On the other hand, antimycin A, rutamycin, valinomycin, or arsenate had no influence on binding.

The finding that binding is present to the same extent for phosphorylating submitochondrial particles and submitochondrial particles deficient in F_1(ATPase) was taken as an indication that the binding sites are located within the inner mitochondrial membrane but not specifically on F_1(ATPase). Another indication of the specificity of interaction is the fact that removal of more than 80% of the mitochondrial lipids with aqueous acetone did not change the extent of analog binding. Preparations of the different electron-transfer complexes did not bind the azido derivative. Since 2,4-dinitrophenol inhibits competitively both the equilibrium and covalent (light dependent) binding to beef heart mitochondria, there is an indication that the same uncoupler binding sites are being labeled under both conditions.

The quantitative relationship of the electron carriers, phosphorylating enzyme, and components being labeled upon photolysis suggests that energy conservation, energy coupling, and dissipation by uncouplers are related to molecular reactions consistent with a chemical coupling mechanism for energy conservation, and is opposed to electrochemical or conformational effects. This latter concept was reinforced by Hatefi's (*166*) finding of a correlation between the uncoupling potencies and the natural affinities of uncouplers for the mitochondrial sites labeled by 2-azido-4-nitrophenol. The 2-azido-4-nitrophenol was bound to a specific uncoupler binding site, a polypeptide of MW 30,000 that fractionates into an enzyme complex called Complex V (or energy-transducing ATPase complex). Complex V is reported capable of oligomycin and uncoupler sensitive ATP–P_1 exchange.

Further characteristics of Complex V have been detailed by Hatefi *et al.* (*167*). It is termed energy-conserving complex on the basis of its ability to catalyze an ATP–P_1 exchange and ATP hydrolysis. The exchange is specific for ATP (ITP, GTP, and UTP are ineffective). The mitochondrial uncoupler-binding sites are located exclusively in Complex V as revealed by photo-affinity studies with 2-azido-4-nitrophenol. In submitochondrial particles, picrate (a membrane-impermeable uncoupler) binds to the same uncoupler binding site as 2-azido-4-nitrophenol. Picrate is a poor protonophore and has very small effect on the proton permeability of phosphorylating sub-mitochondrial particles even at three times the concentration required for complete uncoupling.

A comparative study has been made of the influence of 2-azido-4-nitro-phenol on brown adipose mitochondria from cold-acclimated hamster and rate liver mitochondria (*168*). The binding of 2-azido-4-nitrophenol is competitively inhibited by 2,4-dinitrophenol, palmitic acid, and trifluoro-methylphenylhydrazone carbonyl cyanide. Competitive displacement in-dicates that the uncoupling agents bind to the same site(s) as 2-azido-4-nitrophenol. The number of high-affinity binding sites is the same for both tissues, although binding to brown adipose mitochondria is less tight than to liver mitochondria, probably owing to the presence of residual fatty acids in the former. The analog is a much more effective uncoupler of brown adipose mitochondria than of liver mitochondria.

Brown adipose tissue mitochondria possess an energy-dissipating ion uniport that is inhibited by purine nucleotides. Heaton *et al.* (*169*) have utilized the 8-azido ATP (Compound 36) in a photoaffinity labeling study of the regulatory site of energy dissipation. The regulatory nucleotides bind to the high-affinity sites on the outer face of the inner membrane, independent of the nucleotide translocator. The photoaffinity analog 8-azido adenosine [γ-^{32}P]triphosphate was found to covalently bind following photolysis to two major proteins of the inner membrane. One of these, a 30,000 MW peptide representing 6% of the inner membrane protein, was identified as the carboxyatractylate binding component of the adenine nucleotide trans-locator. A second labeled peptide of 32,000 MW represented 10% of inner membrane protein and was identified as the regulatory site of the energy-dissipating ion uniport. Carboxyatractylate abolished labeling of the 30,000 MW band, but the 32,000 MW band labeling was unaffected. The binding to the 32,000 MW protein was abolished by the presence of 1 m*M* GDP. The GDP nucleotide is known to have a high affinity for the regulatory site of the ion uniport in these mitochondria.

A study of the structural requirements for nucleotides at the regulatory site revealed the necessity for both α and β phosphate groups with a normal oxygen linkage, the absence of bulky substituents on the β or γ phosphate group, and the required absence of bulky substituents on C2 of the 6-membered purine ring and on C2′ of the ribose.

The uncoupling action on oxidative phosphorylation and the binding of $[^3H]$2-azido-4-nitrophenol to the mitochondrial membrane was studied by Kurup and Sanadi (*127*). The uncoupler that bound covalently to the mitochondrial membrane on photoirradiation was four times that bound reversibly in the absence of light. Polyacrylamide gel electrophoresis of photoaffinity label submitochondrial particles in the presence of SDS revealed that a 9000 MW peptide bound high levels of the uncoupler. Other proteins in the molecular weight range of 20,000 to 40,000 and 55,000 contained a very significant proportion of label. Photolysis in the presence of serum albumin or ATP decreased the covalent binding of the uncoupler to all proteins, but primarily to the 20,000 MW component. Soluble ATPase and the mitochondrial proteolipid (i.e., DCCD-binding proteolipid) purified from labeled mitochondria revealed the presence of label.

The investigators found that with AE particles (bovine heart submitochondrial particles prepared by treatment with NH_4OH and EDTA at pH 8.8) the amount of uncoupler bound in the presence of ATP was consistently lower than the amount bound in its absence. Similar data were collected with rat liver mitochondria. The inhibition of binding by ATP was of a partial competitive type. It is of considerable interest that the three proteins presumably labeled (the ATPase, the 20,000 MW peptide, and the DCCD binding proteolipid) all have their binding reduced substantially in the presence of ATP. An outline of the synthesis and pertinent physical data for 2-azido-4-nitrophenol are presented.

The synthesis of another photoaffinity uncoupling agent, 2-nitro-4-azido carbonyl cyanide phenyl hydrazone (N_3CCP), has been reported by Katre and Wilson (*170*). Equilibrium binding studies revealed that there were 1.6 high-affinity binding sites per cytochrome *a* in rat liver mitochondria. The number of high-affinity binding sites were unchanged by extraction of 80% of the mitochondrial phospholipid, whereas much of the low-affinity binding was removed. The nonlinear Scatchard plots obtained for binding of the analog were taken to represent binding sites with different affinities rather than the resultant of cooperative binding.

Binding to the high-affinity sites is not dependent upon active respiration, i.e., it is not the result of energy-driven accumulation of uncoupler. The uncoupling agents 1799, pentachlorophenol, pentachlorothiophenol, dicoumarol, and S-13 all compete with N_3CCP, effectively preventing the azides from binding to the high-affinity sites. The authors advance the proposal that the high-affinity binding sites are in some way involved in the coupling reactions of oxidative phosphorylation and that binding of uncoupler to these sites is directly responsible for uncoupling activity.

The photoaffinity probe, 4-fluoro-3-nitrophenylazide (Compound 35), activates F_1(ATPase); when F_1 and the analog are photolyzed together, strong, irreversible inhibition of the ATPase occurs (*129*). Both the azide analog and 2,4-dinitrophenol appear to stimulate primarily by lowering the K_m for ATP. The photodependent inactivation of F_1 in the presence of 4-fluoro-3-nitrophenylazide is partly protected against by 2,4-dinitrophenol. Dinitrophenol binds initially to a single site on F_1ATPase ($k_d \sim 90\ \mu M$) and at higher concentrations to additional sites. The azido compound appears to prevent binding at the first site competitively. Without photolysis 4-fluoro-3-nitrophenylazide does not influence [^{3}H]ADP binding to F_1, either in the presence or in the absence of Mg^{2+}, and inactivated F_1 after photolysis in the presence of the nitrophenylazide still retains ADP binding capacity.

The authors propose that the catalytic site for ATP hydrolysis is responsive to events occurring at the 2,4-dinitrophenol activation site whereas the tight binding sites (ADP sites) seem unresponsive to interaction at these sites.

Diazido ethidium bromide (Compound 17), 2[3(8), 8(3)]diamino-5-ethyl-6-phenylphenanthridinium bromide, has been used as a specific probe for mitochondrial functions (*138*). This diazido derivative of ethidium bromide was shown to be at least as effective as the parent mitochondrial mutagen. Exposure of mitochondrial of *Saccharomyces cerevisiae* followed by UV irradiation resulted in the specific labeling of a single component identified as the smallest peptide (subunit 9) of the membrane ATPase. A similar labeling pattern was observed with the oligomycin-sensitive ATPase complex, but not with soluble F_1ATPase. Evidence is presented in which diazido ethidium bromide, ethidium bromide, euflavine, *N,N′*-dicyclohexylcarbodiimide, 2,4-dinitrophenol, and 2-azido-4-nitrophenol all appear to compete for the same lipophilic binding site.

Subunit 9 has a high content of apolar amino acids, which within the intact complex appears to be in close association with mitochondrial phospholipids. There is a strong indication that a lipophilic environment is an essential requisite for the binding of diazido ethidium bromide to the oligomycin-sensitive ATPase. The authors describe the synthesis of diazidoeuflavin from euflavin and of azidonitrophenol from 2-amino-4-nitrophenol, i.e., the conversion of phenanthridinium and aoudidinium dyes to related bifunctional photoaffinity agents. Cantley and Hammes (*171*) have demonstrated that at least one of the binding sites for 2,4-dinitrophenol is to be found on the F_1(ATPase) proper.

C. Mitochondrial Energy-Linked ATPase Systems

Regardless of what the molecular mechanism of oxidative phosphorylation may prove to be [chemical coupling, chemiosmotic, conformational, or a combination of these (*17*)], a certainty is that the ATPase enzyme (F_1) associated with mitochondria is a principal, perhaps the fundamental, component of systems responsible for energy conservation. The concept advanced 33 years ago by Lardy and Elvehjem (*172*) that the ATPase activity associated with mitochondria represents an aspect of ATPase synthesis was substantiated some 18 years ago by the isolation and partial purification of the beef heart mitochondrial ATPase (*2,3*).

In view of the central role of this enzyme, an obvious approach to a complete understanding of energy conservation would involve knowledge of the methods by which the ATPase handles the nucleotide products and reactants of oxidative phosphorylation, i.e., ATP, ADP, and inorganic phosphate. In this regard, nucleotide photoaffinity probes offer a powerful and useful approach to understanding.

Hilborn and Hammes (*173*), using equilibrium methods, investigated the nucleotide binding sites of purified F_1ATPase, establishing that there are two types of nucleotide binding sites per enzyme molecule. One type, the "tight" binding site, is specific for ADP and could not be replaced by other nucleotides (K_d with $Mg^{2+} = 0.25\ \mu M$; K_d with EDTA $= 11\ \mu M$). The other class, i.e., the "loose" binding site(s), are not so specific with respect to nucleotide interaction. In addition to ATP (K_d, 220 μM) and ADP (K_d, 47 μM), the nucleotide analogs εADP [3-β-D-ribofuranosylimidazole(2,1-*i*)purine 5′-diphosphate] ($K_d \sim 100\ \mu M$); SHDP (6-mercapto-9-β-D-ribofuranosylpurine 5′-diphosphate) ($K_d \sim 2$ mM), IDP (inosine 5′-diphosphate) ($K_d \sim$ 2 mM), and SHTP (6-mercapto-9-β-D-ribofuranosylpurine 5′-triphosphate) ($K_d \sim 650\ \mu M$) interact with F_1. The loose binding site(s) is assumed to be responsible for the ATPase activity, but the function of the tight binding sites is not yet clear.

The F_1ATPase contains five nucleotide molecules per mole of enzyme (*174*), the tight and loose classes being freely exchangeable under certain conditions. The remaining bound nonexchangeable nucleotide may serve a structural function or may be involved in catalytic activity.

The photoaffinity label arylazido-β-alanyl ATP, (3′-*O*-{3-[*N*-(4-azido-2-nitrophenyl)amino]propionyl} adenosine 5′-triphosphate) (Compound 2) has been shown to bring about a photodependent inhibition of the mitochondrial F_1ATPase and an associated specific covalent labeling of the soluble enzyme (*20*). In these studies the ATPase enzyme, which was first *partially* freed of bound nucleotides by chromatography on Sephadex G-50, was inhibited up to 70% by concentrations of 90 μM analog after a 15-minute photoirradiation. Complete protection against the photoinhibition in the

TABLE V

HYDROLYSIS OF ARYLAZIDO-β-ALANYL ATP (^{32}P) BY F_1 (ATPASE)[a,b]

Time (sec)	$^{32}P_i$ liberated (nmol)	Arylazido-β-alanyl ATP hydrolyzed (%)
15	0.81	40.5
30	1.17	58.5
45	1.25	62.5
60	1.54	77.0

[a] From Russell *et al.* (*20*).

[b] F_1(ATPase) 6.37 μg was incubated at 37°C in 0.5 ml containing 2 nmol of γ-^{32}P-labeled arylazido-β-alanyl ATP (1.36×10^6 cpm/μmol), 83 μmol of glycylglycine, pH 8, and 50 μmol of $MgCl_2$. At the appropriate time 1 ml of a solution containing 2.5% NaCl, 0.85% ammonium molybdate, and 0.25 *M* H_2SO_4 was added to stop the reaction, and the mixture was transferred to a 15-ml glass-stoppered test tube together with a 0.5-ml H_2O wash of the reaction tube. To the combined solutions were added 4 ml of water-saturated isobutanol; the mixture was shaken manually for 60 seconds. The isobutanol phase was washed twice with 1 ml of 1 *M* H_2SO_4 and 2 ml of the isobutanol phase, then mixed in a scintillation vial with 5 ml of Aquasol (New England Nuclear). The radioactivity of the mixture was monitored on a Beckman LS-250 liquid scintillation system.

presence of the analog was achieved by the presence of a 5-fold excess of ATP, indicating a high degree of active-site specificity for the analog. One further indication attesting to active-site interaction with this analog is the fact that in the dark the compound acts as a good substrate for the enzyme. The substrate specificity is indicated in Table V, in which is demonstrated the catalytic liberation of $^{32}P_i$ from ^{32}P-labeled arylazido-β-alanyl ATP as a function of time. In this experiment the "substrate" concentration was at 2 n*M*, much below the K_m(ATP) of 0.2 m*M*. However, calculation of the V_{max} using this K_m shows the analog to be hydrolyzed at 102 μmol mg^{-1} min^{-1}, i.e., at the same rate as the natural compound.

In contrast to the substrate reactivity of arylazido-β-alanyl ATP, 8-azido ATP (Compound 36) is only weakly active as a substrate with a V_{max} of 3 μmol mg^{-1} min^{-1} (*175*). In addition, inhibition due to photolabeling in this latter case is accomplished at millimolar concentrations over a period of an hour of irradiation. Photodependent inhibition with arylazido-β-alanyl ATP is accomplished with micromolar quantities of reagent during 15 minutes or less of irradiation if carried out under the conditions described by Russel *et al.* (*20*). As will be outlined below, such irradiation periods can can be substantially shortened (to the millisecond range) by utilizing higher light intensities.

The high level of substrate reactivity indicates that, if there is any influence of the arylazido-β-alanyl function attached to the nucleotide on the binding of this pseudo substrate to the enzyme, it is minimal. The active site-directed ability of the analog is also indicated by a lack of significant scavenging influence of Tris buffer or *p*-aminobenzoic acid (PABA) on the photodependent inhibition (*176*). The presence of PABA during photolysis up to 15 μM did not influence ATPase activity, and a 1000-fold molar excess of PABA over arylazido-β-alanyl ATP protected only to the extent of 17%. In marked contrast is the fact that equimolar concentrations of ATP together with arylazido-β-alanyl ATP dropped the inhibition from 62% to 22%. With a 5-fold advantage (0.05 mM vs. 0.01 mM) of ATP over the analog, there was no indication of inhibition.

The photodependent inhibition of the enzyme is associated with covalent labeling, 80% inhibition being observed with a concurrent binding of 0.8 μmol of analog per micromole of enzyme, under conditions that would free the enzyme of its loosely bound nucleotides.

Jeng and Guillory (*52*) have reported that analogs of this type in which one has an esterification at the adenine ribose are completely hydrolyzed after 1 hour of incubation at 25°C in 1M NH_4HCO_3. The sensitivity to pH values in excess of 8 has been found to be a general characteristic with other nucleotides (*103*), including the pyridine dinucleotide (*26*) analogs. In the case referred to above, treatment during and subsequent to photoirradiation was accomplished at pH 7.5 with no obvious hydrolysis taking place. In those cases in which basic pH is required for the work-up of the labeled protein such as for proteolytic digestion, the analogs can be prepared with the radioactive label within the β-alanine function, assuring that upon insertion the label is stably attached to the protein (*103*). The position of the label within the analog to be used is thus dictated by the experimental protocol.

1. Photoinhibition of Membrane-Bound F_1ATPase

That the photodependent inhibition of F_1ATPase by the ATP analog is not a unique characteristic of the soluble enzyme is indicated by the data in Table VI. The analog is able to bring about a major degree of photodependent inhibition of the membrane-bound ATPase. Indeed it would appear that

TABLE VI

PHOTODEPENDENT INHIBITION OF PARTICULATE ATPASE[a,b]

Enzyme system	Arylazido-β-alanyl ATP (mM)	C (%)	Control (μmol P_i mg^{-1} hr^{-1})
1. Soluble F_1ATPase (45 μg)	45	48	100
2. ETP (780 μg)	6.5	38	0.82
3. ETP_A (780 μg)	6.5	33	2.00
4. S Particles (780 μg)	6.5	22	4.20

[a] From Russell (*176*).

[b] Irradiation was carried out on 0.2-ml samples containing 0.25 M sucrose, 10 mM Tris-acetate pH 7.5, and the indicated concentration of mitochondrial particles or soluble ATPase. Irradiation was at 10°C for four periods of 1 minute each, spaced by 30 seconds of cooling on ice. Following irradiation, aliquots at several protein concentrations (C) were diluted to 0.76 ml with 0.25 M sucrose, 10 mM Tris-acetate pH 7.5, and the ATPase activity was initiated by addition of 0.24 ml of an ATP regenerating system. The final reaction mixture contained 0.19 M sucrose, 7.6 mM Tris-acetate, 33 mM glycylglycine (pH 8), 2 mM ATP, 2 mM $MgCl_2$, 3 mM phosphoenolpyruvate, 100 μg of pyruvate kinase, and 5 μM carbonyl cyanide-*m*-chlorophenyl hydrazone. The reaction was terminated after 10 minutes at 30°C with 0.2 ml of 70% perchloric acid, and 0.5 ml of protein-free filtrate was assayed for inorganic phosphate. ETP, electron-transport particle.

the particulate (ETP, electron transport particle) bound enzyme is much more sensitive than the soluble (*5*) enzyme.

The time course of inhibition of ETP_AATPase activity as a function of the time of irradiation (Fig. 7) shows that the particulate preparation can be inhibited to greater than 80% and that 50% inhibition is accomplished in less than a 1-minute period of irradiation in the presence of micromolar quantities of the analog.

2. *Photoinhibition of* F_1ATPase *of Heavy Layer Beef Heart Mitochondria*

The ATPase activity of "intact" beef heart mitochondria is comparatively low, within the range of 0.2 to 0.4 μmol P_i liberated per milligram per minute. In comparison to submitochondrial particles, one must consider as well the possibility of an active nucleotide transport system. Consequently, an inhibition of ATPase activity of beef heart mitochondria might represent an inhibition of the nucleotide transport system. In order to test whether ATPase is inhibited relative to transport, samples of mitochondria prior to and following interaction with arylazido-β-alanyl ATP were sonicated to

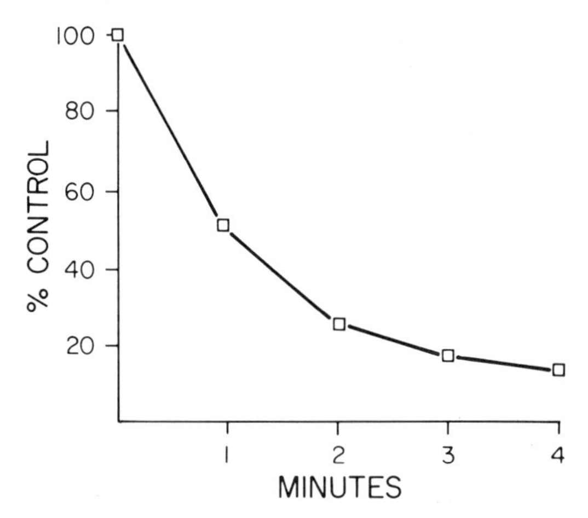

FIG. 7. Time course of photoinactivation of ETP_A particles in the presence of arylazido-β-alanyl ATP. Irradiation was carried out in a 1-mm path length quartz cuvette on a sample volume of 0.2 ml containing 0.25 *M* sucrose, 10 m*M* Tris-acetate (pH 7.5), 0.52 mg of ETP_A, and 31 μ*M* arylazido-β-alanyl ATP. Irradiation was performed at 10°C at the time intervals indicated, spaced by 30-second cooling periods at 0° to 4°C. The light source was provided by two DVY 120 V, 650 W, 3400°K projector lamps located 10 cm on either side of the cuvette. After irradiation, aliquots (at several protein concentrations) were diluted to 0.76 ml with 0.25 *M* sucrose and 10 m*M* Tris-acetate (pH 7.5), and ATPase activity was initiated by the addition of 0.24 ml of an ATP-containing solution. The final reaction mixture contained 0.19 *M* sucrose, 7.6 m*M* Tris-acetate, 33 m*M* glycylglycine (pH 8), 2 m*M* ATP, 2 m*M* $MgCl_2$, 3 m*M* phosphoenolpyruvate, 100 μg of pyruvate kinase, and 5 μ*M* carbonyl cyanide-*m*-chlorophenyl hydrazone. The reaction was terminated after 10 minutes at 30°C with 0.2 ml of 70% perchloric acid. The deproteinated supernatant (0.5 ml) was assayed for inorganic phosphate. ETP, electron-transport particle. From Russell (*176*).

expose the ATPase protein and eliminate membrane barriers. From Table VII it can be seen that the masked ATPase, i.e., that liberated upon sonication, is inhibited to the same extent as that ATPase activity measured prior to sonication. Apparently the translocase is undisturbed by photoirradiation in the presence of arylazido-β-alanyl ATP, although it must be mentioned that the rate of nucleotide transport itself was not specifically determined in this experiment.

3. *Photoinactivation of Energy-Dependent Mitochondrial Reactions*

The energy-dependent transhydrogenase activity of both the electron-transport particles (ETP) and S particles exhibited the same inhibition profile toward photoinactivation in the presence of the nucleotide analog as did the ATPase activity. On the other hand, succinate-driven transhydrogenation

TABLE VII

Arylazido-β-Alanyl ATP Photodependent Inhibition of Heavy Layer Beef Heart Mitochondria and Sonicated Particle ATPase[a,b]

Condition for assay	ATPase[c] Control	+ Arylazido-β-alanyl ATP	% Control
Prior to sonication	0.295	0.124	42
After sonication	1.021	0.360	35

[a] From Russell (*176*).

[b] Samples of mitochondrial protein (1.5 ml containing 6 mg of protein) in 0.25 *M* sucrose, 10 m*M* Tris · HCl pH 7.5 were irradiated for 5 minutes at 0° to 4°C with each minute of irradiation spaced by a 1-minute cooling period. Control samples are compared with samples irradiated in the presence of 49 μmol of arylazido-β-alanyl ATP. Aliquots of 0.8 ml were diluted to 4 ml with 0.25 *M* sucrose and 10 ml of Tris · HCl pH 7.5 and were either assayed directly for ATPase activity or subjected to sonication (1 minute at full output of Model W 185 Heat Systems Ultrasonics, standard $\frac{1}{2}$-inch probe) prior to assay.

[c] ATPase is recorded as micromoles of P_i per milligram per minute.

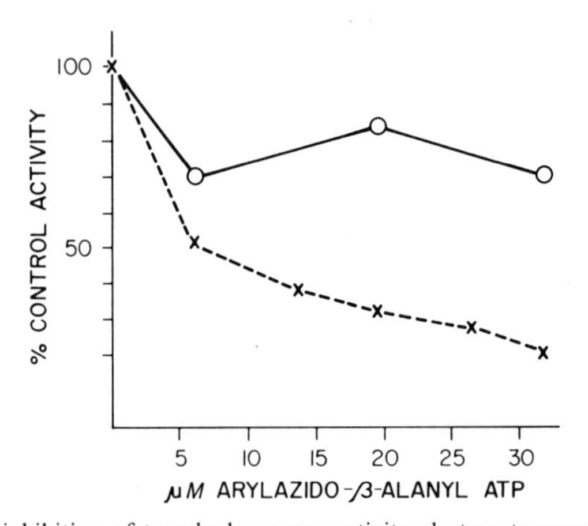

Fig. 8. Photoinhibition of transhydrogenase activity electron-transport particles (ETPs) by arylazido-β-alanyl ATP. For the preparation of arylazido-β-alanyl ATP-modified ETPs, 1 mg of particle protein was subjected to photolysis for 4 minutes in 250 μl of a solution containing 0.25 *M* sucrose, 10 m*M* Tris-Cl (pH 7.5), and the indicated quantity of arylazido-β-alanyl ATP. Aliquots were taken for transhydrogenation assay. Transhydrogenation was assayed in an Eppendorf fluorometer by following the increase in fluorescence due to the formation of NADPH. The assay was carried out in a medium containing 140 m*M* sucrose, 40 m*M* Tris-acetate (pH 7.5), 50 m*M* ethanol, 10 m*M* $MgSO_4$, 0.02 m*M* NAD^+, 0.2 m*M* $NADP^+$, 1 μg of rotenone, 250 μg of yeast alcohol dehydrogenase, and ETPs containing 400 μg of protein. Transhydrogenation was initiated with either 5 m*M* succinate or 3 m*M* ATP. The specific activity of the control preparations (irradiated in the absence of arylazido-β-alanyl ATP) was 8.8 nmol of $NADP^+$ reduced per milligram per minute (ATP-driven, × --- ×) and 19.5 nmol of $NADP^+$ reduced per milligram per minute (succinate-driven, ○—○). From Russell (*176*).

was inhibited maximally only 15% under the same conditions (Fig. 8). In addition, the energy (ATP)-dependent reduction of NAD^+ in ETPs was inhibited to the same extent and with the same concentration dependency as the ATP-driven transhydrogenation. No inhibition of NADH oxidase was observed in the treated preparations, indicating a high degree of specificity of the analog for reactions dependent upon the F_1ATPase protein.

4. *Influence of Arylazido-β-alanyl ATP Photodependent Interaction on the Phosphate-to-Oxygen Ratio*

As can be seen from Table VIII, the ATPase activity of ETP is inhibited to a greater extent than the measured uptake of inorganic phosphate. Since the photoirradiation in the presence of the nucleotide analog does not greatly influence the rate of succinate respiration, this indicates that the phosphate-to-oxygen ratio remains relatively constant in the face of a major inhibition of ATPase activity.

In a large series of experiments using different preparations of submitochondrial particles and different preparations and concentrations of arylazido-β-alanyl ATP, no statistical correlation could be found between the decrease in ATPase activity and any decrease observed in the phosphate-to-oxygen ratio. In each experiment the ATPase activity and the phosphate-to-oxygen ratios were measured on the same samples and on the same day. A product moment correlation coefficient of $r = 0.1168$ was found, which is statistically insignificant (*177*). There is thus no correlation between ATPase inhibition and inhibition of either phosphate uptake or phosphate-to-oxygen ratio (*176*).

5. *Studies on the Effect of Arylazido-β-alanyl ATP on Rat Liver Mitochondrial ATPase*

In view of the finding of a lack of a major arylazido-β-alanyl ATP photodependent inhibition of the phosphate-to-oxygen ratio of beef heart mitochondria, it was decided to investigate a more coupled phosphorylation system to see whether tighter coupling might not reveal an influence of the analog on the phosphorylation mechanism. When rat liver mitochondria were subjected to photolysis in the presence of the nucleotide analog, there was no indication of a photodependent inhibition of ATPase activity. Surprisingly, such an inhibition could, however, be observed in the presence of 2,4-dinitrophenol (Table IX).

As with the beef heart mitochondria, arylazido-β-alanyl ATP had no influence on the phosphate-to-oxygen ratio obtained with succinate (2.0) after photoirradiation. In addition, respiratory control (AMP, values of

TABLE VIII

A. Effect of Arylazido-β-Alanyl ATP on the ATPase and Phosphate Uptake of ETP[a,b]

Expt. No.	Arylazido-β-alanyl ATP (nmol)	ETP protein (mg)	P_i uptake (μmol mg^{-1} min^{-1})		Experimental (% control)	
			Control	Treated	P_i uptake	ATPase
1	28.8	1.68	0.082	0.074	90	20
2	28.8	2.52	0.067	0.064	96	30
3	14.4	3.75	0.311	0.164	48	18
4	24.0	0.75	0.413	0.244	59	28
5	24.0	0.75	0.326	0.373	114	60

B. Effect of Arylazido-β-alanyl ATP on Oxidative Phosphorylation in ETP[a,b]

Expt. No.	Arylazido-β-alanyl ATP (nmol)	ETP protein (mg)	Control ATPase (μmol min^{-1} mg^{-1})	P/O control	Experimental (% control)	
					P/O	ATPase
1	14.4	3.75	1.70	0.502	103	18
2	24.0	0.75	1.76	0.399	93	28

[a] From Russell (*176*).
[b] ETP, electron transport particles.

TABLE IX

PHOTODEPENDENT INHIBITION OF RAT LIVER MITOCHONDRIAL ATPASE BY ARYLAZIDO-β-ALANYL ATP IN THE PRESENCE OF 2,4-DINITROPHENOL[a,b]

Arylazido-β-alanyl ATP	2,4-Dinitrophenol	ATPase activity	% Control
–	–	0.81	100
7.6	–	0.87	108
7.6	+	0.42	53
15.2	–	0.81	100
15.2	+	0.30	38

[a] From Russell (*176*).

[b] Rat liver mitochondria (360 μg of protein) were photolyzed in 1 ml volume of 0.25 *M* sucrose and various concentrations of arylazido-β-alanyl ATP. Where indicated (+) 2,4-dinitrophenol was present at 0.2 m*M*. Photolysis was carried out with five 1-minute irradiation periods spaced by 30-second cooling periods at 0° to 4°C. After photolysis, aliquots of the irradiated samples were assayed for ATPase activity in the presence of 0.2 m*M* 2,4-dinitrophenol. ATPase activity is recorded as micromoles of P_i per milligram per minute.

3.8 to 4.3) was unaffected by arylazido-β-alanyl ATP with or without photoirradiation. While CCCP was able to uncouple mitochondria as evidenced by a loss of respiratory control, it was not able to mimic the effect of 2,4-dinitrophenol in bringing about the nucleotide analog-dependent inhibition of ATPase.

Rat liver mitochondria prepared in 0.25*M* sucrose supplemented with 10 m*M* Tris · HCl pH 7.5 were still refractory to arylazido-β-alanyl ATP-dependent inhibition of ATPase activity; however, in this case, in contrast to that reported above in which mitochondria were prepared in 0.25*M* sucrose, 2,4-dinitrophenol was not able to induce the photodependent inhibition. Although such mitochondria are uncoupled by 2,4-dinitrophenol as evidenced by a loss of respiratory control, the nitrophenol was not able to stimulate a sizable ATPase activity (Fig. 9). The ATPase activity of rat liver mitochondria is only significantly stimulated by 2,4-dinitrophenol in the case of mitochondria prepared with 0.25*M* sucrose, i.e., the same conditions under which arylazido-β-alanyl ATP photoinactivation can occur.

When rat liver mitochondria are subjected to brief sonication, the arylazido-β-alanyl ATP was only partially effective in inhibiting the ATPase activity following photolysis. There was under these conditions still a requirement for 2,4-dinitrophenol presence during photolysis. Apparently neither uncoupling of F_1ATPase nor membrane permeability problems per se are prerequisites for the photoinactivation of rat liver F_1ATPase in the presence of arylazido-β-alanyl ATP. A crude solubilized F_1ATPase preparation from rat liver mitochondria was found to be inactivated after

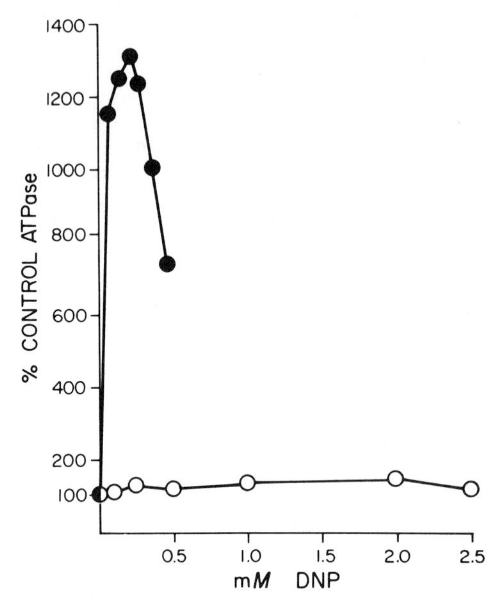

FIG. 9. Effect of the mitrochondrial isolation medium on the 2,4-dinitrophenol stimulation of ATPase activity in rat liver mitochondria. Various amounts of 2,4-dinitrophenol were added to rat liver mitochondria prepared in either 0.25 *M* sucrose or 0.25 *M* sucrose, 10 m*M* Tris-Cl (pH 7.5). The final reaction medium consisted of 33 m*M* glycylglycine (pH 8), 2 m*M* ATP, 2 m*M* $MgCl_2$, 3 m*M* phosphoenolpyruvate, and 100 μg of pyruvate kinase in 1 ml. Concentrations of 2,4-dinitrophenol reported in the graph are final concentrations. The reaction was terminated after 10 minutes at 30°C by the addition of 0.2 ml of 70% perchloric acid, and 0.5 ml of the deproteinated supernatant was assayed for inorganic phosphate. The specific ATPase activity of the mitochondria prepared in sucrose was 0.108 μmol of P_i liberated per milligram per minute, and for those prepared in sucrose-Tris the specific activity was 0.149 μmol of P_i liberated per milligram per minute. From Russell (*176*).

photolysis in the presence of arylazido-β-alanyl ATP in a reaction that was not dependent upon 2,4-dinitrophenol.

6. *Nucleotide Levels of Rat Liver Mitochondria*

In view of the protective effect of ATP and ADP on the photodependent arylazido-β-alanyl ATP inhibition of the soluble and membrane-bound mitochondrial ATPase and the lack of influence on the activity of intact rat liver mitochondrial ATPase, a study was made of the nucleotide levels of mitochondrial preparations. As can be seen from Table X, the ATP level of rat liver mitochondria is much higher than that found in beef heart

TABLE X

NUCLEOTIDE LEVELS OF RAT LIVER MITOCHONDRIA[a,b]

	Control		2,4-DNP Treated	
Nucleotide	Nmol mg^{-1}	% Total	Nmol mg^{-1}	% Total
ATP	2.04	33	0.25	4
ADP	2.32	38	4.17	66
AMP	1.77	29	1.89	30
Total	6.13	100	6.31	100

[a] From Russell (*176*).

[b] Where indicated, 2,4-dinitrophenol was added (final concentration of 0.2 m*M*) to 1 ml of 0.25 *M* sucrose containing 43 mg of rat liver mitochondria. The mixture was incubated at 25°C for 5 minutes followed by addition of 5 ml of 6% perchloric acid (0 to 4°C). The control had H_2O replacing the dinitrophenol. The precipitated protein was centrifuged, and the pellet was reextracted with 1.0 ml of perchloric acid (3%) and recentrifuged. The combined supernatants were neutralized with KOH, and nucleotide levels were determined by enzymic methods [Adam, H. (1965), in Methods of Enzymatic Analysis, 2nd ed., p. 573, H. U. Bergmeyer, ed., Academic Press, New York].

mitochondria. In the latter case, ATP represents but 13% of the total nucleotide, the beef heart mitochondra having a major portion (73%) of nucleotide present as AMP. In the rat liver preparation, incubation with 0.2 m*M* dinitrophenol results in a major shift in nucleotide levels, the ATP level dropping to 4% with a comparable increase in the ADP level to 66% of the total.

Since AMP does not bind to F_1 it is unlikely that it would interfere with arylazido-β-alanyl interaction; thus in beef heart mitochondria the analog inhibits F_1ATPase effectively. Dinitrophenol alters the nucleotide levels in rat liver mitochondria to those approaching the heart mitochondria. Since some 10% of the mitochondrial protein is F_1ATPase amounting to 0.27 nmol mg^{-1} of mitochondrial protein, there is approximately a 7.6-fold excess of ATP over F_1 in the intact mitochondra; 2,4-dinitrophenol reduces this ratio to one. It is known that at least 1 mol of ATP is tightly bound per mole of F_1 (this is presumably not involved in the ATPase or phosphorylative activity), and the effective result of the dinitrophenol action is to reduce the *free* ATP concentration to zero.

It would appear that the nucleotide level in intact rat liver mitochondria is the key factor in determining whether or not the analog will react with the ATPase enzyme. Experiments with beef heart mitochondrial ATPase and reverse electron flow driven by ATP indicate clearly that coupled F_1 is capable of interacting with the analog, i.e., it is not only the respiratory uncoupled ATPase that is capable of interacting with the photoaffinity analog of ATP.

In summary, work with arylazido-β-alanyl ATP and its interaction with mitochondrial ATPase from rat liver and beef heart shows clearly that the analog has the required characteristics of a photoaffinity analog for F_1. The analog is a good substrate for the ATPase in the dark; there is a strong protective effect of ATP and a lack of protective influence of nitrene scavengers. Inhibition as a result of light-dependent reactions takes place at micromolar concentrations of the analog within a range well below the K_m(ATP) for the ATPase. A lower K_m for the membrane-bound enzyme (see below) is consistent with the increased sensitivity of the membrane activity toward arylazido-β-alanyl ATP photodependent inhibition.

The ATP-driven reactions of submitochondrial particles are all equally susceptible to photoinactivation by arylazido-β-alanyl ATP (i.e., succinate-dependent reduction of NAD^+ by ATP-driven NADH–$NADP^+$ transhydrogenation), whereas electron transport-dependent reactions are not specifically inhibited. Although ATP-dependent activities are inhibited by the analog, there is little influence on the respiratory-dependent phosphorylation of ADP in submitochondrial or intact mitochondria.

The possibility that the ATPase activity measured in submitochondrial particles is principally due to uncoupled ATPase and that this is the enzyme that is the major site of photodependent inhibition by arylazido-β-alanyl ATP is unlikely. It would be unlikely for the analog to be capable of inhibiting ATP-driven transhydrogenation and NAD^+ reduction if it were reacting only with the *uncoupled* F_1.

One might well consider, as previously proposed (*178*) and as a logical conclusion from the above data, that ATPase and phosphorylation in mitochondria take place by very different mechanisms on the same enzyme, possibly at different sites within the enzyme's three-dimensional structure. Current work on the arylazido-β-alanyl ATP labeling pattern of F_1ATPase under differing physiological conditions may aid in our understanding of this system.

7. *Interaction of Arylazidobutyryl ADP with Mitochondrial Adenosine Triphosphatase*

Lunardi *et al.* (*118*) have synthesized the photoaffinity analog, arylazidobutyryl ADP, 3′-*O*-3-[*N*-(4-azido-2-nitrophenyl)amino]butyryl adenosine 5′-diphosphate (Compound 29), essentially according to the method of

Jeng and Guillory (52) and have reported on its interaction with isolated F_1ATPase and on the subunits of the enzyme to which it binds after photoirradiation.

While the authors show a clear photodependent inhibition, the data are complicated by an obvious concentration-dependent nonphotolytic inhibition. In the dark and at 50 μM concentration of the photoaffinity label, there is a 20% reduction in ATPase activity, which is further reduced to 40% at 150 μM concentration, the light-dependent inhibition at these two concentrations being 60% and 85%, respectively. Thus the photodependent inhibition amounts to 40 and 45%, respectively. In addition, the author's assertion that the ADP analog is more inhibitory than the arylazidobutyryl adjunct because of the specific interaction of the ADP portion of the analog with the ADP and ATP binding sites of F_1ATPase is difficult to assess because of an observed arylazidobutyryl stimulation of the ATPase. At 150 μM concentration arylazidobutyric acid stimulates F_1ATPase to 120% of the control value. Upon photoirradiation this is reduced to about 65% of the control, resulting in an approximate 45% photodependent inhibition, i.e., comparable to that observed for the nucleotide analog alone.

The photodependent inactivation of F_1ATPase was shown to be accompanied by a covalent binding of the analog to the enzyme. At 80% loss of ATPase activity there was a saturation of binding at 0.95 mol of label per mole of F_1ATPase. This is the same stoichiometry observed by Russell *et al.* (*20*) for photodependent inhibition in the presence of arylazido-β-alanyl ATP. In contrast to the inhibition observed for arylazido-β-alanyl ATP, in which the dark control inhibition was below 15%, that reported for the butyryl ADP analog shows a substantial inhibition in the dark. In the latter case at 300 μM concentration there is a maximal inhibition of the ATPase in the light of 90%; however, in the dark inhibition already approaches 50%. Consequently an assignment of the significance of the quantity of covalently bound nucleotide under these conditions is difficult.

It is clear that, under the conditions of photoirradiation as carried out by Lunardi *et al.* (*118*), both α and β subunits are labeled without significant labeling of the other, (γ, δ, ε) subunits, Preincubation of F_1ATPase with ADP, ATP, or adenylimidodiphosphate prior to addition of the photoreactive nucleotide and irradiation results in a decrease in covalently bound nucleotide. Competitive interaction is taken to represent recognition of a similar site by the four compounds. It is interesting that for 80% protection a 65-fold excess of ADP over the analog is required, whereas only a 3-fold molar excess of adenylimidodiphosphate gives a comparable protection. In a single experiment reported, ATP at 130-fold excess prevents 98% of label incorporation.

The authors suggest that the aryl group may bind to uncoupler sites on the ATPase. If this is true it would mean that the uncoupler sites would of necessity be on the α subunit. Such labeling would amount to about 15% of the total reported incorporation of arylazidobutyryl ADP.

Quercetin, an inhibitor that mimics the effect of the natural ATPase inhibitor (*179,180*), inhibited the binding of arylazidobutyryl ADP to both the α and β subunits of F_1ATPase, whereas pentachlorophenol did not interfere with the photocovalent labeling of the ATPase. The pentachlorophenol did markedly decrease the amount of covalently bound *N*-4-azido-2-nitrophenylaminobutyrate and prevented photoinactivation of F_1ATPase by this compound. Hanstein has rported (*181*) that 2-azido-4-nitrophenol (NAP) and 4-fluoro-3-nitrophenylazide (FNAP), uncouplers of oxidative phosphorylation, bind covalently upon photoactivation to two subunits of the ATPase complex, one the α subunit of F_1ATPase and the other an *uncoupler* binding protein.

8. *Photolabeling of Beef Heart Mitochondrial ATPase by 8-Azido-ATP*

Wagenvoord *et al.* (*175*) have investigated the photolabeling of beef heart mitochondrial ATPase by 8-azido ATP (Compound 36) utilizing the reagent with tritium marker situated in the adenine ring. Although the analog was shown to function as a substrate, its reactivity is quite limited. The K_m is 5.2 times that of the natural substrate, and the maximal velocity is but 1/33.

From a graphic representation of the inhibitory profile presented by Wagenvoord *et al.* (*175*), it can be calculated that at 2 μM 8-azido ATP there is a 70% inhibition of the ATPase activity following a 60-minute photoirradiation. At 4 μM nucleotide concentration there is a comparable degree of inhibition even at 2 hours of irradiation. However, if the enzyme is precipitated with $(NH_4)_2SO_4$ after the first hour of irradiation, photoirradiation of the reisolated preparation for an additional hour in the presence of fresh 8-azido ATP results in almost complete inhibition of enzymic activity. There is no light control recorded for this particular experiment. This is unfortunate, since there is the possibility that the 2-hour irradiation results in some degree of nonspecific inhibition. The authors do state that illumination of F_1 in the absence of 8-azido ATP causes no inhibition. In addition, no dark control, i.e., incubation of the enzyme with the nucleotide analog without irradiation, is recorded. Irradiation for 1 hour at 2 μM 8-azido AMP results in 10–15% inhibition as compared to 80% inhibition with the same concentration of 8-azido ATP. The authors use this experiment as an indication that 8-azido AMP does not bind to F_1. Unfortunately, again no dark control was recorded to indicate possible nonphotodependent inhibition. ATP at 2–5 mM results in a 75% maximum protection of ATPase activity from photodependent inhibition by 2 m*M* 8-azido ATP. AMP at the same concentration has a minimal protective effect.

Most interesting is that the application of 8-azido [2-^{3}H]ATP followed by photoirradiation and gel electrophoresis in the presence of dodecyl sulfate indicates that, of the two major peptide bands of F_1ATPase, only the β band is radioactively labeled. There is no detectable labeling of the minor subunit peptides. In this experiment the authors do not indicate the ratio of azido compound to F_1 used in the labeling experiments. The amount of covalently bound 8-azido [2-^{3}H]ATP per mole of F_1 is proportional to the percentage of light-induced inhibition, and the data indicate that some 1.8 mol of 8-azido ATP are bound per mole of F_1 (MW 360,000) at this level of inhibition. Scheurich *et al.* (*181a*) have examined the binding characteristics of 8-azido ATP to the F_1ATPase from *Micrococcus luteus.*

The difference in the binding specificity of the photoaffinity analogs (i.e., 8-azido ATP to the β subunit and arylazidobutyryl ADP to both α and β subunits) has been explained by Lunardi *et al.* (*118*) on the basis of the structural characteristics of the two analogs and the three-dimensional organization of the F_1ATPase. While there is uncertainty as to the complete subunit stoichiometry of F_1, it is generally accepted that there are at least two α and two β subunits per molecule (*182*).

The proximity of the azido group and the nucleotide base in 8-azido ATP would assure that the nitrene generated upon irradiation is directly at the binding site for nucleotides on F_1, i.e., at the β subunit. On the other hand, in *N*-4-azido-2-nitrophenylaminobutyryl ADP the generated nitrene is isolated from the ribose of ADP by a bridge of four carbons. Assuming binding of this analog to the β subunit, the length of carbon bridges and the resultant flexibility may allow insertion of the activated nitrene not only into the β, but as well into the α, subunits. Satre *et al.* (*182*) investigated the structure of the beef heart mitochondrial subunits by cross-linking reagents (dimethyl suberimidate, dimethyl adipimidate, methylmercaptobutyrimidate) and selective labeling. From their investigations they present evidence that the α subunits are close to each other and to the β subunits, but the β subunits appear to be too far from each other to be able to undergo cross-linking with the reagents used.

9. *Photoinhibition and Labeling of Integrated Forms of Mitochondrial ATPase by Arylazido-β-alanyl ATP*

Cosson *et al.* (*21*) and Cosson and Guillory (*183*) have examined the effectiveness of arylazido-β-alanyl ATP as a photoaffinity probe for the integrated forms of the mitochondrial ATPase, either *in situ* (i.e., submitochondrial particles) or in a form extracted from the mitochondrial membrane but associated with membranous subunits and phospholipid vesicles (ATPase complex). In the first case, the complex enzyme is coupled with electron transport and can produce ATP from ADP and P_i as well as exchange the phosphate of ATP with inorganic phosphate. In the second case, the enzyme is capable of ATP-induced proton translocation as well as ATP–P_i exchange reaction.

The effect of the ATP derivative was detected using either inhibition of enzymic activities (ATP hydrolysis, ATP-$^{32}P_i$ exchange) or by the measurement of binding of arylazido-β-alanyl ATP to native ATPase or its subunits. Photoirradiation was carried out using the flash generated by a xenon lamp (FT 1-BALCAR, 1200 W second U-shaped flash tube) with flashes of less than 1 msec duration. Filtration of the light prevented the passage of more than 99% of irradiation below 300 nm. Under these conditions the ATPase and ATP–$^{32}P_i$ exchange activities of submitochondrial particles were inhibited maximally 5% by irradiation with 15–20 flashes in the absence of arylzaido-β-alanyl ATP.

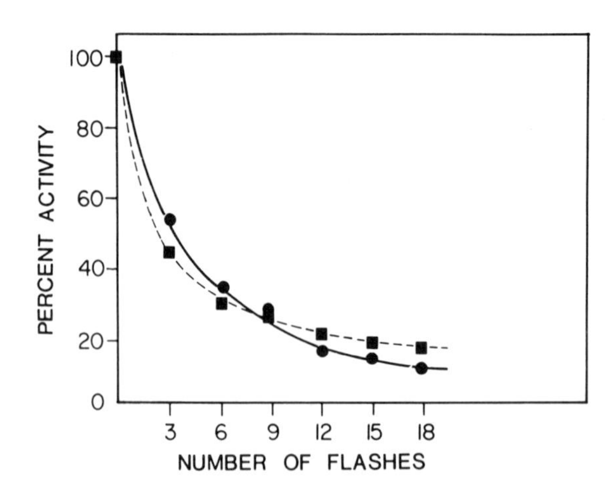

FIG. 10. Comparison of the effect of arylazido-β-alanyl ATP on the ATP–P_i exchange activity and on the ATPase activity of submitochondrial particles. To the submitochondrial particles at 0.723 mg of protein per milliliter in 25 m*M* Tris·HCl, pH 7.5, was added arylazido-β-alanyl ATP to a concentration of 2.5 μ*M*. The mixture was irradiated with 3 flashes; an aliquot was tested for ATPase activity (■---■), and a second aliquot was tested for ATP–P_i exchange activity (●—●). An additional 2.5 μ*M* arylazido-β-alanyl ATP was added, the mixture was again irradiated with three flashes, and aliquots were again taken for measurement of both activities. The procedure was repeated to a final concentration of 15 μ*M* arylazido-β-alanyl ATP and a total of 18 flashes. From Cosson and Guillory (*185*).

10. Photodependent Inhibition of Submitochondrial Particles (SMP) ATPase

Figure 1 shows that at low concentrations of arylazido-β-alanyl ATP (15 μM) greater than 80% inhibition of ATPase is observed. The degree of inhibition is independent of the concentration (4–30 mM) of the ATP used in the subsequent assay for ATPase. Without photolysis there is no inhibition. Addition of the analog in increments followed by flash is more effective than the single addition of a large concentration of the nucleotide analog.

The photodependent inhibition of the ATP–P_i exchange (Fig. 10) has the same inhibitory profile as the inhibition of ATPase activity of SMP. After photodependent inhibition of ATPase activity by (8 μM) arylazido-β-alanyl ATP, there is no change in the K_m(ATP) of the enzyme; there is, however, a decrease in the V_{max} to 21% that of the control (Fig. 11).

Varying the concentration of arylazido-β-alanyl ATP during photo-irradiation shows a clear concentration dependency of the inhibition of

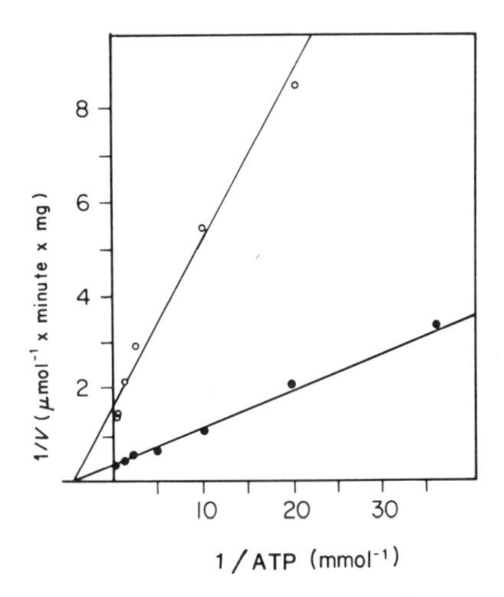

FIG. 11. The photodependent inhibitory profile of submitochondrial ATPase after following arylazido-β-alanyl ATP treatment. Submitochondrial particles at 1.07 mg of protein per milliliter in 20 mM Tris SO_4, pH 7.5, were irradiated with a total of 18 flashes in the presence of 8 μM arylazido-β-alanyl ATP using the successive addition procedure (i.e., 6 flash periods with 1.33 μM increments of analog). The ATPase activity was measured in the presence of various concentrations of ATP using the reaction mixture described in the Methods section. The reciprocal plot of activity versus the ATP concentration gives a K_m (ATP) of 190 μM for both control and analog-treated preparation and V_{max} values of 3.06 μmol of P_i liberated per minute per milligram of protein for the control (●) and of 0.66 μmol of P_i for the analog-treated sample (○). From Cosson and Guillory (*183*).

TABLE XI

COMPARISON OF THE $^{32}P_i$ EXCHANGE ACTIVITY OF SUBMITCHONDRIAL PARTICLES USING ATP OR ARYLAZIDO-β-ALANYL ATP AS SUBSTRATE[a,b]

Substrate	Addition	$^{32}P_i$ Exchange	% Inhibition
ATP	—	3.38	—
ATP	Oligomycin	0.37	89
Arylazido-β-alanyl ATP	—	4.13	—
Arylazido-β-alanyl ATP	Oligomycin	0.39	91

[a] From Cosson and Guillory (*185*).

[b] The $^{32}P_i$ exchange assay was carried out in a total volume of 100 μl with 1.2×10^7 cpm $^{32}P_i$. When added, oligomycin was at the level of 2 μg/ml and ATP or arylazido-β-alanyl ATP at 1.37 m*M*. Activity is recorded as nanomoles of P_i exchanged per minute per milligram of protein.

ATPase with 80% inhibition after irradiation with 2–3 μ*M* arylazido-β-alanyl ATP. Increasing the concentration to 15–20 μ*M* results in an increased inhibition up to 90%. A similar profile is observed for the ATP–^{32}P exchange. In addition to being a substrate for the F_1ATPase, arylazido-β-alanyl ATP is as well a substrate for the ^{32}P–ATP exchange (Table XI). The γ-phosphate exchanged with ^{32}P at a rate similar to that observed for ATP and was sensitive to oligomycin of the same degree as that of the exchange with ATP.

A direct comparison of the mitochondrial reactions (ATPase, ATP–^{32}P exchange) with respect to their response to photoirradiation in the presence of arylazido-β-alanyl ATP showed each to be inhibited to a similar degree (Table XII).

11. Factor-1 ATPase as the Major Protein of SMP to Which Arylazido-β-alanyl ATP Binds

The enzymic reactions shown to be inhibited by arylazido-β-alanyl ATP in SMP are considered to involve directly or indirectly the F_1ATPase protein. The view that this is the component interacting with the nucleotide analog is reinforced by the reported interaction with soluble F_1ATPase of both beef heart (*20*) and rat liver (*176*). In order to test this further and evaluate the specificity of interaction of arylazido-β-alanyl ATP with mitochondrial proteins, the F_1ATPase was extracted from arylazido-β-alanyl ATP phototreated SMP by the chloroform treatment as described by Beechey *et al.* (*184*). The extracted ATPase following photolysis of SMP in the presence of the nucleotide analog was inhibited almost to the same extent as the SMP ATPase. The preferential binding of tritium-labeled

TABLE XII

INHIBITION OF ATPASE AND ATP–P_i EXCHANGE OF SUBMITOCHONDRIAL PARTICLES AFTER PHOTOIRRADIATION IN THE PRESENCE OF ARYLAZIDO-β-ALANYL ATP[a,b]

	Irradiated Control	Irradiated in the presence of arylazido-β-alanyl ATP
ATPase	98%	22%
ATPase (with regenerating system)	101%	-27%
ATP–P_i exchange	94%	25%

[a] From Cosson and Guillory (*185*).

[b] Submitochondrial particles at 1.28 mg of protein per milliliter in 50 m*M* Tris · HCl, pH 7.5, were irradiated with 18 flashes in the presence of a final concentration of 9.4 μ*M* arylazido-β-alanyl ATP using the successive addition procedure (6 additions of 1.57 μ*M* arylazido-β-alanyl ATP with 3 flashes after each addition). A control was irradiated in an identical manner but in the absence of arylazido-β-alanyl ATP. In this case the nucleotide analog was added after the photoirradiation. The ATPase activity was determined in the presence or in the absence of phosphoenolpyruvate and pyruvate kinase. The control activity for ATPase was 11.8 and 10.6 μmol of P_i liberated per minute per milligram of protein, respectively, for the system with and without the ATP regenerating system. For the ATP–$^{32}P_i$ exchange, control activity was assayed at 23.0 μmol ^{32}P exchange per minute per milligram of protein. Enzymic assays were carried out for 5 minutes at 30°C. The data are presented as percent of the control values.

arylazido-β-alanyl ATP to F_1 is seen clearly in Table XIII. The purification of the ATPase protein by the chloroform treatment was accompanied by an increase in the specific radioactivity of the labeled protein. A 3.5-fold enrichment of tritium-labed site was noted during F_1 purification with an 8-fold purification of F_1. This F_1 purification is estimated to be realistically established at 4-fold because of the release of protein inhibitor during purification.

12. Stoichiometry of Covalent Binding of Arylazido-β-alanyl ATP to SMP

That arylazido-β-alanyl ATP is bound with strong affinity can be seen by the concentration dependency of the photodependent inhibition. The concentration of arylazido-β-alanyl ATP during photolysis is 0.10–15 μ*M*; however, even more impressive and added evidence for covalent attachment is the fact that enzymic tests are achieved following a 30- to 50-fold dilution bringing the concentration of the nucleotide analog into the 50- to 500-n*M* range.

The covalent nature of the binding was further examined by experiments involving SDS–Sephadex chromatography. The submitochondrial particles

TABLE XIII

EXTRACTION OF ARYLAZIDO-β-ALANYL PHOTODEPENDENT INHIBITED F_1ATPASE SUBMITOCHONDRIAL PARTICLES[a,b]

Experiment	Preparation assayed	Arylazido-β-alanyl ATP (μM)	Control	Irradiated % Control	Arylazido-β-alanyl ATP bound (nmol/mg)
1	SMP	13.5	3.06	67	0.215
	F_1ATPase (extracted)		24.0	66	0.750
2	SMP	15.0	2.80	78	0.275
	F_1ATPase (extracted)		22.5	66	0.980

[a] From Cosson and Guillory (*185*).

[b] Submitochondrial particles (SMP) (2.45 mg) were irradiated with 18 flashes in the presence of 14 μM [^{3}H]arylazido-β-alanyl ATP (5.8 × 10^6 cpm μmol^{-1}) using the successive addition procedure (six flash periods with the addition of 0.8 μM nucleotide in six increments). The ATPase activity was determined on an aliquot, and another aliquot was treated for F_1 extraction using the procedure described by Beechey *et al.* (*184*). The ATPase activity of extracted F_1 was compared to that of a control prepared in the same manner, but in the absence of arylazido-β-alanyl ATP. The quantity of bound nucleotide was determined after dialysis of the sample and is recorded as nanomoles bound per milligram of protein. ATPase activity of the control sample is recorded as micromoles of P_i liberated per minute per milligram of protein.

were photoirradiated in the presence of [^{3}H]arylazido-β-alanyl ATP, and the preparation was then centrifuged. After denaturation of the pellet by SDS, the SMP was subjected to SDS–Sephadex chromatography. No tritium label was found to be linked to the SDS-treated protein in the absence of photoirradiation. With irradiation, 0.18 nmol of nucleotide was eluted at the void volume of the column per milligram of protein. If, rather than to Sephadex chromatography, the SDS-treated submitochondrial particle preparation was subjected to extensive dialysis against the SDS-containing buffer, 0.165 nmol of label was found to be associated with each milligram of denatured protein.

Attempts were made to measure the stoichiometry of binding of arylazido-β-alanyl ATP by irradiation of submitochondrial particles in the presence of various concentrations of arylazido-β-alanyl ATP. A plateau corresponding to 20–25% of ATPase activity remains when 0.3–0.35 nmol of arylazido-β-alanyl ATP is bound per milligram of SMP. Increasing the nucleotide concentration during irradiation results in further binding with only a slight further decrease in ATPase activity.

The difference in the photodependent inhibition by arylazido-β-alanyl

ATP observed with the successive addition procedure as compared to that in which the number of flashes are varied at an initial high analog concentration (Fig. 1) is quite substantial. Presumably, after a few flashes at high analog concentrations (i.e., 15 μM), the concentration of "free" photoreacted nucleotide analog is high compared to the successive-addition system in which there is a lower analog concentration (1.5 μM) of which a significant proportion is protein bound. In the former case, the accumulated photoreacted free analog can act as a competitor for arylazido-β-alanyl ATP. Thus, with respect to the number of binding sites for arylazido-β-alanyl ATP on the ATPase protein, there may exist but one type, but this becomes more difficult to saturate with photoactive analog in the face of an increasing competitive concentration of the light product of arylazido-β-alanyl ATP as indicated by the inability to inhibit the activity 100%.

13. Influence of Arylazido-β-alanyl ATP on the Partially Purified ATPase Complex Phospholipid Effect

The ATPase activity and the ^{32}P–ATP exchange activity of the ATPase complex are enhanced by the presence of phospholipids. In our hands maximal activating effects were obtained with 0.3–0.5 mg of phospholipid per milliliter. As shown in Fig. 12, the activating effect of phospholipids consists in both an increase in the V_{max} and a decrease in the K_m(ATP) for ATPase. A parallel increase in the affinity and in the inhibitory effect of arylazido-β-alanyl ATP was observed when the ATPase complex was irradiated in the presence of phospholipids (i.e., asolectin). The use of a mixture of purified phospholipids (phosphatidylethanolamine, phosphatidylcholine, cardiolipin in the molar ratio 1 : 1 : 0.02) at 0.3 mg/ml gave a similar increase in the inhibitory effect of arylazido-β-alanyl ATP on both the ATPase and ATP–^{32}P exchange.

14. Covalent Labeling of the ATPase Complex

That the arylazido-β-alanyl ATP is covalently bound to the partially purified ATPase complex can be appreciated from the data illustrated in Fig. 13. After photoirradiation the arylazido-β-alanyl ATP accompanies the ATPase complex proteins eluted at the void volume of a Sephadex G-25 column. Without irradiation the arylazido-β-alanyl ATP remains unbound and is eluted completely following the protein fraction.

At attempt to measure the stoichiometry of arylazido-β-alanyl ATP binding to the ATPase complex was made by a parallel measurement of the degree of inhibition of ATPase and the quantity of arylazido-β-alanyl ATP bound after photoirradiation in the presence of various concentrations of

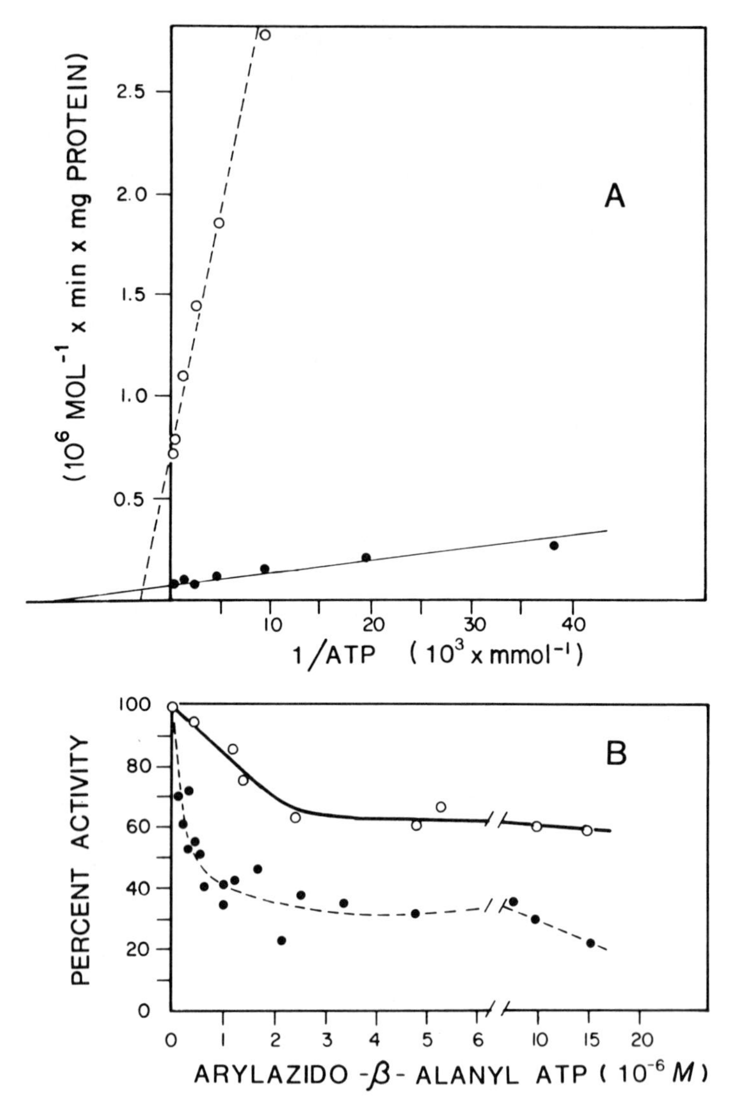

FIG. 12. Effect of phospholipids on the ATPase Activity and on the arylazido-β-alanyl ATP photodependent inhibition in the ATPase complex. The ATPase activity of the partially purified ATPase complex was determined in the absence and in the presence of phospholipids. When used, phospholipids (Asolectin) were at 0.32 mg/ml. This concentration, emulsified by sonication, gave optimal activity when tested at 4 m*M* ATP. (A) The reciprocal plot of activity as a function of the ATP concentration gives respective values of $V_{max} = 1.4$ μmol of phosphate per minute per milligram of protein; K_m(ATP) = 330 μ*M* in the absence of phospholipids (○---○) and $V_{max} = 13.8$ μmol of phosphate per minute per milligram of protein; K_m(ATP) = 77 μ*M* in the presence of phospholipids (●—●). (B) The ATPase complex at 0.53 mg of

tritium-labeled analog (Fig. 14). Some 70–80% of the ATPase activity is inhibited by the binding of 1.6–2.0 nmol of arylazido-β-alanyl ATP per milligram protein of the complex. Extrapolation of the data to complete saturation of binding sites gives a value of 1.6 nmol of arylazido-β-alanyl ATP bound per milligram of ATPase complex protein. However, the irreversible nature of the photodependent inhibition probably renders this approach, i.e., applicability of the Scatchard plot, unreliable.

15. Which Subunits of the ATPase Complex Are Labeled?

The partially purified complex (1.57 mg of protein per milliliter in 50 m*M* Tris-SO_4, pH 7.4) was irradiated with 18 flashes in the presence of phospholipids (0.35 mg/ml) and a total concentration of 6.4 μ*M* [^{3}H]arylacido-β-alanyl ATP (5.8 × 10^6 cpm/μmol) using the successive-addition procedure, i.e., 6 additions of 1.07 μ*M* nucleotides, each addition followed by 3 flashes. The irradiated sample was then dialyzed against buffer containing 0.1% SDS, 150 m*M* DTT, and 50 m*M* Tris·HCl, pH 7.5, lyophilized and resuspended in 1 ml of same buffer.

Aliquots containing 149 μg of protein were applied to the top of 10 different acrylamide gels. Figure 15 shows the gel scanning of the staining profile. A control gel was run in parallel containing the F_1ATPase extracted from the ATPase complex according to Beechey *et al.* (*184*) permitting a comparison for the identification of the F_1 subunits of the ATPase complex (Fig. 15C).

The radioactive pattern (Fig. 15A) when compared to the protein pattern (Fig. 15B) reveals that a restricted number of the subunits of the ATPase complex are labeled. Essentially only the α and β subunits contain significant label. We have no indication that the aryl group associated with arylazido-β-alanyl ATP nor with arylazido-β-alanine binds to independent sites (i.e., uncoupler binding sites on F_1ATPase). When arylazido-β-alanine is photoirradiated with submitochondrial particles, there is no significant inhibition of ATPase activity observed within the concentration ranges below 10 μ*M*. This is true when photoirradiation with arylazido-β-alanine is carried out in the presence or in the absence of free ATP. At much higher concentrations

---

protein per milliliter in 50 m*M* Tris-SO_4, pH 7.5, was irradiated with 18 flashes in the presence of varying concentrations of arylazido-β-alanyl ATP using the successive-addition procedure described in Fig. 1. Irradiation was either in the presence (●---●) or in the absence (○—○) of phospholipids. The ATPase activity was determined in all cases in the presence of phospholipids (0.32 mg/ml) and was compared with controls treated in the same way except the arylazido-β-alanyl ATP was omitted. Control ATPase activity was assayed at 3.02 μmol of phosphate liberated per minute per milligram of protein. From Cosson and Guillory (*183*).

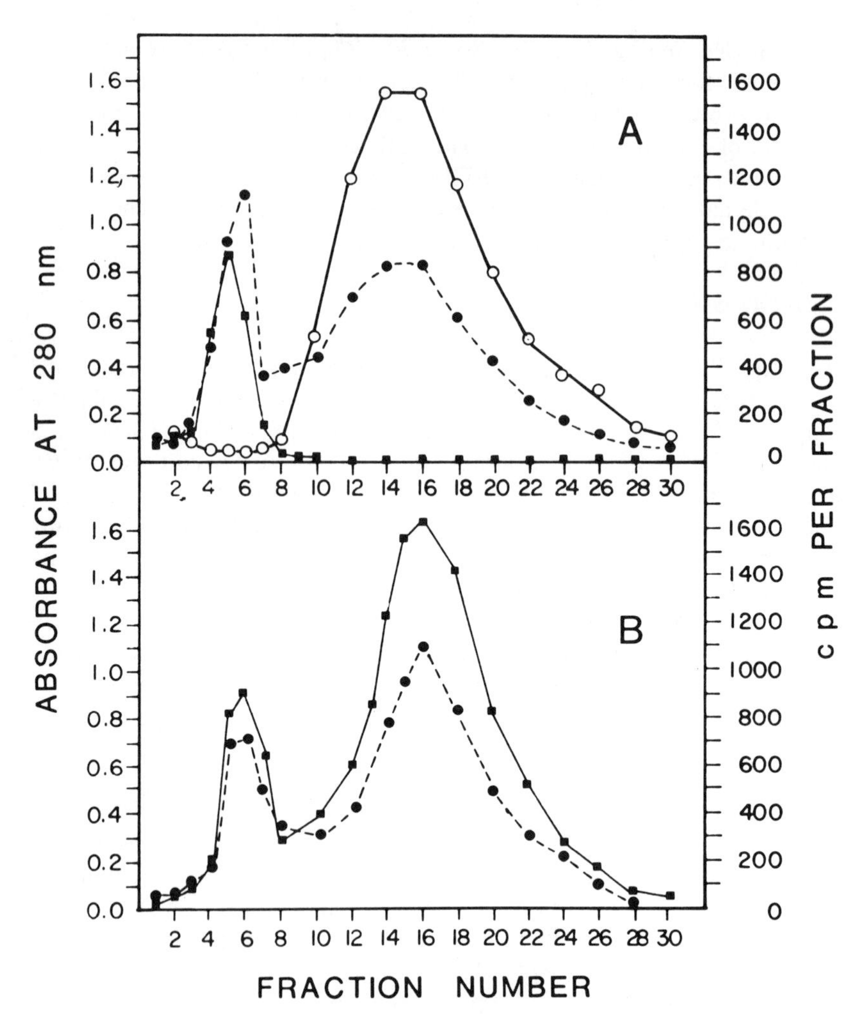

FIG. 13. Chromatography of arylazido-β-alanyl ATP-treated ATPase complex. The partially purified ATPase complex at 1.3 mg of protein per milliliter in 25 mM Tris-SO_4, pH 7.5, was irradiated with 18 flashes in the presence of phospholipids (Asolectin at 0.31 mg/ml) and arylazido-β-alanyl ATP at 5.1 μM using the successive-addition procedure (6 successive treatments of 3 flashes and 0.85 μM analog addition). One sample had in addition 5 mM ADP and 5 mM pyrophosphate; a second control sample was irradiated in the absence of arylazido-β-alanyl ATP. After irradiation the ATPase activity was determined on an aliquot of each sample. In the presence of ADP and pyrophosphate there was a 16% photodependent inhibition by the nucleotide analog, and in the absence of these effectors inhibition was at the level of 62%. The control activity was measured at 13.0 μmol of phosphate liberated per minute per milligram of protein. To the remaining sample sodium dodecyl sulfate (SDS) was added to a final concentration of 0.5%, and the protein solution was placed on the top of a Sephadex G-25 column (0.8 × 20 cm) equilibrated with 25 mM Tris-SO_4, pH 7.5, and 0.5% SDS. The preparation was eluted with

of arylazido-β-alanine (100 μM) a photodependent inhibition (20%) is observed that is not prevented by the presence of ADP and/or pyrophosphate. This latter inhibition might be considered as unspecific, i.e., occurring at sites unrelated to the adenine nucleotide binding site as proposed by Lunardi *et al.* (*118*) for azidonitrophenylaminobutyryl ADP.

The binding of 0.22 mol of arylazido-β-alanyl ATP per milligram of mitochondrial protein is associated with a maximal inhibitory effect. Assuming a molecular weight of 350,000 for F_1 and that F_1 represents 5–10% of the SMP protein, the stoichiometry of binding is between 0.77 and 1.5 mol per mole of ATPase. For the ATPase complex 1.6 nmol of arylazido-β-alanyl ATP is bound per milligram of protein at 100% (extrapolated) inhibition. Considering the preparation to be a 65% "pure" ATPase complex with a molecular weight of 480,000 (*184a*), calculation of binding indicates a stoichiometry of 1.2 mol of analog bound per mole of F_1 complex. These values for the ATPase complex as well as for submitochondrial particles show that arylazido-β-alanyl ATP is bound to a restricted number of sites, close to one per F_1 molecule. Among the 12 to 13 subunits of the ATPase complex by SDS gel electrophoresis, only the α and β subunits carry arylazido-β-alanyl ATP binding sites. No labeling was observed at the level of the subunit corresponding to the adenine nucleotide carrier.

The observation that essentially only the α and β subunits of F_1 integrated into the ATPase complex carry the arylazido-β-alanyl ATP binding site(s) can be compared to the results of Lunardi *et al.* (*118*) showing that a similar derivative, arylazidobutyryl ADP, is also photolytically bound to the α

---

the same buffer. Fractions of 0.25 ml were collected, and their absorbance was determined at 280 nm. The radioactivity content was determined on 0.1-ml aliquots of each of the eluted fractions. Panel A shows the elution pattern obtained for the arylazido-β-alanyl ATP-treated protein (absorbance at 280 nm ■—■; radioactivity, ●---●) and for an irradiated control to which arylazido-β-alanyl ATP was added to a final concentration of 5.1 μM after irradiation (radioactivity, ○—○). Panel B shows the elution pattern obtained for the arylazido-β-alanyl ATP-treated preparation in the presence of ADP and pyrophosphate (absorbance at 280 nm, (■—■; radioactivity, ●---●). The specific radioactivity for the protein fractions 3 to 7 of panel A gave a labeling by arylazido-β-alanyl ATP equal to 1.27 nmol of nucleotide analogs bound per milligram of protein in the absence of ADP and pyrophosphate. In fractions 4 to 8 of panel B the labeling was 0.92 nmol per milligram of protein. An aliquot of each sample was dialyzed for 20 hours in the dark against a buffer containing 25 m*M* Tris-acetate, pH 7.5, and 0.1% SDS (three changes of 1 liter). Five milliliters of Aquasol was then added to each sample for ^3H radioactivity measurement. The specific radioactivity was 1.17 nmol of analog bound per milligram of protein for the sample irradiated in the absence of ADP and pyrophosphate and 0.86 nmol bound per milligram of protein for the sample irradiated with arylazido-β-alanyl ATP in the presence of the effectors. The control sample in which arylazido-β-alanyl ATP was added after irradiation showed a labeling of 0.02 nmol of analog bound per milligram of protein. From Cosson and Guillory (*185*).

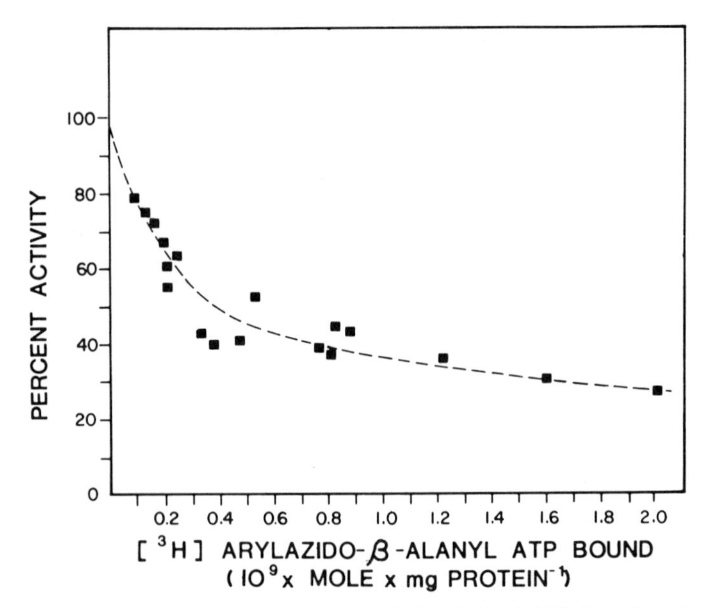

FIG. 14. Determination of the quantity of arylazido-β-alanyl ATP bound to the ATPase complex. The ATPase complex at 1.33 mg of protein per milliliter in 50 m*M* Tris-SO_4, pH 7.5, was irradiated with 18 flashes in the presence of phospholipids (Asolectin at 0.32 mg/ml) and arylizado-β-alanyl ATP at concentrations varying from 0.3 to 8 μ*M*, using the successive-addition procedure described in Fig. 1. For each concentration of arylazido-β-alanyl ATP, the ATPase activity was determined and referred to the control irradiated under the same conditions but without arylazido-β-alanyl ATP. This control ATPase activity tested in the presence of phospholipids was assayed at 11.9 μmol of phosphate liberated per minute per milligram of protein. After irradiation each sample was treated with sodium dodecyl sulfate and dialyzed according to the conditions described in the legend of Fig. 13. From Cosson and Guillory (*183*).

and β subunits of purified F_1. The binding of 0.95 mol of azidonitrophenylaminobutyrl ADP per mole of F_1 resulted in 80% inhibition of the F_1ATPase activity following photoirradiation in the presence of 108 μ*M* of the derivative. These results indicate that the occupancy of the summation of one single nucleotide (ADP or ATP) site by these covalently bound derivatives is sufficient to inhibit a major portion of the ATPase activity. The same level of occupancy with respect to arylazido-β-alanyl ATP results in a similar level of inhibition for the ATPase and the ATP–^{32}P exchange activity of SMP.

A question arises concerning the exact nature of the analog-nucleotide binding site. It has been reported that several classes of sites are present in the F_1 molecule (*186*); there is one class represented by very tightly bound nucleotides, a second class of high-affinity binding sites, and a third class having a relatively low affinity for nucleotides and presumably representing

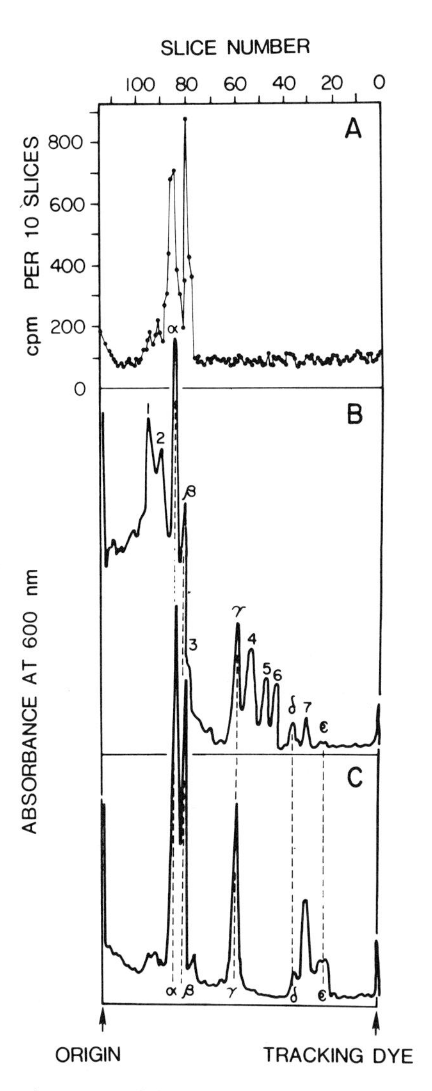

FIG. 15. Electrophoretic pattern of the arylazido-β-alanyl ATP-treated ATPase complex. The partially purified ATPase complex at 1.57 mg of protein per milliliter in 50 mM Tris-SO_4, pH 7.5, was irradiated with 18 flashes in the presence of phospholipids (Asolectin at 0.35 mg/ml) and [^{3}H]arylazido-β-alanyl ATP (5.8×10^6 cpm/μmol) at a total concentration of 6.4 μM using the successive-addition procedure (6 additions of 1.07 μM arylazido-β-alanyl ATP with 3 flashes following each addition). The inhibition of the ATPase activity amounted to 61% of that of the control irradiated in the absence of arylazido-β-alanyl ATP. The control activity was 13.2 μmol of P_i per minute per milligram of protein. When the treated protein was dialyzed under the conditions described in the legend of Fig. 8, a total of 6900 cpm was found associated with 1.49 mg of protein. After lyophilization, the sample was resuspended in 1 ml of 1% sodium dodecyl phosphate (SDS), 50 mM dithiothreitol, and 50 mM Tris-Cl, pH 7.5, and prepared for gel electrophoresis. After the 3-minute treatment at 100°C, glycerol was added to 25%

the ATPase active site (*173,187,188*). It is tempting to conclude from the above results that the arylazido-β-alanyl ATP binds essentially to the active ATPase sites. However, a problem of stoichiometry arises, if each $\alpha + \beta$ subunit pair carries the active site as suggested in the literature (*188*), and apparently confirmed by Lunardi *et al.* (*118*) and by the above results from this laboratory. The fact that presumably more than one $\alpha + \beta$ subunit pair is present per molecule of F_1 (*188*) (or ATPase complex) appears to be in contradiction with the results obtained with the nucleotide derivatives *N*-4-azido-2-nitrophenylaminobutyryl ADP and arylazido-β-alanyl ATP. These later results show that approximately one binding site per molecule of F_1 or ATPase complex is associated with a major inhibitory effect on the ATPase activity.

It may be that the catalytic site for ATPase activity is composed of a combination of the α and β subunits and that the point of insertion of the photoactivated nitrene has an equal probability for interaction with either the α or β subunit region of the catalytic site. Alternatively, one can assume that each active site on the $\alpha + \beta$ subunit pair is not independent of the other, but that the occupancy of one site provokes inhibition of the catalytic activity of the other sites. Such a negative cooperativity could similarly account for the results described in (*189*) showing that the chemical modification of one tyrosine residue blocks the catalytic activity of F_1. Similarly, the modification of one single arginine residue has been shown to result in the complete inactivation of F_1ATPase (*190*). Such a negative cooperativity of occupancy of identical nucleotide binding sites is in good agreement with the alternative site model of oxidative phosphorylation proposed by Boyer and his group (*191*). A quantitative assessment of the amount of photoaffinity nucleotide bound to both the α and β subunit as a function of ATPase inhibition may help to clarify the problem.

Verheijen *et al.* (*192*) have studied the specific labeling of the $(Ca^{2+} + Mg^{2+})$ ATPase of *E. coli* with 8-azido ATP and 4-chloro-7-nitrobenzofurazan (NbfCl). The 8-azido ATP is a substrate for the *E. coli* ATPase with a rate 30-fold lower than that for ATP and similar to that observed with the beef heart F_1ATPase. Photolysis in the presence of the

---

final concentration and bromphenol blue to 0.08 mg/ml. Aliquots (0.1 ml) containing 149 μg of protein were placed on the top of ten different acrylamide gels for electrophoretic separation. After staining the gels were scanned (panel B), sliced, and counted (panel A). A control gel was run in parallel containing the F_1ATPase extracted from the ATPase complex according to (184), permitting the experimental identification of F_1 subunits of the ATPase complex (panel C). From Cosson and Guillory (*183*).

ATP analog (4 mM) results in a 91% inhibition of ATPase, whereas 5 mM 8-azido ADP results in only a 15–20% inhibition.

When radioactive 8-azido ATP was utilized, an analysis of the labeling pattern in the ATPase showed that binding of 1 mol of 8-azido ATP per mole of F_1 results in an inhibition in excess of 90%. Although Verheijen *et al.* state that labeling is predominantly to the α subunit with some label on the β subunit, their calculations show that 72% of the bound activity is on the α subunit and 28% on the β subunit. This is in contrast to the data with beef heart mitochondria, in which only the β subunit has been reported to be labeled (*175*). Under similar conditions, no binding of 8-azido ADP to the *E. coli* enzyme could be detected. The authors mention that, in contrast to the *E. coli* enzyme, the beef heart enzyme is readily inhibited by 8-azido ADP and that 8-azido ADP becomes covalently attached to the enzyme. Unfortunately the α/β distribution of the label was not mentioned.

The initial labeling pattern of NbfCl was similar to that described for the labeling of 8-azido ATP, some 64% of the label being found on the α subunit and 36% on the β subunit following a 30-minute incubation. Upon longer periods of irradiation (24 hours), the NbfCl labeling pattern is reversed, 34% being on the α subunit and 66% on the β subunit. This shift may be related to the photolytic reaction described for NBD-Cl (*176*) and outlined in the next section.

While not a photoaffinity probe the ATP affinity probe described by Budker *et al.* (*193*) ATP-γ-4-(N-2-chloroethyl-N-methylamino)benzyl amidate is of special interest with respect to F_1ATPase interactions. The analog is an inhibitor of the F_1ATPase, the inhibitory influence being greater in the presence of Mg^{2+}. Adenosine triphosphate protects the F_1ATPase against inhibition by the alkylating ATP analog. The radioactive analog was used to show that labeling associated with the inhibition of ATPase takes place at a molar ratio of label bound to the enzyme of 1.4 ± 0.2. The

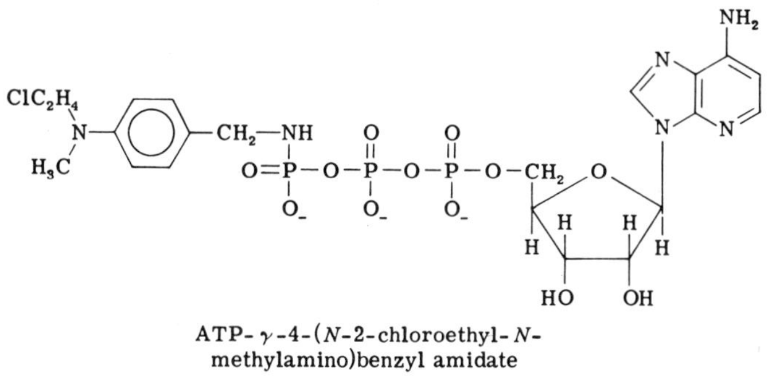

ATP-γ-4-(N-2-chloroethyl-N-methylamino)benzyl amidate

nonnucleotide adjunct 4-(*N*-2-chloroethyl-*N*-methylamino)benzylamine does not inhibit ATPase activity. The ATP analog is reported to bind to the β subunit of F_1ATPase.

16. Photolabeling of F_1ATPase with 7-Chloro-4-Nitrobenzo-2-Oxa-1,3-Diazole (NBD-Cl)

The compound NBD-Cl has been shown to specifically bind to a tyrosine residue near or on the β subunit of F_1ATPase. The evidence that a tyrosine residue is the binding site is based upon spectral evidence, i.e., the appearance of a 385-nm absorption band following reaction of NBD-Cl with F_1ATPase and the spectral similarity to that observed in the reaction of NBD-Cl with *N*-acetyltyrosine ethyl ester (*189*). The F_1ATPase is inactivated by the modification of this single tyrosine residue, an inactivation that can be reversed by a variety of sulfhydryl reagents (*194*). Reversal of the inhibition is accompanied by the transfer of the NBD residue to the added sulfhydryl compound as indicated by a spectral shift from 386 nm (NBD–F_1ATPase complex) to an absorption maxima at 420 nm due to the formation of the R-S-NBD chromophore.

TABLE XIV

INHIBITION OF ETP-ATPASE BY NBD-Cl[a,b]

	− DTT		+ DTT	
NBD-Cl (μM)	ATPase	% Control	ATPase	% Control
—	0.364	100	0.364	100
67	0.175	48	0.306	84
130	0.029	8	0.113	31

[a] From Russell (*176*).

[b] Electron transport particles (ETP) (7.5 mg protein) were incubated with various concentrations of NBD-Cl in a 3-ml volume containing 16 m*M* Tris-Cl (pH 8) and 0.25 *M* sucrose. After incubation for 1 hour at 25°C the samples were centrifuged for 30 minutes at 218,000 *g*, and the particle pellet was resuspended in 0.6 ml of 0.25 *M* surcose, 10 m*M* Tris · HCl (pH 7.5). ATPase activity was assayed using 0.1-ml aliquots of this suspension containing 1.25 mg of protein. The ATPase reaction mixture (1 ml) contained, in addition to particles, 2 m*M* $MgCl_2$, 3 m*M* phosphoenolpyruvate, 2 m*M* ATP, 33 m*M* glycylglycine, 100 μg of pyruvate kinase, and, where indicated, 1.5 m*M* DTT. After a 10-minute incubation at 30°C, 0.2 ml of 70% perchloric acid was added, and inorganic phosphate was determined on the protein free supernatant. ATPase activity is recorded as micromoles of P_i per milligram of protein per minute.

If the NBD-Cl modified enzyme (NBD–F_1ATPase) is incubated at pH 9, an intramolecular transfer occurs resulting in the formation of an *N*-nitrobenzofurazan ATPase. The transfer of the NBD from tyrosine to possibly a ε-NH_2 of lysine is a slow reaction taking several hours and observable as a time-dependent absorption shift from 385 nm to 475 nm. The transfer is felt to be irreversible, since no reactivation could be effected upon the addition of sulfhydryl compounds. Sodium dodecyl sulfate polyacrylamide gel electrophoresis of the *N*-nitrobenzofurazan ATPase showed that only the β subunit of the F_1ATPase was labeled, indicating that an essential tyrosine and a critical group (possibly lysine) on the β subunit when modified by NBD-Cl could inhibit the ATPase activity completely. Nucleotides do not protect the enzyme from this inhibition.

When the ATPase activity of ETP was inhibited by NBD-Cl, it was found that this inhibition was only partially reversible with DTT, in contrast to the complete reversibility found with soluble F_1ATPase (*176*). Inhibition of the enzyme was nearly complete at 130 μM NBD-Cl, and maximal reversal (with 1.5 mM DTT) restored only 31% of the initial activity (Table

TABLE XV

INHIBITION OF ETP_A-ATPASE BY NBD-Cl[a,b]

	− DTT		+ DTT	
Additions	ATPase	% Control	ATPase	% Control
—	0.738	100	0.795	100
NBD-Cl	0.010	1	0.697	88
NBD-Cl + ATP	0.043	6	0.734	92

[a] From Russell (*176*).

[b] ETP_A particles (7.5 mg) were incubated in a total volume of 7.8 ml containing 13 mM Tris-acetate (pH 8), 0.23 M sucrose, 6.5 μM EDTA, and, where indicated, 0.2 mM NBD-Cl and 64 mM ATP. NBD-Cl was added as an ethanol solution, and ethanol (160 μl) was added to the control sample. After incubation at 25°C for 1 hour, the samples were centrifuged for 50 minute at 218,000 g. The protein pellet was resuspended in 1.2 ml of 0.5 M Tris-acetate buffer (pH 8) containing 0.25 M sucrose, and a 0.1-ml aliquot (0.625 mg of protein) was assayed for ATPase activity. The ATPase reaction mixture (1 ml) contained, in addition to the treated ETP_A particles, 2 mM $MgCl_2$, 3 mM phosphoenolpyruvate, 2 mM ATP, 33 mM glycylglycine, 100 μg of pyruvate kinase, and, where indicated, 1.5 mM DTT. Incubation was for 10 minutes at 30°C and was terminated by the addition of 0.2 ml of 70% perchloric acid. Inorganic phosphate was determined on an aliquot of the protein-free supernatant. ATPase activity is recorded as micromoles of P_i per milligram of protein per minute.

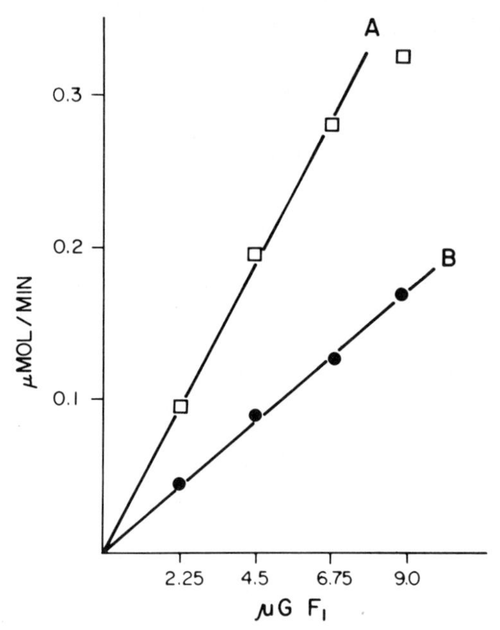

FIG. 16. Effect of order of additions in the reactivation of the NBD-F_1 ATPase–inhibitor protein complex. The NBD-F_1 ATPase (45 μg of protein) was incubated in a volume of 0.8 ml containing 1 μmol of ATP, 1 μmol of $MgCl_2$, and 5 μmol of Tris-Cl (pH 8). To sample A was added 1 μmol of DTT, and, after 5 minutes of incubation at 25°C, 67 μg (50 units) of inhibitor protein. Sample B had 67 μg of inhibitor protein added at 5 minutes, followed by 1 μmol of DTT at 35 minutes. Both samples were precipitated by the addition of 2 ml of saturated $(NH_4)_2SO_4$ after a 40-minute incubation at 25°C. The protein precipitate was centrifuged, and the pellet was resuspended in 1 ml of 50 m*M* Tris-acetate (pH 8), 2 m*M* EDTA, and 4 m*M* ATP. The samples were heated at 64°C for 2 minutes and assayed for ATPase activity. The ATPase reaction mixture (1 ml) contained, in addition to the sample, 2 m*M* $MgCl_2$, 3 m*M* phosphoenolpyruvate, 2 m*M* ATP, 33 m*M* glycylglycine, and 100 μg of pyruvate kinase. Incubation was for 10 minutes at 30°C, and the reaction was terminated by the addition of 0.2 ml of 70% perchloric acid. An aliquot of the deproteinated supernatant was assayed for inorganic phosphate. ATPase activity was recorded as micromoles of P_i liberated per minute. From Russell (*176*).

XIV). On the other hand, ETP_A ATPase in which activity had been inhibited 94% by NBD-Cl was subsequently restored to the extent of 92% when treated with DTT (Table XV). One possible explanation for the difference in the DTT reactivation of NBD-Cl-inhibited ATPase of ETP and ETP_A is that the ATPase inhibitor protein that binds specifically to the F_1ATPase (and is present in ETP but not ETP_A) interferes with DTT reactivation of the enzyme but not with NBD-Cl inhibition.

Prolonged preincubation of the NBD-labeled ETP at 4°C or heating of the labeled ETP at 65°C for 4 minutes (both without DTT) were found to be treatments capable of completely reversing the NBD-Cl inhibition of ATPase upon addition of DTT. The preincubations did not have a direct effect on the NBD–F_1ATPase complex (which was still inhibited), but indirectly allowed DTT to reactivate the inhibited ATPase.

Both prolonged incubation at 4°C and short incubation at elevated temperature are methods utilized to remove inhibitor protein from F_1ATPase (*195,196*). Thus three different physical methods (alkaline pH, heat, and prolonged incubation at 4°C) used to dissociate the inhibitor protein from the enzyme also serve to allow the DTT reactivation of the enzyme following NBD-Cl treatment.

In separate experiments it was demonstrated that DTT itself does not interfer with the binding of inhibitor protein to F_1ATPase. Consequently a test of the influence of inhibitor protein on DTT reactivation of NBD-Cl-inhibited F_1ATPase was made. It was found (Fig. 16) that NBD-Cl-inhibited F_1 treated with DTT prior to incubation with inhibitor protein had more than twice the activity of a preparation incubated with inhibitor protein prior to DTT treatment. It is obvious that, in the reconstituted system, inhibitor protein interferes with DTT activation of NBD-F_1ATPase.

Trypsin–urea particles (TU particles) are submitochondrial particles treated with trypsin and then urea in order to remove F_1ATPase. These particles are capable of binding purified F_1ATPase, rendering the soluble enzyme sensitive to oligomycin and *N,N*-dicyclohexylcarbodiimide (DCCD). When NBD–F_1ATPase was added to TU particles, the ATPase activity remained inhibited; upon addition of DTT, a stimulation of ATPase activity was observed. The fact that this stimulated ATPase activity was sensitive to DCCD indicated that the TU particles had bound the NBD–F_1. In the absence of the TU particles, DCCD had no effect on the ATPase activity of the DTT-activated F_1ATPase. That the NBD-Cl-F_1 bound itself to the TU particles prior to activation by DTT was indicated by an experiment in which the TU particle NBD–F_1 mixture was centrifuged and resuspended before addition to DTT. Activation of the ATPase activity of this preparation by DTT indicated that the inhibited F_1 bound itself to the TU particles.

17. Effect of Irradiation on the NBD-Cl-Labeled ETP$_A$

When NBD-Cl-labeled ETP$_A$ were irradiated with visible light for 4 minutes prior to treatment with DTT, the inhibition of ATPase activity was not reversed by DTT. Photoirradiation in the presence of NBD-Cl resulted in irreversible inactivation. Irradiation had little effect on unlabeled ETP$_A$ and must consequently be a direct effect of light on the NBD–F_1ATPase complex (Table XVI). The amount of inhibition increases with the length of time of irradiation; however, if DTT is present during the irradiation

TABLE XVI

EFFECT OF PHOTOIRRADIATION OF THE REACTIVATION OF NBD–ETA_A–ATPASE BY DTT

Sample	ATPase (μmol/mg^{-1} min^{-1})	% Control
ETP	0.589	100
NBD–ETP_A	0.056	10
NBD–ETP_A + DTT	0.575	98
NBD–ETP_A + light + DTT	0.189	32
ETP_A + light	0.532	90

[a] From Russell (*176*).

[b] The ETP_A particles (7.5 mg) were incubated in a volume of 7.8 ml, containing 13 m*M* Tris-Cl (pH 8), 0.23 *M* sucrose, 6.5 μ*M* EDTA, and 0.2 m*M* NBD-Cl. After incubation for 1 hour at room temperature, the mixture was centrifuged (50 minutes at 218,000 *g*) and the protein pellet was resuspended in 1.2 ml of 0.25 *M* sucrose, 10 m*M* Tris-Cl (pH 7.5). Light treatment consisted of four 1-minute periods of irradiation, with 1-minute cooling intervals at 0°C after each irradiation period. The treatment with DTT consisted of preincubating the sample for 5 minutes at room temperature with 1 m*M* DTT after the irradiation.

period, the inhibition is reversed, indicating that irreversible inhibition is due to the irradiation of bound NBD-Cl.

When NBD–F_1ATPase was prepared in a room exposed to bright sunlight (fluorescent light had no effect), the ATPase of the complex was inhibited relative to that activity of the complex formed in the dark. Comparison of the spectra of the two preparations after Sephadex chromatography showed that the sample prepared in bright sunlight had an absorption peak at 475 nm (see Fig. 17). This corresponds to the peak found by Ferguson *et al.* (*189,194*) when the NBD-Cl-labeled enzyme was left for several hours at pH 9. It appears, therefore, that light is a much more effective catalyst of this reaction than pH, since irradiation for a few minutes brings about an irreversible inhibition of ATPase activity, whereas several hours of incubation at pH 9 are required for the same effect.

Since there is no competition by nucleotides for the NBD-Cl inhibition of F_1ATPase, the nucleotide binding site of the enzyme is most likely not involved with the NBD-Cl binding. As shown here, the labeled enzyme is still capable of reconstitution of DCCD-sensitive ATPase with TU particles. This may indicate that the binding site for NBD-Cl is not involved, fundamentally, with the binding of F_1ATPase to the membrane.

It does appear, however, that the inhibitor protein protects the NBD-F_1ATPase complex from DTT interaction, making DTT a less

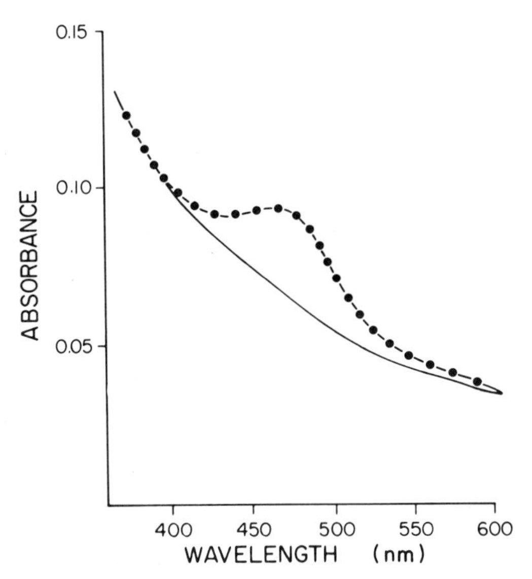

FIG. 17. Spectral characteristics of NBD-F_1. The F_1ATPase (22 mg) in 1 ml of a buffer consisting of 0.25 *M* sucrose, 50 m*M* Tris-Cl (pH 8), m*M* EDTA, and 4 m*M* ATP was brought to 50 μM in NBD-Cl and incubated for 1 hour at room temperature. One sample was prepared in the absence of direct sunlight (indirect fluorescent lighting), and the other in a room exposed to bright direct sunlight. After the incubation period, both samples were precipitated by addition of 1 ml of saturated ammonium sulfate and centrifugation for 20 minutes at 27,000 *g* at 40°C. The precipitated pellets were resuspended in 1 ml of 0.25 *M* sucrose and 20 m*M* triethanolamine-Cl (pH 8) and subjected to Sephadex G-25 chromatography with the same buffer. Light sample (●—●) was exposed to direct sunlight; dark sample (——) was prepared under indirect (fluorescent) lighting. From Russell (*176*).

effective activator. This effect can be eliminated by preparing submitochondrial particles low in inhibitor protein, or by subjecting the labeled mitochondria to treatments that lower the concentration of inhibitor protein. This effect is not limited to submitochondrial particles, but can be reproduced with purified F_1ATPase.

It would appear that the site of inhibition of NBD-Cl is within the same region as the site for binding of inhibitor protein. Since the inhibitor protein does not interfere with oxidative phosphorylation, it is most probably acting at a noncatalytic site, similar to that postulated for NBD-Cl action.

The photostimulation of the NBD-Cl inhibition of ATPase is perhaps not too surprising. NBD-Cl is a chromophore bound initially by a rather unstable ester linkage to tyrosine. In the presence of light, energy absorbed by the ligand would tend to make the ester bonding even less stable. That the shift is most likely to an amine is indicated by the appearance of a 475-nm peak in the absorption spectrum of NBD–F_1ATPase following

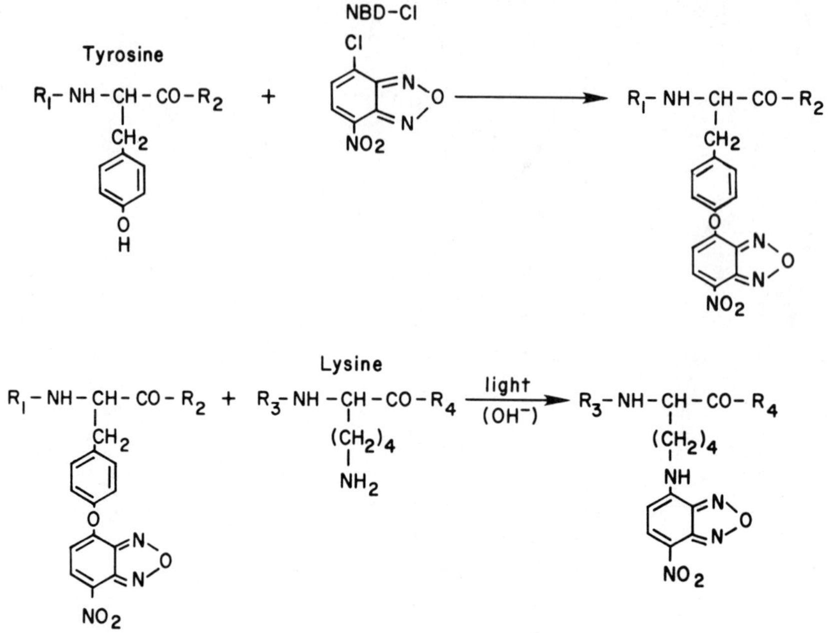

Tyrosine

$$R_1-NH-CH(CH_2-C_6H_4-OH)-CO-R_2 + \text{NBD-Cl} \longrightarrow R_1-NH-CH(CH_2-C_6H_4-O-NBD)-CO-R_2$$

Lysine

$$R_1-NH-CH(CH_2-C_6H_4-O-NBD)-CO-R_2 + R_3-NH-CH((CH_2)_4NH_2)-CO-R_4 \xrightarrow[(OH^-)]{\text{light}} R_3-NH-CH((CH_2)_4-NH-NBD)-CO-R_4$$

FIG. 18. Proposed mechanisms of NBD-Cl binding to F_1ATPase (*176*, *189*, *194*).

exposure to bright sunlight. Figure 18 represents a proposed mechanism for the sequence of events in the NBD-Cl binding to F_1ATPase.

D. Membranes and Membrane Transport Systems

1. *Adeninylate Translocase of Mitochondria*

The adenine nucleotide transporter catalyzes a 1 : 1 exchange of exogenous ADP or ATP for endogenous ATP or ADP. Vignais *et al.* (*197*) have attempted to identify the ADP binding site of the translocator by a series of chemical and physical treatments, such as photooxidation, heat treatment, and controlled trypsin action. A protein fraction reported to have been isolated has the property of binding ADP in an atractyloside-sensitive reaction (*198*). Vignais (*199*) has reviewed the physical and chemical characteristics of the translocator. The inhibitors atractyloside ($K_d \sim 10^{-7}$) and carboxyatractyloside remain bound to the outer mitochondrial membrane surface, while bongkrekic acid ($K_d \sim 10^{-8}$) when partially protonated enters the inner mitochondrial membrane and enhances the affinity of the adenine nucleotide transport system for ADP or ATP. The ability to penetrate (bongkrekic acid) or not penetrate (atractyloside) may explain the inner or outer localization of the action of these inhibitors; alternatively, the inhibitor-binding sites may be asymmetrically distributed on opposite sides of the membrane (*199*).

The preferential uptake of ADP is abolished by uncouplers or by valinomycin + K^+, which bring about an increased membrane conductance to H^+ or K^+.

There is an obvious relationship between the F_1ATPase enzyme and the translocator, since the oxidative phosphorylation of external ADP can be accomplished without an obligatory flux of adenine nucleotides through the mitochondrial membrane. In addition, external ADP is phosphorylated by the ATP synthetase without prior dilution in the matrix pool of nucleotides. The molecular mechanisms for these interactions have yet to be elucidated (*200*).

The application of specific photoaffinity probes to the study of the translocator has added to our knowledge of its structure and relationship to the other proteins of the mitochondrial energy-transducing membrane. Lauquin *et al.* (*119*) have utilized the radiolabeled arylazido analogs of ADP and ATP (*N*-4-azido-2-nitrophenylaminobutyrl ADP and *N*-4-azido-2-nitrophenylaminobutyryl ATP) in an attempt to identify and characterize the adenine carrier in mitochondria and sonic mitochondrial particles. Since the nucleotide carrier is readily damaged by UV irradiation, the use of these analogs has the advantage that they are photolyzed in the visible spectral range. The two analogs bind covalently upon photoirradiation to a 30,000 MW peptide and to a higher molecular weight peptide (50,000–55,000) that probably belongs to F_1ATPase. The addition of bongkrekic acid specifically decreases the photolabeling of the 30,000 MW polypeptide in submitochondrial particles. In *Saccharomyces cerevisiae* the adenine nucleotide binding protein, which is protected by atracytyloside, is by contrast a 37,000 MW polypeptide. The K_i for the nucleotide translocator exhibited by arylazido nitrophenylaminobutyryl ADP is of the order of 10 μM.

The 8-azido ADP analog was shown by Schafer *et al.* (*201*) to be a weak inhibitor of adenine nucleotide transport in rat liver mitochondria in the dark; upon UV irradiation irreversible inhibition occurred. In a continuation of the work with 8-azido ADP, Schafer and Penades (*202*) have described the synthesis of $[^{14}C]8\text{-}N_3$ ADP prepared from $[^{14}C]$8-Br ADP. With the radiolabeled material 0.5 mol of the ADP analog was bound per mole of translocator subunit (MW 31,000) as judged by SDS gel electrophoresis of the labeled peptide. The labeling pattern suggests that *in situ* the carrier possesses a dimeric structure. The labeling is specific for 8-N_3 ADP, 8-N_3

AMP being inactive. Binding is inhibited by ADP, ATP, atractylate, or carboxyatractylate when added prior to photolysis.

The 8-N_3 ADP inhibits the normal state 4 to 3 respiratory transition induced by ADP in rat liver mitochondria, and maintains the redox state of both cytochrome *b* and pyridine nucleotides, which indicates that 8-N_3 ADP does not induce phosphorylation cycles (*201*). Thus it is either not translocated or the analog does not serve as a phosphate acceptor or serves very inefficiently. Since inhibition is released by uncouplers, the authors feel that the nucleotide analog does not necessarily penetrate the inner membrane, but interacts with the carrier, decreasing the rate of translocating of normal ADP.

Neither the ATPase activity of rat liver mitochondria and submitochondrial particles nor that of beef heart submitochondrial particles was inhibited by 8-N_3 ADP with or without photolysis.

The stoichiometry of labeling and the proposed membrane impermeability of 8-N_3 ADP lead to the conclusion that only half of the binding sites of the intact carrier system are exposed to the outer surface of the inner membrane, i.e., accessible to the photolabel. The lack of efficient translocation of the analog is explained on the basis of the anticonformation of the nucleotide, since the conformation needed for a mobile carrier substrate complex is unlikely in the case of 8-N_3 ADP. Schlimme *et al.* (*203*) have outlined the structural properties of a nucleotide substrate which are essential for binding and transport over the carrier system. Substitution of C8 hydrogen by a bulky group does not result in loss of carrier specific binding although the syn $\rightleftharpoons$ anti isomerization of the nucleotide base is prevented in favor of a preferred syn conformation.

The K_i(8-N_3 ADP) for nucleotide translocation in a dark reaction is 400 μM compared to 10 μM for azidonitrophenylaminobutyryl ADP. The K_m(ADP) varies from 1 to 10 μM and that for ATP from 1 to 150 μM. Estimation of the ratio of translocator molecules to cytochrome *a* ranges from 1.5 to 2 for rat liver mitochondria and from 2.5 to 3 for rat heart mitochondria (*199*).

Another photoaffinity approach to the study of the adenine nucleotide translocator is the preparation and utilization of an arylazido atractyloside (*117*). In synthesis of the analog the primary alcohol function of the glucose molecule in atractyloside was linked to the carboxyl group of the propionyl or butyryl derivatives of 4-fluoro-3-nitrophenylazide in the presence of carbonyldiimidazole in dry dimethylformamide. Both analogs were found to inhibit ADP transport, an inhibition that was relieved by excess ADP or by uncouplers.

For the aminobutyryl azido analog of atractyloside NAP_4-ATR (Compound 28) there were 1.2 nmol of high affinity sites evaluated with a k_d of 40 mM. Under photolytic conditions, 0.6 nmol was covalently bound to the ADP carrier per milligram of mitochondrial protein. The covalent binding of NAP_3-[^3H]ATR was less efficient than that of NAP_4-[^3H]ATR.

The radioactive profile of the mitochondrial membrane photolyzed in the presence of NAP_4-[^3H]ATR showed that two proteins, a major atractyloside binding polypeptide (MW 30,000) and a smaller one (MW 42,000–45,000), were labeled. Bongkrekic acid and atractyloside protect the 30,000 MW peptide from labeling by the azido analog. When the tritium-labeled NAP_4 residue was used in place of the atractyloside analog, there were again two peptides labeled—a major radioactive peak of 42,000–45,000 MW and a minor one of 56,000 MW.

Hanstein (*181*) has reported that NAP binds to two polypeptides of the inner mitochondrial subunit 1 of mitochondrial ATPase (MW 56,000) and to a 31,000 MW protein. The latter is believed to be the specific uncoupler binding protein.

2. *Ouabain-Sensitive* (Na^+ *plus* K^+) *ATPase*

Cardioactive digitalis glycosides, such as ouabain, specifically inhibit the ($Na^+ + K^+$) ATPase, an enzyme component of the Na^+ pump. This inhibition is related to the positive ionotropic action of these drugs by mechanisms still to be elucidated. The enzyme itself undergoes a cycle of phosphorylation in which the γ phosphate of ATP is transferred to, and then released from, the active center. Na^+ and Mg^{2+} stimulate this phosphorylation while K^+ enhances dephosphorylation. Ouabain binds preferentially to a transient phosphorylated form of the enzyme. Most of the reported purified preparations of ($Na^+ + K^+$) ATPase contain at least two different subunits (*204*). The enzyme contains two 95,000 MW polypeptide chains and two glycoprotein molecules ($\sim$50,000 MW). The large chains are thought to be involved in the formation of phosphate anhydride bond with ATP as an intermediate during hydrolysis.

Cardiac glycosides and aglycons, such as digoxin, ouabain, and strophanthidin, are well known therapeutic agents for the treatment of heart disease. Their biochemical action appears to be a specific inhibition of this ($Na^+ + K^+$) ATPase, which is involved in the transport of Na^+ and K^+ in opposite directions across the cell membrane.

The question of which of the two large peptides contains the cardiac glucose binding site has recently been explored with the aid of photoaffinity probes. Ruoho and Kyte (*205*) have synthesized an ethyldiazomalonyl derivative of the cardiac glycoside cymarin, 4′-(ethyldiazomalonyl)cymarin, (DAMN-cymarin) (Compound 43).

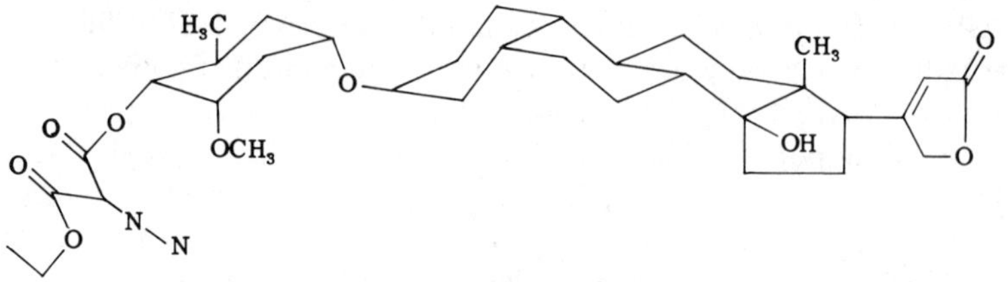

Compound 43:
4′-(Ethyldiazomalonyl)cymarin

Using purified preparations obtained from microsomes of canine renal medulla, three radioactive proteins were resolved by electrophoresis following photolysis in the presence of the radiolabeled analog. Two polypeptides of the major pe tide of the ($Na^+ + K^+$) ATPase and a covalently cross-linked dimer of the same chain (see below). Thus the long-chain ($Na^+ + K^+$) ATPase is labeled by DAMN-cymarin and, as would be anticipated, label was also found in the cross-linked dimers; no radioactivity was found associated with the small chain. Cymarin itself prevented incorporation; also there was no photoaffinity labeling when ATP was replaced by its phosphorimidate analog AMP-PNP. This is consistent with the observation that cardiac glycosides are known to bind most tightly to the ($Na^+ + K^+$) ATPase when it is in the phosphorylated form (*206*).

Since the cardiac glucoside binding site, known to be accessible from the outer surface of the plasma membrane, and the phosphorylation site, accessible from the inside surface, are both on the larger polypeptide subunit, a logical conclusion advanced is that the polypeptide must have sequences exposed to both sides of the membrane.

There are a number of problems that prevent a clear interpretation of the photoaffinity effect. While incorporation of the probe is definitely light dependent, the fact is that another analog of the cardiac glycoside, CM-cymarin, 4′-(ethylchloromalonyl)cymarin, can photolabel the enzyme preparation in a manner identical to that resulting from DAMN-cymarin interaction. Since the CM-cymarin lacks the diazo group, it cannot form a carbene by the normal N_2 dissociation mechanism. One proposal advanced is that a residue present on the protein is the photoactive species and that a "reverse" photoaffinity labeling of the reagent occurs. The observation that photochemical damage induces a degree of protein polymerization is consistent with the possibility of a photoactive species being present on the protein subunit.

An important consequence of the study, irrespective of the mechanism by which covalent bond formation occurs between the azido analog and the protein binding site, is that the large subunit has reactive characteristics that are explained by having one surface exposed to the outside of the cell and another exposed to the inside; i.e., that the protein spans the membrane. What appears to be certain is that the "large" subunit is the binding site for ouabain.

Hegyvary (*207*) explored the possibility that other membrane units distinct from the subunits of the ($Na^+ + K^+$) ATPase might be responsible for the binding of digitalis glycosides. Periodate oxidation of the vicinal hydroxyl groups of the rhannose in ouabain resulted in the formation of the dialdehyde. The binding of such oxidized [^{3}H]ouabain to renal plasma membranes (outer medulla of sheep kidney) was accomplished by borohydride reduction of the presumed Schiff bases formed between the aldehyde groups and the primary amino groups of the protein.

More than 90% of the oxidized ouabain was bound by the ($Na^+ + K^+$) ATPase component of the membrane. Such binding required the simultaneous presence of Na^+, Mg^{2+}, and ATP, K^+ inhibited binding, and there was no binding in the absence of Mg^{2+}. The covalent nature of the binding was evident from the stability against dissociation by acid, by 90% methanol, or by heating.

In the presence of Na^+, ATP, and Mg^{2+}, the oxidized [^{3}H]ouabain was bound preferentially to an electrophoretically single membrane protein of MW 89,000. This peptide is thought to be part of the ($Na^+ + K^+$) ATPase because it bound ouabain in the presence of Na^+, Mg^{2+}, and ATP, but not in the presence of Tris-EDTA. Second, the protein was phosphorylated by [^{32}P]ATP in the presence of Na^+ and Mg^{2+}, but not in the presence of K^+ and Mg^{2+}.

The finding that 90% of the digitalis glycoside (ouabain) is bound to a single membrane protein that was a part of the ($Na^+ + K^+$) ATPase supports the concept that, at the level of the cell membrane, ($Na^+ + K^+$) ATPase is the main receptor for digitalis glycosides.

The synthesis of 2-nitro-5-azidobenzoyl derivatives of ouabain have been reported by Forbush *et al.* (*208*). The synthetic material inhibits the Na^+-K^+ ATPase of pig kidney outer medulla and is competitive with ouabain. Examination of the labeled protein revealed 45% of the bound material to be present on the large polypeptide (95,000 MW), some 10% with the 53,000 MW peptide, and 45% with a proteolytic component.

A photochemical analog of strophanthidin, 3-azidoacetylstrophanthidin (Compound 22), was synthesized and tested as a cardiotonic steroid site-directed photoaffinity label for ($Na^+ + K^+$) ATPase (*109*). The analog inhibited rat brain ATPase 50% at a concentration of $10^{-6} M$ and displaced ouabain from its binding sites on the enzyme. In the presence of UV light and acetylphosphate [which supports glucoside binding on the ($Na^+ + K^+$) ATPase] the analog produced 15% irreversible inhibition of ATPase. Another photoactivatable analog of strophanthidin, strophanthidin-3,5-bis-*p*-benzoylbenzoate, inhibits rat brain ($Na^+ + K^+$) ATPase at the same level as 3-azidoacetylstrophanthidin and, like the latter, displaces ouabain from its binding site. Methyl-*p*-benzoylbenzoate alone was not effective. Ultraviolet photoactivation produced 35% irreversible inhibition of enzyme activity; however, methyl-*p*-benzoylbenzoate of itself also produced marked inhibition. It would appear that a major fraction of the inhibitory action is due in this case to other than interaction at the cardiotonic steroid binding site.

Haley and Hoffman (*130*) observed that, in the absence of irradiation, 8-azido ATP (Compound 36) is a substrate for both Mg^{2+} ATPase and ouabain sensitive ($Na^+ + K^+$) ATPase of human red cell membranes. With photolysis, covalent incorporation into the red cell membrane occurs with a labeling of but three peptide components of the membrane. This labeling is abolished at sufficient concentrations of ATP. Also ATP competes more effectively with 8-azido ATP for the active site of the ($Na^+ + K^+$) ATPase than it does for the active site of the Mg^{2+} ATPase.

The reaction of NBD-Cl with the purified eel electroplax Na^+ and K^+ ATPase was monitored by changes in ($Na^+ + K^+$) ATPase, K^+ stimulated *p*-nitrophenyl phosphate hydrolysis, and the protein UV absorption spectrum (*209*). The NBD-Cl reacts with two tyrosine residues per mole of enzyme. The modified tyrosines are located on the 95,000 MW polypeptide chain, and only one tyrosine modification is necessary for complete inhibition of ATPase activity. Mercaptoethanol restores 65% of the inhibited activity. A model is proposed that suggests an asymmetric arrangement of two 95,000 MW polypeptide chains with a single tyrosine residue at the ATP site.

3. *Labeling of Bilayers of Lipid Membranes: Lipophilic Photogenerated Reagents for Membrane Labeling*

One potential use of photoaffinity probes containing lipophilic portions is the possibility of labeling portions of membrane proteins in contact with the lipid bilayer. The photoactivatable apolar 1-azidonaphthalene (Compound 24) and 1-azido-4-iodobenzone (Compound 25) have been shown to label covalently the apolar components present within sarcoplasmic reticulum membranes (*112,210*). Upon their addition to sarcoplasmic reticulum vesicles in the dark, they partitioned rapidly and almost quantitatively into the lipid hydrocarbon regions of the membrane lipids. There was an incorporation of more than 70% of radioactivity into vesicles, some 99% of which was removed following equilibrium with bovine serum albumin.

After exposure to light, some 30% of the radioactivity was not removed by the same procedure. Repeated cold acetone precipitation removed some 60% of the membrane phospholipids, with a resultant retention of 15% of radioactivity in the precipitate. Direct analysis of the distribution of label indicated that the fatty acids contain most of the radioactivity.

The azides were found to effectively quench the fluorescence of perylene present in liposomes and the sarcoplasmic reticulum membranes. Since the quenching is due to molecular encounter, it is reasoned that a significant portion of the azides must dissolve in sites similar or identical to those of perylene (*211–213*). Radioactivity not extracted into methanol–chloroform was found to be associated principally with two main "integral" proteins, a 150,000 MW polypeptide and a 100,000 MW polypeptide, presumably the Ca^{2+} dependent ATPase of the sarcoplasmic reticulum. Pronase, which removed 79% of the protein from the membrane, liberated only ~19% of the radioactivity; total incorporation into the proteins, however, was low. From the fatty acid labeling, the inaccessibility to Pronase digestion, and the quenching of perylene fluoresence, it would appear that a significant fraction of the membrane proteins are labeled from within the lipid core. Either the nitrene species responsible for insertion is formed within the lipid hydrocarbon core or migrates there during the time period involved in photoactivation.

The pyrene azide (1-azidopyrene) has been used to label *E. coli* membranes (*214*), and, although the authors present data showing labeling to take place upon photolysis, quantitative data and characterization of the products are lacking. On the other hand, the use of 5-[^{125}I]iodonaphthyl 1-azide as a reagent for the investigation of the extent of penetration of proteins into the lipid bilayer looks promising (*215*). Upon photoactivation it inserts covalently with high efficiencies (20–55%) into membrane components. The azido derivative was found to insert from the lipid bilayer mainly into intrinsic membrane proteins. In the sarcoplasmic reticulum the main insertion occurs into the Ca^{2+}-sensitive ATPase. Mild tryptic cleavage of the 100,000 MW labeled protein results in the formation of two fragments of 52,000 and 46,000 MW. Since both these fragments are equally labeled, it is suggested that the protein is in contact with the bilayer minimally at two segments of its polypeptide chain. The ability to prepare rather highly labeled reagent and the lack of formation of polymeric products upon irradiation offer advantages when compared to labeling with the 1-azidonaphthalene derivation. On the other hand, when 5-iodonaphthyl 1-azide was used to label a preparation of the Na-K^+ ATPase, not all the membrane-bound peptides were labeled (*216*), and the large 100,000 MW subunit appears to be selectively labeled. When membranes containing the labeled enzyme were treated with trypsin, less than 10% of the label was released, although up to 55% of the protein was.

Gel electrophoresis indicated that most of the radioactivity was at the 12,000 MW level, implying that selective labeling of only a portion of the largely hydrophobic section occurred.

The photolytic effect of [^{3}H]phenylazide on the lipid bilayer of egg phosphatidylcholine vesicles and bovine myelin membranes has been studied by Abu-Salah and Findlay (*217*). The azide partitions into the lipid bilayer of egg phosphatidylcholine vesicles, and upon photoactivation some 80% of the covalently attached label is associated with the fatty acyl residues of the phospholipids, while the remainder is in the polar head groups. The cholesterol component of myelin membranes is also heavily labeled. The authors propose it as an effective potential covalent label for membrane proteins.

Bayley and Knowles (*218*) have also investigated the interaction of phenylnitrene formed from phenylazide with artificial phospholipid vesicles. The reagent labeled highly unsaturated soybean lecithin vesicles to the extent of 3.3%, while dimyristroyllecithin (completely saturated) vesicles were labeled to 0.25%. The extent of labeling correlated roughly with the degree of unsaturation and the number of doubly allylic methylene groups present in the fatty acyl side chain. The labeling is completely eliminated by reduced glutathione in the aqueous phase. On the basis of this study the authors hold that nitrenes are unsatisfactory reagents for the labeling of either lipids or hydrophobic portions of membrane proteins.

When prephotolyzed phenylazide was mixed with egg yolk lecithin, only 90% of the radioactivity could be removed by exhaustive dialysis, whereas there was effective removal of all unphotolyzed phenylazide. There was obviously a great deal of noncovalently bound photolyzed products of phenylazide present. A large fraction of the radioactivity derived from the photolysis products cochromatographs with the fatty acid methyl esters. Consequently the authors caution that estimates of the amount of label phenylazide attached to fatty acids on the basis of chromatography of fatty acid methyl esters could be in error. The amount of labeling of egg yolk lecithin by the interaction of phenylnitrene with synthetic vesicles was of the order of 12%.

The low effectiveness of nitrenes with respect to labeling hydrophobic regions of membranes arises first from their long half-life, favoring dimerization or reactions with the water-buffer components at the bilayer surface, and, second, from their electrophilic character. Thus longevity and chemical selectivity may result in the labeling of membrane components that are exposed to solvent even though the reagent may have been generated within the bilayer. Bayley and Knowles thus consider that photolabile groups bound to amphipathic molecules such as lipids may present difficulties with respect to labeling in a random fashion and that previous work with such reagents should be reevaluated considering the problems of nitrene electrophilicity (*218*). They point out as well that the danger of misinterpretation of labeling patterns obtained with aryl nitrenes is not necessarily avoided by the use of more hydrophobic precursors, since the site of nitrene generation may bear little or no relation to the ultimate site(s) of reaction owing to the electrophilic nature of the photolytic product. Such considerations make it impossible to rule out the specific arylnitrene modification of a few nucleophilic groups on the externally situated parts of membrane systems.

In an accompanying paper Bayley and Knowles (*219*) provide evidence for the thesis that carbenes generated from diazirines are superior reagents for photolabeling of lipids. Phenylcarbene and adamantylidene were formed from the corresponding diazirines [phenyldiazirine (Compound 44) and adamantanediazirine (Compound 45)] photochemically within lipid bilayers. The carbene products inserted into C—H bonds of saturated fatty acids

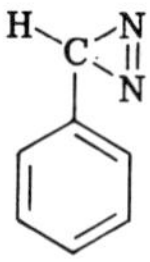

Compound 44:
Phenyldiazirine

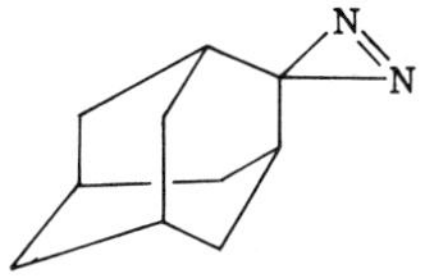

Compound 45:
Adamantanediazirine

and into C—C double bonds of unsaturated fatty acids. The lipid labeling pattern was not reduced by the water-soluble scavenger glutathione. Phenylcarbene is reported to be 10–20 times more efficient in labeling fatty acids than is phenylnitrene. The properties of diazirines offer as well an increased stability of products as compared to those derived from nitrenes.

The reagent *N*-(4-azido-2-nitrophenyl)-2-aminoethylsulfonate [NAP-taurine (Compound 13) was synthesized by Staros and Richards (*78*) and used to study those proteins situated on the surface of the erythrocyte cell. In ghost preparations the reagent was incorporated into all detectable proteins. The NAP-taurine is considered to be membrane impermeable, since there was a complete absence of label in hemoglobin isolated from labeled intact erythrocytes as well as from protein bands 1 and 2 (spectrin).

Labeling of a group of proteins on the outer membrane previously not labeled by small molecular reagents or inaccessible to enzyme probes is felt to be due to their low content of nucleophilic groups and represents a newly discovered class of external proteins. The selectivity of NAP-taurine in labeling the cell surface components of human red cells could not, however, be completely explained on the basis of the degree of exposure of the various proteins that were labeled (*97*). It is suggested that different local concentrations of the reagent at various parts of the cell arise owing to the varying electrostatic interaction.

Matheson *et al.* (*220*) have demonstrated the value of NAP-taurine in the labeling of essentially all exposed residues of ribonuclease A. On photolysis of the azido compound with ribonuclease, there was an almost random modification of "surface" amino acids including hydrophobic residues. Residues buried in the interior of the molecule did not react with the nitrene. However, when exposed by denaturation such residues underwent reaction.

4. *Cyclic Nucleotide Binding Sites*

The 8-azido adenosine-3′,5′-monophosphate (8-N_3-cAMP) has been shown to be a photoaffinity label for the cAMP binding site present in the partially purified soluble protein kinase from bovine brain (*132*). The binding protein has a molecular weight of 49,000 and comigrates with the single protein from the complex, which is endogenously phosphorylated by ATP. This latter protein has been shown to be a regulatory subunit of the protein kinase. Haley and his co-workers (*221,222*) have shown that the 8-azido cAMP can act as a probe for labeling the membrane-bound cAMP binding proteins of the erythrocyte membrane. In this case it labels two specific membrane proteins following photolysis. This work has been extended into more complex membrane systems showing the utility of the nucleotide analog as a photoaffinity probe for the cAMP binding site of whole nucleated cells (*133*). In the latter study a linear gradient SDS–polyacrylamide gel system was used to separate possible cAMP binding sites as monitored by the binding of labeled [^{32}P]8-azido cAMP. There are at least four sites within the sarcoma 37 cell that bind 8-azido cAMP. While these all saturate at 1 μM concentration of the analog, they have measurably different affinities for 8-azido cAMP. The cyclic adenosine monophosphate analog, N^6-(ethyl-2-diazomalonyl) adenosine 3′,5′-cyclic monophosphate, has been incorporated photochemically into a single protein of intact human erythrocyte ghosts (*223*). The protein labeled is suggested to be the regulatory subunit of membrane-bound cAMP-dependent protein kinase.

Synthesized β-$^{32}P_i$-labeled 2-azido ADP is a substrate for creatine kinase, apyrase, and pyruvate kinase (*224*). A high-affinity binding to human platelets is blocked by cold 2-azido ADP, ADP, ATP, AMP, 8-BrADP, and diphosphates of other nucleotides. At acid pH 2-azido ADP is photolabile, and at alkaline pH it becomes fluorescent and photostable. The compound is 2- to 5-fold more potent than ADP as an aggregating agent for human platelets and 20-fold more potent as an inhibitor of PGE-stimulated adenylate cyclase. Substantiation of the preliminary data could indicate that 2-azido ADP is potentially an attractive nucleotide photoaffinity labeling agent, perhaps without the conformational limitations of 8-azido ADP.

A light-induced covalent coupling of unmodified [^{3}H]cAMP and [^{3}H]cGMP to high-affinity receptors in testicular tissue has been described by Antonoff and Ferguson (*225*) and Antonoff *et al.* (*226*). Both nucleotides labeled a similar group of peptides. In a continuation of their interesting work Antonoff and Ferguson (*227*) have shown that when [^{3}H]cAMP and [^{3}H]cGMP were irradiated at various UV wavelengths in the presence of cyclic nucleotide receptor protein (testicular parenchyme), the photoincorporation is maximal in the region of 280 nm. The action spectra for the incorporation more closely resemble the absorption spectrum of the reacting protein than that of the nucleotide. It is suggested that the absorption of radiant energy by components of the protein receptor leads to the photoaffinity labeling. It is suspected that aromatic amino acids in the receptor proteins are the entities that are activated.

Similarly, the reported covalent joining of deoxyribonucleic acid to bovine serum albumin (*228,229*) may occur via photoactivation of bovine serum albumin.

5. *Transport Systems*

The azido compound, 2-nitro-4-azidophenyl-1-thio-β-D-galactopyranoside (azidophenylgalactoside) (Compound 11), is a competitive inhibitor of lactose transport in membrane vesicles from *E. coli* (K_i 75 nM) (*94*). Azidophenylgalactoside-dependent irreversible inactivation of the lac transport system following exposure to visible light requires the presence of D-lactate (KD77 nM). The azido analog does not inactivate amino acid transport, and membrane vesicles without the lac transport do not take up significant amounts of the azido compound. An absolute dependence upon an energy source (D-lactate) for azidophenylgalactoside-dependent photodependent inhibition is added evidence for the proposal that the lac carrier protein is inaccessible to substrate in the absence of energy coupling. Lactose protects against photodependent inactivation (*93,230*).

Intact adipocyte plasma membranes have been allowed to react with the photoreactive glucose derivative, *N*-(4-azido-2-nitrophenyl)-2-amino-2-D-deoxyglucose (NAP-D-glucose) (Compound 23), as a means of identifying the glucose transport components in the plasma membrane (*231*). The glucose derivative was taken up by the plasma membranes with K_t and V_{max} values similar to those for glucose under identical conditions. The azido analog inhibits uptake of D-glucose and 3-*O*-methylglucose.

Photolysis of the analog adipose cell complex results in inhibition of transport and the covalent labeling of four membrane components, two of which correspond in electrophoretic mobility to the major plasma membrane glycoproteins. The labeled membrane components with molecular weights of 100,000 and 75,000 may be directly involved in the process of glucose translocation (*232*). The labeling pattern was significantly altered when the membrane system was subjected to disruptive homogenization. A large number of membrane protein components that were inaccessible to NAP-D-glucose in the intact adipocyte system were now labeled.

6. *Photochemical Cross-Linking of Cell Membranes*

A problem in routine cross-linking experiments is the ability to distinguish between those cross-linked products that represent true near-neighbor coupling and those due to secondary collisions between independent molecules. Kiehm and Ji (*233*) have attempted to eliminate this uncertainty by chemically cross-linking membrane with cleavable photosensitive heterobifunctional reagents, ethyl-4-azidobenzimidate (EABI) and 4,4′-dithiobisphenylazide (DPA), utilizing flash discharges of a millisecond duration. This is comparable to the time period of the cross-linking itself if one considers the lifetime for arylnitrenes to be within the range of 0.1 to 1 msec (*77*). During this period of cross-linking it is assumed that it is unlikely that a significant number of collisions will occur among intramembranous particles. Consequently, cross-linking under these conditions represents either naturally occurring stable complexes or randomly colliding particles that remain associated during the period of cross-linking.

The photosensitive heterobifunctional cross-linking reagents produce well defined cross-linked complexes, making the interpretation of cross-linking results less ambiguous with respect to possible random collisions between membrane molecules. One would consequently expect to see further application of flash photolysis in such cross-linking experiments.

E. Photoaffinity Probes of Ion Channels in Excitable Membranes

Electrically excitable tissue is characterized by ion channels, structures that regulate the ion permeability of the membrane as a consequence of a particular chemical or physical signal. Within such structures one can recognize operationally two components: a gating device that acts as a door opening and closing the channel allowing for the regulation of the ion flux; and, second, a selective filtering device that determines which specific ion may pass. The gating property of different tissues responds to different signals; by way of example, while the sodium and potassium channels of nerve axons are gated by the action potential of the cell, i.e., by depolarization, the postsynaptic channels of the neuromuscular junction are gated by a chemical rather than an electrical signal. In the latter case, gating is controlled by a transmitter molecule released from the presynaptic membrane and bound at specific receptor sites at the postsynaptic membrane. One of the fundamental problems of these systems resides in the answer to the question: Are the molecular component(s) of the membrane system responsible for the selectivity of ion transport the same as those responsible for the electrical or chemically controlled gate?

One of the principal experimental obstacles for the investigation of such systems is identical to that involved in the study of the biochemical mechanism of energy transduction in mitochondria, i.e., the lack of a functional assay of the components of the system after extracting them from their membrane environment. It is anticipated that the method or resolution and reconstitution used to advantage with mitochondrial investigation offers some avenues of experimental approach. One aid is the fact that ion channels are, in excitable membranes, the site of action of numerous ligands of small molecular weight. The most important of these from the standpoint of potential photoaffinity probes may be the neurotoxins. Some of the neurotoxins are bound with such a selectivity and with such a high affinity that they can be used for *in vitro* binding assays. By way of example, the inward movement of sodium ions in the initial phase of the action potential has been shown to be blocked by the butterfish neurotoxin, tetrodotoxin (TTX) (*234*). Using radioactive TTX, Reed and Raftery (*235*) have shown that the binding of the toxin to the sodium channels in the electroplaque of *Electrophorus electricus* takes place with an apparent dissociation constant of $6 \times 10^{-9} M$. This binding was restricted to a single population of noninteracting receptor sites. On the other hand, the most efficient blocking agent of potassium conductance in nerve and muscle is tetraethylammonium (TEA); however, its affinity is too low [$K_d = 0.4$ mM (*236*)] for its use in the isolation of its binding component. Utilizing reversible binding ligands as a means of identifying membrane binding sites will always be open to some question, especially with regard to reequilibration of ligands among membrane subunits during the resolution of the membrane components.

As evidenced by a number of recent experimental reports, photoaffinity labels offer an exciting and potentially straightforward means for the investigation of the molecular constitution of the ion pores of excitable tissues. The synthetic approach in which already known active pharmacological reagents are modified into photoaffinity probes would appear to be a particularly attractive experimental approach.

1. *Acetylcholine Receptor and the* Na^+/K^+ *Channels of Postsynaptic Membranes*

The acetylcholine receptor is believed to be the gating device in the postsynaptic membrane of cholinergic synapses (*239*). At the neuromuscular junction it regulates directly or indirectly the membrane permeability for sodium and potassium ions. After binding of a transmitter molecule to the acetylcholine receptor of the postsynaptic membranes, the channel opens for a short period of time (all or none response) dependent upon physical conditions.

The purification of the nicotinic acetylcholine receptor protein in milligram quantities has allowed the initiation of a detailed biochemical investigation of its structure and mechanism of function (*237,238*). Kiefer *et al.* (*100*) reported in 1970 on the use of two potential photoaffinity probes, 4-azido-2-nitrobenzyltriethylammonium (HK68) (Compound 15) and 4-azido-2-nitrobenzyltrimethylammonium (HK83) (Compound 14), as photoaffinity labeling agents for acetylcholinesterase of the red blood cell and the acetylcholinesterase receptor of the intact frog sartorius muscle. In the absence of photolysis the arylazides were reversibly bound to both types of acetylcholine binding sites, and upon photolysis both types of sites were irreversibly inactivated. An interesting aspect was that the frog sartorius acetylcholine receptor of intact muscle was inactivated despite the presence of overlying tissue.

The irreversible loss of receptor activity upon photolysis was inhibited by protectors of the receptor site and photoproducts of the two photoaffinity probes. In a subsequent study (*86*) utilizing ^3H-labeled 4-azido-2-nitrobenzyltrimethylammonium, it was found that, while the probe appeared to be capable of photoaffinity labeling the acetylcholinesterase of the erythrocyte membrane, it was also reacting nonspecifically and extensively with other membrane components. In addition, *p*-aminobenzoate and soluble proteins, such as bovine serum albumin and ribonuclease, were effective scavengers of the photoaffinity probe. Consequently, nonspecific reaction with other proteins of the membrane could not be selectively prevented.

The photoaffinity labeling was considered to be a pseudophotoaffinity process in which the activated compound (nitrene) present in free solution was capable of binding reversibly to and dissociating from the binding site many times before it reacted to form a covalent bond within the active site. Such reactivity, it was stressed, is in effect similar to ordinary affinity labeling except that the labeling reagent is a long-lived intermediate generated by photolysis.

Hucho *et al.* (*240*) have used 4-azido-2-nitrobenzyltrimethylammonium fluoroborate in their examination by photoaffinity labeling of the quaternary structure of the acetylcholine receptor from *Torpedo californica.* Membrane fragments from the electric tissue containing the nicotinic acetylcholine receptor were found to be composed of four different polypeptide chains with molecular weights of 40,000 (α), 48,000 (β), 62,000 (γ), and 66,000 (δ). It is suggested that these four different types of polypeptide chains comprise all the proteins of this membrane as an integrated protein complex. Consequently they would be responsible for all its functional characteristics, i.e., transmitter binding and ion translocation.

All the components of the receptor structure were found to react covalently with the photoaffinity label, and agonists and antagonists such as flexedil, decamethonium, and hexamethonium, containing a quaternary ammonium group, protected all the chains against the photodependent labeling. Since the 4-azido-2-nitrobenzyltrimethylammonium fluoroborate reacts with all the major subunits of the receptor complex, it allowed the investigation of the influence of cholinergic ligands on the components of the complex. When the neurotoxin from *Naja naja siamensis* was incubated with the preparation and the photoaffinity probe during irradiation, the toxin was found to protect completely the α chain against the photodependent labeling. The other chains were protected to a much lesser extent. It is postulated that β chain protection is caused by an overlapping and steric hindrance derived from the binding of the neurotoxin to the α chain. It is concluded as well that the α chain binds the neurotoxin and that the α and β chains are both involved in the binding of ligands with quaternary ammonium groups.

The 40,000 MW subunit of the acetylcholine receptor from *Torpedo californica* in addition to binding α-bungarotoxin has as well binding site(s) for other small ligands. The cholinergic antagonist 4-(*N*-maleimido)benzyltrimethylammonium iodide (MBTA), acts as an affinity label in being bound to the receptor when reduced by DTT. Two α-bungarotoxin binding sites have been found to exist for each of the affinity label binding sites of MBTA. The α-bungarotoxin binds to two populations of receptor sites in purified acetylcholine receptors of which half can bind ligands such as acetylcholine, *d*-tubocurarine, and decamethonium with high affinity. This "half-site" phenomenon is demonstrated as well for the fluorescent probe DAP [bis(3-aminopyridinium)-1,10-decanediiodide, a decamethonium analog (Compound 46)]. Under certain conditions DAP occupies half of the toxin binding sites almost exclusively (*241*). Witzemann and Raftery (*242*) synthesized from DAP a bisazido derivative (Compound 47) for use as a photoaffinity labeling agent for the acetylcholine receptor of *Torpedo californica.* The photoaffinity probe was found to bind specifically to the purified acetylcholine receptor with physiological characteristics resembling the nonazido precursor.

In the purified "soluble" preparation of the acetylcholine receptor, two

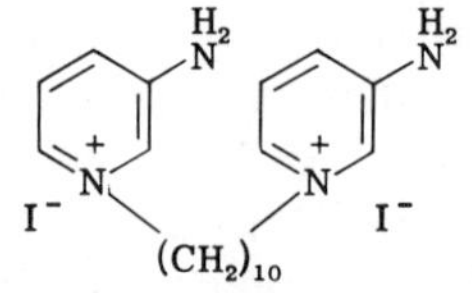

Compound 46:
DAP; bis(3-aminopyridinium)-1,10-decanediiodide

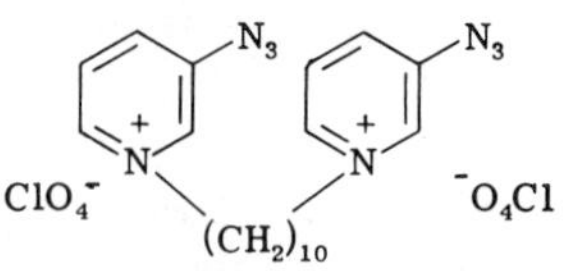

Compound 47:
DAPA; bis(3-azidopyridinium)-1,10-decanediperchlorate

of the four subunits, specifically the α (40,000 MW) and γ (60,000 MW) subunits, contained 90% of the bound analog. In contrast with the membrane-housed receptor system, the subunits of molecular weight 40,000 (α) and 50,000 (β) were found to be labeled. It would appear that the structural arrangement of the acetylcholine subunits of the membrane-bound receptor differ from the purified solubilized form. The labeling results are interpreted as indicating also that the specific binding sites are located on the 40,000 MW subunit and that labeling of the other subunits reflects the proximity to this binding site.

Two mechanisms for changes in ionic permeability have been proposed. First, that the binding of a transmitter brings about a conformational change in the receptor protein; it is this conformation transition which then regulates the gating of the ionic channels. Conformational changes as a result of ligand binding have in fact been observed for the acetylcholine receptor system (*243,244*). However, the relationship of the conformational change to gating movement is still a subject of debate. Second, it has been proposed that the receptor regulates an enzymic process, such as a kinase one, which influences directly the ion permeability of the membrane.

Gordon *et al.* (*245*) have shown that there is an endogenous membrane protein kinase activity in membrane fractions enriched in the acetylcholine receptor of *Torpedo californica*. The phosphorylation of the four polypeptides was stimulated 9-fold by K^+. The cholinergic ligand carbachol inhibited phosphorylation of the four polypeptides. Endogenous phosphorylation of the acetylcholine receptor is stimulated maximally by 100 m*M* K^+. The same preparation contains phosphatase activity (*246*). The substrate for the kinase appears to be a membrane component of MW 65,000. In the acetylcholine receptor from *Torpedo californica* the α chain has been shown to contain the acetylcholine receptor site (*247*); the function of the other chains is unknown. By means of the photoaffinity label arylazido-β-alanyl ATP, the ATP binding sites have been shown to be present in polypeptides of MW 45,000 and MW 55,000, which are distinct from the polypeptide chains of the acetylcholine receptor (*248*). It is consequently unlikely that the major component of the acetylcholine receptor represents a protein kinase or pyrophosphatase enzyme. With the radioactive ATP photolabel it was possible to show that the ATP-binding proteins are a part of the complete receptor complex, radioactivity being bound by an affinity column of *Naja naja siamensis* toxin, a ligand for the α chains of the complex. The two labeled bands, MW 45,000 and MW 55,000, would appear to represent ATP-binding proteins of the acetylcholine receptor-rich membrane, possibly the protein kinase and/or phosphatase enzymes.

The synthesis of diazido propidium (Compound 16) and diazido ethidium (Compound 17) has been described together with the application of these compounds as photoaffinity labels for cholinergic proteins (*249–251*).

Diazido propidium inhibits neuromuscular transmission in a reversible manner; when photoirradiated, the inhibition becomes irreversible. Evidence is presented that the reagent is acting postsynaptically by blocking the acetylcholine receptors. Both compounds bind to a peripheral acetylcholine binding site on acetylcholinesterase. The binding of labeled α-neurotoxin from *Naja naja siamensis* to purified acetylcholine receptor membranes of *Torpedo californica* is decreased after incubation and irradiation with diazido propidium. About half of the toxin binding sites are blocked by the photoaffinity label.

A number of attempts to reconstitute the biological function of the acetylcholine receptor have been reported. Karlin *et al.* (*252*) and Michaelson and Raftery (*253*) have reported recombination of detergent-extracted acetylcholine receptor purified by affinity chromatography with lipid vesicles formed from lipid extracted from *Torpedo californica.* Since the *in vivo* characteristics of the acetylcholine receptor were not consistently restored, further work is required for a truly biological reconstitution. Schiebler and Hucho (*254*) characterized the acetylcholine receptor-enriched membranes from *Torpedo californica* showing a protein : lipid ratio of 70% : 30%, with 70% of the lipids consisting of phospholipids. The lipid-depleted receptor complex was recombined with artificial lipid vesicles. Such receptor vesicles were chemically excitable by 10 μM carbamoylcholine as measured by efflux of $^{22}Na^+$ from the vesicles. The excitability was blocked by the α toxin from *Naja naja siamensis* venom and by reduction with 5 m*M* dithioerythritol.

2. Sodium Channels

The voltage-dependent sodium channels of axonal and muscle membranes can be examined by means of a large number of apparently specific and non-cross-reacting neurotoxins (*32*). Veratridine and batrachotoxin increase the resting sodium permeability of the membrane whereas tetrodotoxin and saxitoxin specifically block sodium flux. A third class of neurotoxins, i.e., scorpion toxins and the anemone toxins, prevent inactivation of the sodium channel (*255,258*). The anemone toxins are small polypeptides and in view of their structure may be potentially useful precursors for the preparation of photoaffinity analogs.

The molecule tetrodotoxin has been modified by the methods used by Jeng and Guillory for the preparation of the arylazido-β-alanyl derivatives of nucleotides (*52*). This method resulted in the isolation of a product that is a biologically active arylazido analog of tetrodotoxin (*30*). The analog acts as an active site-directed reagent for covalent insertion at the tetrodotoxin binding site of excitable membranes.

Amiloride (Compound 48) is a potassium-sparing diuretic that inhibits sodium transport across amphibian skin as well as through other transport tissues at concentrations as low as $10^{-5} M$. The reagent is effective in frog skin only when present in the external solution (*259*). Benos and Mandel (*260*) have recently reported that one of the analogs of amiloride, i.e., bromoamiloride (Compound 49), can irreversibly inhibit sodium transport across frog skin after irradiation with ultraviolet light. The mechanism for the inhibition is considered to be either a free-radical generation of photonucleophilic substitution. There is indication that absorption of ultraviolet light by the covalently labeled material results in rearrangement or breakage of the pyrazine ring.

3. *Potassium Channels*

No specific neurotoxin comparable to those for the sodium channel is available for the potassium channel. Potassium flow through nerve and muscle membranes is inhibited by tetraethylammonium (TEA) and by aminopyridine with apparent dissociation constants of the order of 0.4 mM, an amount insufficient for their application in resolution of this specific membrane site. Hucho *et al.* (*261*) have attempted to label the site associated with the voltage-dependent potassium conductance changes covalently with 4-azido-2-nitrobenzyltriethylammonium fluoroborate. This was shown to block reversibly and specifically the potassium current of the nodal membrane of frog sciatic nerve. The inhibition becomes irreversible when the nerve is incubated with the label in UV light.

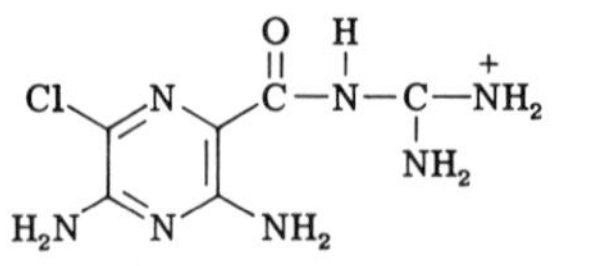

Compound 48: Amiloride

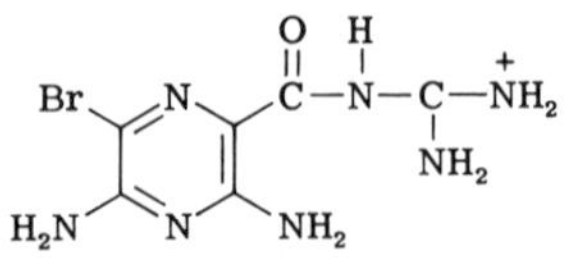

Compound 49: Bromoamiloride

In these experiments application of the label by perfusion from the inside of the axon caused in the dark a blocking effect similar to the one observed with application from the outside. During irradiation this effect was decreased and abolished, while the label applied from the outside with irradiation caused an irreversible block of the sodium conductance. Hucho *et al.* (*261*) postulate that this is due to the binding site for the quaternary ammonium group being situated close to the internal opening of the potassium channel. In this model, during irradiation in the case of application from the internal side of the axon, the "reactive tail" of the photoaffinity label can react only with sites outside the channel, i.e., with proteins from the axoplasm that are being removed by diffusion from the channel. When applied from the external side, the label enters the channel far enough for a covalent attachment to sites inside the channel. See the excellent review by Hucho and Schiebler (*29*) for a clear pictorial representation of this model. The tritiated derivative of the TEA analog was able to bind with saturation kinetics to unmyelinated crayfish walking nerve (*262*). The nerve incorporated an amount of radioactivity indicating that there were about 220 binding sites per micrometer. The total radioactivity was extracted from the membrane with chloroform–methanol 2 : 1 indicating that the labeled molecule is a phospholipid or a small proteolipid.

F. Contractile Systems

A central dogma of muscle biochemistry is that the hydrolysis of ATP by myosin is the key chemical reaction in muscle contraction (*263*). The work by Yount and his colleagues (*264*) has shown that ATP analogs capable of forming disulfide bonds with myosin react at sites distinct from the substrate-binding site, but do produce inhibition of ATP hydrolysis and actin binding.

Jeng and Guillory (*52*) have shown that myosin ATPase activity is substantially inhibited when the protein is photoirradiated in the presence of arylazido-β-alanyl ATP. Control samples, i.e., myosin, incubated with the nucleotide analog in the dark were found not to undergo a sizable inhibition of their ATPase activity. The photodependent chemical interaction between myosin and arylazido-β-alanyl ATP was shown to be covalent in nature.

Jeng (*265*) has shown that in addition to the Ca^{2+}-dependent ATPase activity of myosin, the heavy meromyosin (HMM) subfragments prepared with papain [i.e., S_1(P)] or with chymotrypsin [i.e., S_1(CT)], as well as the EDTA-stimulated ATPase of S_1(P) were all inhibited in a photodependent manner by arylazido-β-alanyl ATP. In all cases the presence of ATP during the irradiation period was able to protect against the nucleotide analog-dependent inhibition. In addition, the analog is able to function as a substrate for the

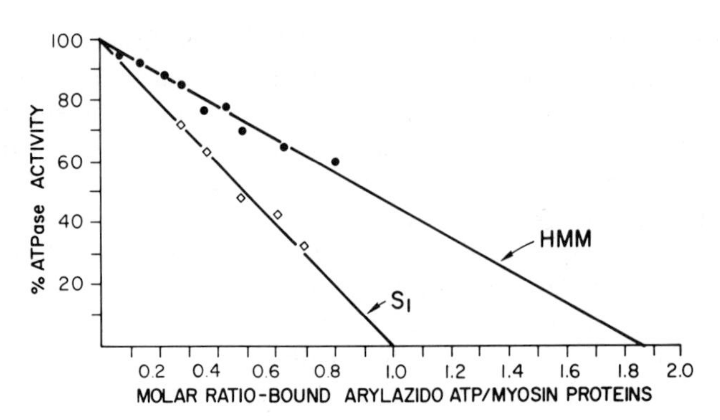

FIG. 19. Interaction of arylazido-β-alanyl ATP with subfragment 1 (S_1P) and heavy meromyosin (HMM) ATPase. See text for description. From Jeng and Guillory (*265*).

ATPase active site as indicated by the catalytic hydrolysis of arylazido-β-alanyl [^{32}P]ATP. Using the radioactive analog as a substrate, a K_m of 6.15 μM was obtained for S_1(P).

Arylazido-β-alanyl [^{3}H]ATP is bound photolytically to HMM and S_1(P) at a maximum molar ratio of protein to analog of 1 : 2 and 1 : 1, respectively. The ratio was evaluated after a stepwise procedure of irradiation, assaying for activity loss, filtration through Sephadex gel, and measuring of ^{3}H radioactivity as a function of protein concentration. The binding ratios were obtained by extrapolating the percentage of ATPase activity inhibited to 100% (Fig. 19). On the basis of these data it would appear that the arylazido-β-alanyl ATP does not distinguish the two heads of HMM as being different with respect to nucleotide binding. The SDS gel pattern of S_1(P) photoirradiated in the presence of tritium-containing arylazido-β-alanyl ATP demonstrated that the light chains of myosin are not labeled by the nucleotide analog. The fact that only the subfragment 1 heavy chains become photolytically labeled does not exclude the possible involvement of the light chain in the enzymic activity of the myosin molecule.

Szilagyi *et al.* (*266*) reported on the labeling of proteolytic fragments containing the active site(s) regions of myosin by arylazido-β-alanyl ATP. At a 3- to 4-fold molar excess of the analog per mole of binding sites, 0.45–0.65 mol were incorporated covalently after a 3-minute irradiation with UV light. This incorporation was accompanied by a 45–60% loss of Ca^{2+} ATPase activity. The label was found to be associated with the heavy chain, as shown by SDS gel electrophoresis. The label is reported to be in the N-terminal 25,000 MW peptide, which is separated by an at least 50,000 MW peptide from the region that contains SH_1 and SH_2. If these thiols are near the ATP binding sites (*267*), it means that portions of the two regions of the heavy chain, widely separated in the primary structure, come together in the tertiary structure to form the nucleotide binding site.

A photoaffinity analog of GTP (8-azido guanosine triphosphate, 8-N_3GTP, Compound 50) has been shown to be able to support the polymerization of

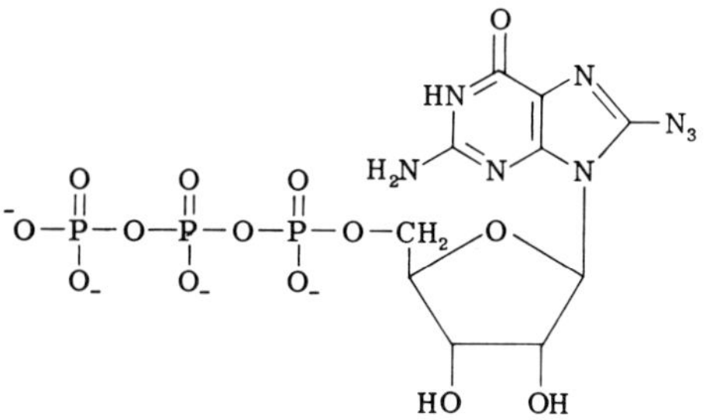

Compound 50:
8-Azido guanosine triphosphate

tubulin dimers isolated from brain tissue (*268*). The tubulin dimers are known to contain two GTP binding sites, and GTP is normally required for polymerization. Photolysis of tubulin with 0.01 or 0.1 mM [β, γ ^{32}P]8-N_3GTP yielded label only on the β-monomer, indicating that the exchangeable GTP binding site is present on the β subunit. Only GTP, not ATP, was effective in preventing the photodependent incorporation, indicating that the analog is reacting with specific GTP binding sites.

The arylazido-β-alanyl GTP photoaffinity reagent has been prepared (Compound 51) (*24*). In the dark this GTP analog is able to mimic the allosteric effect of inhibiting glutamic dehydrogenase in a reversible manner.

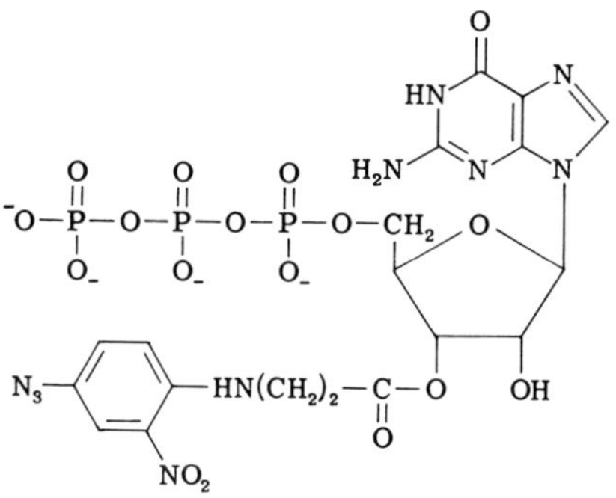

Compound 51:
Arylazido-β-alanyl GTP;
{3′-[*N*-(4-azido-2-nitro-phenyl)amino]propionyl} GTP

Characterization of the inhibitory profile indicates that it is as effective as that of the natural cofactor GTP. Photoirradiation of arylazido-β-alanyl GTP with the enzyme results in an irreversible inhibition of enzymic activity associated with a covalent binding of the radioactive nucleotide. Tryptic and chymotryptic peptides prepared from the photoirradiated enzyme indicates that but two peptides contain the photodependent inserted compound. Current experiments are directed toward elucidating the particular peptide sequences and evaluating their position(s) within the three-dimensional structure of the enzyme.

An interesting aspect of the interaction of the arylazido-β-alanyl GTP analog and the arylazido-β-alanyl ADP analog (*24*) with glutamic dehydrogenase is the fact that photoinsertion of both into what is presumably their allosteric binding sites appears to freeze the enzyme into its specific configuration. After photoirradiation with the nucleotide analogs the enzyme was cleared of residual nucleotide by gel filtration; in the case of the arylazido-β-alanyl GTP-treated preparation, the enzyme had undergone irreversible inhibition consistent with a GTP effect while the arylazido-β-alanyl ADP-treated preparation retained a specific characteristic of the ADP activated state.

G. Membrane Components for Fatty Acid and Lipid Binding

1. Phospholipid Reagents

Chaknabarti and Khorana (*110*) have reported on the synthesis and characterization of a variety of saturated and nonsaturated fatty acids containing photosensitive groups in different positions of the alkyl chain. It is anticipated that these compounds and their phospholipid derivatives will be exceedingly useful in the study of hydrophobic interactions between proteins and phospholipids in biological membranes. In one group of fatty acids an azido function is positioned at various positions along the alkyl chain (i.e., 16-azidopalmitelaidic acid, 12-azidoaleic acid (Compound 52), 12-azidostearic acid, 9-azidostearic acid, 6-azidostearic acid); in a second group an arylazido function (4-azido-2-nitrophenol) is positioned at the 12 position, i.e., 12-*O*-(4-azido-2-nitrophenol)oleic acid (Compound 53) for the unsaturated and 12-*O*-(4-azido-2-nitrophenyl) stearic acid for the saturated acid. Two other derivatives have azomalonyl adjuncts, i.e., 12-*O*-(ethyl-2-diazomalonyl)stearic acid and 12-*O*-ethyl-2-diazomalonyl)-oleic acid (Compound 54). Some representative examples are given:

$$CH_3(CH_2)_5\underset{\substack{|\\N_3}}{C}HCH_2\overset{H}{C}{=}\overset{H}{C}(CH_2)_7COOH$$

Compound 52:

12-Azidoaleic acid

$$\begin{array}{l} \quad\quad\quad\quad\quad\quad\quad\;\; H\;\; H \\ CH_2(CH_2)_5CHCH_2C{=}C(CH_2)_7COOH \\ \quad\quad\quad\quad\;\; | \\ \quad\quad\quad\quad\; O \end{array}$$

(4-azido-2-nitrophenyl ring: NO_2, N_3)

Compound 53:
12-*O*-(4-Azido-2-nitrophenyl)oleic acid

$$\begin{array}{l} \quad\quad\quad\quad\quad\quad\quad\quad\; H\;\; H \\ CH_3(CH_2)_5CHCH_2C{=}C(CH_2)_7COOH \\ \quad\quad\quad\quad\quad\; | \\ \quad\quad\quad\quad\quad O{-}\underset{\underset{O}{\|}}{C}{-}\underset{\underset{N_2}{|}}{C}{-}\underset{\underset{O}{\|}}{C}{-}O{-}C_2H_5 \end{array}$$

Compound 54:
12-*O*-(Ethyl-2-diazomalonyl)oleic acid

A number of the synthesized fatty acids were then used to acylate the 2-position in the preparation of a number of mixed acyl phosphatidylcholines (e.g., Compound 55) and mixed acyl phosphatidylethanolamines (e.g., Compound 56).

$$\begin{array}{l} \quad\quad\quad\quad\quad\quad\quad\quad\quad\quad\quad\quad\quad\quad\quad\quad\quad\quad\quad\quad\quad O \\ \quad\quad\quad\quad\quad\quad\quad\quad\quad\quad\quad\quad\quad\quad\quad\quad\quad\quad\quad\quad\quad \| \\ \quad\quad\quad\quad\quad\quad\quad\quad\quad\quad\quad\quad\quad\quad CH_2{-}O{-}C{-}C_{15}H_{31} \\ \quad\quad\quad\quad\quad H\;\; H \quad\quad\quad\quad\quad\;\; O \quad\quad\quad | \\ N_3CH_2(CH_2)_5C{=}C(CH_2)_7\overset{\|}{C}{-}O{-}CH \quad\quad\quad O \\ \quad\quad\quad\quad\quad\quad\quad\quad\quad\quad\quad\quad\quad\quad\quad | \quad\quad\quad\quad\quad\quad \| \quad\quad\quad\quad\quad\quad + \\ \quad\quad\quad\quad\quad\quad\quad\quad\quad\quad\quad\quad\quad\quad CH_2{-}O{-}P{-}OCH_2CH_2N(CH_3)_3 \\ \quad\quad\quad\quad\quad\quad\quad\quad\quad\quad\quad\quad\quad\quad\quad\quad\quad\quad\quad\quad\quad | \\ \quad\quad\quad\quad\quad\quad\quad\quad\quad\quad\quad\quad\quad\quad\quad\quad\quad\quad\quad\quad\quad O^- \end{array}$$

Compound 55:
Palmitoyl-2-(16-azidopalmitelaidyl)-
S_n-glycero-3-phosphorylcholine

$$\begin{array}{l} \quad\quad\quad\quad\quad\quad\quad\quad\quad\quad\quad\quad\quad\quad\quad\quad\quad\quad\quad\quad\quad O \\ \quad\quad\quad\quad\quad\quad\quad\quad\quad\quad\quad\quad\quad\quad\quad\quad\quad\quad\quad\quad\quad \| \\ \quad\quad\quad\quad\quad\quad\quad\quad\quad\quad\quad\quad\quad\quad CH_2{-}O{-}C{-}C_{15}H_{31} \\ \quad\quad\quad\quad\quad H\;\; H \quad\quad\quad\quad\quad\;\; O \quad\quad\quad | \\ N_3CH_2(CH_2)_5C{=}C(CH_2)_7\overset{\|}{C}{-}O{-}CH \quad\quad\quad O \\ \quad\quad\quad\quad\quad\quad\quad\quad\quad\quad\quad\quad\quad\quad\quad | \quad\quad\quad\quad\quad\quad \| \quad\quad\quad\quad\quad\quad\quad\quad\quad + \\ \quad\quad\quad\quad\quad\quad\quad\quad\quad\quad\quad\quad\quad\quad CH_2{-}O{-}P{-}O{-}CH_2CH_2{-}NH_3 \\ \quad\quad\quad\quad\quad\quad\quad\quad\quad\quad\quad\quad\quad\quad\quad\quad\quad\quad\quad\quad\quad | \\ \quad\quad\quad\quad\quad\quad\quad\quad\quad\quad\quad\quad\quad\quad\quad\quad\quad\quad\quad\quad\quad O^- \end{array}$$

Compound 56:
1-Palmitoyl-2-(16-azidopalmitelaidyl)-
S_n-glycero-3-phosphorylethanolamine

Upon sonication the synthetic phospholipids formed sealed vesicles as judged by retention of [^{14}C]glucose. Photolysis of the vesicles resulted in the formation of high molecular weight products in which intermolecular cross-linking of fatty acid chains was demonstrated.

In a continuation of their synthetic program to develop reagents for the study of phospholipid protein interactions in biological membranes, Gupta *et al.* (*269*) have reported on an improved method for the synthesis of phospholipid analogs containing photoactivable groups. Utilizing *N*,*N*-dimethyl-4-aminopyridine as catalyst and moderate amounts of fatty acids, Khorana and his associates have been able to prepare diazyl or 1,2 mixed diacylphosphatidylcholines, N-protected phosphatidylethanolamines, and phosphatidic acids in 75–90% yield. These contained the photoactivable groups: 2-nitro-4-azidophenoxy, *m*-azidophenoxy and α,β unsaturated keto groups in the fatty acyl chains. In the first published application of the above compounds to biological systems, Greenberg *et al.* (*270*) demonstrated that a variety of fatty acids containing azido or 4-azido-2-nitrophenoxy groups support growth of an auxotroph of *E. coli* that requires unsaturated fatty acids for growth. The labeled fatty acids are incorporated with their photosensitive groups intact. With the exception of the 6-azido stearate analog, all were incorporated at levels ranging from 31% to 43% of total fatty acids in phospholipids. The incorporated fatty acids had the photosensitive groups at different positions on the paraffin chain, from 6-, 9-, 11-, and 12-azidostearic acids to 12-azidooleic and 16-azidopalmitelaidic and finally 12-(4-azido-2-nitrophenoxy)stearic and oleic acids. Those fatty acids with quite bulky azidonitrophenoxy groups still showed good incorporation. The authors have pointed out that, stearically, the latitude for the acceptance of the substituted fatty acid into the biosynthetic pathways forming phospholipids and into the resulting membrane appears to be surprisingly large. The azido stearic acids appear to meet the requirements for a functional membrane better than do methyl derivatives. The large azido group ($=N=N^{+}=N^{-}$) would seem to provide more appropriate packing in the membrane bilayer and allows necessary interaction in specific protein–phospholipid complexes.

Unfortunately azido groups have relatively low extinction values and their absorption band in the UV overlaps those of protein and nucleic acid. Care must be taken, consequently, in the type and degree of irradiation required for activation of these reagents when present within membrane.

2. *Surface Membrane Receptors, Estrogen and Steroid Binding Sites*

Wolff *et al.* (*271*) have reported that the 21-diazosteroids are very suitable reagents as photogenerated probes of the mineral corticoid receptors of the toad bladder system. In this tissue the 21-diazo derivatives had significant functional activity and augmented active Na^{+} transport. By competition experiments the relative affinities of diazosteroids to rat kidney slice systems (aldosterone receptor) was 9α-fluoro-21-diazo-21-deoxycorticosterone > 9α-bromo-21-diazo-21-deoxycorticosterone > 21-diazoprogesterone. Affinities toward the corticosterone binding site were 21-diazo-21-deoxycorticosteron > 21-diazoprogesterone. Photodependent binding of [1,2-^{3}H]9-α-bromo-21-diazo-21-deoxycorticosterone to plasma proteins was detected.

The synthesis of estrogen photoaffinity labels has been reported (*272*). Both hexestrolazide (Compound 58) and hexestroldiazoketopropyl ether

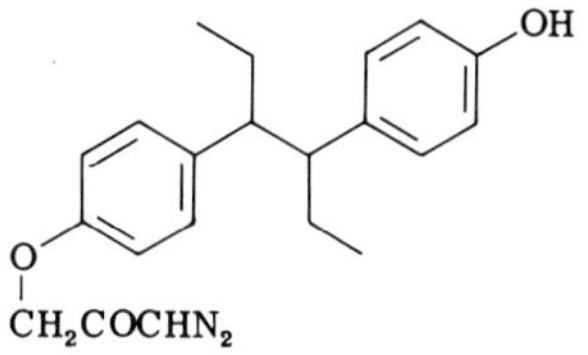

CH_2COCHN_2

Compound 57:
Hexestroldiazoketopropyl ether

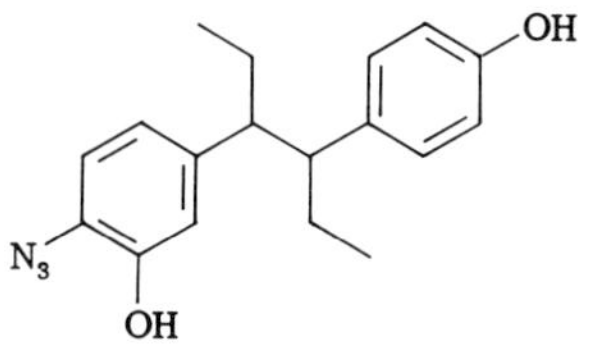

N_3 OH OH

Compound 58:
Hexestrolazide;
3,4-bis(4′-hydroxyphenyl)hexane

(Compound 57) were prepared in good yields, are stable, and undergo photochemical decomposition in solution. The application of the labeled compound resulted in the first example of the labeling of a steroid hormone receptor by photoaffinity labeling. The covalent labeling process is estrogen-site specific, and the efficiency of labeling is 15–20%.

Blackburn *et al.* (*273*) investigated the photochemical linkage of diethylstilbestrol (DES) to DNA. Diethylstilbestrol is an established synthesic estrogenic hormone and has been found to be bound covalently to DNA *in vitro* as a result of the action of long-wavelength UV irradiation. A level of binding of one molecule of DES per 1000 bases was achieved by photochemical means with an apparent 2 : 1 selectivity for purine base binding. Blackburn *et al.* suggest that binding is a result of a photochemical oxidation involving electron expulsion from the stilbestrol followed by covalent process linking the DES to a purine base.

The affinity labeling of the active site of a Δ^5-ketosteroid isomerase (*Pseudomonas testosteroni*) using photoexcited natural ligands has been accomplished (*274*). The order of effectiveness of steroid ketones in promoting photoinactivation of the enzyme is 3-oxo-4-estren-17-β-yl acetate (Compound 59) > 17-β-hydroxy-4-estren-3-one > 17-β-hydroxy-4-androsten-3-one ≫ 1-cyclohexen-2-one. This order parallels that for the affinity of substrate analogs to the enzyme's active site. Inactivation by the steroid analog is slowed down in the presence of competitive inhibitors.

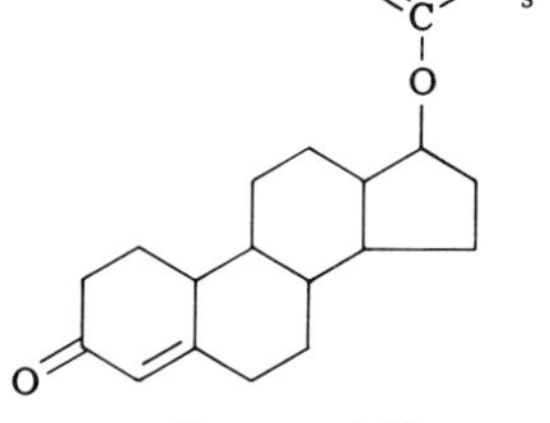

Compound 59:
3-Oxo-4-estren-17-β-yl acetate

Photoinactivation is assumed to take place by means of excitation of the keto group of the active site-bound 3-ketosteroid followed by chemical reactions between enzyme functional groups and the electronically excited ketone. Presumably the formation of the triplet state of the ketene is followed by covalent modification.

Sweet *et al.* (*275*) have utilized 21-bromoacetylaminoprogesterone in the study of 20-β-hydroxysteroid dehydrogenase.

The design of the photoaffinity reagent *N′*-(2-nitro-4-azidophenyl)-ethylenediamine (NAP-propranolol) (Compound 60), was based on its potential use as a photoaffinity probe for the β-adrenergic receptor (*276*).

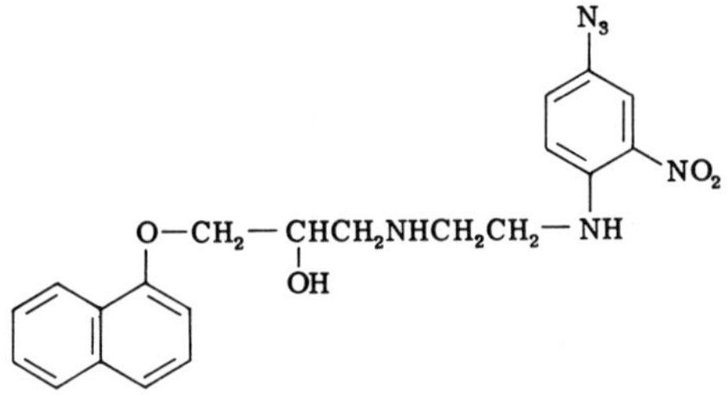

Compound 60:
N-(2-Hydroxy-3-naphthoxy-propyl)-*N*-(2-nitro-4-azidophenyl)ethylenediamine

The reagent inhibits the stereospecific binding of a β-adrenergic antagonist (K_i, 100 nM) and the l-isoproterenol-stimulated adenylate cyclase of turkey erythrocytes (K_i, 19 nM).

Klaasner *et al.* (*277*) designed a photolabile peptide, L-methionyl-L-tyrosyl-*p*-azido-L-phenylalaninamide as a label for the hormone binding site of neurophysin. The compound binds reversibly and specifically in the dark to bovine neurophysin II ($K_d = 1.4 \times 10^{-4} M$). Under photolytic conditions the compound is responsible for an irreversible inhibition of the noncovalent ligand binding activity of neurophysin II. Indications are that the photoaffinity probe is reacting at the hormone binding site of the protein.

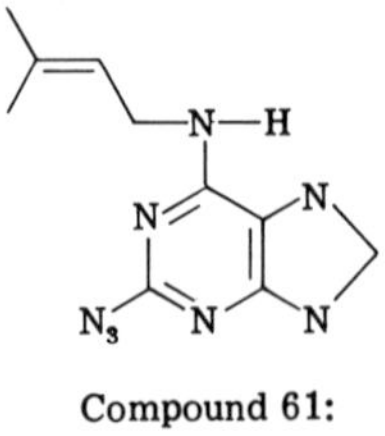

Compound 61:
2-Azido-N^6-(Δ^2-isopentenyl)adenine

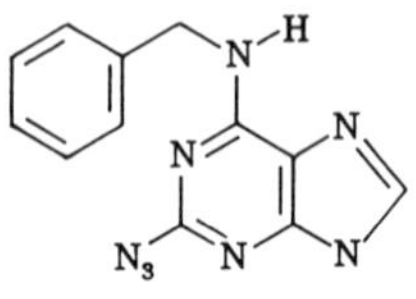

Compound 62:
2-Azido-N^6-benzyladenine

Two potent photoaffinity labels for probing cytokinin binding sites have been reported. The two azidopurine derivatives, 2-azido-N^6-(Δ^2-isopentenyl)adenine (Compound 61) and 2-azido-N^6-benzyladenine (Compound 62) possess high cytokinin activity (*278*).

The synthesis of two heterobifunctional cross-linking reagents that can be used to attached photoactivatible nitroarylazides to primary amino groups of proteins has been reported (*279*). The photoactive *N*-5-azido-2-nitrobenzoyloxysuccinimide (ANS-NOS) (Compound 63) and ethyl-*N*-5-azido-2-nitrobenzoylaminoacetimidate (ANB-AL) (Compound 64) were attached to lysine residues of cobra venom phospholipase A_2 without loss

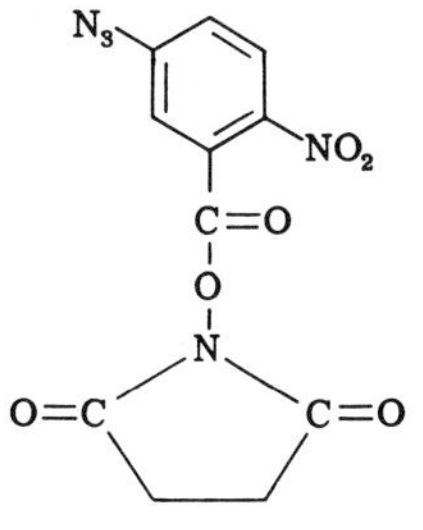

Compound 63:

ANS-NOS; *N*-5-azido-2-nitrobenzoyloxysuccinimide

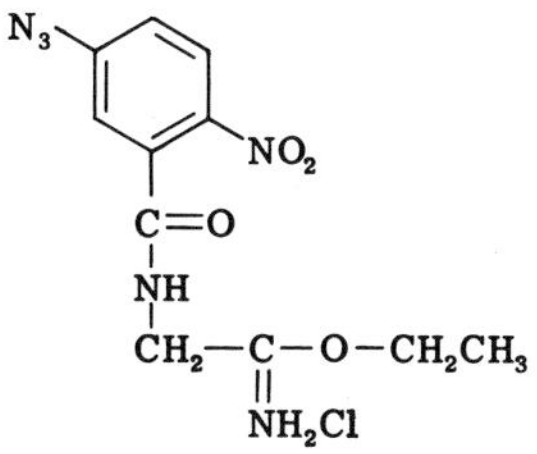

Compound 64:

ANB-AI; Ethyl-*N*-5-azido-2-nitrobenzoylaminoacetimidate

in enzymic activity. Photoirradiation of the modified enzyme resulted in the covalent linking of the enzyme into dimers and larger aggregates. The reagents offer certain definite advantages over homobifunctional reagents, such as dimethylsuberimidatesine. The initial labeling position is specific for amino groups, and the secondary (photodependent) insertion is nonspecific. ANS-NOS applied in cross-linking studies is the first reagent to show an effective cross-linking of soluble proteins.

IV. Concluding Comments

One obvious conclusion concerning the current status of photoaffinity labeling in bioenergetic systems is that to date the technique has been most successful in identifying particular ligand-binding-site peptides in complex mixtures. However, even here the technique has decided limitations, which must be recognized; in particular, the cautionary advice of Cooperman (*44*) should be mentioned as a guideline. Because of the chemical nature of the arylazido analogs the possibility of nonspecific interactions due both to the hydrophobic nature of the aromatic moiety and the electrophilic character of the nitrene should be always considered. Examination of the specificity of interaction at a particular ligand site by a number of analogs and under a variety of experimental conditions is always recommended as a means of eliminating artifacts (*44*).

Advances in the use of photoaffinity labeling in the investigation of the structure of ligand sites on proteins became possible only in conjunction with the development of techniques for the isolation of peptides and their sequence determination. Advances in the area of peptide chemistry will continue to add to the usefulness of the photoaffinity techniques.

However, for further investigations of biological transducing systems, a critical need is the establishment of techniques that would allow for the isolation of membrane components, which have been specifically tagged, in nondestructive biologically reconstitutable forms. Sodium dodecyl sulfate–polyacrylamide gel electrophoresis in combination with cross-linking reagents is a powerful tool in the analysis of the structure of complex multicomponent systems. This will still be utilized in conjunction with specific photoaffinity probe studies, but should be augmented by nondestructive methods for resolution and reconstitution of membrane systems. Perhaps the use of chaotropic reagents or organic solvent extraction can be combined with the labeling by photoaffinity probes as a means of resolving specific and biologically active membrane components.

The types of photoaffinity probes that can be synthesized are limited only by the imagination of the chemist doing the synthesis and the willingness of the ligand binding site to accept the molecule. It should be expected that the photoaffinity technique will be utilized or attempts will be made to utilize it in as many different ways as there are specific systems to study. A portent for the future is the development of combined photoaffinity–fluorescence (or spin label) probes that could be used in the investigation of the dynamics of bioenergetic systems. The labeling of the tetrodotoxin binding site with such a dual probe is envisioned as a potential aid in the study of the gating channels of excitable membranes. In such cases the investigator need not necessarily be concerned with direct insertion at the specific binding site, but only that covalent insertion carries the fluorescent portion close enough to ensure that the binding site region is being monitored.

The development of fatty acid and phospholipid photoaffinity probes for both nitrene and carbene precursors offers the most exciting potential for examination of the relationship of membrane proteins to the membrane structure that holds them. In addition, those components that are as well necessary for their functional activity can now come under scrutiny. It is a certainty that this will become a most active avenue of research in the not too distant future.

It is possible that photoaffinity labeling of active sites could be used to localize proteins by electron microscopy. Active-site labeling reagents containing heavy atoms have been suggested as being capable of achieving resolving powers of 30 Å or better, i.e., sufficient for electron microscopic investigations. Such labels would have adequate electron-scattering power and would have a high degree of specificity of labeling.

Another possible future development is the possibility of bringing about specific photodependent insertion at restricted regions, possibly by laser beams. The use of microbeams of light to localize photoactivation morphologically has been suggested (*86*).

Concerning the photoaffinity probes in current use, there is still little known of their reactivity with specific sites on membranes or proteins. A concerted effort is required with model systems to delineate the specific reactivity of the photogenerated species in order to substantiate the often postulated nonspecificity of reactivity (*75*).

The variety of methods utilized for the photolytic generation of excitable species, the time period of irradiation and energy flux, the solvent system utilized during irradiation, and other potential variables are different from one laboratory to another. This is a deficiency in this rapidly expanding field, and without proper experimental safeguards it could result in some misconceptions and much confusion.

A most significant contribution of photoaffinity labeling will be in the elucidation of active-site structure and the mechanistic details of the catalytic process. With care and the use of a variety of photoaffinity probes, one should be able to draw conclusions concerning the amino acid sequence in or around the active site (*52*). It should be possible by this method to collect data on families of proteins for comparative studies of active-site peptides in cases in which the complete sequence determination of the native proteins, for one reason or another, is not possible.

Finally, photoaffinity probes should also be of importance in deciding as to aspects of solution chemistry as opposed to X-ray structural analysis in crystal structures. The timed insertion of a photoaffinity probe as a function of the physiological condition of an enzyme (i.e., presence or absence of allosteric activators or inhibitors) may be of great aid in resolving questions of conformational (structure function) analysis.

ACKNOWLEDGMENTS

I wish to express my appreciation to my colleagues of the past few years, who have contributed in no small way to the development of the ideas and experimental procedures of photoaffinity probes currently being utilized in our laboratory. In particular, thanks are extended to Dr. S. J. Jeng, Dr. S. Chen, Dr. J. Russell, and Dr. J. Cosson. The experimental work described in this review originating from our laboratory has been supported by the American

Heart Association, the United States Public Health Service (NIH), and the National Science Foundation (NSF).

REFERENCES

1. Lipmann, F. (1941). *Adv. Enzymol.* **1**, 99–121.
2. Pullman, M., Penefsky, H. S., Datta, A., and Racker, E. (1960). *J. Biol. Chem.* **235**, 3322–3329.
3. Penefsky, H. S., Pullman, M. E., Datta, A., and Racker, E. (1960). *J. Biol. Chem.* **235**, 3330–3336.
4. Quiocho, F. A. and Lipscomb, W. N. (1971). *Adv. Protein Chem.* **25**, 1–59.
5. Kirsch, J. (1973). *Ann. Rev. Biochem.* **42**, 205–234.
6. Mitchell, P. (1977). *Ann. Rev. Biochem.* **46**, 996–1005.
7. Boyer, P. D. (1977). *Ann. Rev. Biochem.* **46**, 957–966.
8. Chance, B. and Williams, G. R. (1956). *Adv. Enzymol.* **17**, 65–134.
9. Slater, E. C. (1974). *In* "Dynamics of Energy-Transducing Membranes" (L. Ernster, R. W. Estabrook, and E. C. Slater, eds.) BBA Libr. **13**, 1–20. Elsevier, Amsterdam.
10. Bisson, R., Gutweniger, H., Montecucco, C., Colonna, R., Zanotti, A., and Azzi, A. *FEBS Lett.* **81**, 147–150.
11. Bisson, R., Azzi, A., Gutweniger, H., Colona, R., Montecucco, C., and Zanotti, A. (1978). *J. Biol. Chem.* **253**, 1874–1880.
12. Kagawa, Y. and Racker, E. (1966). *J. Biol. Chem.* **241**, 2467–2474.
13. Jost, P. C., Griffith, O. H., Capaldi, R. A., and Vanderkooi, G. (1973). *Proc. Nat. Acad. Sci. U.S.A.* **70**, 480–484.
14. Brierley, G. P. and Merola, A. J. (1962). *Biochim. Biophys. Acta* **64**, 205–217.
15. Olsen, W. L., Gupta, C. M., Mann, C., Radhakrishnam, R., Schaechter, M., and Khorana, H. G. (1977). *Am. Soc. for Microbiology Annual Meeting Abstracts* K139.
16. McConnell, H. M. (1977) and McFarland, B. G. (1970). *Quart. Rev. Biophys.* **3**, 91–136.
17. Boyer, P. D., Chance, B., Ernster, L., Mitchell, P., Racker, E., and Slater, E. C. (1977). *Ann. Rev. Biochem.* **46**, 955–1026.
18. Hanstein, W. G. (1973). *Fed. Proc.* **32**, Abstract 1663, p. 515.
19. Hatefi, Y. and Hanstein, W. G. (1974). *In* "Membrane Proteins Transport and Phosphorylation," Proc. Int. Symp. (G. F. Azzone, M. E. Klingenberg, E. Qualiariello, and N. Silipraudi, eds.), p. 287. North-Holland, Amsterdam.
20. Russell, J., Jeng, S. J., and Guillory, R. J. (1976). *Biochem. Biophys. Res. Commun.* **70**, 1225–1233.
21. Cosson, J., Jeng, S. J., and Guillory, R. J. (1978). *Fed. Proc.* **37**, Abstract 2526, p. 1728.
22. Russell, J. and Guillory, R. J. (1976). Abstr. Am. Chem. Soc. 172nd Meeting, Biol. 102, San Francisco.
23. Russell, J. and Guillory, R. J. (1975). Abstr. *Pacific Slope Bioch. Conf.* Honolulu, Hawaii.
24. Jeng, S. J., Guillory, R. J., and Sund, H. (in preparation).
25. Racker, E. (1967). *Fed. Proc.* **26**, 1335–1340.
26. Chen, S. and Guillory, R. J. (1977). *J. Biol. Chem.* **252**, 8990–9001.
27. Chen, S. and Guillory, R. J. *J. Biol. Chem.* (submitted).
28. Deformamido azido antimycin A was synthesized from deformyl antimycin A. See reference 113.
29. Hucho, F. and Schiebler, W. (1977). *Molecular and Cellular Biochemistry* **18**, 151–172.
30. Guillory, R. J., Rayner, M. D., and D'Arrigo, J. S. (1977). *Science* **196**, 883–888.

31. Hafemann, D. R. and Housten, A. H. (1971). *Fed. Proc.* **30**, Abstract 349, p. 255.
32. Narahashi, J. (1975). *In* "The Nervous System" (D. B. Tower, ed.), Raven Press, New York.
33. Bergman, C., Dubois, J. M., Rojas, E., and Rathmayer, W. (1976). *Biochim. Biophys. Acta* **455**, 173–184.
34. Beress, L., Beress, R., and Wunderer, G. (1975). *FEBS Lett.* **50**, 311–314.
35. Means, G. E. and Feeney, R. E. (1971). "Chemical Modification of Proteins." Holden-Day, San Francisco.
36. Wold, F. (1977). *In* "Methods in Enzymology," XLVI (W. B. Jakoby and M. Wilchek, eds.), pp. 3–14. Academic Press, New York. See also reference 51.
37. Singh, A., Thornton, E. R., and Westheimer, F. H. (1962). *J. Biol. Chem.* **237**, p. 3006–3008.
38. Fleet, G. W. J., Porter, R. R., and Knowles, J. R. (1969). *Nature* **224**, 511–512.
39. Shafer, J., Baronowsky, P., Laursen, R., Finn, F., and Westheimer, F. H. (1966). *J. Biol. Chem.* **241**, 421–427.
40. Hexter, C. S. and Westheimer, F. H. (1971). *J. Biol. Chem.* **246**, 3928–3933.
41. Vaughan, R. J. and Westheimer, F. H. (1969). *J. Am. Chem. Soc.* **91**, 217–218.
42. Knowles, J. R. (1972). *Acct. Chem. Res.* **5**, 155–160.
43. Creed, D. (1974). *Photochemistry and Photobiology* **19**, 459–462.
44. Cooperman, B. S. (1975). *In* "Aging, Carcinogenesis and Radiation Biology" (K. C. Smith, eds.). Plenum, New York.
45. Jakoby, W. B., and Wilchek, M. (eds.), (1977). "Methods in Enzymology," Vol. XLVI. Academic Press, New York.
46. Wofsy, L., Metzger, H., and Singer, S. J. (1962). *Biochem.* **1**, 1031–1039.
47. Vallee, B. and Riodean, F. (1969). *Ann. Rev. Biochem.* **38**, 733–794.
48. Cohen, L. A. (1968). *Ann. Rev. Biochem.* **87**, 695–726.
49. Metzger, H., Wofsy, L., and Singer, S. J. (1963). *Biochem.* **2**, 979–988.
50. Singer, S. J. and Doolittle, R. F. (1966). *Science* **153**, 13–25.
51. Singer, S. J. (1967). *Adv. Protein Chem.* **22**, 1–54.
52. Jeng, S. J. and Guillory, R. J. (1975). *Supramolecular Structure* **3**, 448–468.
53. Tonomura, Y. (1973). "Muscle Proteins, Muscle Contraction and Cation Transport." University Park Press, Baltimore.
54. Azuma, N., Ikegara, M., Otsuka, F., and Tonomura, Y. (1962). *Biochim. Biophys. Acta* **60**, 104–111.
55. Yamashita, T., Soma, Y., Kobayashi, S., Sekine, T., Titani, K., and Narita, K. (1964). *J. Biochem* **55**, 576–577.
56. Hiratsuka, T. and Uchida, K. (1973). *Biochim. Biophys. Acta* **320**, 635–647.
57. Meloche, H. P. (1967). *Biochem.* **6**, 2273–2280.
58. Cosson, J. and Guillory, R. J. *J. Biol. Chem.* (in press).
59. Laidler, K. J. (1973). "The Chemical Kinetics of Enzyme Action," 2nd ed. Oxford Univ. Press, New York. (1962). See also Lawson, W. B., and Schramm, H. J. *J. Am. Chem. Soc.* **84**, 2017–2018.
60. Cohen, J. A. and Warringa, M. G. P. (1953). *Biochim. Biophys. Acta*, 1953 **11**, 52. See also Phillips, A. T. (1977). *In* "Methods in Enzymology" (W. B. Jakoby and M. Wilchek, eds.), Vol. XLVI, pp. 59–69. Academic Press, New York.
61. Silhavy, T., Szmeleman, S., Boos, W., and Schwartz, M. (1975). *Proc. Nat. Acad. Sci. U.S.A.*, 1975 **72**, 2120–2124.
62. Pressman, D. and Roholt, O. (1961). *Proc. Nat. Acad. Sci. U.S.A.* **47**, 1606–1610. See also Roholt, O. A., Radzimski, G., and Pressman, D. (1965). *Proc. Nat. Acad. Sci. U.S.A.* **53**, 847–853.

63. Crestfield, A. M., Stein, W. H., and Moore, S. (1963). *J. Biol. Chem.* **238**, 2413–2420.
64. Crestfield, A. M., Stein, W. H., and Moore, S. (1963). *J. Biol. Chem.* **238**, 2421–2428.
65. Heinrikson, R. L., Stein, W. H., and Moore, S. (1965). *J. Biol. Chem.* **240**, 2921–2934.
66. Bayley, H. and Knowles, J. R. (1977). *In* "Methods in Enzymology," (W. B. Jakoby and M. Wilchek, eds.), Vol. XLVI, pp. 69–114. Academic Press, New York.
67. Doering, W. V. E., Buttery, R. G., Laughlin, R. G., and Chaudhuri, N. (1956). *J. Am. Chem. Soc.* **78**, 3224.
68. Gilchrist, T. L. and Pres, C. W. (1969). *In* "Carbenes, Nitrenes and Arynes," (Gilchrist, T. L. and Rees, C. W., eds,), pp. 10–27. Pitman Press, London.
69. Chaimovich, H., Vaughan, R. J., and Westheimer, F. H. (1968). *J. Am. Chem. Soc.* **90**, 4008–4093.
70. Doering, W. V. E. Odmn, R. A. (1966). *Tetrahedron* **22**, 81–93.
71. Wentrup, C. and Crow, W. D. (1970). *Tetrahedron* **26**, 3965–3981.
72. Brungwick, D. J. and Cooperman, B. S. (1973). *Biochem.* **12**, 4074–4084.
73. Rosenstein, R. W. and Richards, F. F. (1972). *J. Immunol.* **108**, 1467–1469.
74. Jeng, S. J. and Guillory, R. J. (1974). *Fed. Proc.* Abstract 893, p. 1381.
75. Staros, J. V., Bayley, H., Standring, D. N., and Knowles, J. R. (1978). *Biochem. Biophys. Res. Commun.* **80**, 568–572.
76. Reiser, A., Willets, R. W., Williams, G., and Harley, R. (1971). *Faraday Soc. Trans.* **64**, 3265–3275.
77. Chowdhry, R., Vaughan, R. J., and Westheimer, F. H. (1976). *Proc. Nat. Acad. Sci. U.S.A.* **73**, 1406–1408.
78. Staros, J. V. and Richards, F. M. (1974). *Biochem.* **13**, 2720–2726.
79. Staros, J. V., Haley, B. E., and Richards, F. M. (1971). *J. Biol. Chem.* **249**, 5004–5007.
80. Reiser, A. and Leyshorn, L. (1970). *J. Am. Chem. Soc.* **92**, 7487.
81. McRobbie, I. M., Meth-Cohn, O., and Suschitzky, N. (1976). *Tetrahedron Letters* **12**, 926–928.
82. Doering, W. V. E. and Odmn, R. A. (1966). *Tetrahedron* **22**, 81–93.
83. Wentrup, C. and Crow, W. D. (1970). *Tetrahedron* **26**, 3965–3981.
84. Fleet, G. W. J., Knowles, J. R., and Porter, R. R. (1972). *Biochem. J.* **128**, 499–508.
85. Breslow, R., Feiring, A., and Herman, F. (1974). *J. Am. Chem. Soc.* **96**, 5937–5939.
86. Ruoho, A. E., Kiefer, H., Roeder, D., and Singer, S. J. (1973). *Proc. Nat. Acad. Sci. U.S.A.* **70**, 2567–2571.
87. Yoshioka, M., Lifter, J., Hew, C. L., Converse, C. A., Armstrong, M. Y. K., Koningsberg, W. H., and Richards, F. F. (1973). *Biochem.* **12**, 4679–4684.
88. Hew, C. L., Litter, J., Yoshioka, M., Richards, F. F., and Koningsberg, W. H. (1973). *Biochem.* **12**, 4685–4689.
89. Lifter, J., Hew, C. L., Yoshioka, M., Richards, F. F., and Koningsberg, W. H. (1974). *Biochem.* **13**, 3567–3571.
90. Richards, F. F., Lifter, J., Hew, C. L., Yoshioka, M., and Koningsberg, W. H. (1974). *Biochem.* **13**, 3572–3575.
91. Cannon, L. E., Woodard, D. K., Woehler, M. E., and Lovins, R. E. (1974). *Immunology* **26**, 1183–1194.
92. Rudnick, G., Kaback, H. R., and Weil, R. (1975). *J. Biol. Chem.* **250**, 6847–6851.
93. Rudnick, G., Kaback, H. R., and Weil, R. (1975). *J. Biol. Chem.* **250**, 1371–1375.
94. Yariv, J., Kalb, A. J., and Yariv, M. (1972). *FEBS Lett.* **27**, 27–29.
95. Perry, M. B. and Heung, L. L. W. (1972). *Canad. J. Biochem.* **50**, 510–575.
96. Saman, E., Clacyssens, M., Kersters-Hilderson, H., and DeBruyne, C. K. (1973). *Carbohydr. Res.* **30**, 207–210.
97. Staros, J. V., Richards, F. M., and Haley, B. E. (1975). *J. Biol. Chem.* **250**, 8174–8178.

98. Rothstein, A., Cabantchik, A. I., and Knauf, P. (1976). *Fed. Am. Soc. Exp. Biol.* **35**, 3–12.
99. Knauf, P. A., Brenen, W., Davidson, L., and Rothstein, A. (1976). *Biophys. J.* **16**, 107a.
100. Kiefer, H., Lindstrom, J., Lennox, E. S., and Singer, S. J. (1970). *Proc. Nat. Acad. Sci. U.S.A.* **67**, 1688–1694.
101. Hucho, F., Bandini, G., Layer, P., Stengelin, S., and Suarez-Isla, B. A. (1977). *Hoppe-Seyler's Z. Physiol. Chem.* **358**, 253.
102. Hucho, F., Bergman, C., Dubois, J. M., Rojas, E., and Kiefer, H. (1976). *Nature* **260**, 802–804.
103. Guillory, R. J. and Jeng, S. J. (1977). *In* "Methods in Enzymology" (W. B. Jakoby and M. Wilchek, eds.), Vol. XLVI, pp. 259–288. Academic Press, New York.
104. Katzenellenbogen, J. A., Johnson, H. J., and Myers, H. N. (1973). *Biochem.* **12**, 4085–4092.
105. Katzenellenbogen, J. A., Johnson, H. J., Carlson, K. E., and Myers, H. N. (1977). *Biochem.* **16**, 2986–2994.
106. Katzenellenbogen, J. A., Carlson, K. E., Johnson, H. J., and Myers, H. N. (1977). *Biochem.* **16**, 1970–1976.
107. Winter, B. and Goldstein, A. (1972). *Mol. Pharmacol.* **8**, 601–611.
108. Levy, D. (1973). *Biochim. Biophys. Acta* **322**, 329–336.
109. Tobin, T., Akera, T., Brady, T. M., and Taneja, H. R. (1976). *Eur. J. Pharmacol.* **35**, 69–76.
110. Chakrabarti, P. and Khorana, H. G. (1975). *Biochem.* **14**, 5021–5033.
111. Greenberg, G. R., Chakrabarti, P., and Khorana, H. G. (1976). *Proc. Nat. Acad. Sci. U.S.A.* **73**, 86–90.
112. Klip, A. and Gitler, C. (1974). *Biochem. Biophys. Res. Commun.* **60**, 1155–1162.
113. Das Gupta, U. and Rieske, J. S. (1973). *Biochem. Biophys. Res. Commun.* **54**, 1247–1254.
114. Wilson, D. F., Mukai, Y., Erecinska, M., and Vanderkooi, J. M. (1975). *Arch. Biochem. Biophys.* **171**, 104–107.
115. Erecinska, M., Vanderkooi, J. M., and Wilson, D. F. (1975). *Arch. Biochem. Biophys.* **171**, 108–116.
116. Erecinska, M. (1977). *Biochem. Biophys. Res. Commun.* **76**, 495–501.
117. Lauquin, G., Brandolin, G., and Vignais, P. (1976). *FEBS Lett.* **67**, 306–311.
118. Lunardi, J., Lauquin, J. M., and Vignais, P. V. (1977). *FEBS Lett.* **80**, 317–323.
119. Lauquin, G., Brandolin, G., Lunardi, J., and Vignais, P. V. (1978). *Biochim. Biophys. Acta* **501**, 10–39.
120. Browne, D. T., Hixson, S. S., and Westheimer, F. H. (1971). *J. Biol. Chem.* **246**, 4477–4484.
121. Hixson, S. S. and Hixson, S. H. (1973). *Photochemistry and Photobiology* **18**, 135–138.
122. Koberstein, R. (1976). *Eur. J. Biochem.* **67**, 223–229.
123. Hanstein, W. G. (1973). *Fed. Proc. Fed. Am. Soc. Exp. Biol.* **32**, Abstract 1663, p. 515.
124. Hatefi, Y. and Hanstein, W. G. (1974). *In* "Membrane Proteins Transport and Phosphorylation; Proc. Int. Symp." (G. F. Azzone, M. E. Klingenberg, E. Qualiariello, and N. Silipraudi eds.), p. 187. North-Holland, Amsterdam.
125. Hanstein, W. G. and Hatefi, Y. (1974). *J. Biol. Chem.* **249**, 1356–1362.
126. Hanstein, W. G. (1976). *Trends Biochem. Sci.* **1**, 65–67.
127. Kurup, C. K. R. and Sanadi, D. R. (1977). *J. Bioenergetics Biomembranes* **9**, 1–15.
128. Cyboron, G. W. and Dryer, R. L. (1977). *Arch. Biochem. Biophys.* **179**, 141–146.
129. Senior, A. E. and Tometsko, A. M. (1975). *In* "Electron Transfer Chains and Oxidative Phosphorylation" (E. Quagliariello, S. Pope, F. Palmieri, E. C. Slater, and N. Silipraudi, eds.), pp. 155–170. North-Holland, Amsterdam.
130. Haley, B. E. and Hoffman, J. F. (1974). *Proc. Nat. Acad. Sci. U.S.A.* **71**, 3367–3371.
131. Koberstein, R., Cobianchi, L., and Sund, H. (1976). *FEBS Lett.* **64**, 176–180.

132. Pomerantz, A. H., Rudolph, S. A., Haley, B. E., and Greengard, P. (1974). *Biochem.* **14**, 3858–3862.
133. Share, K., Black, J. L., Pancoe, W. L., and Haley, B. E. (1977). *Arch. Biochem. Biophys.* **180**, 409–415.
134. Lau, E. P., Haley, B. E., and Barden, R. E. (1977). *Biochem. Biophys. Res. Commun.* **46**, 843–849.
135. Lau, E. P., Haley, B. E., and Barden, R. E. (1977). *Biochem.* **16**, 2581–2585.
136. Hixon, S. C., White, W. E., and Yielding, K. L. (1975). *J. Mol. Biol.* **92**, 319–329.
137. Hixon, S. C., White, Jr., W. E., and Yielding, K. L. (1975). *Biochem. Biophys. Res. Commun.* **66**, 31–35.
138. Bastos, R. de Norbrega. (1975). *J. Biol. Chem.* **250**, 7739–7746.
139. Nielsen, P. E., Leick, V., and Buchardt, O. (1975). *Acta Chem. Scand.* **B29**, 662–666.
140. Seela, F. and Cramer, F. (1975). *Hoppe-Seyler's Z. Physiol. Chem.* **356**, 1185–1186. See also Sonenberg, N., Zamir, and A., Wilchek, M. (1977). *In* "Methods in Enzymology" W. B. Jacoby and M. Wilchek, eds.), Vol. XLVI, pp. 702–707. Academic Press, New York.
141. Bispink, L. and Matthaei, H. (1973). *FEBS Lett.* **37**, 291–294.
142. Aggarwal, B. B., Quintanilha, A. T., Cammack, R., and Packer, L. (1978). *Biochem. Biophys. Acta* **502**, 367–382.
143. Penzer, G. R., Radda, G. K., Taylor, J. A., and Taylor, M. B. (1970). *Vitam. Horm.* **28**, 441–446.
144. Massey, V. and Hammerich, P. (1977). *J. Biol. Chem.* **252**, 5612–5614.
145. Jeng, S. J. and Guillory, R. J. (in preparation).
146. Rydstrom, J., Panov, A. V., Paradies, G., and Ernster, L. (1971). *Biochem. Biophys. Res. Commun.* **45**, 1389–1397.
147. Hatefi, Y. and Hanstein, W. G. (1973). *Biochem.* **12**, 3515–3422.
148. Hatefi, Y. and Galante, Y. M. (1977). *Proc. Nat. Acad. Sci. U.S.A.* **74**, 846–885.
149. Hatefi, Y. and Stempel, K. E. (1967). *Biochem. Biophys. Res. Commun.* **26**, 301–308.
150. Chen, S. and Guillory, R. J. *J. Biol. Chem.* (submitted).
151. Tzagoloff, A. and MacLennon, D. H. (1965). *Biochim. Biophys. Acta* **99**, 476–485.
152. Chuang, T. F. and Crane, F. L. (1973). *J. Bioenerg.* **4**, 563–578.
153. Bisson, R., Gutweniger, H., Montecucco, C., Colonna, R., Zanotti, A., and Azzi, H. (1977). *FEBS Lett.* **81**, 147–150.
154. Birchmeier, W., Kohler, C. E., and Schatz, G. (1976). *Proc. Nat. Acad. Sci. U.S.A.* **734**, 4334–4338.
155. Vanderkooi, J., Erecinska, M., and Chance, B. (1973). *Arch. Biochem. Biophys.* **157**, 531–540.
156. Dickerson, R. C., Takano, T., Eisenberg, D., Kallai, O. B., Samson, L., Cooper, A., and Margoliash, E. (1971). *J. Biol. Chem.* **246**, 1511–1535.
157. Folin, M., Azzi, A., Tamburro, A. M., and Jori, G. (1972). *Biochim. Biophys. Acta* **285**, 337–345.
158. Jori, G., Tamburro, A. M., and Azzi, A. (1944). *Photochemistry and Photobiology* **19**, 337–345.
159. Folin, M., Gennari, G., and Jori, G. (1974). *Photochemistry and Photobiology* **20**, 357–370.
160. Ivanetich, K. M., Bradshaw, J. J., and Kaminsky, L. A. (1976). *Biochem.* **15**, 1144–1153.
161. Vanderkooi, J. M. (1976). *Biochem. Biophys. Res. Commun.* **69**, 1043–1049.
162. Yu, C. A., Chiang, X. L., Yu, L., and King, T. E. (1975). *J. Biol. Chem.* **250**, 6218–6221.
163. Loomis, W. F. and Lipmann, F. (1948). *J. Biol. Chem.* **173**, 807–808.
164. Greville, G. D. and Needham, D. M. (1955). *Biochim. Biophys. Acta* **16**, 284. See also Chappell, J. B., Perry, S. V. (1955). *Biochim. Biophys. Acta* **16**, 285.
165. Hanstein, W. G. and Hatefi, Y. (1974). *J. Biol. Chem.* **249**, 1356–1362.

166. Hatefi, Y. (1975). *J. Supra. Mol. Struct.* **3**, 201–213.
167. Hatefi, Y., Hanstein, W. G., Galante, Y., and Stiggall, D. L. (1975). *Fed. Proc.* **34**, 1699–1706.
168. Cyboron, G. W. and Dryer, R. L. (1977). *Arch. Biochem. Biophys.* **179**, 141–146.
169. Heaton, G. M., Wagenvoord, R. J., Kemp, A., and Nicholis, P. (1978). *Eur. J. Biochem.* **82**, 515–521.
170. Katre, N. V. and Wilson, D. F. (1977). *Arch. Biochem. Biophys.* **184**, 578–585.
171. Cantley, L. C. and Hammes, G. (1973). *Biochem.* **12**, 4900–4904.
172. Lardy, H. A. and Elvehjem, C. A. (1945). *Ann. Rev. Biochem.* **14**, 1–30.
173. Hilborn, D. H. and Hammes, G. G. (1973). *Biochem.* **12**, 983–990.
174. Harris, D. A., Rosing, J., and Slater, E. C. (1974). *Biochem. Soc. Trans.* **2**, 86–87.
175. Wagenvoord, R. J., Vanderkeaan, T., and Kemp, A. (1977). *Biochim. Biophys. Acta* **460**, 17–24.
176. Russell, J. (1977). Ph.D. Dissertation, Univ. of Hawaii.
177. Campbell, R. C. (1974). "Statistics for Biologists," 2nd ed. Cambridge University Press, London.
178. Chang, T., and Penefsky, H. S. (1973). *J. Biol. Chem.* **248**, 2746–2754.
179. Lang, D. R. and Racker, E. (1974). *Biochim. Biophys. Acta* **333**, 180–186.
180. DiPietro, A., Godinot, C., Bouillant, M. L., and Gautheron, D. C. (1975). *Biochimie* **57**, 959–967.
181. Hanstein, W. G. (1976). *Biochim. Biophys. Acta* **456**, 129–148.
181a. Scheurich, P., Schafer, H. J., and Dose, K. (1978). *Eur. J. Biochem.* **88**, 253–257.
182. Satre, M., Klein, G., and Vignais, P. V. (1976). *Biochim. Biophys. Acta* **453**, 111–120.
183. Cosson, J. and Guillory, R. J. *J. Biol. Chem.* (in press).
184. Beechey, R. B., Hubbard, S. A., Linnett, P. E., Mitchell, A. D., and Munn, E. A. (1975). *Biochem. J.* **148**, 533.
184a. Tzagoloff, A. and Meagher, P. (1972). *J. Biol. Chem.* **247**, 6624–6630.
185. Cosson, J. and Guillory, R. J. (unpublished observations).
186. Harris, D. A., Rosing, J., Van de Stadt, R. J., and Slater, E. C. (1973). *Biochim. Biophys. Acta* **314**, 149–153.
187. Hammes, G. G. and Hilborn, D. A. (1971). *Biochim. Biophys. Acta* **233**, 580–590.
188. Senior, A. E. (1973). *Biochim. Biophys. Acta* **301**, 249–277.
189. Ferguson, S. J., Lloyd, W. J., Lyons, M. H., and Radda, G. K. (1975). *Eur. J. Biochem.* **54**, 117–126.
190. Marcus, F., Schuster, S. M., and Lardy, H. A. (1976). *J. Biol. Chem.* **251**, 1775–1780.
191. Kayalar, C., Rosing, J., and Boyer, P. D. (1977). *J. Biol. Chem.* **252**, 2486–2491.
192. Verheijen, J. H., Postma, P. W., and Van Dam, K. (1978). *Biochim. Biophys. Acta* **502**, 345–353.
193. Budker, V. G., Koslov, I. A., Kurbatou, V. A., and Hilgrom, Y. M. (1977). *FEBS Lett.* **81**, 11–14.
194. Ferguson, S. J. M., Lloyd, W. J., and Radda, G. K. (1975). *Eur. J. Biochem.* **54**, 127–133.
195. Van de Stadt, R. J., De Boer, B. L., and Van Dam, K. *Biochim. Biophys. Acta* **292**, 338–349.
196. Warshaw, J. B., Lam, K. W., Nagy, B., and Sanadi, D. R. (1968). *Arch. Biochem. Biophys.* **123**, 385–396.
197. Vignais, P. V., Vignais, P. M., and Defaye, G. (1973). *Biochem.* **12**, 1508–1519.
198. Egan, R. W. and Lehninger, A. L. (1974). *Biochem. Biophys. Res. Commun.* **59**, 195–201.
199. Vignais, P. V. (1976). *Bioenergetics* **8**, 9–17.
200. Harris, D. A., Radda, G. K., and Slater, E. C. (1977). *Biochim. Biophys. Acta* **459**, 560–572.

201. Schafer, G., Schroder, E., Rowohl-Quisthoudt, G., Penades, S., and Rimpler, M. (1976). *FEBS Lett.* **64**, 185–189.
202. Schafer, G. and Penades, S. (1977). *Biochem. Biophys. Res. Commun.* **78**, 811–818.
203. Schlimme, E. and Stahl, K. W. (1974). *Hoppe-Seyler's Z. Physiol. Chem.* **335**, 1139–1142.
204. Kyte, J. (1972). *J. Biol. Chem.* **247**, 7642–7649.
205. Ruoho, A. and Kyte, J. (1974). *J. Proc. Nat. Acad. Sci. U.S.A.* **71**, 2352–2356.
206. Matsui, H. and Schwartz, A. (1966). *Biochim. Biophys. Acta* **156**, 655–663.
207. Hegyvary, C. (1975). *Mol. Pharmacol.* **11**, 588–594.
208. Forbush, B., Kaplan, J., and Hoffman, J. F. (1978). *Fed. Proc.* **37**, Abstract 142, p. 239.
209. Cantley, L. C., Gelles, J., and Josephson, L. (1978). *Biochem.* **17**, 418–425.
210. Klip, A., Darszon, A., and Montal, M. (1976). *Biochem. Biophys. Res. Commun.* **72**, 1350–1358.
211. Pownall, H. J. and Smith, L. C. (1974). *Biochem.* **13**, 2594–2597.
212. Papageorgion, G. and Argoudelis, C. (1973). *Arch. Biochem. Biophys.* **156**, 134–142.
213. Rudy, B. and Gitler, C. (1972). *Biochim. Biophys. Acta* **288**, 235–236.
214. Nieva-Gomoz, D. and Gennis, R. B. (1977). *Proc. Nat. Acad. Sci. U.S.A.* **74**, 1811–1815.
215. Bercovici, T. and Gitler, C. (1978). *Biochim. Biophys. Acta* **17**, 1484–1489.
216. Kaelish, S. J. D., Jorgensen, P. L., and Gitler, C. (1977). *Nature* **269**, 715.
217. Abu-Salah, K. M. and Findlay, J. B. C. (1977). *Biochem. J.* **161**, 223–228.
218. Bayley, H. and Knowles, J. R. (1978). *Biochem.* **17**, 2414–2419.
219. Bayley, H. and Knowles, J. R. (1978). *Biochem.* **17**, 2420–2423.
220. Matheson, Jr., R. R., Van Wart, H. E., Burgess, A. W., Weinstein, L. I., and Scheraga, H. A. (1977). *Biochem.* **16**, 396–403.
221. Haley, B. (1975). *Biochem.* **14**, 3852–3857.
222. Owens, J. R. and Haley, B. E. (1976). *J. Supramol. Struc.* **5**, 91–102.
223. Guthrow, C. F., Rasmussen, H., Brunswick, D. J., and Cooperman, B. S. (1973). *Proc. Nat. Acad. Sci. U.S.A.* **70**, 3344–3346.
224. Mills, D. C. B. and Macfarlane, D. E. (1978). *Fed. Proc.* **37**, Abstract 1754, p. 546.
225. Antonoff, R. S. and Ferguson Jr., J. J. (1974). *J. Biol. Chem.* **249**, 3319–3321.
226. Antonoff, R. S., Ferguson Jr., J. J., and Idelkope, G. (1976). *Photochemistry and Photobiology* **23**, 327–329.
227. Antonoff, R. S. and Ferguson Jr., J. J. (1978). *Photochemistry and Photobiology* **27**, 499–501.
228. Braun, A. and Merrick, B. (1975). *Photochemistry and Photobiology* **21**, 243–247.
229. Schott, H. N. and Schotler, M. D. (1977). *Biochem. Biophys. Res. Commun.* **59**, 1112–1116.
230. Rudnick, G., Weil, R., and Kaback, R. (1975). *J. Biol. Chem.* **250**, 6847–6851.
231. Trosper, T. and Leoy, D. (1977). *J. Biol. Chem.* **252**, 181–186.
232. Lin, S. and Spudich, J. A. (1974). *J. Biol. Chem.* **249**, 5778–5783.
233. Kiehm, D. J. and Ji, T. H. (1977). *J. Biol. Chem.* **252**, 8524–8531.
234. Moore, J. W., Blaustein, M. P., Anderson, W. C., and Narahashi, T. (1967). *J. Gen. Physiol.* **50**, 1401–1411.
235. Reed, J. K. and Raftery, M. A. (1976). *Biochem.* **15**, 944–953.
236. Hille, B. (1967). *J. Gen. Physiol.* **50**, 1287–1302.
237. Meunier, J. C., Sealock, R., Olsen, R., and Changeux, J. R. (1974). *Eur. J. Biochem.* **45**, 371–394.
238. Hucho, F., Bandini, G. A., and Suarez-Isla, B. (1978). *Eur. J. Biochem.* **83**, 335–340.
239. Nachmansohn, D. (1959). "Chemical and Molecular Basis of Nerve Activity." Academic Press, New York.
240. Hucho, F., Layer, P., Kiefer, H. R., and Bandini, G. (1976). *Proc. Nat. Acad. Sci. U.S.A.* **73**, 2624–2628.

241. Martinez-Carrion, M. and Raftery, M. A. (1973). *Biochem. Biophys. Res. Commun.* **55**, 1156–1164.
242. Witzemann, V. and Raftery, M. A. (1977). *Biochem.* **16**, 5862–5868.
243. Grunhagen, H. H. and Changeux, J. P. (1976). *J. Mol. Biol.* **106**, 497–516.
244. Grunhagen, H. H. and Changeux, J. P. (1976). *J. Mol. Biol.* **106**, 517–535.
245. Gordon, A. S., Davis, C. G., and Diamond, I. (1977). *Proc. Nat. Acad. Sci. U.S.A.* **74**, 263–267.
246. Gordon, A. S., Davis, C. G., Miltay, D., and Diamond, I. (1977). *Nature* **267**, 539–540.
247. Weill, C. L., McNamee, M. G., and Karlin, A., (1974). *Biochem. Biophys. Res. Commun.* **61**, 997–1003.
248. Hucho, F., Gordon, A. S., Jeng, S. J., and Guillory, R. J. (unpublished observations).
249. Parisi, M., Reader, T., and de Robertis, E. (1972). *J. Gen. Physiol.* **60**, 454–470.
250. Kemp, G., Dolly, J. A., Bernard, E. A., and Wenner, C. E. (1973). *Biochem. Biophys. Res. Commun.* **54**, 607–613.
251. Bradley, R. J., Howell, J. H., Romine, W. O., Carl, G. F., and Kemp, G. E. (1976). *Biochem. Biophys. Res. Commun.* **68**, 577–584.
252. Karlin, A., Weill, C. L., McNamee, M. G., and Valderrana, R. (1975). *Cold Spring Harbor Symp. Quant. Biol.* **40**, 211–230.
253. Michaelson, D. M. and Raftery, M. A. (1974). *Proc. Nat. Acad. Sci. U.S.A.* **71**, 4768–4772.
254. Schiebler, W. and Hucho, F. (1978). *Eur. J. Biochem.* **85**, 55–63.
255. Bergman, C., Dubois, J. M., Rojas, E., and Rathmayer, W. (1976). *Biochim. Biophys. Acta* **455**, 173–184.
256. Romoy, G., Abita, J. P., Schweitz, H., Wunderer, G., and Lazdunski, M. (1976). *Proc. Nat. Acad. Sci. U.S.A.* **73**, 4055–4059.
257. Wunderer, G. (1976). *Eur. J. Biochem.* **68**, 193–198.
258. Martinez, G., Kopeyan, C., Schweitz, H., and Lazdunski, M. (1977). *FEBS Lett.* **84**, 247–252.
259. Biber, T. U. L. (1971). *J. Gen. Physiol.* **58**, 131–144.
260. Benos, D. J. and Mandel, L. J. (1978). *Science* **199**, 1205–1206.
261. Hucho, F., Bergman, C., Dubois, J. M., Rojas, E., and Kiefer, H. (1976). *Nature* **260**, 802–804.
262. Hucho, F. (1977). *Nature* **267**, 719–720.
263. Gergely, J. (1976). *In* "Chemical Mechanisms in Bioenergetics" (D. R. Sanadi, ed.), pp. 221–262. American Chemical Society, Washington, D.C.
264. Wagner, P. D. and Yount, R. G. (1975). *Biochem.* **14**, 5156–5162.
265. Jeng, S. J. and Guillory, R. J. (1976). Abst. *Am. Chem. Soc. 172nd Meeting*, Abstr. Biol. 135, San Francisco.
266. Szilagyi, L., Balint, M., Sreter, F. A., and Gergely, J., (1978). *Fed. Proc.* **37**, Abstract 2348, p. 1695.
267. Elzinga, M. and Collins, J. H. (1977). *Proc. Nat. Acad. Sci. U.S.A.* **74**, 4281–4284.
268. Geahlen, R. L. and Haley, B. E. (1977). *Proc. Nat. Acad. Sci. U.S.A.* **74**, 4375–4377.
269. Gupta, C. M., Radhakrishnan, R., and Khorana, H. G. (1977). *Proc. Nat. Acad. Sci. U.S.A.* **74**, 4315–4319.
270. Greenberg, G. R., Chakrabarti, P., and Khorana, H. G. (1976). *Proc. Nat. Acad. Sci. U.S.A.* **73**, 86–90.
271. Wolff, M. E., Feldman, K., Catsoulacos, P., Funder, J. W., Hancock, C., Amano, Y., and Edelman, I. S. (1975). *Biochem.* **14**, 1750–1759.
272. Katzenellenbogen, J. A., Myers, H. N., Johnson, H. J., Kempton, R. J., and Carlson, K. E. (1977). *Biochem.* **16**, 1964–1970.

273. Blackburn, G. M., Flavell, A. J., and Thompson, M. H. (1974). *Cancer Research* **34**, 2015–2019.
274. Martyr, R. J. and Benisek, W. F. (1973). *Biochem.* **12**, 2172–2178.
275. Sweet, F., Strickler, R. C., and Warren, J. C. (1978). *J. Biol. Chem.* **253**, 1385–1392.
276. Dartler, F. S. and Martinetti, G. V. (1977). *Biochem. Biophys. Res. Commun.* **79**, 1–7.
277. Klausner, Y. S., McCormick, W. M., and Chaiken, I. M. (1978). *Int. J. Peptide Protein Res.* **11**, 82–90.
278. Theiler, J. B., Leonard, N. J., Schmitz, R. Y., and Skoog, F. (1976). *Plant Physiol.* **58**, 803–805.
279. Lewis, R. V., Roberts, M. F., Dennis, E. A., and Allison, W. S. (1977). *Biochem.* **16**, 5650–5654.

Subject Index

M

N

O

P

R

S

T

U

V

W